SECOND EDITION

ORIGINS OF INTELLIGENCE

Infancy and Early Childhood

SECOND EDITION

ORIGINS OF INTELLIGENCE

Infancy and Early Childhood

EDITED BY

MICHAEL LEWIS

University of Medicine and Dentistry of New Jersey
Rutgers Medical School
New Brunswick, New Jersey

PLENUM PRESS • NEW YORK AND LONDON

Library of Congress Cataloging in Publication Data

Main entry under title:

Origins of intelligence.

Includes bibliographical references and indexes.
1. Intellect—Social aspects. 2. Infant psychology. I. Lewis, Michael, 1937 Jan. 10–
[DNLM: 1. Intelligence tests—In infancy and childhood. 2. Intelligence—In infancy and
childhood. BF 431 069]
BF431.O67 1983   155.4′13   83-11013
ISBN 0-306-41225-X

© 1983 Plenum Press, New York
A Division of Plenum Publishing Corporation
233 Spring Street, New York, N.Y. 10013

Printed in the United States of America

Contributors

BEVERLY BIRNS, Social Science Interdisciplinary Program, State University of New York at Stonybrook, Stonybrook, New York

JEANNE BROOKS-GUNN, Institute for the Study of Exceptional Children, Educational Testing Service, Princeton, New Jersey

EARL C. BUTTERFIELD, Department of Education, University of Washington, Seattle, Washington

RICHARD D. EWY, Department of Psychology, University of California, Berkeley, California

MARK GOLDEN, Department of Psychiatry, Albert Einstein College of Medicine, 1300 Morris Park Avenue, Bronx, New York

JEANNETTE HAVILAND, Department of Psychology, Rutgers—The State University, New Brunswick, New Jersey

MARJORIE P. HONZIK, Institute of Human Development, University of California, Berkeley, California

JANE V. HUNT, Institute of Human Development, University of California, Berkeley, California

SHARON LANDESMAN-DWYER, Department of Psychiatry and Behavioral Sciences, University of Washington, Seattle, Washington

JACQUELINE V. LERNER, College of Human Development, The Pennsylvania State University, University Park, Pennsylvania

RICHARD M. LERNER, College of Human Development, The Pennsylvania State University, University Park, Pennsylvania

ROBERT A. LEVINE, Laboratory of Human Development, Harvard Graduate School of Education, Cambridge, Massachusetts

MICHAEL LEWIS, Department of Pediatrics, Rutgers Medical School—University of Medicine and Dentistry of New Jersey, New Brunswick, New Jersey

CATHERINE LUTZ, Department of Anthropology, State University of New York at Binghamton, Binghamton, New York

ROBERT B. McCALL, Communications and Public Service, The Boys Town Center, Boys Town, Nebraska

DAVID J. MESSER, Psychology Division, The Hatfield Polytechnic, Hatfield, Herts, England

SANDRA SCARR, Department of Psychology, Yale University, New Haven, Connecticut

INA Č. UŽGIRIS, Department of Psychology, Clark University, Worcester, Massachusetts

JOHN S. WATSON, Department of Psychology, University of California, Berkeley, California

MARSHA WEINRAUB, Department of Psychology, Temple University, Philadelphia, Pennsylvania

LOUISE CHERRY WILKINSON, Department of Educational Psychology, University of Wisconsin—Madison, Madison, Wisconsin

LEON J. YARROW, Late of the Child and Family Research Branch, National Institute of Child Health and Human Development, Bethesda, Maryland

Preface to the Second Edition

Since the first edition of this volume was published in 1976, interest in the problem of intelligence in general and infant intelligence in particular has continued to grow. The response to the first edition was heartening: many readers found it a source of information for the diverse areas of study in infant intelligence. Because of the success of that volume, we have decided to issue a second edition. This edition is in many ways both similar to and different from the first. Its similarity lies in the fact that many of the themes and many of the contributors remain the same. Its difference can be found in the updating of old chapters and the addition of several new ones.

Taken together, the chapters present a rounded picture of the central issues in infant intelligence. Because the aim was to present a picture of the issues, no attempt, other than the selection of authors and themes, can be made to integrate these chapters into a single coherent whole. In large part, this reflects the diversity of study found in the area of early intellectual behavior. Rather than having a comprehensive theory of infant intelligence, the field abounds with a series of critical questions. To unite these chapters into some coherence, it will be necessary to articulate what these issues might be. Five major themes run throughout the field of infant intelligence and thus through this volume.

Theme one: The nature of intelligence. This theme addresses the questions surrounding the definition of intelligence. How is intelligence to be measured? Is it a general competence or a specific set of skills (Lewis, McCall, Landesman-Dwyer and Butterfield)? What are the appropriate instruments of measures? What tests should one look at? Is it more than simple learning (Watson and Ewy)? Does one consider psychometric tests (Brooks-Gunn and Weinraub, Honzik) or does one consider social competence, affect, and motivation (Wilkinson, Haviland, Yarrow and Messer)? Should one look to advances in cognitive development?

Theme two: Stability and change in intellectual ability. Are individuals consistent or variable, and is development a series of transformations or the accumulation of more of the same abilities? The major issues under this theme are addressed through discussions of the organization of infant intelligence (Užgiris), the notion of stages versus continuous development (McCall), and the consideration of individual differences and predictability (Honzik, McCall, Lewis). Finally the role of infant intelligence tests in predicting subsequent abilities needs consideration (Honzik, Hunt, McCall, Lewis).

Theme three: Factors affecting intellectual development. Risk factors, including teratogens as well as perinatal and prenatal factors, need to receive increased attention (Hunt), as do temperament differences (Lerner and Lerner). From a more environmental view, factors such as social class and cultural differences should affect performance as well as define the levels and the type of performance observed (Lutz and LeVine, Golden and Birns, Landesman-Dwyer and Butterfield).

Theme four: The nature–nurture controversy. In the study of intelligence, this theme crops up over and over. Throughout this volume, the theme is addressed, in particular by Scarr, Lutz and LeVine, McCall, and Lewis.

Theme five: The sociopolitical nature of intelligence tests. This theme addresses the broad issues surrounding the role of the concept and measurement of intelligence within and across cultures. In the present volume, Lutz and LeVine, Lewis, and Wilkinson address this theme.

The volume consists of 15 chapters, each of which addresses some particular aspect of the topic of infant intelligence. Lewis, in the introductory chapter, *On the Nature of Intelligence: Science or Bias?* discusses the political and social issues that are pertinent when one talks about the nature of infant intelligence. Specifically, the relationship between sociopolitical views and the nature of intelligence is considered, taking into account some of the more recent information available in this area. Chapter 2, by Brooks-Gunn and Weinraub, reviews the *Origins of Infant Intelligence Testing* and presents an up-to-date account of some of the more recent tests used to assess infant intelligence. *Measuring Mental Abilities in Infancy: The Value and Limitations* are reviewed by Honzik in Chapter 3. She considers the relationship of early infant intelligence scores to later behavior. Honzik reviews and challenges the notion of whether test scores accurately describe the growth of mental abilities. McCall, in *A Conceptual Approach to Early Mental Development* (Chapter 4), attempts to review studies of infant intelligence and offers a road map to the study of some of the central issues in intellectual development by considering the topics of continuity, consistency, and change as well as the issues of individual differences and nature–nurture. In Chapter 5, the *Organization of Sensorimotor Intelligence* is discussed by Užgiris, who presents a review of the organization of intelligence from a Piagetian perspective. Research with both normal and handicapped children as well as with diverse cultures is considered. Sandra Scarr, in Chapter 6, considers *An Evolutionary Perspective on Infant Intelligence: Species Patterns and Individual Variations* and presents a view of canalization as a way of looking at the development of intelligence as well as a way of looking at individual differences. Watson and Ewy discuss the topic of *Early Learning and Intelligence* in Chapter 7. Here they consider the relationship between early learning and intelligence, asking whether intelligence determines learning, learning determines intelligence, or learning expresses intelligence. In Chapter 8, Hunt considers *Environmental Risks in Fetal and Neonatal Life as Biological Determinants of Infant Intelligence.* The effects of drugs, malnutrition, and prematurity are described as they affect infant intellectual development. Wilkinson, in *Social Intelligence and the Development of Communicative Competence* (Chapter 9), challenges the notion of what is commonly considered intellectual behavior. She suggests that intelligence has to be considered from a "competence" point of view, in particular, from that aspect of competence called *social intelligence.* One specific form of social intelligence that she considers is communicative competence. Lutz and LeVine, in *Culture and Intelligence in Infancy* (Chapter 10), review the issue of intelligence from a cultural perspective. Cultural beliefs and theories about the nature of the developing child and of mental and social abilities vary among cultures. In a sense, by observing intelligence as it relates to culture, they provide additional support for the possibility that intelligence is a cultural construct rather than an attribute of a person. Golden and Birns's *Social Class and Infant Intelligence* (Chapter 11), reviews the effects and influences of social class on intelligence. They document the view that class effects appear to make the most noticeable effect around the first to second year of life.

Factors affecting intellectual performance become relevant for understanding individual differences; temperament differences have been identified as such factors. In Chapter 12, *Temperament–Intelligence Reciprocities in Early Childhood: A Contextual Model,* Lerner and Lerner explore the relationship between individual differences in temperament and intelligence and offer a model linking these domains. Chapter 13, *Looking Smart: The Relationship between Affect and Intelligence in Infancy*, by Haviland, attempts to demonstrate that the distinction drawn between intellectual and affective abilities is without justification. In Chapter 14, Yarrow and Messer explore the relationship between *Motivation and Cognition in Infancy*, pointing out that the connection between competence in general and intellectual behavior in particular cannot be considered as only intellectual action but must include the constructs of motivation and social competence. The authors, by exploring the concepts of motivation, efficacy, and cognition, go a long way toward defining the important connection among these aspects of infant behavior.

In the last chapter, *Mental Retardation: Developmental Issues in Cognitive and Social Adaptation*, Landesman-Dwyer and Butterfield explore a variety of issues related to mental retardation. First, they consider some general issues, for example, the causes for and the number of mentally retarded children; second, the authors describe the cognitive as well as social behavior of retarded infants; in the final section, they consider the reasons for studying mental retardation.

The volume consists of a broad sweep of the major issues in infant intelligence. The collection represents an important set of papers, each addressing some critical issue in the field. To conclude, I repeat what I said in the first edition: "As the contents of this volume make clear, the perspective is broad, the views personal and educated." It is for the reader to integrate these perspectives. Our hope is that this collection will continue to stimulate theory and research on the topic of infant intelligence.

New Brunswick, New Jersey   MICHAEL LEWIS

Preface to the First Edition

A preface is an excellent opportunity for an editor to speak directly to the reader and share with him the goals, hopes, struggles, and production of a volume such as this. It seems to me that I have an important obligation to tell you the origins of this volume. This is no idle chatter, but rather an integral part of scientific inquiry. It is important before delving into content, theory, and methodology to talk about motivation, values, and goals. Indeed, it is always necessary to explicate from the very beginning of any intellectual and scientific inquiry the implicit assumptions governing that exercise. Failure to do so is not only an ethical but a scientific failure. We learn, albeit all too slowly, that science is a moral enterprise and that values must be explicitly stated, removing from the shadows those implicit beliefs that often motivate and determine our results. No better or more relevant example can be found than in the review of the implicit assumptions of the early IQ psychometricians in this country (see Kamin's book, *The Science and Politics of IQ*, 1975). What might have been the result had we known their biases? What might have been the result had we known from the very beginning that their scientific quest was not one in which their values were removed, but in fact their hypotheses and their values were highly integrated? The comments that are to follow are an attempt to elucidate the reason for this volume.

The thought of this book occurred nearly fourteen years ago when I first became interested in the subject of infant mental activity. In the early 1960s, there was relatively little work on the subject of infancy. When Jerome Kagan and I first set out to look at infant behavior, we had in mind the idea of studying infant mental abilities. The major question that we confronted was, "What do we mean by mental abilities?" Was there such a thing or things, and how might they be measured? We have both struggled with these questions.

Although I knew relatively little about infant intelligence tests, the notion of intelligence testing, specifically infant intelligence testing, was never one that particularly interested me. I chose instead to study particular cognitive functions of the infant and sought through this means to come to an understanding of mental activity. Specifically, I chose to explore infants' attentional behavior with the hope of understanding through the infant's transactions with the environment what might be the structure and processes guiding some mental activities. From the outside, an important guiding premise has been to study the infant's changing behaviors as a function of demand characteristics of a situation in which he might find himself. The specific model that I chose was the attention paradigm. I chose to look at the organism's attending behavior—defined in a variety of ways—through the presentation of redundant information. From the results of these studies and others, one finds that the infant's attentive behavior declines as it interacts with redundant information. Moreover, when that information is altered, attentive behavior recovers. I saw in that paradigm and in the organism's changing transaction with its environment the basic feature of intelligence behavior, namely, the adaptation of the organism to its environment. Having undertaken these studies, it appeared important to determine the relationship between attending behavior and other measures of infant intellectual capacity. Supported by the National Science Foundation, I undertook a longitudinal study in which attentive behavior, object permanence as measured on the Corman-Escalona Sensorimotor Scales, infant intelligence tests as represented by the Bayley, and language capacity as represented by the Peabody Picture Vocabulary Test were all administered to groups of children in the first two years of life. The aim, quite frankly, was to show that attentive behavior to redundant and changing information was related to other measures of infant intelligence. To my surprise, several results emerged and it was in the emergence of these findings and their significance that the seeds of this volume were sown.

First, to my dismay, I found that the infant's attentive behavior bore no relationship to the infant's performance on the other intellectual tasks. However, when I looked carefully at the infant's performance on the Bayley, Object Permanence test, and language test, I found (1) within the first 2 years of life these tests were not highly correlated with one another. Thus, a notion of a unitary concept of intelligence which could be tapped over a variety of different tasks was seriously questioned. (2) Within any particular task there was little individual stability over the first 2 years of life. Thus, a child who performed well on object permanence at 3 months was not necessarily the child at 18 months or 24 months who likewise performed well. The same was true for the Bayley Mental Development Index. Parenthetically, the infant's attending task did show more individual stability than did either of these two other infant tasks.

These results confused and then shocked me. What did I mean by infant intelligence? What in fact did others mean by infant intelligence? The reviews of the literature on infant intelligence quickly revealed that our findings were not unique, and in fact Bayley herself had written:

> The findings of these early studies of mental growth of infants have been repeated sufficiently often so that it is now well established that test scores earned in the first year or two have relatively little predictive validity.

With this rather late but startling insight, I began to explore the issue of infant intelligence. This exploration has led to the present volume.

The creation of this volume is motivated by the desire to come to understand what people think and study about when they think and study about infant intelligence and intelligence scores. What I wanted to do was a volume which would look at infant intelligence from a wide variety of perspectives—a biological perspective, a social perspective, a cognitive and affective perspective. By viewing infant intelligence from a multi-perspective in this way, the end result should be the emergence of a picture of a construct which could not possibly be obtained by its examination from any particular single perspective. Thus, it seemed absolutely essential from the very beginning that a multi-dimensional perspective be given, because it was only through this perspective that one could come to view clearly this conceptualization. Simply stated, I wished to get the best people there were—in terms of their effort, interest, and knowledge—to examine the concept from the perspective they were most comfortable with. In that way, each perspective would have an advocate. There is no summary statement to be found in this volume. No one will do the work for the reader—it must be the reader himself interacting with each of these perspectives (and their sum) that will enable the emergence, successful or otherwise, of the concept that is being grappled with here. Thus, in some sense, it is a truly interactive process between the perspectives of the various authors and the mind of the reader.

As the contents of this volume make clear, the perspective is broad—the views personal and educated. Thus, it is left to the interac-

tion between the reader and this volume; the hope is for a clearer understanding of the concept of infant intelligence.

Finally, to Rhoda, Benjamin, and Felicia, who molded reason with
love and who altered knowledge with experience, I dedicate this
volume.

Princeton, New Jersey   MICHAEL LEWIS

Contents

1   On the Nature of Intelligence

Science or Bias?

MICHAEL LEWIS

The concept of intelligence—subsuming the belief that intelligence is relatively easy to measure and that, as a monolithic construct, it is a useful predictor of subsequent human behavior—is firmly entrenched in the mind of Western man.

In any discussion of the construct, it is necessary to define as precisely as possible what we mean when we say "intelligence." This essay will attempt to make such a definition. As a consequence of the discussion, it will be found that this construct is rather frail. It lacks the strength usually associated with it in theory and therefore fail to support the elaborate superstructure based on the premise of its existence. Such a discussion must lead the reader into a serious consideration of the uses and misuses of the IQ score in a technological society.

THE CONCEPT OF INTELLIGENCE

In common with many others (e.g., Galton, 1884; Goddard, 1912; Spearman, 1904; Terman, 1906) Burt, Jones, Miller, and Moodie (1934) expressed a view of intelligence that is a good starting point for the discussion. Burt *et al.* viewed intelligence as a finite potential with which the individual is endowed at conception, that is, subject neither to qualitative change nor to environmental influence. Finally, they believed intelligence is easily measured. This definition possesses a wide

MICHAEL LEWIS • Department of Pediatrics, Rutgers Medical School—University of Medicine and Dentistry of New Jersey, New Brunswick, New Jersey 08903.

assortment of features that should be carefully explicated. Specifically, these features include:

1. There is a *single* factor called g that subsumes all mental activity.
2. *All* performance in mental activity can be predicted by this factor.
3. It is an easily measured factor.
4. This factor can be measured by the measurement of a subset of behavior.
5. This factor is innate.

Since we are interested here in a rapidly developing organism, we must consider a final feature derived from Burt's view:

6. Intelligence is not subject to qualitative change.

While these features can be discussed for organisms at any age, the discussion will be restricted here to the opening years of life, since our subject is infant intelligence.

A SINGLE FACTOR

Probably no one feature is more central to the construct of intelligence than that it is a single potential, a single factor—often referred to as the g factor. While such theorists as Spearman, Goddard, and Burt believed in a single-factor notion of intelligence, others believed that, rather than having a single g factor, intelligence was a composite of factors and that there was a finite set of mental abilities (Thurstone, 1938). Such a view of intelligence—that of a set of abilities—does considerable damage to the more simple view outlined in the above six features. Nevertheless, it still enables one to believe in the concept of abilities as some finite set, to reify this set as having a real location in the brain, and to believe that these abilities are controlled by heredity. Intelligence, conceived as a single factor is not like cognitive activity, since cognitive activity has never been considered a single capacity but rather a wide and varied set of skills.

Is there any basis in fact for this single-factor view of intelligence, especially in infancy? First, in order to understand the multifaceted nature of the question, it is necessary to consider how tests of intelligence are constructed. One central feature in test construction is the production of items and subtests that are related to one another and to the score on the test as a whole. Items are so constructed and eliminated that this must be the case. Thus, if there are 10 test items, 9 of which are highly related to one another and to the total test score, the tenth item will be eliminated. It is no wonder that these tests have high inter-item agreement as well as high split-half reliability (consistency). They are designed that way. Thus, test construction perpetuates the notion of a single factor by the manipulation of items designed to produce just such an outcome.

Even more to the point is the issue of data analysis. It is possible to derive either a single factor or multiple factors depending on the type of factor analytic technique employed. For example, the use of a principal component factor analysis, by design, generates a single factor or g through projecting a single axis (others being at right angles), accounting for the major amount of variance, while more complex factor analysis, using rotating and oblique solutions, allows for multiple factors or abilities by projecting axes (not at right angles) and rotating them for maximal solution. Spearman and Burt invented and used the principal component analysis since they believed that intelligence could be represented by a single factor, while Thurstone invented more complex solutions since he believed that intelligence was not a single factor but a set of factors. It is important to note here the relationship between analytic techniques for data analysis and the belief system of the scientist. Clearly, the nature of the concept of intelligence, the measurement system, and analytic technique used are not independent. Prior belief and the nature of inquiry are often related, and this appears especially true in the study of intelligence (Gould, 1981; Hearnshaw, 1979; Kamin, 1974).

This indirect evidence demonstrated that the notion of a single factor, on its face, is quite related to the measurement system. Thus, the demonstration of direct evidence supporting the single-factor view from studies of infant performance on standardized tests of infant IQ does not appear warranted. A review of them should be undertaken if only to point out the empirical failure to demonstrate among infants a single g factor. McCall, Hogarty, and Hurlburt (1972) took great pains to find individual or factor-item stability across tests and age; nevertheless, they were forced to conclude that even with this type of analysis and the use of a variety of other multivariate techniques, the correlation between different ages "remains modest and of minimal practical utility" (p. 746). In conclusion, they rejected the simple conceptualization of a g factor in infancy:

> The search for correlational stability across vastly different ages implies a faith in a developmentally constant, general conception of intelligence that presumably governs an enormous variety of mental activities. Under that assumption, the nature of the behavioral manifestations of g would change from age to age, but g itself is presumed constant, and this mental precocity at one age should predict mental precocity at another. Confronted with the evidence reviewed above, this g model of mental development must be questioned. (p. 736)

In much of McCall's work (McCall et al., 1972; McCall, Eichorn, & Hogarty, 1977), a component factor analysis is used as the primary analytic device. Thus, while McCall rejects the notion or utility of g, the analytic technique would appear to support such a concept. In fact, although recognizing that the nature of the principal component is both unstable (vis-à-vis an individual's score in comparison to the group) and in fact changeable over age (first appearing as an attention factor, then by an active exploration of objects, and finally a verbal factor), McCall's work would appear to support the view that infant intelligence is primarily a single *but changing* structure.

Given that various factor analytic techniques lead to different outcomes (see Thurstone, 1938), Lewis and Enright (1982) have recently analyzed a set of infant test scores using an oblique rotation solution. Their results indicate that for each of the three ages studied (3, 12, and 24 months), a set of infant skills can be generated. In Figure 1 this set and the relationship over age is presented. These factors tend to agree with much of the work of past research utilizing factor analysis that could generate multiple factors (Bayley, 1970; Richards & Nelson, 1939; Stott & Ball, 1965).

Figure 1 indicates that there are at least four major abilities at 3 months, which include a *search factor* composed of orientation and attention items; an *auditory factor* that centers on vocalization and noisemaking; a *social factor*, including smiling at mirror image and frolic play; and finally a *manipulation factor* containing items relating to holding or reaching for objects. By 12 months, three major abilities can be identified, including a *verbal*, an *imitation*, and a *means–end* factor. Finally, by 24 months another set of abilities appear, including a *verbal-symbolic*, a *lexical*, an *imitation*, and a *spatial* factor. The sequence indicates three distinct

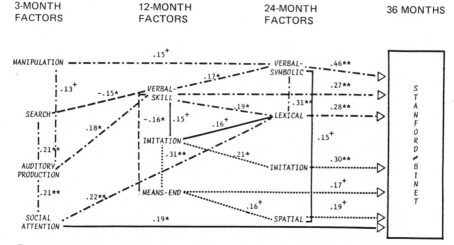

FIGURE 1. Diagram of the multiple paths of mental development characteristic of the sample of infants as a whole. Important within-age and cross-age correlations among Bayley factors abstracted at 3, 12, and 24 months and between the Bayley factors and Stanford-Binet scores at 36 months are illustrated ($**p<.01$, $*p<.05$, $+p<.10$). The various paths of mental development are illustrated with distinctively patterned lines. See text for further elaboration.

developmental paths. (1) a verbal, (2) a social, and (3) an imitation and means–ends.

Such a multiple-abilities analysis indicates that (1) some abilities appear early and then disappear or become only minimally related (manipulation), (2) some appear and are transformed (auditory), (3) some appear later, and (4) some appear and remain the same (imitation). Without specific analyses designed to elicit multiple abilities, such observations would go unnoted.

Perhaps if we turn from the standardized kinds of test, such as the Bayley or the Gesell, to the more recent approaches suggested by the Genevan school, we could find a single factor of infant mental ability. It may be necessary to utilize Piagetian theory and explore tasks more closely related to sensorimotor development to find this factor. King and Seegmiller (1971) applied the Užgiris and Hunt sensorimotor scales to 14-, 18-, and 24-month-old infants. The consistency of scores on these seven scales was compared across three ages, as was the relationship at 14 months across the different scales. Not only did the authors find relatively little consistency in terms of the correlations of scores (only 4 out of 24 possible correlations were significant), but they also found relatively little consistency across the various scales at a single age. Užgiris (1973), in trying to understand the patterns of sensorimotor intelligence, measured a limited number of subjects' performances on seven subscales of the Užgiris and Hunt (1975) sensorimotor intelligence scale. The most parsimonious explanation of her results was that there was almost no agreement between performance on one scale and performance on the others. Thus, even when we consider the nonstandard intelligence tests and look at sensorimotor development, at least as measured by the Užgiris and Hunt Scales, we find no evidence for a g factor.

Lewis and McGurk (1972) obtained and related three different types of infant intelligence test. Infants were seen longitudinally from 3 to 24 months, at which time they received the Bayley Scales of Infant Development (1969) and the object permanence scale from the Corman and Escalona sensorimotor scales (1969). In addition, at 24 months the children received a modified Peabody Picture Vocabulary Test in which both comprehension and production language scores were obtained. For the Bayley scales and the object permanence scales, the interage correlations proved to be relatively weak. Lewis and McGurk also observed the correlation between the Bayley and the object permanence scales at each age and between language development at 24 months and the Bayley and the object permanence scores at each age. In general, the results failed to indicate any consistent pattern across the tests that might be likened to a g factor.

There is little consistency across different measures of intellectual functioning—for example, between the Bayley scales and the sensorimotor scales—and little consistency within the sensorimotor scales or across different factors such as those found by McCall *et al.* (1972) for the Gesell scales. The data, therefore, offer little support for the notion of a single g factor in infant intelligence.

THE PREDICTABILITY OF BEHAVIOR

An important source of the glamour of intelligence scores is the belief that by knowing an organism's IQ score, it is possible to know a great deal about the organism's potential performance in all activities. IQ scores might not be important for some activities, but which ones? Presumably those activities not involving intellectual capacity. Here one encounters a difficulty: We believe on the one hand that IQ is an underlying general capacity predicting performance in some activities, but on the other hand, we have no good theory to tell us which activities. Thus, if carpentry is believed to involve intellectual activity, the natural conclusion is that IQ scores will predict carpentry ability. On the other hand, if carpentry has no intellectual components, then one's IQ should not predict one's performance as a carpenter. Thus, to a large extent, what we believe falls within the domain of intellectual activity will determine how broadly IQ can predict behavior.

Historically, the IQ test was designed around school performance (Binet, 1898, 1900). Thus the high relationship between IQ and school performance is not a measure of the validity of the IQ construct but rather only an example of how a test can be constructed to predict performance in a particular type of activity. If carpentry or hockey-playing are not considered intellectual activities, there is no reason for IQ and performance in these skills to be related.

For the first time in this discussion it becomes apparent that we must choose specific criteria to define *intellectual activity* (a choice not based on any apparent scientific theory). These criteria should be as explicit as possible. Toward such a goal one might well ask why school performance has been and continues to be the intellectual activity most related to IQ scores.

Doubts about the predictability of behavior from IQ scores are further reinforced by the knowledge that intelligence has not been shown to be a unitary factor. If this is indeed the case, by what logic could IQ test performance be related to all intellectual activity (given that one could logically define the domain)? It could well be that only a certain portion of that intellectual domain has been tapped by the test and related to those intellectual activities which, although subsumed within the larger domain of intellectual behavior, are independent of the skills tapped by the test. In such a case, there would be no relationship between IQ score and performance. Our everyday experience lends at least some face validity to the belief that this is indeed the case.

Indeed, the review of the relationship between IQ scores (or school performance, both highly related) and subsequent life "success" has been questioned. There is little support for the general belief that high IQ scores will lead to any of the goals we hold important—for example, wealth, happiness, physical and mental health, or success. McClelland (1973) concluded that "the poorer students in high school or college did as well in life as the top students" (p. 2). This seems to be true for jobs ranging from factory worker, bank teller, and air traffic controller (Berg, 1970) to scientific researcher (Taylor, Smith, & Chiselin, 1963). Jencks, Smith, Acland, Bane, Cohen, Gintis, Heyns, and Michalson (1972) also noted that, while a person's occupation may be related to the amount of education received (insofar as educational background enables one to enter the occupation initially), the relationship between childhood IQ and later occupation disappears when the effects of education are statistically controlled. Aptitude test scores seem to be no better at predicting than school grades or IQ (Holland & Richards, 1965; Elton & Shevel, 1969; McClelland, 1973; Thorndike & Hagen, 1959).

In studies showing that scores on general intelligence tests correlate with proficiency across jobs (e.g., Ghiselli, 1966), the effects of social class background or of personal credentials are seldom taken into account. It is fairly well established that social class background is related to getting higher scores on ability tests (Nuttal & Fozard, 1970) as well as to having the right personal credentials for success. Consequently, the relationship between intelligence and job success is likely to be illusory, a product of their joint association with social class. Intelligence measures do not seem to predict successful job performance and they bear little relationship to later earnings (Duncan, Featherman, & Duncan, 1972; Jencks et al., 1972). While studies of longitudinal data (Jensen, 1972; Kohlberg, LaCross & Ricks, 1970) do indicate that gifted children experience more material and social success and have lower incidences, for example, of alcoholism and homosexuality than do nongifted individuals, the data were based on a study in which no attempt was made to equate for the opportunity to achieve success. The gifted came from higher socioeconomic backgrounds. Again, a failure to take into account social class and all that it entails may result in spuriously high correlations between intelligence and these outcomes.

Such results should lead to some question as to why intelligence is used as the chief standard by which to gauge the development of children and the efficacy of intervention programs. Indeed, such a view has been questioned (Lewis & Michalson, 1983; Zigler & Trickett, 1978).

In terms of infant behavior, the relationship between IQ performance and alternative intellectual behavior has received almost no attention. In one study, Lewis and Lee-Painter (1974) related the Bayley intelligence performance scores of 100 12-week-old infants from a wide variety of socioeconomic backgrounds to their behavior in interaction with their mothers in a naturalistic home situation. The results showed that there were no significant relationships between performance on the Bayley intelligence scores and the infant behaviors as measured by the infant–mother interactions. Studies relating infant behavior to other social, cognitive, and affective behaviors also fail to find any correlation with IQ scores. Studies of infants' responses to strangers (Lewis & Brooks, 1975) or self-recognition (Lewis & Brooks-Gunn, 1979) have failed in general to find any relationship between emerging social skills and IQ scores. Moreover, although infant IQ performance has been related to infant trauma (see Chapter 8, by Hunt), there is no evidence that infant IQ test performance is related to other infant intellectual activities. Finally, the findings indicating a lack of relationship between performance on a variety of difficult tests, all reporting to measure intellectual activity, have already been discussed.

Of central importance to the issue of predictability is the relationship of test performance over age within infancy and test performance in infancy as related to older ages. Here, although there are some studies that show high across-age stability of individuals vis-à-vis groups (Siegel, 1981; Wilson, 1973), the findings have generally been quite consistent. Bayley, for one, has concluded, "The findings of these early studies of mental growth have been repeated sufficiently often so that it is now well established that test scores earned in the first year or two have relatively little predictive validity" (1970, p. 1174). Stott and Ball (1965), Thomas (1970), McCall et al. (1972, 1979), and Lewis (Lewis & Brooks-Gunn, 1981; Lewis & Enright, 1983; Lewis & McGurk, 1972) have all reported relatively little predictive validity between the early scores of infant IQ and later measures of intelligence.

One might argue that intelligence is stable but the behaviors in its service may vary in an ontogenetic fashion. Thus, there may be no reason to predict consistency from epoch to epoch. McCall et al. (1977), in their structural consideration, give support to such a view although it has been demonstrated for full scale scores that even within an epoch—within the first two years of infancy—there is little consistency in IQ performance.

From logical as well as from empirical considerations there are reasons to question the feature of intelligence that allows prediction of performance in activities from scores on IQ tests, even when the activities in question are themselves labeled as IQ tests. The difficulties are only compounded when one considers activities not directly related to school performance. That school performance has been singled out as the criterion for intelligence is a historical accident with sociopolitical implications. Since, in our culture, success is so interwoven with school performance, it is no wonder that IQ would appear to be predictive of a wide range of human activity. Even so, studies do not necessarily support this belief. Unfortunately, the relationship of intellectual activity to activities in general has not been seriously considered. This being the case, one cannot possibly review the predictability of IQ performance in terms of all human activity. Until we are able to develop a taxonomy of human activity and are able to see the functional significance, intellectual or otherwise, of the activities organisms perform at each age, it will not be possible for us to talk about how accurately IQ scores predict performance in other activities involving intellectual capacity. Until theory has caught up with out bias, it may still be necessary to select carpenters for their carpentry skills rather than their IQ scores, as it may be important to select teachers for their teaching performances rather than their IQ scores.

MEASUREMENT OF IQ

The measurement of IQ is based on a series of assumptions that must be reviewed. First, the criterion for intellectual behavior is school performance. Thus, the tasks and materials are designed around the tasks and materials used in school. This design would be of little concern if all that was of interest was, in fact, school performance. Unfortunately, the discussion has not stopped at this point, but theorists have chosen instead to generalize test performance to all mental activity, not just school performance. As has already been discussed, it must be recognized that *mental activity* has not been properly defined. This logical slippage is rather interesting, and it is necessary to question its origin. While we have no answers to this problem, it is important to determine historically why a test used as a measure of school performance became a test of general mental ability. Clearly the association between school performance and mental ability is tenuous. The question of what constitutes mental activity is by no means easy; perhaps this explains the ready acceptance of school performance. If so, our scientific inquiry has again given way to a nonscientific assumption.

This problem, difficult to solve for both adults and children, is made more difficult when we consider infant mental activity. Simply put, how is infant mental ability to be measured? Reaching for a red ring hardly seems to qualify as a mental activity, yet a careful survey of infant tests reveals a wide variety of such items. As Yarrow and Messer (see Chapter 14) and Haviland (see Chapter 13) have suggested, and as Lewis and Enright's analysis of infant performance indicates (see Figure 1), one might consider many of the items in infant tests of mental activity to be motivational or affectional in nature rather than mental. In fact, the finding that social attention is related to subsequent other factors in infant performance (Figure 1) lends support to this view. Moreover, recent studies of infant abilities in communication (Lewis & Rosenblum, 1977), social cognition (Lamb & Sherrod, 1981), and affect (Lewis & Michalson, 1983) should provide information related to infants' motivational, affective, and social abilities. Such information needs to be utilized in order to generate new definitions of mental ability or intelligence. The work of Zigler & Trickett (1978) and of Scarr (1981) in attempting to find new and expanded definitions of intelligence, and the use of such terms as *social competence,* is a movement toward this goal.

Moreover, test items have been freely borrowed from test constructor to test constructor, giving these items a high reliability, if no validity. Again the theory and empirical evidence of what constitutes mental activity lag behind test construction. A notable exception is Piaget, who has tried to explicate mental activity. Unfortunately he has constructed a model that both ignores and utilizes behaviors that he chooses to call nonmental. Thus, Haviland (see Chapter 13) quite rightly calls attention to the fact that Piaget uses affectional responses to infer mental activities. It is interesting to note that with the growing influence of Darwin and ethology, there is a renewed attempt to observe a wide variety of the young organism's activities and responses. Such a methodology may well give rise to a taxonomy of mental activity in the very young. This taxonomy (since school performance cannot be used as its criterion) must be more encompassing than the one constituted for school-age children. Thus, for the very young it may be possible to free ourselves from the constraint of a school performance criterion of mental activity. As a result, however, test items for infant mental activity may take on a rather strange appearance, such as nose-directed behavior in mirrors (see Lewis & Brooks-Gunn, 1979) or surprise affects in a new social situation (see Lewis & Rosenblum, 1974).

It is clear that school performance cannot be the sole measure of intelligence. As we are able to rid ourselves of this bias, we will be able to include a wider range of test items. Historically, the measurement of mental activity went from sensory capacities or even reaction time to activities related to school performance. It is now necessary that school performance give way to a large set of activities just beginning to be considered. Thus, the sociopolitical influences that dictate our scientific requirements (item selection) at any one point should be explored in order to broaden (or limit) the definition of intelligence. It may be that an authoritarian ideology requires tasks of rote memory, while an ideology of personal freedom requires tasks of creativity. This connection between ideology and the nature of tests of IQ needs to be explored.

NATURE, NURTURE, AND SOCIAL NEEDS

The literature on the role of innate factors versus environmental influences is elaborate. We will not go into detail on this issue since it has already received more space than it requires. The nature–nurture argument is an across-paradigm discussion resting on quite different views of the nature of man. Following Overton (1973), it appears that across-paradigm discussions do not lend themselves to easy solution and that the evidence for one paradigm does not negate the evidence for the other. More to the point, such paradigmatic views as the nature–nurture controversy address important sociopolitical issues as well as scientific ones.

The nature–nurture controversy centers on the issue of heritability coefficients. The index of heritability coefficients for performance on an IQ test has been estimated, at least for whites, to be as high as 0.80. Since a great deal of time has been spent arguing that IQ scores may have relatively little to do with the domain of mental activity, it would be foolish to argue that even if IQ scores are solely determined by heredity, mental activity in general is so determined.

Scarr-Salapatek (1971) has presented ample evidence that a 0.80 heritability index for whites tells little about the index for nonwhites and that it is not possible to talk of white–nonwhite differences as a function of heritability. However, the nature of this position still supports the heritability argument; the position merely argues that there is not enough information to comment on individual or group differences. Implicit in this argument is another—that if the heritability index of nonwhites were known, one in fact could regard IQ score differences as a function of race. This heritability discussion has led to a belief that IQ performance, as well as mental activity, is possessed in the genes and that individual differences are also so determined. Let us quote from an article by Lee Cronbach (1975):

> The evidence is that differences among American or British whites in the past generation have been due in part to genetic differences. The precise proportion, but not the principal, of hereditary influence can be debated. If the statistic is an index of social-cultural conditions, not a biological inevitability, change the distribution of nutrition, home experience, and schooling in the next generation, and the heritability index will change. Findings on heritability within white populations tells us nothing whatsoever about how white and black groups would compare if their environments had been equalized. (p. 2)

Thus, we are talking about sociocultural and political conditions, not issues of biology. In case this is not clear, let us continue the quotation:

> Note also that a high degree of heritability does not imply that in truth environment can have no effect. Even if the heritability index is as high as .80, two children with the same genotype may differ by as much as 25 IQ points if one is reared in a superior environment and the other in an unstimulating one. (p. 2)

Thus, given heritability coefficients and given an index as high as .80, two children can differ by 25 points on a standard test of IQ. Simply put, a person with an IQ of 95 (just average) and a 120-point child (a college graduate) may have the same genetic potential. It would seem, then, that IQ differences as differential measures of mental activity are as often as not a function of sociopolitical and cultural issues. Thus, to argue for heritability differences between individuals and groups (given an index of 0.80) and to base social policy on these indexes is to deny that differences can be accounted for by cultural as well as biological factors. A heritability index or difference indexes still have nothing to tell us about the performance differences between two people or two groups. Why then our interest in heritability? The sociopolitical consequences of heritability appear to be more important than its scientific value, since to know a heritability index (even one as high as .80) is not to know the phenotypic outcome.

Such an anlysis appears warranted, given the recent revelation about Burt's work and the conclusion that the heritability coefficient, as least as estimated through the use of his data, is no longer valid (Hearnshaw, 1979; Kamin, 1975). Of more importance than the falsity of Burt's findings is the conclusion that research on the nature of intelligence and on individual differences in intelligence may be generally biased in terms of the beliefs of investigators. The evidence for this more general conclusion is considerable. Already mentioned is the fact that analytic techniques used in the study of the structure of intelligence remain connected to particular ideologies. Thus, belief in a g factor is associated with a principal component analysis, while belief in multiple factors is associated with a complex factor analysis (Burt, 1940; Thurstone, 1938).

Most recently, Gould (1981) has raised the more general issue of belief/ideology and the scientific study of human intelligence. The general errors of the study of intelligence as Gould sees them are (1) the error of reification—the belief that intelligence is a real thing, having a location and being easily available for measure; and (2) the error of measurement—that is, the systematic biases of data through falsification, simplification, omission, and particular analytic techniques. Gould demonstrates that Burt's falsification is but one of a long list of human "errors" in the study of intelligence, most of which stem from the inability to separate out personal belief and ideology from empirical investigation. The historical trail of such error and bias is long, and Gould traces it carefully as he shows error piled upon error in the work of Morton, Broca, Goddard, Terman, Yerkes, and the Army Mental Tests (Gould, 1981). Two examples should suffice: In the measurement of intelligence, Morton, like others, has argued for a relationship between brain size and IQ: "Other things equal, there is a remarkable relationship between the development of intelligence and the volume of the brain" (1849, as cited in Gould, 1981, p. 51). Although it was known that brain size was highly related to such factors as body size and age, investigators often neglected them in their analyses of group differences or systematically eliminated or included brains that fit or did not fit the *a priori* view (Gould, 1981, pp. 59–60). Gould reanalyzed Morton's data with corrections and found no group differences, although Morton had claimed there were such differences.

The second example brings us closer in time, to Yerkes and the Army Mental Tests. World War I brought the opportunity to introduce the notion of intelligence into the society at large and to utilize test results for distribution of the war activity to be performed. The results from such testing were (1) the average mental age of white males was 13, just above the edge of moronity when set against the standard of 16. This meant that nearly 37% of whites and 89% of blacks were morons. (2) European immigrants, especially the darker peoples of southern Europe and the Slavs, were less intelligent than the western and northern Europeans. Russians' mental age was 11.34; Italian, 11.01; and Poles, 10.74. (3) Blacks are at the bottom of the scale, with a mental age of 10.41 (Gould, 1981, p. 197). There is insufficient space to recount Gould's description and critique of the contents and procedure for taking the test, but one or two of the multiple-choice test items should alert us to the possible bias inherent in the selection of items used to measure intelligence: "Crisco is a: patent medicine, disinfectant, toothpaste, food product" or "Christy Mathewson is famous as a: writer, artist, baseball player, comedian." Is it any wonder that recent immigrants scored lower than long-term residents? The use of such items further serves to reinforce concern for the definition of intelligence—that is, the items used in its measurement. The interaction of this definitional problem with individual and group difference score is made clear by such examples.

The reification of intelligence and the attempt at measuring it have historically been connected to the belief system that (1) the capacity exists, (2) it can be measured, (3) it is inherited, and (4) groups differ, with darker people having less of it than white people. These remain *a priori* assumptions to which studies have been addressed (Gould, 1981). The possibility of studying mental abilities is a valid one that requires the disengagement of such assumptions.

THE ISSUE OF INTERVENTION

It has been argued that the success or failure of intervention programs in early childhood and infancy is an indicator of the effect of environment on the intellectual growth and capacity of the child. If a variety of intervention programs are shown to be ineffective, then intervention or environment *per se* is ineffective in altering the intellectual performance capability of the infant. If, on the other hand, the intervention procedures are effective, then changes in environment are a useful tool in altering intellectual performance. The use of intervention becomes highly relevant to the issue of infant intelligence and its ability or inability to change with time or situation. Indeed, one might argue that this is the one important method for getting at the effect of environment on the infant's intellectual capacity.

Already discussed is some problem of measuring the effect of intervention, notably the difficulty in considering infant intelligence as a unitary construct measured by a single instrument. Moreover, it is necessary to match the evaluation of the intervention with the appropriate instrument. Thus, if one were affecting the object permanence capability in the young infant by such interventions as a peekaboo game and showing the child how to find hidden objects, then the type of measurement should not be a Peabody Picture Vocabulary Test or some other verbal task but rather a specific measure of sensorimotor capacity. Thus, if intervention programs are used as a means of assessing whether the infant's intellectual ability is fixed and unalterable by environment, it is necessary to match the nature of the intervention procedure to the criteria of effectiveness. This, unfortunately, is rarely done.

Of even more importance is the notion that all children can benefit from the same type of intervention procedure. This is a naive view, yet, unfortunately, it is also widely held. Under this model every infant in an intervention procedure must receive the same treatment, for the sake of either "scientific objectivity" or technical simplicity. Thus, every child is to watch a particular TV program or be instructed about a particular concept using a particular set of instructional materials. The popularity of this method is surprising, since both the educational experience in the classroom and, more recently, the educational psychology literature have increasingly shown that children need different types of intervention programs to arrive at the same goal. Both children and infants, after all, come into the intervention procedure with different kinds of experience. The educational literature refers to the aptitude–treatment interaction or the subject–treatment interaction in discussing this issue. In order to reach the same goals, it is often necessary to apply different kinds of intervention (have various curricula), dependent on the characteristics of the child. In a review of this subject, Berliner and Cahen (1973) make a strong argument for this position in educational programs and, more importantly, in the evaluation of their effectiveness. The evaluation of environmental effectiveness is one important way to consider the innate quality of infant intellectual capacity.

How we create difficulty by not considering the subject–treatment interaction and by not evaluating the effect of the child's experience on intervention programs can be seen in this example. Assume that 100 children are in intervention program A and that 10% of these children show increases in some measure of the effectiveness of intervention A. On the other hand, 90% show no effect or even show some negative effect of intervention A. When looking at the data of the experimental group averaged over all 100 children, it must be concluded that Intervention A was not a success. If there were 10 different intervention programs, each of them helping a different 10% of the experimental group, it would be concluded that each of these programs failed to affect the measured capacity of these 100 children. Thus, intervention *per se* seems ineffective in influencing the child's intellectual capacity. In fact, this is not the case. Each program succeeded in affecting the children's intellectual capacity but did not do so for the group as a whole. Thus, across all 10 programs, the intellectual capacity of all 100 children, at least of those that were supposed to be affected by the intervention procedures, did show improvement. However, when looking at mean data, no significant positive effect can be located. This example argues most powerfully for a subject–treatment interaction design, both in terms of the nature of the program to be used and in terms of the evaluations of the effectiveness of that program. It becomes, then, the function of the experimenter, the curriculum developer, the evaluator, and the theoretician to find what conditions individual children need to optimize their intellectual capacities. Until such programs are initiated, it is not far wrong to conclude that intervention *per se* is ineffective or, further—and central to the present argument—that intellectual capacity cannot be altered by environmental change. This discussion makes it obvious that the methods of evaluation and the methods of intervention have not yet reached sufficient sophistication to be used as tests of the effect of the environment on infant intellectual capacity.

A final issue has to do with the nature and effort of infant intervention programs. Remember that the hypothesis underlying intervention is that IQ can be affected by environmental factors. How is this general hypothesis tested? We wish to argue from the outset that those supporting an environmental position do so within conditions that restrict the opportunity for a significant effect. If intervention is, in fact, designed to test the malleability of infant intelligence, why then are infant intervention programs so limited in their scope? It would seem that in order fully to test the hypothesis under discussion, large amounts of time, energy, and money *per* child should be invested in the intervention program. Thus a child and his/her family should receive as much money as necessary to change their total life style—for example, the housing, medical care, nutrition, and educational experiences for all the family. In other words, a complete and total program of environmental change. Instead, what is normally undertaken? Several hours a day, at the most, of a limited type of intervention program are provided, usually just for the infant. How can one argue for the hypothesis of malleability of IQ through environmental influence when the total environment is hardly affected, at least in terms of what we could think of doing? What underlies this limited strategy is the question of how IQ can be affected for the least amount of money. While such a question is an important one for any society to ask, it is a second order question, one to be asked after malleability has been demonstrated. This second order question pertains to sociopolitical issues having to do with fiscal spending priorities of the society and the distribution of resources. To confuse this question with that of malleability is to fail to address the scientific issue. The first order of inquiry must be to test the general hypothesis, and to do so, great quantities of time, effort, and money must be expended. Once it has been demonstrated that with the expenditure of unlimited resources and with a complete change of environment it is possible to alter intellectual capacity, the next question—and a sociopolitical one, at that—must be the cost of this change and the test of practicality for the society. However, to ask such questions prior to the test of the general hypothesis is both to interfere with the test of the hypothesis and to insert sociopolitical factors into the discussion of environmental influences on intellectual capacity.

RECAPITULATION: A CONCEPT OF INTELLIGENCE

Recall that at the outset, we outlined several features that needed to be reviewed when the concept of intelligence is being considered. The discussion of these features revealed that none of them appears free from considerations that are not solely scientific. This is not surprising, for it would be naive to believe that science in general and certainly psychology in particular is devoid of moral considerations. Science, although it may try, is not valueless, but value-laden. The task of the scientist is to expose the values so that their consequences will be apparent. When science fails to recognize the values associated with the particular scientific effort, scientists fall prey to bias. This point should be clear from Kamin's (1975) and Gould's (1981) analyses of the values of many of the scientists responsible for the current views and data on IQ. If it is true that all the features inherent in the concept of IQ have sociopolitical aspects, it may not be too strong to suggest that the IQ score up until now has been more of a sociopolitical than a scientific construct. In the section to follow the function of the IQ score is explored.

The Use and Function of Infant IQ Scores

A review of the empirical research on infant intelligence tests supports the notion there is no consistency across or within age in a wide variety of tests purported to measure infant mental functioning. Therefore, the concept of a developmentally constant, general, unitary concept of intelligence is not very tenable. What do these conclusions imply for the notion that intelligence is "inherent or at least innate, not due to teaching or training and remains uninfluenced by industry or zeal" (Burt et al., 1934, p. 124)? Such a model of human capacity must clearly be dealt a severe blow by a review of the infancy literature. And yet such a conception of people remains and tests continue to be sold and taken. While these intelligence scales have been acknowledged to have limited function, they still are widely used in clinical settings in the belief that, although lacking in predictive validity, they provide a valuable aid in assessing the overall health and developmental status of babies at the particular time of testing. This procedure is justified only if the scores are regarded solely as measures of present performance, not as indicators of future potential. What this "present performance" may mean is questionable, since superior performance may be followed by poor performance. For example, Bayley (1955) shows a negative correlation of .30 between males' early test behavior and their IQs at 16–18 years. Just as infant scales are quite invalid as measures of future potential, it is also unlikely that they properly assess a child's current performance vis-à-vis the other children, except when extreme samples of dysfunction are utilized.

Concurrently, intelligence test scores are widely used as criterion measures in the evaluation of infant intervention or enrichment programs. The experimental subjects are compared with the control subjects in terms of their performance on intelligence tests. If the scores of the experimental group are higher than those of the control group, the program is evaluated positively; if not, it is evaluated negatively. Implicitly assumed is that infant intelligence is a general, unitary capacity and that mental development can be enhanced as a result of an enrichment experience in a few specific areas. Zigler and Trickett (1978) suggest that the use of IQ as a measure in evaluation of childhood intervention programs became popular because standard IQ tests are well-developed instruments that are easily administered. Similarly, it is assumed that infant scales are adequate to reflect an improvement that occurs in competence as a consequence of a specific enrichment experience. However, as our previous discussion has made clear, infant intelligence—or intelligence at any age, for that matter—considered as a general unitary capacity is highly questionable. Moreover, that infant skills are adequate to reflect improvement from specific enrichment experiences must also be questioned, since a variety of infant skills tested shows relatively little interscale or intrascale consistency. In part, the need for different measures has been recognized for some time. IQ scores need to give way to measures such as social competence (Anderson & Messick, 1974; Zigler & Trickett, 1978) or intellectual competence (Scarr, 1981).

The data on infant intelligence tests also cast doubt on whether the scores can be generalized beyond the particular set of abilities or factors sampled at the time of testing. An infant who showed dramatic gains in testing that involved sensorimotor function would not necessarily manifest such gains in tests involving verbal skills. The implication of these conclusions for a wide variety of evaluation policies concerning infant intervention must be considered. For example, infant intelligence scales, no matter how measured, are quite unsuitable instruments for assessing the effects of specific intervention procedures, primarily because infant intelligence is not a general unitary trait but is rather a composite of skills and abilities that do not necessarily covary. Such a view of intelligence is by no means new (see for example, Guilford, 1959; Thurstone, 1938), but it is one that must be repeatedly stated in order to counteract the tendency to utilize simple and single measures of infant intelligence. An example can be used to clarify this issue.

Consider an intervention procedure that is designed primarily to influence sensorimotor intelligence—for example, the development of object permanence. An appropriate curriculum must involve training infants in a variety of peekaboo and hide-and-seek tasks. According to the data presented, a standard infant intelligence scale would be the wrong instrument to use in assessing the efficiency of such a program and is likely to lead to erroneous conclusions concerning the program's efficacy. Even more serious is the possibility that by using the wrong instrument of evaluation over a large number of programs, one would erroneously conclude that intervention in general is ineffective in improving intellectual ability, thus supporting the genetic bias that environment is ineffective in modifying intelligence. There are few who would suggest that schoolchildren be administered a standard intelligence test after a course in geography. Yet such a procedure would be exactly analogous to using an intelligence test to measure the success of teaching the object permanence concept to the infant. The success of a geography course is best assessed by a test of geographical knowledge; by the same token, the success of a program stressing sensorimotor skills is best assessed by specific tests of sensorimotor ability. In both cases there may in some instances be improvement in intelligence test scores; but such improvement has to be regarded as fortuitous.

The general view that intelligence is a unitary construct and easily measured cannot be supported by the data. Why, then, should this view of intelligence hold such a dominant position in the thinking of contemporary scientists and public alike? The answer to such a question may be found in a consideration of the functional use of the IQ score in a complex and technological society. The analysis of Gould (1981) on the misuse of the measurement of intelligence suggests that one of the more important functions of the IQ score remains its use in stratifying society into a hierarchy having those with high IQ considered better in all ways than those with low IQ scores. Moreover, the selection of items and procedures for such tests, as the example of the Army tests demonstrates, is to benefit those already at the top of that hierarchy. The test then serves to justify the existing order. Moreover, by arguing for the innate quality of the measured difference, the society or those members at the top of the hierarchy can both justify their position and resist any reordering. Such a belief system, supported by the set of *a priori* beliefs (which are always confirmed "scientifically"), has been labeled "social Darwinism." The purpose of these self-confirming beliefs is in the preservation of the social hierarchy. In complex societies, it serves to maintain a caste system and to create a division of labor within the culture— that is, to determine who will go to school in the first place, who will get into academic programs that lead to college, etc. These divisions, in turn, determine the nature of labor the children will perform as adults and, in turn again, will determine the children's socioeconomic status and social, economic, and political position in the society. This division of labor and, as a consequence, the division of goods and services available to those who succeed are a necessity in a complex society. This stratification is then justified by scores on a test designed to produce just such a stratification. If we cannot make the claim that IQ differences are genetically determined, then we must base them on differences in cultural learning. But these differences in cultural learning, for the sake of the division of labor, are exactly what the IQ tests are intended to produce. The hierarchy produces the test differences and the test differences are used to maintain the hierarchy. Thus, IQ scores have come to replace, in part, the class systems or feudal systems that previously had the function of stratifying society and distributing the goods and wealth of that society. Whereas these earlier systems were supported by the invoking of constructs having to do with the Almighty, the present system invokes Mother Nature instead. In any social system in which stratification is necessary in order to distribute the wealth of that society in a disproportionate fashion, some sort of stratifying device is necessary. The twentieth century technological society's stratification device has become the intelligence test. As such, the intelligence test rests more on its function for distribution of wealth than on its scientific merit.

REFERENCES

ANDERSON, S., & MESSICK, S. Social competency in young children. *Developmental Psychology*, 1974, *10*, 282–293.

BAYLEY, N. On the growth of intelligence. *American Psychologist*, 1955, *10*, 805.

BAYLEY, N. *Bayley scales of infant development: Birth to two years*. New York: Psychological Corp., 1969.

BAYLEY, N. Development of mental abilities. In P. Mussen (Ed.), *Carmichael's manual of child psychology*. New York: Wiley, 1970.

BERG, I. *Education and jobs: The great training robbery*. New York: Praeger, 1970.

BERLINER, D. C., & CAHEN, L. S. Trait-treatment interactions and learning. In L. Kerlinger (Ed.), *Review of research in education*. Itasca: Peacock, 1973.

BINET, A. Historique des recherches sur les rapports de l'intelligence avec la grandeur et la forme de la tête. *L'Année Psychologique*, 1898, *4*, 245–298.

BINET, A. Recherches sur la technique de la mensuration de la tête vivante. *L'Année Psychologique*, 1900, *5*, 314–429.

BURT, C. *The factors of the mind*. London: University of London Press, 1940.

BURT, C., JONES, E., MILLER, E., & MOODIE, W. *How the mind works*. New York: Appleton-Century-Crofts, 1934.

CORMAN, H. H., & ESCALONA, S. K. Stages of sensorimotor development: A replication study. *Merrill-Palmer Quarterly*, 1969, *15*, 351.

CRONBACH, L. J. Five decades of public controversy over mental testing. *American Psychologist*, 1975, *30*, 1.

DUNCAN, O. D., FEATHERMAN, D. L., & DUNCAN, B. *Socioeconomic background and achievement*. New York: Seminar Press, 1972.

ELTON, C. F., & SHEVELL, L. R. *Who is talented? An analysis of achievement* (Research report No. 31). Iowa City, Ia.: American College Testing Program, 1969.

GALTON, F. *Hereditary genius*. New York: D. Appleton, 1884.

GHISELLI, E. E. *The validity of occupational aptitude tests*. New York: Wiley, 1966.

GODDARD, H. H. *The Kallikak family: A study in the heredity of feeble-mindedness*. New York: Macmillan, 1912.

GOULD, S. J. *The mismeasurement of man*. New York: W. W. Norton, 1981.

GUILFORD, J. P. Three faces of intellect. *American Psychologist*, 1959, *14*, 469.

HEARNSHAW, L. S. *Cyril Burt psychologist*. London: Hodder & Stoughton, 1979.

HOLLAND, J. L., & RICHARDS, J. M., JR. *Academic and nonacademic accomplishment: Correlated or uncorrelated?* (Research Report No. 2). Iowa City, Ia.: American College Testing Program, 1965.

JENCKS, C., SMITH, M., ACLAND, H., BANE, M. J., COHEN, D., GINTIS, H., HEYNS, B., & MICHALSON, S. *Inequality: A reassessment of the effect of family and schooling in America*. New York: Harper & Row, 1972.

JENSEN, A. R. The heritability of intelligence. *Saturday Evening Post*, 1972, *244*, 9, 12, 149.

KAMIN, L. F. *The science and politics of IQ*. New York: Halsted Press, 1974.

KING, W., & SEEGMILLER, B. *Cognitive development from 14 to 22 months of age in black, male, first-born infants assessed by the Bayley and Hunt–Użgiris scales*. Paper presented at the meeting of the Society for Research in Child Development meetings, Minneapolis, 1971.

KOHLBERG, L., LACROSSE, J., & RICKS, D. The predictability of adult mental health from childhood behavior. In B. Wolman (Ed.), *Handbook of child psychopathology*. New York: McGraw-Hill, 1970.

LAMB, M. E., & SHERROD, L. R. (Eds.). *Infant social cognition: Empirical and theoretical considerations*. Hillsdale, N. J.: Erlbaum, 1981.

LEWIS, M., & BROOKS, J. Infants' social perception: A constructivist view. In L. Cohen & P. Salapatek (Eds.), *Infant perception: From sensation to cognition*. New York: Academic Press, 1975.

LEWIS, M., & BROOKS-GUNN, J. *Social cognition and the acquisition of self*. New York: Plenum Press, 1979.

LEWIS, M., & BROOKS-GUNN, J. Visual attention at three months as a predictor of cognitive functioning at two years of age. *Intelligence*, 1981, *5*, 131–140.

LEWIS, M., & ENRIGHT, M. K. *The development of mental abilities: A multidimensional model of intelligence in infancy*. Manuscript submitted for publication, 1982.

LEWIS, M., & LEE-PAINTER, S. *Mother-infant interaction and cognitive development*. Paper presented at the meeting of the Eastern Psychological Association, Philadelphia, 1974.

LEWIS, M., & McGURK, H. The evaluation of infant intelligence: Infant intelligence scores—true or false? *Science*, 1972, *178*(4066), 1174.

LEWIS, M., & MICHALSON, L. *Children's emotions and moods: Developmental Theory and measurement*. New York: Plenum Press, 1983.

LEWIS, M., & ROSENBLUM, L. (Eds.). *The origins of behavior* (Vol. 2). *The origins of fear*. New York: Wiley, 1974.

LEWIS, M., & ROSENBLUM, L. *The origins of behavior* (Vol. 5). *Interaction, conversation, and the development of language*. New York: Wiley, 1977.

McCall, R. B., Hogarty, P. S., & Hurlburt, N. Transitions in infant sensorimotor development and the prediction of childhood IQ. *American Psychologist*, 1972, *27*, 728.

McCall, R. B., Eichorn, D. H., & Hogarty, P. S. Transitions in early mental development. *Monographs of the Society of Research in Child Development*, 1977, *42* (3, Serial No. 171).

McClelland, D. C. Testing for competence rather than for "intelligence." *American Psychologist*, 1973, *28*, 1–14.

Morton, S. Observations on the size of the brain in various races and families of man. *Proceedings of the Academy of Natural Science*, 1849, *4*, 221–224.

Nuttall, R. L., & Fozard, T. L. Age, socioeconomic status and human abilities. *Aging and Human Development*, 1970, *1*, 161–169.

Overton, W. On the assumptive base of the nature-nurture controversy: Additive versus interactive conceptions. *Human Development*, 1973, *16*, 74.

Richards, T. W., & Nelson, V. L. Abilities of infants during the first eighteen months. *Journal of Genetic Psychology*, 1939, *55*, 299–318.

Scarr, S. Testing for children. *American Psychologist*, 1981, *36*, 1159–1166.

Scarr-Salapatek, S. Race, social class and IQ. *Science*, 1971, *174*, 1285.

Siegel, L. S. Infant tests as predictors of cognitive and language development at two years. *Child Development*, 1981, *52*, 545–557.

Spearman, C. General intelligence objectively determined and measured. *American Journal of Psychology*, 1904, *15*, 201–293.

Stott, L. H., & Ball, R. S. Infant and preschool mental tests: Review and evaluation. *Monographs of the Society for Research in Child Development*, 1965, *30* (Serial No. 101).

Taylor, C., Smith, W. R., & Chiselin, B. The creative and other contributions of one sample of research scientists. In C. W. Taylor & F. Barron (Eds.), *Scientific creativity: Its recognition and development*. New York: Wiley, 1963.

Terman, L. M. Genius and stupidity. A study of some of the intellectual processes of seven "bright" and seven "stupid" boys. *Pedagogical Seminary*, 1906, *13*, 307–373.

Thomas, H. Psychological assessment instruments for use with human infants, *Merrill-Palmer Quarterly*, 1970, *16*, 179.

Thorndike, R. L., & Hagen, E. *10,000 careers*. New York: Wiley, 1959.

Thurstone, L. L. *Primary mental abilities*. Chicago: University of Chicago Press, 1938.

Užgiris, I. Č. Patterns of cognitive development in infancy. *Merrill-Palmer Quarterly*, 1973, *19*, 181.

Užgiris, I. Č., & Hunt, J. McV. *Assessment in infancy: Ordinal scales of development*. Unpublished manuscript, University of Illinois, Urbana, Ill.: 1975.

Wilson, R. S. Testing infant intelligence. *Science*, 1973, *182*, 734–737.

Zigler, E., & Trickett, P. K. IQ, social competence, and evaluation of early childhood intervention programs. *American Psychologist*, 1978, *33*, 789–798.

2   *Origins of Infant Intelligence Testing*

JEANNE BROOKS-GUNN AND MARSHA WEINRAUB

Infant intelligence tests, like all other psychological tests, have their roots in the intelligence testing movement of the late 19th and early 20th centuries. Infancy was never the focus of the early test developers and was studied only because *idiots*, the lowest classification of the mentally retarded, were thought to exhibit the mental abilities of a 2-year-old (Binet & Simon, 1905a,b). However, infants were not entirely neglected, since the child study movement led by Darwin and Preyer increased interest in the early development of the species. As the testing movement gained momentum and branched out into more areas in the 1920s, intensive studies of infants and preschoolers were initiated, leading to the development of normative scales and intelligence tests. These tests of the 1920s and 1930s were rapidly accepted and were used to investigate stability, reliability, and predictive validity from infancy to childhood. Two other waves of tests were developed, in the 1960s and in the 1970s, with the focus of the 1970s tests being somewhat different from those preceding them.

In this chapter, the origins of infant intelligence tests will be examined and the developments that followed will be traced. It is our hope that this historical account of infant testing will provide a framework in which to interpret the issues that will be addressed in this decade.

JEANNE BROOKS-GUNN • Institute for the Study of Exceptional Children, Educational Testing Service, Princeton, New Jersey 08541.   MARSHA WEINRAUB • Department of Psychology, Temple University, Philadelphia, Pennsylvania 19122.   The preparation of this chapter was supported by the Department of Education, Office of Special Education, under the aegis of an Early Childhood Research Institute.

ORIGINS OF INTELLIGENCE TESTING: THE NINETEENTH CENTURY

The late nineteenth century was an exciting time for scientists. Advances in medicine, the surge of interest in evolution, the humane concern for the mentally deficient and the insane—all were widely discussed in western Europe. The variety of scientific interests were favorable for mental testing in that the topic had an eclectic base. The testing involvement in different countries sprang from different concerns. As cross-fertilization and adaptation occurred, the strong national flavor subsided, as illustrated by the rapid acceptance of the Binet tests.

Mental Retardation and Diagnosis: France

That there are differences among persons with respect to intellect as well as physiognomy has long been recognized. Mythology abounds with references to such differences in beasts and man, and the Greeks, as well as others, were said to have destroyed so-called inferior individuals in infancy (Peterson, 1925). Physiognomy and especially phrenology were believed to be related to both intellectual and emotional deficiencies. Relationships between intellectual and physiological characteristics are receiving renewed interest today, as neurologists talk about "soft signs" such as webbed fingers, hair swirls, and attached ear lobes as being indicative of neurological insult and perhaps predictive of learning disabilities (Waldrop, Bell, McLaughlin & Halverson, 1978).

Prior to the 1800s, mental deficiency and insanity were not differentiated. The first attempt to do so was made by Esquirol, a French doctor who, in his 1838 treatise on mental illness, stated that

> Idiocy is not a disease but a condition in which the intellectual facilities are never manifested or have never been developed sufficiently to enable the idiot to acquire such amount of knowledge as persons of his own age reared in similar circumstances are capable of receiving.

Not only did Esquirol differentiate between mental retardation and insanity, but he also anticipated several later developments in the study of mental retardation. First, understanding that retardation was not an all-or-none phenomenon, he identified a continuum from normalcy to idiocy. Second, in his attempt to classify mental retardates (which was the first attempt to do so), language capability was used as the classification criterion rather than measurements of sensations or physiognomy. That speech was probably the best indicator of intellectual functioning was not mentioned again until Terman's 1918 statement (Goodenough, 1949; Terman, Kohs, Chamberlain, Anderson, & Henry, 1918).

At the same time that the French medical community was distinguishing between the mentally deficient and the insane, a controversy arose about the educability of the former. Esquirol himself did not believe that the retarded would benefit from training, which was the stance taken by most of the Paris medical community. Some 30 years before Esquirol's treatise appeared, Itard had published his accounts of the Wild Boy of Aveyron (1801/1938), whom he had spent five painstaking years trying to train. Although the boy improved dramatically, he was not able to function in society and Itard felt he had failed.

However, the French scientific community did not believe that Itard had failed (Goodenough, 1949; Seguin, 1896/1907). In fact, Itard's student, Seguin, started a school for the mentally deficient in 1837, in the belief that training was possible. The program was a success in that students did learn, albeit within certain limitations. Visitors came from all over the world and, on returning to their countries, opened similar schools. By 1870, there were 80 schools for the mentally retarded throughout the western world (Goodenough, 1949).

Although schools were being opened at a rapid rate, no universal or empirically based criteria for the evaluation of mental deficiency existed. Therefore, admission procedures were quite arbitrary. As Binet said, on surveying the admission criteria at a number of schools and institutions in France, "We have compared several hundreds of these certificates, and we think we may say without exaggeration that they looked as if they had been drawn by chance out of a sack" (Binet & Simon, 1907/ 1914). Thus, the need for a reliable and valid diagnostic procedure was recognized by the French medical and pedagological community in the late nineteenth century. However, nothing was done until 1904, when the Minister of Public Instruction in Paris appointed a commission to study the question of special education.

This commission authorized the development of a mental test to be given to school children in Paris to identify those who would benefit from special education. Binet, a member of the commission, set out to develop this screening test. Binet was not a clinician, but an experimental psychologist who had studied mental functioning in normal children for at least 10 years before the appearance of his now famous test. However, Binet had always been interested in retardation and had even written a book on the subject (Peterson, 1925), although he was not involved in any retardation research until 1900. In 1900, Simon, under Binet's direction, gave anthropometric tests to both mentally retarded and normal boys, but these tests failed to distinguish between the two groups (Peterson, 1925; Simon, 1900).

Evolutionary Theory and the Testing Movement in England

Concern about individual differences in mental functioning was not limited to France, however. In the same time in England, Charles Darwin was developing his revolutionary theory of evolution, culminating in *The Origin of Species* (1859) and *Expression of the Emotions in Man and Animals* (1872/1973). The ideas that intra- as well as interspecies variation might be due to evolutionary forces and be intimately tied to the biological make-up of individuals were a catalyst for Galton's work. Galton, who happened to be a cousin of Darwin's, noticed that eminence ran in families; this observation was easily related to the notion of evolution and human variation. Galton's book *Hereditary Genius*, published in 1869, traced the inheritance of genius in families. When Galton's interests in both intelligence and individual differences became coupled with a belief in the efficiency of quantitative measurement, he opened a testing laboratory in the South Kensington Museum in 1882 where anyone could take a variety of anthropometric tests for a small fee. It was thought that such measurements might be related to mental functioning. Galton's contributions to the testing movement were enormous, for he pioneered work regarding the genetic inheritance of intelligence, individual differences in general, and the quantification of psychophysiological abilities.

In addition, the work of Darwin and Galton generated interest in the development of certain abilities and behaviors in children. They believed that the origins of behavior might be elucidated by a study of ontogenesis. One of Darwin's theses, of course, was that facial emotions were not learned but were biological in nature (1872/1973). What better way to study this than to watch the development of emotions in young children. A number of scientists, including Darwin himself, did just that, and numerous diaries and records of infant development were published (e.g., Darwin, 1877; Preyer, 1882/1888; Shinn, 1900; Stern, 1914/1924). These studies clearly showed that there were developmental behavior sequences and that at the same time there were individual differences in rates of development (Goodenough, 1949). These diaries provided many ideas for test items in the infant intelligence tests developed in the 1930s (Buhler, 1930; Buhler & Hetzer, 1935; Shirley, 1933).

The Psychology of Sensation: Germany

The psychology of sensation and perception originated in Germany. Individual differences were seen as sources of errors rather than legitimate areas of study, as in France and England. However, even in the most controlled conditions, individual differences still appeared. Cattell, while working under the direction of Wundt, became interested in this phenomenon. Although Wundt was said to disapprove (Cattell & Farrand, 1896; Peterson, 1925), Cattell did his dissertation on individual differences in college students.

The fact that the Germans could not eliminate individual differences had an important effect on the testing movement. As Goodenough (1949) stressed,

> the work of the psychophysicists served to point out the need for a different kind of psychological research by their very failure to account for all the sources of variation in responses. (pp. 30–31)

EARLY TESTS OF INTELLIGENCE

Even though nineteenth century German psychology was not conducive to the testing of individual differences, in the 1890s there was a flurry of test development in Germany as well as in the rest of Europe. Series of tests were developed in America and Germany both for school-age children (Gilbert, 1894; Munsterberg, 1891; Boas, 1891, cited in Peterson, 1925) and for college students (Cattell, 1890; Jastrow, 1892; Kraepelin, 1895). It was Cattell who first introduced the term "mental tests" in his article "Mental Tests and Measurements" (1890). The tests developed at this time measured sensory, anthropometric, and memory functions as well as reaction times and imagery. An underlying assumption was that intelligence was related to the simpler sensory and memory functions. Cattell spoke for many when he said that the "measurements of the body and of the senses come as completely within our scope as the higher mental processes" (Cattell & Farrand, 1896).

There were critics of Cattell's view, the most vocal being Binet. In the first issue of *L'Année Psychologique*, a journal begun by Binet in 1895, Binet criticized the German and American tests as being too sensory in nature. Instead of choosing the most simple functions as measures of intelligence, he argued that higher mental functions should be measured since they are more likely to differentiate among individuals (Binet & Henri, 1895, 1896). Binet and Henri listed memory, attention, and comprehension as 3 of the 10 mental functions that they felt should be studied.

Binet's advice was not followed, however, and the testing movement outside of France lost its momentum. Peterson (1925) suggested that disillusionment with testing was due in part to the fact that two widely publicized studies found no relationship between the current mental tests and school performance (Wissler, 1901; Sharp, 1898–1899), even though the failures were due to the homogeneity of the students tested at Cornell (all were graduate students) and to the sensory nature of the tests at Columbia.

Development of the 1905 Binet Scale

Binet, however, continued his work on the development of mental tests. His studies, which were usually conducted with normal, school-age children, were published in the journal *L'Année Psychologique*. A study of the journal from its inception in 1895 to the early 1900s shows Binet gradually moving away from the measurement of "simple" tasks and toward the measurement of more complex mental tasks, as study after study indicated that sensory and physical performance did not increase systematically with age and were not related to school performance. Binet, in the French tradition of practicality, found it easier to reject concepts that proved fruitless than did his German and American counterparts. His well-known definition of intelligence, for example, reflects his increasing interest in complex mental function. "To judge well, to comprehend well, to reason well, these are the essentials of intelligence. A person may be a moron or an imbecile if he lacks judgment; but with good judgment he could not be either" (Binet & Simon, 1905a, p. 196).

The 1905 scale, then, reflected the importance of complex mental tasks and the inability of frequency tests or sensory items to differentiate among students. In addition, the scale introduced more pass/fail items and included items increasing in difficulty. The 1905 scale included 30 items that were suitable for children ranging from idiocy to normalcy in intellectual ability. Although not age-normed, the test distinguished among the three classes of retardation and between retarded and normal intellectual ability. The first six items covered the competencies of idiots and normal infants up to 2 years of age. The first item was following a lighted match with the eyes and head, the fourth was making an appropriate choice between a piece of wood and a piece of chocolate, and the sixth was following simple orders and imitating gestures. Fifty normal children aged 3 to 11 and an unspecified number of retarded children in both schools and institutions were used in the standardization sample.

The 1905 scale revolutionized the testing movement. Intelligence would no longer be measured in terms of simple sensory tests, and mental-age levels and unitary test scores became firmly established (despite the fact that Binet himself was not totally enamored of the concept of a single score).

Tests Following the 1905 Scale

A number of revisions made by both Binet and others, including Binet's own 1908 and 1911 revisions, appears (Binet & Simon, 1908, 1914). In America, Goddard translated the 1908 scale in 1910 and Kuhlmann modified the same scale in 1911.

Mental testing was quickly adopted by the special education community. Goddard, at the Vineland Training School in New Jersey, had been searching for a diagnostic test when he discovered the 1908 Binet scale. He gave his translation to hundreds of children in his institution and in the public schools in Vineland (1910a,b, 1911). This test was widely used in special education until Terman's Stanford–Binet test appeared in 1916.

At the same time, conditions in the educational system also led to a rapid acceptance of testing. At the turn of the century, compulsory education up to the age of 13 or 14 was prevalent in many states. Although progressive in nature, these laws affected the slow learner adversely. The child with limited intellectual capacity was left in the lower grades for longer and longer periods of time as the compulsory age increased. In 1908, the United States Bureau of Education reported that the percentage of children who were more than three years below age grade level ranged from over 20% in Birmingham to 6%–10% in Detroit, Philadelphia, and Los Angeles (Strayer, 1911, in Goodenough, 1949). Concern with these conditions led to the establishment of special classes within the public school system and of diagnostic testing for placement.

Testing also had advocates in the scientific community, as the tests developed by Terman and Kuhlmann indicate. Much of the interest in the testing of children was generated at Clark University under G. Stanley Hall. Goddard, Terman, and Kuhlmann all studied with him, although it was said that Hall opposed the study of mental tests (Peterson, 1925).

As the concept of intelligence was redefined and reexamined, so too were notions regarding the constancy of intelligence through the lifespan. According to Goodenough (1928), it had been generally assumed that the IQs of preschool children were as stable over time as the IQs of school-age children. In the 1920s, the interest in the preschool child and the effects of nursery school prompted an investigation of this assumption. The tests developed in the 1920s were further revisions and extensions of Binet's original test to provide for the measurement of the mental development of preschool children. Burt (1921) produced an English revision that measured intelligence in children as young as 3 years of age. The Yerkes, Bridges, and Hardwick Point Scale, originally introduced in 1915 for children aged 4 and older, was revised and extended downward in 1923.

The Earliest Infant Tests

Testing of school-age children was firmly established by 1910, but what about testing of children under 5 years of age? Pedagogical demands were for diagnostic instruments for the school-age, not the preschool, child. Although interest in the preverbal child was not high, a small cadre did pursue the issue.

Binet, for example, included items for children in the first years of life and for idiots in the 1905 scale, but excluded them in the 1908 and 911 revisions. Goddard's revision did not include items for young children although Kuhlmann's did. The 1916 Stanford–Binet, which became the standard mental test, had no items for the preverbal child.

Even though no infant intelligence tests were developed in the early 1900s, the diaries of men such as Darwin or early development suggest that infancy was seen as an important and valid topic of study. In addition, the social welfare movement, coming of age in the twentieth century, was concerned with early diagnostic instruments for the placement of adoptive children.

The general lack of interest in infant testing is illustrated by the fact that the first infant test never received any attention. In 1887, an article on the testing of children up to 3 years of age appeared in the *New Orleans Medical and Surgical Journal.* Chaille, the physician who wrote the article, devised a test that predated Binet's by eight years on a number of important developments, such as the concept of mental age. The test was probably too advanced for its time.

Three infant tests were published in the 1920s. In 1922, Trabue, an educator, and Stockbridge, an author and journalist, introduced the *Mentimeter.* It was designed to measure inherited mental ability at all ages. It was to be used by parents, employers and teachers, as well as by anyone interested in self-knowledge and self-improvement. What was unusual about the Mentimeter was that it included tests for infants. There were three items each at the 3- and 6-month level, three items at 1 year, and six items each at 3 and 4 years. These items tapped various aspects of muscular control, imitation, language production, and language comprehension. Administration was informal, the tests were not standardized, and reliability and validity were not assessed. These tests were not influential in the infant-testing movement, probably because they were intended for a lay audience. Nevertheless, their appearance demonstrated the public interest in early testing.

The first serious professional attempt to measure mental development in infancy was Kuhlmann's second revision (1922) of the Binet–Simon Scale. Kuhlmann both revised the scale and extended it downward to include five items for each of the following age groups: 2, 6, 12, 18, and 24 months. The new test included measures of coordination, imitation, recognition, and speech. As with other intelligence tests of the day, a mental age and intelligence quotients were calculated. Only a small number of children were tested at the preschool and infant levels;

these small samples were probably unrepresentative, judging from the high IQs that were obtained.

The Kuhlmann infant tests were never widely used. For example, in Goodenough's (1928a) validation study of the test, no children under 18 months were included in the sample of nearly 500. Although Goodenough recognized that infant testing was an important and promising area, she believed that the currently available tests were not "sufficiently refined to render them serviceable for use in the solution of problems for which a high degree of precision is necessary" (p. 127).

In 1928, Linfert and Hierholzer produced the first well-standardized scale of intellligence for infants. The test consisted of two series: 29 items for infants from 30 to 132 days old, and 35 items for infants from 153 to 365 days. The tests were standardized on 300 infants, and 50 of the infants were seen longitudinally. Although the best infant test to date, it never received widespread attention.

Infant Testing in the 1930s

By the end of the 1920s and the early 1930s, the infant testing movement had built up steam, so that many tests were being developed. Several now-renowned investigators—Gesell at the Yale Clinic of Child Development, Buhler in Vienna, Shirley at the University of Minnesota, Bayley in the Berkeley Growth Project, and Fillmore at the University of Iowa—began to publish reports of scales they and their colleagues had been working on for some time.

Gesell Developmental Schedules

The most extensive series of investigations of infants and preschool children was carried out at the Yale Clinic of Child Development under the direction of Arnold Gesell, a pediatrician. These investigations continued for nearly 40 years, refining methods of observation and collecting extensive normative data. The tests originated out of the need for normative data on behavioral development. Working in association with the pediatric and well-baby clinics of the New Haven Dispensary, doctors had been increasingly called upon to examine children psychologically as well as physically. Finally, in 1916, a project was undertaken to observe and record infant and preschool behavior at home and in the clinic in order to develop comprehensive norms.

In the preliminary investigation, 50 children were observed at each of 10 age levels—birth, 4, 6, 9, 12, and 18 months and 2, 3, 4, and 5 years. Data on *motor development* (including posture and prehension), *language development* (including comprehension and imitation), *adaptive behavior* (including eye–hand coordination, recovery of objects, alertness and manipulation of objects), and *personal-social behavior* (including reactions to people, initiative, independence, and play behavior) were collected. Gesell and his colleagues were the first to observe the child's behavior in naturally occurring situations (Buhler & Hetzer, 1935).

The norms collected in this preliminary survey, which covered about 150 behavioral items, were codified into a "Developmental Schedule." Normative levels were referred to, using the concept of "developmental age." However, Gesell did not regard this schedule as an intelligence test in any way. In fact, Gesell's only references to the concept of intelligence were generally phrased in the negative. Interested in a more general concept of "mental growth," Gesell assumed that mental growth, like physical growth, followed an orderly pattern of maturation. Since growth was regarded as lawful, reliable diagnoses were viewed as possible so long as accurate and extensive observations could be made.

In 1927 Gesell initiated a larger, more systematic program to investigate "the wealth and variety of behavioral patterns" of the infant and preschooler (Gesell & Thompson, 1934, p. 4). A homogeneous group of 107 boys and girls from carefully selected white, middle-class families were seen longitudinally. Between 26 and 49 children at each age were observed monthly throughout the first year, at 18 months, and at each birthday until 5 years of age. Protocols on each child included a home record, a medical history, a daily record, anthropometric measurements, maternal observations, detailed behavioral reports of the child's day at the clinic, a normative exam, and ratings in each developmental category.

Several years later, after they had examined hundreds of children, Gesell and his colleagues introduced the *Gesell Development Schedules* (from 1940 to 1947). These schedules provided standardized procedures for observing and evaluating the child's behavioral development between the ages of 1 month and 6 years. The items were grouped into four subtests: *motor items*, including postural reactions, balance, sitting, and locomotion; *adaptive items*, including "alertness, intelligence, and various forms of constructive exploration" (Gesell & Amatruda, 1954, p. 338); *language items*, including facial expressions, gestures, and vocalizations; and *personal-social items*, including feeding, dressing, toilet-training, and play behavior. Age placements were determined by the percentage of subjects who passed each item. Gesell cautioned against the presentation of a composite score—an all-inclusive developmental quotient—and suggested instead that clinical judgments and the developmental quotients on each of the subtests be used to evaluate the child's developmental status (Gesell & Amatruda, 1962).

Today, Gesell's schedules are considered less standardized and more subjective than the many other psychological tests (Anastasi, 1961). The restricted size and homogeneous nature of the sample described (the 107 infants in 1927) further limit its value. Moreover, some of the items are considered to have been rather arbitrarily placed (Stott & Ball, 1965). Gesell and his co-workers never reported a careful statistical analysis of the reliability or validity of the schedules. Nevertheless, the wealth of thorough observations have made the Gesell schedules the main source of material for almost all subsequent infant tests.

The Viennese Test Series

Buhler's Baby Tests, which were first released in German in 1928 and then in English in 1930, resulted from intensive observations of children aged 2 months to 2 years. Like Gesell's Schedule, the Baby Tests were to provide standards by which normality of development could be assessed. Children at different age levels were observed for 24-hour periods. From these observations, test items were developed to measure physical and mental control, behavior in social relationships, and manipulation of objects. The test included 10 items for every month in the first year and for each 3-month period in the second year.

In 1932 (translated in 1935), Buhler, Hetzer, and their colleagues presented the Viennese Series, a thorough revision of the earlier Baby Tests. The development of these tests was guided by a desire both to study the child's personality and development in all its "fundamental dimensions" and to develop tests that would cover all of the "essential steps of development" from birth to 5 years. Six fundamental categories of development were identified: sensory reception, bodily movements, social behavior (including language), learning (including imitation), manipulation of materials, and mental productivity or thinking processes.

Each test in the Viennese Series had 10 items; tests were available monthly until 8 months, bimonthly through 18 months, and yearly from

2 through 5 years. Testing was done in a familiar environment for the child; interest was in the child's natural, not optimal, performance. Developmental age, age within each of the six categories, and the overall pattern of development were considered important in the diagnosis of the child. It was hoped that these tests could be used to determine acceleration or retardation in the child's development in order to assess the normalcy of the child's personality, to predict further development, and to suggest the way for treatment, if necessary. Although some evidence of validity was reported (Buhler & Hetzer, 1935), the generality of the tests was limited by the fact that most of the more than 500 children in the standardization group were drawn from orphanages.

Shirley's Study of the First Two Years

In the late 1920s, Shirley was collecting data for the Minnesota Infant Study, one of the first longitudinal infant studies. Twenty-five infants from birth to 2 years were observed first in the hospital and then at home. Observations were made daily in the first week, once a week from two weeks to one year, and biweekly thereafter. Although Shirley's sample was too small to provide normative information on babies, her study was one of the first to describe comprehensive behavioral development in infancy. The study was initiated to trace the "roots of intellectual behavior . . . into the simplest acts of the infant" (Shirley, 1933, p. 3). Like Gesell, Shirley did not focus on intelligence *per se*. Shirley felt that in infancy, intellect had to be "used as a blanket term to cover almost anything the infant does" (1933, p. 3). In addition, since it would be difficult to determine how much each behavior type contributed to the development of intelligence, it was decided to study all behaviors as carefully as possible.

Unlike Buhler, who was observing infants continuously for 24 hours at a stretch, Shirley gave her infants weekly timed exams lasting about 30 minutes each. It was felt that these exams allowed for more controlled observations. A wide variety of physical and behavioral tests were given. Tests for the youngest levels concentrated on physical development and on observations of the infant's reactions to sounds, tastes, and postural positions. Tests from 3 to 9 months concentrated on the infant's reaching and grasping behaviors. Tests for older infants considered children's behaviors toward pictures, mirrors, odors, various objects, and instructions. Some test items were original, and others were borrowed from Kuhlmann (1922), Gesell (1926), Watson and Watson (1921), Stutsman (1931), and Wallin (1918).

The tests were scored by a point system in which different values were assigned to each item, making scoring rather cumbersome and introducing erratic rates of development, especially in the first year. Although Shirley's tests never gained widespread use, her work is valuable for its detailed, longitudinal description of developmental sequences.

California First Year Mental Scale

Also in the late 1920s, Bayley was beginning to develop scales of infant mental and motor development for use in the Berkeley Growth Study. In reviewing previous work, Bayley noted that "sufficient norms of infant behaviors were already available; what was needed was exact determination of those behaviors which were 'significant as criteria of development'" (Bayley, 1933b, p. 10). Theoretical issues guided Bayley's research. She was interested in determining to what extent mental development was consistent over age—how patterns of behavior changed within the short but crowded period of infancy. Questions such as "What behavior is 'mental?' What specific behaviors in infancy precede later achievements? How dependent are later achievements on earlier ones? What effect does environment have on development?" naturally led to the development of an infancy test designed to predict future intelligence.

In 1933, Bayley published the California First Year Mental Scale (1933a). It included 185 test items from birth through 3 years. However, administration and scoring instructions and norms were provided only for the 115 items covering the first 18 months. The test contained original items as well as items from other tests, and tapped motor maturation, eye–hand coordination, adaptive behavior, response to sound, visual maturation, language comprehension and production, and social responsiveness. Each item was age-placed to the nearest tenth of a month.

The test was standardized on a longitudinal sample from the Berkeley Growth Study. The children were mainly from upper-middle-class homes. Between 52 and 61 infants were tested during each of the months through the first year. In the second year, 46 to 53 infants were tested monthly until 15 months, and then every 3 months until 3 years.

Bayley's scales were the most well-researched scales of their time. Bayley provided extensive split-half reliability data (ranging from 0.51 to 0.95) and predictive validity information. According to Thomas (1970), the scales were not used extensively outside the Berkeley Growth Study at the time of their development because of their inaccessibility and the fact that the standardization group represented only upper-middle-class children. However, Bayley seemed less bent on "selling" her test as a diagnostic tool than she was on the implications of her results for the understanding of mental development. She concluded:

> Behavior growth of the early months of infant development has little predictive relation to the later development of intelligence. . . . We have measured at successive ages varying components of more or less independent functions; not until the age of 2 years do these composites exhibit a significant degree of overlapping with the aggregations of traits constituting 'intelligence.' (1933b, pp. 74 & 82)

The failure to secure high predictive validity, said Bayley, "can be explained by a series of shifts from one type of function to another as the child grows older" (1933b, p. 74), by rapid and inconsistent developmental changes, and possibly even by difficulties inherent in infant testing (e.g., temperamental problems, restricted behavioral repertoire). Instead of Bayley's findings adversely affecting the infant-testing movement, they resulted in a reconsideration of the nature of infant tests.

Iowa Tests for Young Children

The *Iowa Tests for Young Children* are less well known than the tests discussed previously. Like other researchers in the late 1920s, those in Iowa were working to establish a scale of mental development for children from birth to 3 years (Fillmore, 1936). The scale's purpose was to facilitate ongoing research at the Iowa Child Welfare Research Station.

To test early forms of the scale, home-reared children were brought to the laboratory about three times a year for mental and physical measurements; in some cases, medical examinations and nutritional advice were given. Many children were seen at each bimonthly age level; exact numbers range between 60 and 200. Rigorous attempts were made to ensure consistent testing conditions across subjects. The final scale consisted of 49 items for use between the ages of 4.5 and 23.4 months. The items were similar to those of other previously published tests. Test reliability and predictive validity reports were rather low.

Unlike other test developers, Fillmore (1936) attempted to include only those test items that predicted later IQ. Unfortunately, this did not result in much higher reliability or predictability.

EARLY INVESTIGATIONS OF INFANT INTELLIGENCE SCORES

After the flurry of infant test development in the early 1930s, few new tests were introduced, notable exceptions being the Griffiths Abilities of Babies Scale (1954) and the Cattell Infant Scale (1940), both of which will be discussed later. Instead, a consolidation period began, as research on reliability, stability, and predictability flourished. This led to the gradual elimination of some if not most of the earlier tests from general usage. In fact, in Stott and Ball's 1963 (Stott & Ball, 1965) survey of the use of various infant and preschool tests, only the Cattell Infant Scale and the Gesell Developmental Schedules were used by a number of the respondents. The work of Shirley, Buhler, and Fillmore became largely referential, as test developers used them to identify test items. Bayley's test was not used extensively in the 25 years following its inception, remaining a research instrument until its revision in 1960.

Stability of Infant Intelligence Test Scores

With the availability of infant tests, researchers became concerned with the extent to which measurements of mental abilities were stable. Two kinds of stability measures were considered: those dealing with stability or consistency of test measure within a testing session—internal consistency—and those dealing with stability or consistency of scores within a particular infant test from one testing session to another—test–retest reliability.

Internal Consistency. To measure internal consistency, split-half coefficients are used. With corrections for attenuation by the Spearman-Brown formula, correlation coefficients ranged from .51 to .95 for tests given within the first 18 months of life (Werner & Bayley, 1966). Split-half coefficients averaged .81 to .84 for tests given within the first year.

Test–Retest Reliability. Test–retest reliability reports varied widely, depending on the particular test, the interval between tests, and the age at testing. For two Buhler Baby Tests given a day apart, Herring (1937) reported that correlation coefficients ranged from .40 to .96 over the first 12 months of life. Conger (1930, cited in Shirley, 1933) found little agreement between Linfert–Hierholzer Test scores obtained two days apart for very young infants. For 1-, 2-, and 3-month-olds, correlation coefficients were −.24, .44, and .69 respectively. Shirley (1933) reported that week-to-week test scores correlated from −.16 to .69 across the first 3 to 11 weeks of life.

Studies of the California First Year Mental Scale (Bayley, 1933b), Kuhlmann– Binet (Driscoll, 1933), Minnesota tests (Shirley, 1933), Iowa Tests (Fillmore, 1936), Buhler Baby Tests (Herring, 1937), and Gesell Schedules (Richards & Nelson, 1939) all seemed to agree that repeated testing correlations over the first 18 months were too low for any kind of individual prediction. The longer the interval between test sessions, the lower the reliability was. In Thorndike's (1940) review of reliability studies, monthly test–retest reliability coefficients for tests given after the first four months of life were reasonable, although long-term prediction was low.

Many explanations for the unreliability of early mental scores have been advanced. First, since development is so very rapid, it might not proceed at a uniform rate from birth through 2 years, resulting in little stability (Wilson, 1973). Second, those behaviors that might be used to measure "intelligence" at some stages of infancy might be very different from those used to measure "intelligence" at other stages of infancy. In fact, the quality of intelligence might be different at different periods within infancy (Lewis & Brooks-Gunn, 1981a,b; McCall, Eichorn, & Hogarty, 1977). Third, the growing impact of environment and environmental fluctuations might account for the unreliability of the tests (Hunt, 1961). Fourth, difficulties in motivating infants to perform and the limitations of the child's behavioral repertoire further increased the unreliability of the tests.

Predictive Validity of Early Infant Tests

A common principle of tests and measurement is that without retest reliability, there is little hope for predictive validity. One would have thought that the lack of retest reliability would have been sufficient to erode confidence in the tests. In this case, however, it was the predictive validity studies that ultimately determined each test's worth. In fact, concern with predictive validity was so important that no matter how well-standardized or reliable the infant test was, if it did not predict to later measurements of intelligence, the test was doomed to obscurity.

The Linfert–Hierholzer test was standardized on 300 infants (50 were tested repeatedly), while the Stanford–Binet Test was standardized on only 10 3-year-olds and 51 4-year-olds (Stott & Ball, 1965, p. 133). Nevertheless, because the Linfert–Hierholzer test did not correlate with the 1916 Stanford–Binet test for 4-year-olds (Furfey & Muehlenbein, 1932), it never received widespread attention.

The 1930s were a period of growing disillusionment with infant tests, primarily because there were very few reports of high predictive validity. Driscoll (1933) observed that children's Kuhlmann–Binet scores from 12 to 24 months and from 24 to 36 months correlated only .58 and .57, respectively. Bayley (1933b) found that California First Year Mental Scales scores averaged over months 7, 8, and 9 of the first year correlated only .22 with scores at 2 years. Using ratings, Fillmore (1936) reported that children's scores on the Iowa Tests for Young Children at $5\frac{1}{2}$ months correlated only .26 with scores at $18\frac{1}{2}$ months.

Even the studies that did report high predictive validity were called into question. Cunningham (1934) found that 12-month scores on the Kuhlmann–Binet predicted to 2- and 8-year IQ scores, but her study was generally regarded as inconclusive due to the small sample (Stott & Ball, 1965). Two studies of the Buhler Baby Tests (Buhler, 1930; Hetzer & Jenschke, 1930, cited in Buhler, 1930) that also reported high predictive validities suffered from similar problems. Small samples were used (24 and 25 infants), ages of subjects varied from 3 months to over 2 years, and test intervals varied from 3 to 14 months. In addition, predictive validities may have been amplified by the fact that classifications—normal, retarded, or advanced—were used instead of numerical scores.

Hallowell (1925–1927) reported that intellectual assessments of 142 children were reliable enough in 95% of the cases to be used for adoption placement; Gesell (1928) felt that his studies also showed consistency of mental growth. Bayley (1933b), however, criticized these studies on the grounds that age groups were not treated separately, time intervals between tests were not considered, and statistical treatments were unclear.

The only successful predictive validity measures were obtained by using sophisticated item analysis and multiple regression techniques. Nelson and Richards (1938) correlated 6-month Gesell scores, 2-year Merrill–Palmer scores, and 3-year Stanford–Binet scores. Although the intertest correlations were less than .50, certain items correlated highly with later intelligence scores. For example, awareness items, such as "regards pellet" and "splashes in tub," had low correlations with the overall 6-month Gesell score but acceptable correlations with the 2-year Merrill–Palmer score. In contrast, motor items, such as "secures cube" and "reacts to mirror," had high correlations with the 6-month Gesell test but lower correlations with later test scores. Observing that the predictive validity for several single items was higher than the predictive validity of the entire 6-month test, Nelson and Richards combined several highly predictive items in order to obtain the highest predictive validity possible. Multiple correlations using several highly predictive items from the 6-month test yielded a multiple R of .80 with 3-year Stanford–Binet scores.

In a similar study, Anderson (1939) repeatedly tested nearly 100 infants in the first two years of life with a test composed of original items and items from the Gesell, Buhler, and Linfert–Hierholzer tests (Anderson, 1939). The infants also were given a Stanford–Binet at 5 years. In general, Anderson found low or insignificant relationships between the 5-year-old IQ score and the earlier infant tests. Use of item analysis techniques helped increase the correlations to between .20 and .55. Nevertheless, only two items were useful in predicting later intelligence—alertness in the early months and language in the second year. Multiple correlations considering test scores through 24 months and parental education yielded the most reliable prediction—a coefficient of .71.

Thus, the reliability and validity of infant intelligence tests developed in the 1930s were disappointingly low. Among the explanations offered were the infant's limited behavioral repertoire, the failure to consider the infant's gestational age, difficulties in maintaining the attention and motivation of the infant, interference of temperamental characteristics such as activity level and alertness, and failure to make accurate clinical assessments of the infant's overall responsiveness to the testing session. More important, however, was the fact that several theoretical assumptions had to be reevaluated. People began to question whether intelligence was, as had been hoped, a reliable stable characteristic that developed continuously and at a constant rate for all. In particular, the results of more sophisticated studies lent support to arguments that intelligence varied not only quantitatively but also qualitatively across the first years of life.

TABLE I
Mirror Test Items in Five Infant Tests

Test items	Age in months at which item occurs																
	4	5	6	7	8	9	10	11	12	13–14	15–17	18–20	21–23	24–26	27–29	30–32	36
Approaches image	Bayley																
Regards own image		Gesell			Buhler Griffiths												
Smiles at image		Gesell Bayley		Cattel	Buhler		Griffiths										
Vocalizes to image			Gesell Bayley														
Pats, feels image			Bayley	Catell Gesell					Buhler								
Leans forward							Gesell										
Plays with image			Bayley				Griffiths										
Searches							Griffiths					Buhler					
Reaches toward mirror for object								Gesell	Buhler								
Reaches toward adult, not reflection											Buhler						
Refers to self by name														Gesell	Gesell		
Refers to self by pronoun																Gesell	
Refers to self by sex																	Gesell

Interest Item Consistency

It seems that the various test developers looked over each other's tests quite carefully. However, such cross-fertilization of ideas and borrowing of items did not result in the use of identical items. Even when similar items were used, they were placed at different ages. For instance, the fact that mirrors were of great interest to infants and elicited pleasurable responses was noted by those in the child study movement (Darwin, 1877; Preyer, 1882/1888) and by those who were developing infant intelligence tests. An examination of the tests developed by Bayley, Cattell, Buhler, Griffiths, and Gesell reveals that the mirror items themselves and the ages at which they occur are somewhat divergent across the five infant tests. The items and age ranges are presented in Table I.

The lack of consensus as to when various mirror responses occur is surprising and suggests that conclusive normative ages were not established by the test developers. The passing or failing of a similar item on different scales would often result in different intelligence scores. Regarding one's own image in the mirror "typically" occurs at either 5 or 8 months, while smiling at the image occurs at either 5, 7, 8, or 10 months of age. Searching behind the mirror is placed at 10 months by Griffiths and at 18 to 20 months by Buhler.

Some of the divergence can be accounted for by the type of mirror used and the situation in which the behavior is elicited (Lewis & Brooks-Gunn, 1979). However, large inconsistencies in the age at which certain behaviors were placed by the test developers still remain. This was true even though all of the later tests were standardized on large numbers of infants.

Environmental Effects on Intelligence

Another major issue that generated a great deal of research concerned environmental effects on intelligence. From the beginning of the testing movement in America, intelligence was seen as fixed in nature or predetermined. For example, Goddard, who was the first to translate the Binet Scale into English and the first to use it extensively, was a strong proponent of fixed intelligence. Of the infant test developers, Gesell was the best known for his adherence to a notion of predetermined maturation. So prevalent was this view that even when an early study found that nursery school attendance resulted in an increase in IQ (when compared with an appropriate control group), these differences were attributed to the inadequate standardization of the preschool scales (Goodenough, 1928a,b).

However, not everyone believed that intelligence was impervious to environmental stimulation. From 1932 to 1939, a series of 11 papers from the Iowa Child Welfare Station appeared, reporting some highly controversial studies (Wellman, 1932a,b). In these studies, children in a variety of settings (e.g., nursery schools, longitudinal studies, adoption settings, foster children of feebleminded women, children in orphans' homes) were tested and retested in order to determine the effects of environmental circumstances on IQ scores. It was reported that participation in the Iowa Studies' nursery schools resulted in IQ gains (Wellman, 1932a,b), that children placed in adoptive homes in the first six months of life had higher IQ scores than those placed from 2 to 5 years of age (Skodak, 1939), that institutionalized children classified as feebleminded had IQ scores that declined (Crissey, 1937), and that children of feebleminded mothers, when placed in another home, had normal IQ scores (Skodak, 1939).

These findings were bitterly attacked. Goodenough (1939, 1940), one of the advocates of fixed intelligence, dismissed the Iowa findings for methodological and statistical reasons. As she pointed out, the control groups in many of the studies were inadequate.

Other attacks on the Iowa group were less rational. In what Stott and Ball (1965) termed the most violent article, it was said:

> But to claim miracle working at Iowa that cannot be duplicated in other parts of the country . . . is much worse than nonsense and ought to be exposed as such. (Simpson, 1939, p. 366)

As a result of the controversy generated by the Iowa Studies, a series of better-controlled studies on the impact of environment on IQ emerged in the 1940s (e.g., Bradway, 1945; Frandsen & Barlow, 1940; McHugh, 1943). However, even these data were inconclusive, satisfying neither the so-called environmentalists nor the nativists. The effects of environment on intellectual ability are still being debated today, especially in early school enrichment programs, although the present argument takes a different form (Brooks-Gunn & Hearn, 1982).

INFANT INTELLIGENCE TESTING IN THE 1940s AND 1950s

New Tests

Child psychologists, although disillusioned by the failure of infant tests to show predictive validity, did not give up easily. Two approaches to the depressing validity studies of the 1930s were taken. The less popular approach was to allow more room for clinical interpretation and subjective assessment instead of focusing narrowly on the child's performance skills. Hallowell (1941) argued that the qualitative aspects of performance—complexity and variety of responsiveness, attention span and discrimination ability—had to be considered. McGraw (1942) developed a series of three laboratory tests in which a piece of plate glass was placed between the child and a lure so that the style quality of children's reactions rather than their particular behavior could be appraised. In both instances, ratings were necessarily subjective and reliability was not assessed.

The second and more popular approach was to improve and modify the older infant tests. This tack was taken by Cattell (1940), Shotwell and Gilliland (1943), and Griffiths (1954). Their attempts are briefly considered below.

Cattell's Infant Intelligence Scale

In an extensive study at Harvard University on the development of normal children from birth to several years of age, Cattell was called upon to select psychological tests for measurement of mental development in infants and young children. In the early years of the study, existing infant tests were used (in particular the Gesell tests). Cattell (1940) was disappointed with these tests, listing seven problems with the tests available at that time:

1. They lacked objective procedures for administration and scoring items.
2. They included many personal-social items that were heavily influenced by home training.
3. They contained measures of large motor control that Cattell believed were not related to later mental development.

4. Scaling was often inadequate.
5. The tests covered limited age ranges.
6. Test items were unequally distributed.
7. The tests were often poorly standardized.

Cattell developed her own scale for infants from 2 to 30 months. She drew heavily from the Gesell test items, but included items only if they: (1) showed a regular increase in percentage of passage from one age to another, (2) were easy to administer and were interesting to the child, (3) did not require cumbersome equipment, (4) measured performance not appreciably influenced by home training, and (5) were not dependent on the use of large muscles. At least five items were included at each age level—monthly in the first year, bimonthly in the second year, and quarterly in the third year. Precise objective directions for administering and scoring the test were written.

The tests were standardized on 274 middle-class children tested at 3, 6, 9, 12, 18, 24, and 30 months. Split-half reliabilities for tests from 6 through 30 months were high (from .71 to .90); the split-half reliability for the 3-month test was only .56. Although the Cattell test was designed to be a downward revision of Form L of the Stanford–Binet test, predictive validities for the younger infants were not terribly high. For 3, 6, 9, 12, 18, 24, and 30 months, the correlations were, respectively, .10, .34, .18, .56, .67, .71, and .83.

Other studies of the Cattell test have reported high test–retest correlations (Gallagher, 1953; Harms, 1951). However, predictive validity studies have been disappointing. Escalona and Moriarty (1961) reported no relationship between either the Cattell or Gesell test at 7 to 8 months of age and school-age WISC scores, and Cavanaugh, Cohen, Dunphy, Ringwall, and Goldberg (1957) found low correlations between the Cattell and the preschool Stanford–Binet tests.

Northwestern Intelligence Test

In the mid-1940s, Shotwell and Gilliland (1943) developed a test to screen very young infants for adoption. The results were two tests (Gilliland, 1949, 1951): Test A, measuring infants from 4 to 12 weeks, and Test B, measuring infants from 13 to 36 weeks. The tests were very similar to earlier tests in administration and scoring. Many of the 40 items on each test were borrowed from the Gesell scales, although only those items that seemed to measure adaptability rather than physical coordination or reflexes were included.

Gilliland (1949, 1951) reported high split-half reliability coefficients for each test. However, predictive validity information was inadequate. Gilliland (1949) reported small mean differences between Kuhlmann–Binet or Cattell Test Scores and Test A; thus, he concluded that Test A was as valid as either of these other tests. However, the sample was small (n = 29) and composed entirely of mentally retarded children. There is only one other validity study. Infants ranging in age from 13 to 35 weeks of age were given both Test B and the Cattell, and at 18 months were again given the Cattell. Early Cattell scores were correlated .39 with later Cattell scores; early Test B scores correlated .38 with later Cattell scores (Braen, 1961).

In reviewing the literature, Thomas (1970) concluded that the Northwestern scales "have not been adequately researched, nor do they appear to offer any advantages over existing scales" (p. 199). Although the goal of the tests—very early screening of infants for adoption—was an admirable one, it may have been unattainable considering both the difficulties involved in testing infants so young with traditional types of items and the intra- and interindividual variation in the first year of life.

Griffiths's Mental Development Scale

Griffiths's Mental Development Scale (1954), covering the ages from 2 weeks to 2 years, was designed as an improvement on existing infant tests. Griffiths hoped that her assessment of what she regarded as innate general cognitive ability would help in early psychological diagnoses of infants.

The scale was divided into five subscales: locomotor, personal-social, hearing and speech, hand and eye, and performance. Each scale had 52 items; there were 3 items for each week in the first year and 2 items for each week in the second year. Although borrowing heavily from other tests, particularly the Gesell, Griffiths criticized earlier tests for their lack of speech items, especially in the first year. Believing that speech was a uniquely human intellectual task, Griffiths included twice as many speech items in her infant scale as anyone else. First-year measures included imitation, listening to sounds and conversation, repetition of single sounds, and babbling. Also emphasized were the social nature of many of the items and the importance of social relationships for the infant's intellectual development. Despite Griffiths's stress on the importance of the infant's relationship with others, her test included measures of the infant's orientation toward, interaction with, and preference for other people only during the first year; for infants in the second year, all of the personal-social items involved self-help skills that tapped the child's relationship to objects, not people.

Griffiths standardized the tests on 571 London children using about 25 children for every month of testing. According to 1940 U.S. Census figures, Griffiths's English standardization sample seemed very similar to the U.S. urban population. The test–retest reliability obtained from 60 children tested at various ages with test intervals ranging from 7 to 70 weeks was .87 (Griffiths, 1954). Griffiths did not report predictive validity data, although others have. Hindley (1960) reported rather low test–retest correlations—ranging from .13 to .56—for infants tested at 3, 6, 12, and 18 months. Hindley (1960, 1965) and Roberts and Sedgley (1965) correlated the Griffiths scale in infancy with Stanford–Binet scores at 5 years. Both studies reported similar coefficients—from .30 to .46. Roberts and Sedgley reported that infant scores taken at about 18 months of age were correlated less than .40 with Binet scores at 7 years of age. Thus, while the Griffiths test was viewed by many as an improvement over other infant tests because of equal or better standardization efforts, higher test–retest reliability, and higher predictive validity, the test was still inadequate for individual predictions. Although Griffiths's test scores are highly correlated (.80s to .90s) with Cattell scores (Caldwell & Drachman, 1964), the Griffiths has rarely been used in the United States, probably because, according to Horrocks (1964, reported in Thomas, 1970), Griffiths provided test materials only to those trained by her.

Uses of the Infant Tests

Despite the questions concerning reliability, predictive validity, and the effects of environment being raised by researchers, infant intelligence tests increased in popularity as diagnostic aids. Indeed, although infant test research was widely reported in the journals, the tests were used primarily by practitioners. Mental retardation, brain damage, physical and sensory disabilities, and even personality problems were all assessed using the current tests.

Adoption agencies found infant tests especially useful. Throughout the 1940s and 1950s, there was a dramatic increase in the number of adoptive placements in the United States. The number of adoptive petitions in Connecticut doubled in the decade from 1943 to 1952, numbering over 1000 by 1952 (Wittenborn, 1957). At the same time, the placement age was steadily lowering; for example, in 1947–1948 only 9% of the cases in Connecticut were adopted before 6 months of age, while in 1951–1952, 25% of the cases were placed in the first 6 months. Thus, the status of a child eligible for adoption had to be determined early, and the infant tests were the only diagnostic instruments available.

To obtain a clearer picture of who used the infant tests and how they were used, Stott and Ball (1965) conducted a survey of infant and preschool test users in the early 1960s. Of the 750 questionnaires sent to individuals and centers, 330 were returned and 217 of respondents were involved with children under 6 years of age. Of the respondents, 93% had used at least one of the early tests; 81% had used the tests for diagnostic purposes. One-third had also used the tests for preadoptive screening, for admission to special schools or programs, and for testing physically handicapped children. Not surprisingly, only 7% had used the tests for research, 5% for general development or longitudinal studies, and 2% for validation studies. The groups most frequently tested for research purposes were autistic and anoxic young children.

INFANT INTELLIGENCE TESTING IN THE 1960s AND 1970s

After the appearance of new tests in the 1940s and 1950s, a new round of evaluation and criticism began. Remember that the new tests were more carefully designed than the tests of the 1930s: Testing procedures were standardized, item selection was more rigorous, an adequate number of infants was used for normative purposes, split-half reliability was adequate, and retest reliabilities, although often low, at least were reported. However, predictive validity lingered as an unresolved theoretical and practical issue. Multivariate techniques and multideterminal models were employed in an attempt to enhance prediction.

Predictive Validity and Related Issues

Study after study revealed that test scores from the first two years could not be used to predict preschool or school-age intelligence in normal samples (Bayley, 1949; Bowlby, 1952; Cavanaugh et al., 1957; Hindley, 1965; MacFarlane, 1953; McCall, Hogarty, & Hurlburt, 1972; Wittenborn, Astrachan, DeGouger, Grant, Janoff, Kugel, Myers, Riess, & Russell, 1956). However, results from studies of retarded infants were somewhat more encouraging (Ames, 1967; Hallowell, 1941; Illingworth, 1960/1972, 1961; Simon & Bass, 1956; Symmes, 1933). In general, prediction of intellectual functioning of those who were suspected to be retarded or neurologically impaired, was much better than prediction for a normal sample alone. For example, using the Gesell Developmental Schedules in infancy and the Stanford–Binet at later ages with large, heterogeneous samples, Knobloch and Pasamanick (1960, 1966) reported moderate correlations between first- and third-year ($r = .48$, 1960) and first- and eighth- to tenth-year ($r = .70$, 1966) IQ scores. However, correlations in both studies were considerably higher in the abnormal than in the normal infant groups. For example, in the 1960 study, the correlation over age for normal infants was .43 while for the below-normal infants it was .74.

Caution should be used in interpreting studies of low-scoring infants. Many may be criticized on methodological grounds (cf. Thomas, 1970). For example, in MacRae's study of adoptive children (1955), the early test scores may have influenced adoptive placement, thus exaggerating the correlations between early and later test scores. In addition, studies of known handicapped infants report low predictability from infancy to school age (Brooks-Gunn & Lewis, 1983).

Attempts to Increase Predictive Validity

A variety of methods for increasing intelligence tests' predictive validity was suggested and tested. One technique, introduced by Escalona, involved testing under optimal conditions and the use of clinical appraisals in conjunction with the infant tests (Escalona, 1950; Escalona & Moriarty, 1961; Simon & Bass, 1956). Persuasive as the argument for clinical assessment was, predictions were still less than .70. The failure to obtain higher correlations was probably influenced by methodological problems such as the deletion of large numbers of infant-testing sessions, failure to describe objectively the basis on which clinical assessments were made, and smallness of sample sizes.

The use of multiple regression was a more promising technique for increasing predictive validity (Knoblock & Pasamanick, 1966; Werner, Honzik, & Smith, 1968). Werner and her colleagues were able to predict IQ at 10 years for normal children by using 20-month Cattell scores, pediatricians's ratings, social ratings, prenatal stress scores, and parental socioeconomic class ($r = .58$). When the infant's Cattell score at 20 months had been under 80, the multiple correlation coefficient increased to .80. As can be seen, these investigators were relying less and less on individual IQ scores in infancy to predict later IQ.

Item Analysis and Predictive Validity

Item and factor analyses also were performed on various infant intelligence tests. Although item analyses revealed that some individual items (in particular, verbal items) had higher predictive validity to later IQ than did overall test scores (Cameron, Livson, & Bayley, 1967; Fillmore, 1936; Hunt & Bayley, 1971; Nelson & Richards, 1938; Richards & Nelson, 1939), these correlations were still not high enough for reliable individual predictions. In an exploratory investigation, Catalano and McCarthy (1954) found a multiple correlation of .52 between three phonological characteristics of infant speech at 13 months and Stanford–Binet scores at 45 months.

Factor analytic studies by Nelson and Richards (1938), Richards and Nelson (1939), and Stott and Ball (1965) were especially helpful in revealing reasons for differences in scores between various infant tests given at the same age and at different ages, and for differences in scores between the same tests given at different ages. Considering the wide range in content across tests, Stott and Ball remarked, "It is almost surprising that there is a degree of correlation between two simultaneously administered scales as seems to be present" (p. 221). According to Stott and Ball, infant scales were "more diversified" and covered a wider range of intellectual abilities than later tests. Whereas infant scales were judged to include measures of divergent production—creative abilities—later tests, such as the Stanford–Binet and Merrill-Palmer tests, had very few measures of creative abilities. The Stanford–Binet emphasized memory; the Merrill-Palmer, manual skills.

More recently, McCall and his colleagues (1972, 1977) have performed principal component analyses of items on the Gesell Schedule and the California First Year and Preschool Scales (the precursors to the Bayley Scales) from the Berkeley and Fels Longitudinal Studies. Both samples were tested longitudinally so that the ability of factors identified in the first year of life to predict factors in the later years could be explored. Changes in the item composition of the first principal component over age, and discontinuities in the pattern of correlations between the first principal components extracted at various ages were used to identify stages or qualitative shifts in the nature of mental abilities in infancy. Given this epigenetic view of mental development, individual consistency in performance was not expected since the skills characterizing different stages are qualitatively different. In fact, no important correlations between performance on the first principal components in the first 6 months of life and on the first principal component at 24 months were found. In addition, there was no long-term predictability from performance in the first 18 months of life to intelligence scores attained betwen 6 and 36 years (McCall *et al.*, 1977).

New Methods of Assessing Intelligence

The concern about prediction and related issues led to a third wave of tests being developed in the 1960s and 1970s. These tests were to serve a new purpose, as the old notions of intelligence as a unitary factor became obsolete.

Bayley Scale of Infant Development

By the 1960s, investigators were becoming reconciled to the fact that even improvement of existing tests would not lead to high predictive validity for normal children (Illingworth, 1960/1972). Qualitative differences in intelligence across different age periods, the effects of varied environmental experiences, and the incidence of delayed maturation and unanticipated degenerative disease were recognized as significant impediments to prediction. Thus, Bayley's California First Year Scale was revised and restandardized (Bayley, 1969), not so much, it seems, in the interests of increasing predictive validity, but more to allow for a comparative assessment of current infant abilities. The Bayley Scale of

Infant Development is the most widely used and rigorously constructed of the American infant tests. The Mental Scale consists of 163 items appropriate for infants approximately 2–30 months of age. The competencies assessed include sensory-perceptual abilities, object constancy and memory, language comprehension and production, cognitive abilities related to abstract thinking, and social abilities associated with interaction and imitation (Bayley, 1969). The Bayley Scale, like the tests preceding it, is not particularly predictive of childhood functioning until 18 to 24 months of age.

Tests to Assess Specific Abilities

The majority of new tests of the 1960s and 1970s have not been designed to test all aspects of mental development in infancy; they have been designed to fulfill particular purposes and measure specific types of behaviors. Some tests are designed primarily to screen for developmental delay such as the Developmental Screening Inventory (Frankenberg & Dodds, 1967), which is discussed by Brooks-Gunn and Lewis (1983). Other tests were devised as research instruments for the assessment of specific competencies, such as sensorimotor functioning attentional processes and social competency. Finally, the most activity has been in the area of neonatal assessment.

Assessing Specific Functional Domains. Those interested in more theoretical issues regarding mental development have developed tests that would tap specific areas of development. Both cognitive and social skills have been translated into infant tests, the most interest having been generated in the areas of sensorimotor functioning, social competency, and attentional processing.

The importance of sensorimotor skills for later development had been proposed by Piaget in the 1940s and 1950s but did not gain widespread acceptance in America until the 1960s. At least five tests have been developed to measure sensorimotor skills (Bell, 1970; Decarie, 1965; Escalona & Corman, 1969; Golden & Birns, 1968; Užgiris & Hunt, 1966). Ricciuti's (1965) Object Grouping and Selective Ordering Tasks assess the development of categorizing skills in 1- to 3-year-old children, while the Fantz–Nevis Visual Preference Test (Fantz & Nevis, 1967) looks specifically at visual preferences from birth to 6 months of age. There are several tests that focus on early language development: the Communicative Evaluation Chart from Infancy to Five Years (Anderson,

Miles, & Matheny, 1963), the Prelinguistic Infant Vocalization Analysis (Ringwall, 1965), the Verbal Language Developmental Scale (Mecham, 1971), the Shield Speech and Language Developmental Scale (Shield Institute for Retarded Children, 1968), and the Bzoch–League Receptive-Expressive Language Scale for the Measurement of Language Skills in Infancy (Bzoch & League, 1970–1971).

Renewed interest in social development has led to the search for measures of social competency. Two early tests on social skills are the Measurement of Social Competence (Doll, 1953) and the Ring and Peg Tests of Behavior Development (Banham, 1964). Several tests were developed to assess the social competencies of children in day-care settings (Day Care and Child Development Council of America, Inc., 1973; Lewis & Michalson, 1983).

Finally, specific early cognitive abilities have been targeted as indicative of intellectual functioning. Two separate sets of investigators have found attentional processing in the 3- to 7-month-old to be predictive of IQ scores later on (Lewis & Brooks-Gunn, 1981a; Fagan & McGrath, 1981).

The infants' responses to a novel stimuli—specifically, a comparison of fixation times to familiar and novel visual stimuli—were positively related to intelligence test scores at 2, 5, and 7 years of age; these correlations ranged from .35 to .65. These remarkable findings (in light of the failure of early intelligence tests to predict childhood functioning) have been replicated in several studies across laboratories; attentional processing may turn out to be the intellectual measure used for diagnostic purposes in the 1980s. In fact, Fagan is developing the Infant Test of Attention (1982).

Neonatal Tests. The three most well-known neonatal tests are the Prechtl Neurological Examination (Prechtl & Beintema, 1964), the Graham/Rosenblith Behavioral Examination of Neonates (Rosenblith, 1961), and the Neonatal Behavioral Assessment Scale (NBAS; Brazelton, 1973). The NBAS will be described in some detail.

In contrast with many other neonatal examinations, which focus primarily on the assessment of reflexive behavior as an index of CNS integrity, the Brazelton NBAS was designed to evaluate the quality and organization of higher-level functions in the newborn (Brazelton, Als, Tronick, & Lester, 1979). A guiding principle underlying the NBAS is that an infant's capacity for eliciting, participating in, and sustaining interactions with others in the environment will affect development. Thus, in addition to reflexes, temperament and state are assessed, since the quality and organization of reflexes are thought to vary with state and because the range and lability of states is likely to affect the way that others respond to an infant.

The NBAS is composed of three types of items—elicited reflexes, general behavioral dimensions, and specific behavioral items. A total of 20 elicited reflexes (e.g., Moro, Babinski, rooting) typical of most neonatal exams are included to assess neurological integrity. Two general behavioral dimensions, attractiveness and need for stimulation, describe the infant's overall organization. "Attractiveness" refers to the infant's ability to elicit and sustain social interactions with the examiner, while "need for stimulation" measures whether the examination procedures and the concomitant stimulation have an organizing or disorganizing effect on the infant's behavior.

The 26 specific behavioral items are the core of the examination and measure the infant's interactive capabilities. Included among the items are response decrement measure, orientation to social and nonsocial stimuli, motor maturity and tone, activity level, self-quieting abilities, and measures of smiling and cuddling. Conceptually, these behavioral items can be separated into four behavioral categories (Brazelton *et al.*, 1979): (1) interactive capacities, (2) motoric capacities, (3) organizational capacities with respect to state, and (4) organizational capabilities in response to stress items.

Test items are not administered in any specific order, although there are specified constraints on when different items are to be administered with respect to the infant's state. The examiner's handling of the infant mimics typical infant–caretaker interactions, and the examiner seeks to bring the infant from an initial light sleep to more active states including crying and then to return the infant to a less aroused state. The reliability of independent testers ranges from .85 to 1.00 (Horowitz & Brazelton, 1973), and test–retest reliability from 3 to 28 days of age has been reported to be .80 for males and .85 for females (Horowitz, Aleksandrowicz, Ashton, Tims, McCluskey, Culp, & Gallas, 1973).

Because items on this examination do not have comparable scaling properties (i.e., for some items, midrange scores represent optimal performance, while others have more traditional linear scales), analysis of performance on the NBAS is complex and a variety of approaches have been used. These have included item-by-item analyses (e.g., Brazelton, Koslowski, & Tronick, 1976; Freedman & Freedman, 1969), summary scales for subscales (e.g., Brackbill, Kane, Manniello, & Abramson, 1974; Powell, 1974), factor analyses (e.g., Aleksandrowicz & Aleksandrowicz,

1974; Scarr & Williams, 1973), and typological-profile analysis (Adamson, Als, Tronick, & Brazelton, 1975; Lester & Zeskind, 1978). Although the lack of a standardized scoring system may make comparisons between studies difficult, the scales were designed to present a multidimensional characterization of neonates that would capture their individuality and complexity better than a single developmental score could.

Long-term predictability is still being assessed. Several studies have looked at the predictive validity of the scale over the first year of life (Brazelton *et al.*, 1979). Powell (1974) found that a summary scale, "responsivity," correlated moderately with the Bayley behavior record at 4 months (.67) and 6 months (.64) as well as with the Bayley motor score at 6 months (.67). In addition, another summary scale, "head control," had a correlation of .47 with the 4-month Bayley behavior record. Correlations between measures of infant temperament at 4 months and factors of alertness, motor-maturity, tremulousness, habituation, and self-quieting abstracted from the Brazelton Scale have been reported by Bakow, Sameroff, Kelly, and Zax (1973). Finally, Scarr and Williams (1971) reported that nine of the scales administered on day 7 correlated .30 or better with Cattell IQ scores at 1 year.

Tronick and Brazelton (1975) tested a sample of 53 at-risk infants with both the Brazelton Scale and a standard neurological examination. While both examinations were equally successful in predicting abnormality at 7 years of age, the Brazelton scale led to fewer misclassifications of normal infants as abnormal (24%) than did the standard neurological exam. Thus the Brazelton Scale may be a more sensitive discriminator of normal and abnormal infants than the standard neurological examination, although others' research (e.g., Horowitz, 1977) suggests that the issue of predictability remains unresolved.

Two modified versions of the NBAS are of note. The *Assessment of Preterm Infants' Behavior* was developed by Als and her colleagues (Als, Lester, & Brazelton, 1979) for use with preterm infants. The *NBAS-K* was developed by Frances Degen Horowitz and her colleagues (Horowitz, Sullivan, & Linn, 1978) to include the newborn's typical as well as optimal performance and ratings of the examiner's responses to the infant.

Many researchers have used the NBAS and its modifications to study at-risk infants, cross-cultural and socioeconomic differences in neonates, effects of prenatal and perinatal variables such as obstetric medication and maternal drug addiction on newborns, and relationships between neonatal behaviors and mother–infant interactions (see review by Brazelton *et al.*, 1979). The NBAS characterization of the infant as an active participant in the world from the moment of birth and its emphasis on individuality and change are congruent with the current zeitgeist and make the scale an attractive research tool. Thus, though the long-term predictive validity of the scales is not clear at present, it is likely that the scale will continue to be widely used.

SUMMARY

The infant-testing movement seems to be moving away from a notion of general intelligence toward emphasis on specific behaviors and skills as possible predictors of later development. Such tests hold great promise for a more precise analysis of the course of cognitive development in the early years of life, although predictability of these more specific tests (with the exception of attention) is still at issue. However, as theoretical models of infant development move further and further from a linear model of infant development and become more concerned with *change* and the infant's *transactions* with its environment, predictive validity may play a diminishing role in evaluating new tests. Predictions of infant variability and particular kinds of changing relationships at key points in development may emerge as a criteria of greater importance as new tests are introduced and evaluated.

ACKNOWLEDGMENTS

We wish to thank Mary Enright for her assistance.

REFERENCES

ADAMSON, L., ALS, H., TRONICK, E., & BRAZELTON, T. B. *A priori profiles for the Brazelton Neonatal Assessment.* Mimeo, Child Development Unit, Children's Hospital, Boston, 1975.

ALEKSANDROWICZ, M., & ALEKSANDROWICZ, D. Pain-relieving drugs as predictors of infant behavior variability. *Child Development*, 1974, *45*, 935.

ALS, H., LESTER, B. M., & BRAZELTON, T. B. Dynamics of the behavioral organization of the premature infant: A theoretical perspective. In T. M. Field, A. M. Sostek, S. Goldberg, & H. H. Shuman (Eds.), *Infants born at risk: Behavior & development.* Jamaica, N.Y.: Spectrum Press, 1979.

AMES, L. B. Predictive value of infant behavior examinations. In J. Hellmuth (Ed.), *Exceptional infant* (Vol. 1): *The normal infant*. Seattle, Wash.: Straub & Hellmuth, 1967.

ANASTASI, A. *Psychological testing*. New York: Macmillan, 1961.

ANDERSON, D. The predictive value of infancy tests in relation to intelligence at five years. *Child Development*, 1939, *10*(3), 203–212.

ANDERSON, R. M., MILES, M., & MATHENY, P.A. *Communicative evaluation chart from infancy to five years*. Cambridge, Mass.: Educators Publishing Service, 1963.

BAKOW, H., SAMEROFF, A., KELLY, P., & ZAX, M. *Relation between newborn and mother-child interactions at four months*. Paper presented at the biennial meeting of the Society for Research in Child Development, Philadelphia, 1973.

BANHAM, K. M. *Ring and peg tests of behavior development*. Munster, Ind.: Psychometric Affiliates, 1964.

BAYLEY, N. *The California first year mental scale*. Berkeley: University of California Press, 1933. (a)

BAYLEY, N. Mental growth during the first three years. *Genetic Psychology Monographs*, 1933, *14*, 1–92. (b)

BAYLEY, N. Consistency and variability in the growth of intelligence from birth to 18 years. *Journal of Genetic Psychology*, 1949, *75*, 165–196.

BAYLEY, N. *Bayley scales of infant development*. New York: The Psychological Corporation, 1969.

BELL, S. B. The development of the concept of the object as related to infant-mother attachment. *Child Development*, 1970, *41*, 191–211.

BINET, A., & HENRI, V. La mémoire des phrases. *L'Année Psychologique*, 1895, *1*, 24–59.

BINET, A., & HENRI, V. La psychologie individuelle. *L'Année Psychologique*, 1896, *2*, 411–465.

BINET, A., & SIMON, T. Méthodes nouvelles pour le diagnostic du niveau intellectuel des anormaux. *L'Année Psychologique*, 1905, *11*, 191–244. (a)

BINET, A., & SIMON, T. Application des méthodes nouvelles au diagnostic du niveau intellectuel chez les infants normaux et anormaux d'hospice et d'école primaire. *L'Année Psychologique*, 1905, *11*, 245–266. (b)

BINET, A., & SIMON, T. Le développement de l'intelligence chez les enfants. *L'Année Psychologique*, 1908, *14*, 1–94.

BINET, A., & SIMON, T. [*Mentally defective children*] (W. B. Drummond, trans.) London: Edward Arnold, 1914. (Originally published, 1907.)

BOWLBY, J. Maternal care and mental health. *World Health Organization Monograph Series*, 1952 (No. 2).

BRACKBILL, Y., KANE, J., MANNIELLO, R. L., & ABRAMSON, M. D. Obstetric meperidine usage and assessment of neonatal status. *Anesthesiology*, 1974, *40*, 116.

BRADWAY, K. P. An experimental study of the factors associated with Stanford-Binet IQ changes from the preschool to the junior high school. *Journal of Genetic Psychology*, 1945, *66*, 107–128.

BRAEN, B. B. An evaluation of the Northwestern infant intelligence test: Test B. *Journal of Consulting Psychology*, 1961, *25*, 245–248.

BRAZELTON, T. B. *Neonatal behavioral assessment scale*. Children's Hospital Medical Center, Boston, Mass., n.d.

BRAZELTON, T. B. Neonatal behavioral assessment scale. *Clinics in Developmental Medicine*, No. 50. Philadelphia, Pa.: Lippincott, 1973.

BRAZELTON, T. B., KOSLOWSKI, B., & TRONICK, E. Study of the neonatal behavior in Zambian and American neonates. *Journal of the American Academy of Child Psychiatry*, 1976, *15*, 97.

BRAZELTON, T. B., ALS, H., TRONICK, E., & LESTER, B. M. Specific neonatal measures: The Brazelton neonatal behavior assessment scale. In J. D. Osofsky (Ed.), *Handbook of infant development*. New York: Wiley, 1979.

BROOKS-GUNN, J., & HEARN, R. Early intervention and developmental dysfunction. Implications for pediatrics. *Advances in Pediatrics*, 1982, *29*, 326–350.

BROOKS-GUNN, J., & LEWIS, M. Assessing the handicapped young: Issues and solutions. *Journal of the Division for Early Childhood*, 1981, 84–95.

BROOKS-GUNN, J., & LEWIS, M. Screening and diagnosing handicapped infants. *Topics in Early Childhood Special Education*, 1983, *3*(1), 14–28.

BUHLER, C. *The first year of life*. New York: John Day, 1930.

BUHLER, C., & HETZER, H. *Testing children's development from birth to school age*. New York: Farrar & Rinehart, 1935.

BURT, C. *Mental and scholastic tests*. London: King, 1921.

BZOCH, K. R., & LEAGUE, R. *Bzoch-League receptive—expressive language scale for the measurement of language skills in infancy*. Gainesville, Fla.: Tree of Life Press, 1970–1971.

CALDWELL, B. M., & DRACHMAN, R. H. *Comparability of three methods of assessing the developmental level of young infants*. Syracuse, N.Y.: Department of Pediatrics—SUNY, Upstate Medical Center, 1964.

CAMERON, J., LIVSON, N., & BAYLEY, N. Infant vocalizations and their relationship to mature intelligence. *Science*, 1967, *157*, 331–333.

CATALANO, F. L., & McCARTHY, D. Infant speech as a possible predictor of later intelligence. *The Journal of Psychology*, 1954, *38*, 203–209.

CATTELL, J. Mental tests and measurements. *Mind*, 1890, *15*, 373–381.

CATTELL, J., & FARRAND, L. Physical and mental measurements of the students of Columbia University. *Psychological Review*, 1896, *3*, 618–648.

CATTELL, P. *The measurement of intelligence of infants and young children*. New York: The Psychological Corporation, 1940, 1960, 1966.

CAVANAUGH, M. C., COHEN, I., DUNPHY, D., RINGWALL, E. A., & GOLDBERG, I.D. Prediction from the Cattell infant intelligence scale. *Journal of Consulting Psychology*, 1957, *21*, 33–37.

CHAILLE, S. E. Infants: Their chronological process. *New Orleans Medical and Surgical Journal*, 1887, *14*, 893–912.

CRISSEY, O. L. Mental development as related to institutional residence and educational achievement. *University of Iowa Studies on Child Welfare*, 1937, *13*(1).

CUNNINGHAM, B. V. Infant IQ ratings evaluated after an interval of seven years. *Journal of Experimental Education*, 1934, *11*(2), 84–87.

DARWIN, C. *The origin of species*. London: John Murray, 1859.

DARWIN, C. *Expression of the emotions in man and animals*. New York: D. Appleton, 1973. (Originally published, 1872.)

DARWIN, C. A biographical sketch of an infant. *Mind*, 1877, *2*, 285–294.

DAY CARE AND CHILD DEVELOPMENT COUNCIL OF AMERICA, INC. *Evaluating Children's progress: A rating scale for children in day care*. Washington, D.C.: Author, 1973.

DECARIE, TH. G. *Intelligence and affectivity in early childhood*. New York: International Universities Press, 1965.

DOLL, E. A. *Measurement of social competence*. Educational Test Bureau, 1953. (Available from American Guidance Service, Circle Pines, Minnesota.)

DRISCOLL, G. P. The development status of the preschool child as a prognosis of future development. *Child Development Monographs*, 1933, No. 13.

ESCALONA, S. The use of infant tests for predictive purposes. *Bulletin of the Menninger Clinic*, July 1950.

ESCALONA, S. K., & CORMAN, H. *Albert Einstein scales of sensori-motor development.* New York: Albert Einstein College of Medicine of Yeshiva University, 1969.

ESCALONA, S. K., & MORIARTY, A. Prediction of school-age intelligence from infant tests. *Child Development*, 1961, *32*, 597–605.

ESQUIROL, J. D. *Des maladies mentales considerées sous les rapports médical, hygienique et médico-légal.* Paris: J. B. Baillière, 1838.

FAGAN, J. F. *A visual recognition test of infant intelligence.* Paper presented at the International Conference on Infant Studies, Austin, Texas, March 1982.

FAGAN, J. F., & MCGRATH, S. K. Infant recognition memory and later intelligence. *Intelligence*, 1981, *5*, 121–130.

FANTZ, R. L., & NEVIS, S. *Fantz-Nevis visual preference test.* Cleveland: Case-Western Reserve, 1967.

FILLMORE, E. A. Iowa tests for young children. *University of Iowa Studies on Child Welfare*, 1936, *11*, No. 4.

FRANDSEN, A., & BARLOW, F. P. Influence of the nursery school on mental growth. *39th Yearbook of the National Society of Education*, 1940, Part II, 143–148.

FRANKENBURG, W. K., & DODDS, J. B. *Denver developmental screening test.* Denver, Colo.: Ladoca Publishing Foundation, 1967.

FREEDMAN, D. G., & FREEDMAN, N. Behavioral differences between Chinese-American and European-American newborns. *Nature*, 1969, *224*, 122.

FURFEY, P. H., & MUEHLENBEIN, J. The validity of infant intelligence tests. *Journal of Genetic Psychology*, 1932, *40*, 219–223.

GALLAGHER, J. J. Clinical judgment and the Cattell intelligence scale. *Journal of Consulting Psychology*, 1953, *17*, 303–305.

GALTON, F. *Hereditary genius.* London: MacMillan, 1869.

GESELL, A. *The mental growth of the preschool child.* New York: MacMillan, 1926.

GESELL, A. *Infancy and human growth.* New York: MacMillan, 1928.

GESELL, A. The ontogenesis of infant behavior. In D. Carmichael (Ed.), *Manual of child psychology.* New York: Wiley, 1954.

GESELL, A., & AMATRUDA, C. *Developmental diagnosis.* New York: Paul B. Holber, 1954.

GESELL, A., & AMATRUDA, C. *Developmental diagnosis: Normal and abnormal child development, clinical methods and practical applications.* New York: Harper, 1962.

GESELL, A., & THOMPSON, H. *Infant behavior, its genesis and growth.* New York: McGraw-Hill, 1934.

GILBERT, J. A. Research on the mental and physical development of school children. *Studies of Yale Psychological Laboratory*, 1894, *2*, 40–100.

GILLILAND, A. R. *The Northwestern intelligence tests. Examiner's manual. Test A: Test for infants 4-12 weeks old.* Boston: Houghton Mifflin, 1949.

GILLILAND, A. R. *The Northwestern intelligence tests. Examiner's manual. Test B: Test for infants 13-36 weeks old.* Boston: Houghton Mifflin, 1951.

GODDARD, H. H. The Binet and Simon tests of intellectual capacity. *The Training School*, 1908, *5*, 3–9.

GODDARD, H. H. Four hundred feeble minded children classified by the Binet method. *Pedagogical Seminary*, 1910, *17*, 387–397. (a)

GODDARD, H. H. A measuring scale for intelligence. *The Training School*, 1910, *6*, 146–155. (b)

GODDARD, H. H. Two thousand normal children measured by the Binet measuring scale of intelligence. *Pedagogical Seminary*, 1911, *18*, 232–259.

GOLDEN, M., & BIRNS, B. *Piaget object scale.* New York: Albert Einstein College of Medicine of Yeshiva University, 1968.

GOODENOUGH, F. L. *The Kuhlmann-Binet tests for children of preschool age.* Minneapolis: University of Minnesota Press, 1928. (a)

GOODENOUGH, F. L. A preliminary report on the effects of nursery school training upon intelligence tests scores of young children. *27th Yearbook of the National Society of Education*, 1928, 361–369. (b)

GOODENOUGH, F. L. Look to the evidence: A critique of recent experiments on raising the IQ. *Education Methods*, 1939, *19*, 73–79.

GOODENOUGH, F. L. New evidence on environmental influence on intelligence. *39th Yearbook of the National Society of Education*, 1940, Part I, 307–365.

GOODENOUGH, F. L. *Mental testing.* New York: Rinehart, 1949.

GRIFFITHS, R. *The abilities of babies.* London: University of London Press, 1954.

HALLOWELL, D. K. Mental tests for preschool children. *Psychological Clinic*, 1925–1927, *16*, 235–276.

HALLOWELL, D. K. Validity of mental tests for young children. *Journal of Genetic Psychology*, 1941, *58*, 265–286.

HARMS, I. E. *A study of some variables affecting the reliability of intelligence test scores during late infancy.* Unpublished doctoral dissertation, State University of Iowa, 1951.

HERRING, R. M. An experimental study of the reliability of the Buhler baby tests. *Journal of Experimental Education*, 1937, *6*(2), 147–160.

HINDLEY, C. B. The Griffiths scale of infant development: Scores and predictions from 3 to 18 months. *Journal of Child Psychology and Psychiatry*, 1960, *1*, 99–112.

HINDLEY, C. B. Stability and change in abilities up to 5 years: Group trends *Journal of Child Psychology and Psychiatry*, 1965, *6*, 85–99.

HOROWITZ, F. D., & BRAZELTON, T. B. Research with the Brazelton Neonatal Scale. In T. B. Brazelton (Ed.), Neonatal Behavioral Assessment Scale. *National Spastics Society Monograph*. Philadelphia: Lippincott, 1973.

HOROWITZ, F. D., ALEKSANDROWICZ, M., ASHTON, L. J., TIMS, S., McCLUSKEY, K., CULP, R., & GALLAS, H. *American and Uruguayan infants: Reliabilities, maternal drug histories and population difference using the Brazelton Scale.* Paper presented at the biennial meeting of the Society for Research in Child Development, 1973.

HOROWITZ, F. D., SULLIVAN, J. W., & LINN, P. Stability and instability in the newborn infant: The quest for elusive threads. In A. J. Sameroff (Ed.), *Organization & stability of newborn behavior: A commentary on the Brazelton Neonatal Behavior Assessment Scale.* (With commentary by Robert N. Emde.) *Monographs of the Society for Research in Child Development*, 1978, *43*(5–6, Serial No. 177).

HORROCKS, J. E. *Assessment of behavior.* Columbus, Oh.: C. E. Merrill, 1964.

HUNT, J. McV. *Intelligence and experience.* New York: Ronald, 1961.

HUNT, J. McV., & BAYLEY, N. Explorations into patterns of mental development and prediction from the Bayley Scales of infant development. *Minnesota Symposium on Child Psychology*, 1971, *5*, 52–71.

ILLINGWORTH, R. S. *The development of the infant and young child, normal and abnormal.* Baltimore, Md.: Williams & Wilkins, 1972. (Originally published, 1960.)

ILLINGWORTH, R. S. The predictive value of developmental tests in the first year, with special reference to the diagnosis of mental subnormality. *Journal of Child Psychology and Psychiatry*, 1961, *2*, 210–215.

ITARD, J. G. [*The wild boy of Aveyron*] (G. Humphrey & M. Humphrey, trans.) New York: Appleton-Century-Crofts, 1938. (Originally published, 1801.)

JASTROW, J. Some anthropological and psychological tests on college students—a preliminary survey. *American Journal of Psychology*, 1892, *4*, 420–427.

KNOBLOCH, H., & PASAMANICK, B. An evaluation of the consistency and predictive value of the forty week Gesell development schedule. In C. Shagass & B. Pasamanick (Eds.), *Child development and child psychiatry*. Psychiatric Research Report of the American Psychiatric Association, 1960, No. 13, 10–31.

KNOBLOCH, H., & PASAMANICK, B. *Prediction from assessment of neuromotor and intellectual status in infancy*. Paper presented at American PsychopathologicalAssociation meeting, February 1966.

KRAEPELIN, E. Der psychologische Versuch in der Psychiatrie. *Psychologische Arbeiten*, 1895, *1*, 1–91.

KUHLMANN, F. Binet and Simon's system for measuring the intelligence of children. *Journal of Psycho-Asthenics*, 1911, *15*, 76–92.

KUHLMANN, F. *A handbook of mental tests*. Baltimore, Md.: Warwick & York, 1922.

LESTER, B., & ZESKIND, P. Brazelton Scale and physical size correlates of neonatal cry features. *Infant Behavior and Development*, 1978, *4*, 393.

LEWIS, M., & BROOKS-GUNN, J. *Social cognition and the acquisition of self*. New York: Plenum Press, 1979.

LEWIS, M., & BROOKS-GUNN, J. Attention and intelligence. *Intelligence*, 1981, 5(3). (a)

LEWIS, M., & BROOKS-GUNN, J. Visual attention at three months as a predictor of cognitive functioning at two years of age. *Intelligence*, 1981, *5*, 131–140. (b)

LEWIS, M., & BROOKS-GUNN, J. Developmental models and assessment issues. In N. Anastasiow, W. Frankenburg, & A. Fandal (Eds.), *Identifying the developmentally delayed child*. Maryland: University Park Press, 1982.

LEWIS, M., & MICHALSON, L. *Children's emotions and moods: Theory and measurement*. New York: Plenum Press, 1983.

LINFERT, H. E., & HIERHOLZER, H. M. *A scale for measuring the mental development of infants during the first years of life*. Baltimore, Md.: Williams & Wilkins, 1928.

MACFARLANE, J. W. The uses and predictive limitations of intelligence tests in infants and young children. *Bulletin WHO*, 1953, *9*, 409–415.

MACRAE, J. M. Retests of children given mental tests as infants. *Journal of Genetic Psychology*, 1955, *87*, 111–119.

MCCALL, R. B., HOGARTY, P. S., & HURLBURT, N. Transitions in infant sensorimotor development and the prediction of childhood IQ. *American Psychologist*, 1972, *27*, 728–748.

MCCALL, R. B., EICHORN, D. H., & HOGARTY, P. S. Transitions in early mental development. *Monographs of the Society for Research in Child Development*, 1977, 42(3, Serial No. 171).

MCGRAW, M. B. Appraising test responses of infants and young children. *The Journal of Psychology*, 1942, *14*, 89–100.

MCHUGH, G. Changes in IQ at the public school kindergarten level. *Psychology Monographs*, 1943, *55*, No. 2.

MECHAM, M. J. *Verbal language development scale*. Circle Pines, Minn.: American Guidance Service, 1971.

MUNSTERBERG, H. Zur individual Psychologie. *Centralblatt für Nervenheilkunde und Psychiatric*, 1891, *14*.

NELSON, V. L., & RICHARDS, T. W. Studies in mental development: I. Performance on Gesell items at 6 months and its predictive value for performance on mental tests at 2 and 3 years. *Journal of Genetic Psychology*, 1938, *52*, 303–325.

PETERSON, J. *Early conceptions and tests of intelligence*. New York: World Book, 1925.

POWELL, L. F. The effect of extra stimulation and maternal involvement on the development of low birth weight infants and on maternal behaviors. *Child Development*, 1974, *45*, 106.

PRECHTL, H., & BEINTEMA, D. *The neurological examination of the full term newborn infant.* London: Heineman, 1964.

PREYER, W. *The mind of the child.* New York: D. Appleton, 1888. (Originally published, 1882.)

RICCIUTI, H. N. Object grouping and selective ordering behavior in infants 12 to 24 months old. *Merrill-Palmer Quarterly,* 1965, *11,* 129–148.

RICHARDS, T. W., & NELSON, V. L. Studies in mental development: II. Analyses of abilities tested at age of 6 months by the Gesell schedule. *Journal of Genetic Psychology,* 1938, *52,* 327–331.

RICHARDS, T. W., & NELSON, V. L. Abilities of infants during the first 18 months. *Journal of Genetic Psychology,* 1939, *55,* 299–318.

RINGWALL, E. A. *Prelinguistic infant vocalization analysis.* Buffalo, N.Y.: State University of New York, 1965.

ROBERTS, J. A. F., & SEDGLEY, E. Intelligence testing of full-term and premature children by repeated assessments. In C. Banks & P. L. Broadhurst (Eds.), *Studies in psychology.* London: University of London Press, 1965.

ROSENBLITH, J. F. *Manual for behavioral examination of the neonate as modified by Rosenblith from Graham.* Providence, R.I.: Brown Duplicating Service, 1961.

SCARR, S., & WILLIAMS, M. L. The effects of early stimulation on low birth weight infants. *Child Development,* 1973, *44,* 94.

SEGUIN, E. *Idiocy: Its treatment by the physiological method.* New York: Bureau of Publications, Teachers College, Columbia University, 1907. (Reprinted from the original edition of 1886.)

SHARP, S. E. Individual psychology: A study in psychological method. *American Journal of Psychology,* 1898–1899, *10,* 329–391.

SHIELD INSTITUTE FOR RETARDED CHILDREN. *Early identification and treatment of the infant retardate and his family.* New York: Author, 1968.

SHINN, M. *The biography of a baby.* Boston: Houghton Mifflin, 1900.

SHIRLEY, M. *The first two years.* Minneapolis: University of Minnesota Press, 1933.

SHOTWELL, A. M., & GILLILAND, A. R. A preliminary scale for the measurement of the mentality of infants. *Child Development,* 1943, *14,* 167–177.

SIMON, T. Recherches anthropométriques sur 223 garçons anormaux agés de 8 à 23 ans. *L'Année Psychologique,* 1900, *6,* 191–247.

SIMON, A. J., & BASS, L. G. Toward a validation of infant testing. *American Journal of Orthopsychiatry,* 1956, *26,* 340–350.

SIMPSON, B. R. The wandering IQ: Is it time to settle down? *Journal of Psychology,* 1939, *7,* 351–367.

SKODAK, M. Children in foster homes; A study of mental development. *University of Iowa Studies on Child Welfare,* 1939, *16*(1).

STERN, W. *Psychology of early childhood up to the sixth year of age.* New York: Henry Hal, 1924. (Originally published, 1914.)

STOTT, L. H., & BALL, R. S. Evaluation of infant and preschool mental tests. *Monographs of the Society for Research in Child Development,* 1965, *30*(3), Serial No. 101.

STRAYER, G. D. *Age and grade census of schools and colleges* (Report of the U.S. Bureau of Education, Bulletin #5). Washington, D.C.: Government Printing Office, 1911.

STUTSMAN, R. *Mental measurement of preschool children.* New York: World Book, 1931.

SYMMES, E. An infant testing service as an integral part of a child guidance clinic. *American Journal of Orthopsychiatry,* 1933, *3,* 409–430.

TERMAN, L. M. *The measurement of intelligence.* Boston: Houghton Mifflin, 1916.

TERMAN, L. M., KOHS, S. C., CHAMBERLAIN, M. B., ANDERSON, M., & HENRY, B. The vocabulary test as a measure of intelligence. *Journal of Educational Psychology*, 1918, *9*, 452–466.

THOMAS, H. Psychological assessment instruments for use with human infants. *Merrill-Palmer Quarterly*, 1970, *16*, 179–224.

THORNDIKE, R. L. Constancy of the IQ. *Psychological Bulletin*, 1940, *37*, 167–186.

TRABUE, M. R., & STOCKBRIDGE, F. P. *Measure your mind*. New York: Doubleday, 1922.

TRONICK, E., & BRAZELTON, T. B. Clinical uses of the Brazelton Neonatal Behavioral Assessment. In B. Z. Friedlander, G. M. Steritt, & G. E. Kirk, (Eds.), *Exceptional infant* (Vol. III). New York: Brunner/Mazel, 1975.

UŽGIRIS, I. C., & HUNT, J. McV. *An instrument for assessing infant psychological development*. Mimeographed paper, Psychological Development Laboratories, University of Illinois, 1966.

WALDROP, M. F., BELL, R. Q., McLAUGHLIN, B., & HALVERSON, C. F. Newborn minor physical anomalies predict short attention span, peer aggression and impulsivity at age 3. *Science*, 1978, *179*, 563–564.

WALLIN, J. E. W. The peg form boards. *Psychological Clinician*, 1918, *12*, 40–53.

WATSON, J. B., & WATSON, R. R. Studies in infant psychology, *Scientific Monthly*, 1921, *13*, 493–515.

WELLMAN, B. L. Some new bases for interpretation of the IQ. *Journal of Genetic Psychology*, 1932, *41*, 116–126. (a)

WELLMAN, B. L. The effects of preschool attendance upon the IQ. *Journal of Experimental Education*, 1932, *1*, 48–69. (b)

WERNER, E., & BAYLEY, N. The reliability of Bayley's revised scale of mental and motor development during the first year of life. *Child Development*, 1966, *37*, 39–50.

WERNER, E. E., HONZIK, M. P., & SMITH, R. S. Prediction of intelligence and achievement at 10 years from 20-month pediatric and psychological examinations. *Child Development*, 1968, *39*, 1063–1075.

WILSON, R. S. Testing infant intelligence. *Science*, 1973, *182*, 734–739.

WISSLER, C. L. The correlation of mental and physical tests. *Psychology Review Monograph Supplement*, 1901, *3*(6).

WITTENBORN, J. R. *The placement of adoptive children*. Springfield, Ill.: Charles C Thomas, 1957.

WITTENBORN, J. R., ASTRACHAN, M. A., DeGOUGAR, M. W., GRANT, W. W., JANOFF, I. E., KUGEL, R. B., MYERS, B. J., RIESS, A., & RUSSELL, E. C. A study of adoptive children: II. The predictive validity of the Yale developmental examination of infant behavior. *Psychological Monographs*, 1956, *70*(2).

3 Measuring Mental Abilities in Infancy
The Value and Limitations

MARJORIE P. HONZIK

The development of mental abilities during infancy is impressive and measurable. The pediatrician measures growth in head size, which reflects the growth of the brain. The neuropathologist measures cerebral DNA to estimate cell number and possible damage from malnutrition or other causes (Winick, 1970). The psychologist measures behavioral change by means of careful observations of responses to specific tasks. In this chapter we shall review critically infant tests and their contribution to the understanding of mental growth in the first months of life.

Mental development in infancy is rapid and not *easily* measured. The adequacy of the tests, the skill of the examiner, and, above all, the state of the infant affect mental measurement. Despite these limiting factors, pediatricians, psychologists, neurologists, and parents have found the results of testing to be crucial in the diagnosis of specific abilities and deficits, and research workers are turning to tests with increasing frequency to evaluate mental growth and its relevant determinants. We shall discuss sequentially the limitations imposed by the triad of infant, test, and examiner; the available tests, and then the contribution of these tests to diagnosis of disability and to research on the development of infant intelligence.

MARJORIE P. HONZIK • Institute of Human Development, University of California, Berkeley, California 94720.

THE INFANT

The growth of cognitive functions is intertwined with somatic growth, which makes the testing of the infant a challenge. The baby triples body weight in the first year, but the greatest changes occur in the brain, where there is some increase in cell number but marked increase in cell size, in number of dendrites, and in the myelin sheath covering the nerve axons. The weight gain of the brain during early infancy is greater than that taking place in any other somatic area (Dodgson, 1962). Accompanying structural growth are the marked changes in cognitive development that are revealed in the infant's changing perception of and reactions to his or her world.

The rapidly changing infant requires skillful management by the mental tester, who must be continuously sensitive to the marked developmental changes as well as to the continuously changing state of the baby during the testing period.

The requirements of a good test are exacting. Not only does it have to yield reliable and valid scores, but it has to cover the repertoire of cognitive behaviors that are developing, and clearly it must include materials that will elicit maximum responses at all developmental levels. The perfect assessment, then, takes place when the squalling, sucking, chewing, ever-moving baby is relatively quiescent, attentive, wide-eyed with interest, and above all, responsive to the tester and the toys. This idyllic situation is achieved only with effort on the part of all concerned.

Information processing begins at birth or earlier, but the baby has the greater problem during the paranatal period of accommodating to a life of independent breathing and assimilation of food. Somatic development and change create difficulties in testing at many stages of adjustment during the first years of life. In the immediately postnatal period, the infant is often drowsy and unresponsive, and the time periods available for testing are very short. Careful observations indicate that infants in the postnatal period are, on the average, alert only 10% of the time. The great ingestion of food relative to body size during the early postnatal months frequently leads to discomfort and colic, which may interfere with the baby's responses. As this stage passes and he or she is more consistently comfortable, the baby's awareness of strange people and situations becomes apparent. "Awareness of a strange situation" or "fear of strangers" may themselves be used as test items (Bayley, 1969), but this apprehensive stage may make for less than optimal responses on certain other test items. The teething child is often more interested in chewing the test equipment than in responding to it. The child who is learning to crawl and walk wants to practice these skills and is proportionately less interested in responding to form boards and problem situations. Skilled testers, using all the patience, perseverance, and charm that they can muster, usually can elicit the baby's best responses, but it is not always easy. This descriptive account suggests aspects of the baby's development that may result in errors of measurement, but it also suggests that we are considering the mental development of a living, vibrant, growing organism. The metamorphosis may not be as great as in some species, but the sturdy, almost-verbal 2-year-old has come a long way from the immature state at birth. Do the scores on infant tests document these changes? One of the purposes of this chapter is to answer this question.

INFANT TESTS

Infant tests reflect the concerns of their authors. Diagnosis was the primary objective of the Gesell (Gesell & Amatruda, 1941), the Griffiths (1954), and the Brunet–Lézine (1951) tests. Cattell (1940/1960) and Bayley (1933, 1969) were more interested in devising tests that could be used to study the *development* of mental abilities during the first years of life. However, all five tests have been used in major research projects, and all have been and are being used in the diagnosis of individual children. More recently, Užgiris and Hunt (1966) have assembled test items that test Piaget's stages more specifically than the tests mentioned above. Another trend is the use of abbreviated screening tests. The most notable is the Denver Developmental Screening Test, which is being widely used by pediatricians in the United States and in a number of other countries.

Arnold Gesell of Yale was the pioneer in infant testing. His first degree was in psychology, but his approach to infant testing was that of the pediatric neurologist, as indicated by the title of his handbook *Developmental Diagnosis* (with Amatruda, 1941). Griffiths (1954) expressed her indebtedness to Gesell, stating that the justification for a new test is the "urgency of the need for early diagnosis of mental condition in special cases" (p. 1). However, she showed greater concern about test standardization and the reliability of the scores than did Gesell.

All infant tests include many items first described by Gesell. The Brunet–Lézine test (1951) is a translation into French of the Gesell schedules, but the test items are arranged by age levels. The two tests constructed by the American psychologists Bayley (1933, 1969) and Cattell (1940/1960) were developed as research instruments. These investigators were far more attentive than Gesell to the problems of test construction, sampling, and adequate standardization.

The original five tests have been widely used. Their distinctive characteristics will be summarized before we discuss the contribution of infant tests to our understanding of intellectual development.

The Gesell Developmental Schedules

Gesell and Amatruda wrote that "developmental diagnosis is essentially an appraisal of the maturity of the nervous system with the aid of behavior norms" (1941, p. 15). Many of the test items in Gesell's Developmental Schedules stem from this point of view. For example, he described the stimulating value of the 1-inch cube at successive ages: "grasps when placed in hand (4 weeks); ocular fixation when cube is placed on table (16 weeks); prehension on sight by palmar grasp (28 weeks); prehension by digital grasp (40 weeks)" (p. 19). These behaviors are classified as *adaptive behavior*. Other major fields of behavior, according to Gesell, are *motor language,* and *personal-social.* The motor sequence includes both gross skills, such as sitting and walking, and fine motor skills, such as picking up a small sugar pellet or building with blocks. *Language* behavior is assessed by comprehension as well as vocalizations and use of words. Gesell stated that the *personal-social* behavior sequence, consisting of responses to people and self-help items, is "greatly affected by the kind of home in which the child is reared" (p. 14). In contrast, he wrote of the *motor* scale that it "is of special interest because it has so many neurological implications" (p. 5), and of *language* behavior, that it furnishes clues to the organization of the infant's central nervous system. These distinctions as to the determinants of the subscale scores have never been adequately tested. In fact, cross-cultural research on motor development suggests that child-rearing practices may in part determine gross motor skills (Geber & Dean, 1964). Experiential factors in the home affect not only *personal-social* behaviors but also *language* development and *adaptive* behavior, according to Bernstein (1961), Honzik (1967), and others.

The Gesell Schedules yield developmental quotients (DQs) obtained by the division of the maturity age by the chronological age of the child. Gesell wrote that the "DQ represents the proportion of normal development present at a given age" (p. 113).

The Gesell Schedules are based on extensive observations, but Gesell did not present statistical evaluations of the normative findings nor is there any attempt to assess the reliability or the validity of the DQs. Gesell's perceptive and almost eloquent descriptions of infant behavior have seldom been equaled:

> The baby can reach with his eyes before he can reach with his hand; at 28 weeks the baby sees a cube; he grasps it, senses surface and edge as he clutches it, brings it to his mouth, where he feels its qualities anew, withdraws it, looks at it on withdrawal, rotates it while he looks, looks while he rotates it, restores it to his mouth, withdraws it again for inspection, restores it again for mouthing, transfers it to the other hand, bangs it, contacts it with the free hand, transfers, mouths it again, drops it, resecures it, mouths it yet again, repeating the cycle with variations—all in the time it takes to read this sentence. (p. 49)

Gesell and Amatruda's *Developmental Diagnosis* has been enlarged and revised by Knobloch and Pasamanick (1974), and Knobloch, Stevens, and Malone (1980) are the authors of a new manual for this test.

All subsequent authors of infant tests raided the Gesell Schedules for test items, with appropriate acknowledgment of indebtedness. The test materials from the Gesell Schedules found most frequently in other tests are the red ring and string, the red 1-inch cubes, the sugar pellet, and the small dinner bell with handle. These materials are used in the early months to test visual and auditory responses, eye–hand coordinations, and problem solving.

A screening test that not only uses Gesell items but is modeled on it is the Denver Developmental Screening Test (DDST). This test deserves mention because of the care with which it was assembled and because of its wide use. Frankenburg and Dodds (1967) developed this test to "aid in the early detection of delayed development in young children" (p. 181). Frankenburg, Camp, and Van Natta (1971) reported that the reliability and validity of this scale is fairly high. They concluded that this screening test when used with infants in the first year of life, misses approximately 13% of the cases that would obtain an abnormal rating on the Bayley test. Infants and young children earning an abnormal rating on the DDST achieve psychomotor development indexes (PDIs) averaging between 50 and 73 on the Bayley.

The Griffiths Scale of Mental Development

The Griffiths Scale, published in 1954, resembles the Gesell Schedules in some important dimensions. Ruth Griffiths was concerned with diagnosis; a perceptive observer of the behaviors of infants, she wanted to avoid any pretense of measuring "intelligence" in infancy. Her interest was in assembling a good, reliable test that detected deficits in the following areas of functioning: *locomotion, personal-social, hearing and speech, eye and hand,* and *performance.* These categories resemble those of Gesell, but Gesell's motor scale is divided into "Locomotion" and "Performance," his language scale is called more precisely "Hearing and Speech," and the adaptive scale is termed simply "Eye and Hand." An effort was made to standardize this scale on a representative sample of infants. The distribution of the paternal occupations of the 552 children tested was similar to that of employed males in the United States. A test–retest correlation of 0.87 was reported for 60 infants who were retested after an average interval of 30 weeks. Griffiths wrote that there is nothing necessarily fixed or permanent about an intelligence quotient as such. In addition, she wrote:

> It has often puzzled the writer why psychologists should for so long have expected that the results of a single test, applied in some one particular hour of a child's life, under particular circumstances, should necessarily carry within it any particular implication of finality, or suggestion that if repeated the ratio of the child's performance to his age should remain unchanged. In no other field of diagnostic work is such a condition expected. (p. 43)

Brunet–Lézine Test

This test follows the Gesell Schedules in dividing the test items into the four categories of motor, adaptive, language, and personal-social. However, it differs in that it is a point scale with six test items at each month level, followed by four questions that are asked of the mother or caretaker. The number of test items measuring each of the four behavior categories varies at the different age levels, but for the most part there are two *motor* and four *adaptive* items, and one *motor*, one *language*, and two *personal-social* questions. The test is scored according to the four categories and according to whether the score is based on an observation item or on a question. The manual for this test is not available in the United States. The test is used extensively in Europe for diagnosis and was the test of choice in the Stockholm longitudinal study described by Klackenberg-Larsson and Stensson (1968) and Brucefors (1972).

Cattell's Infant Intelligence Scale

Psyche Cattell assembled this scale to assess the mental develop-ment of a group of normal children in a longitudinal study conducted in the School of Public Health at Harvard. Cases were selected from prena-tal clinics, which meant that well-to-do families were excluded from the standardization sample, as were those of the lowest economic levels. Cattell acknowledged "the vast amount of pioneer work of Gesell" (p. 19) and stated that Gesell's battery of tests was used as the foundation for her scale. However, Cattell excluded from her test items of a *personal-social* nature, since they are influenced to a marked degree by "home training," and *motor* items involving the large muscles. Cattell provided objective procedures for administering and scoring the tests. Her test is an age scale with five items listed at each month level in the first year. The standardization was based on a total of 1,346 examinations admin-istered to 274 children, who were tested at the ages of 3, 6, 9, 12, 18, 24, 30, and 36 months.

The Cattell test has many good points. The items interest children. The instructions for administering and scoring are clear. Odd–even reli-ability is .56 for the 3-month test, but for the 6-month test it is .88, and at subsequent age levels it varies from .71 to .90.

There are two disadvantages to this test as compared with the Bay-ley series. Because standardization testing was done at 3-month inter-vals, the item placements at ages 2, 4, 5, 7, 8, 10, and 11 months are interpolated values. As a result, the transitions from age to age are less smooth than when standardization testing is done at monthly intervals. A second disadvantage is that the total number of test items covering the age period 2–12 months is 55, in contrast to over 100 items in both Bayley's California First Year Mental Scale and in the Brunet-Lézine.

Griffiths's scale, which includes both locomotor and performance items, has 155 test items for the age period 1–12 months. Because the Cattell test is shorter than the other tests, it seldom tires the babies. It can usually be completed within a half hour.

The Bayley Scales of Infant Development

These scales, which include both mental and motor tests, are based on more than 40 years of research. Bayley published the California First Year Mental Scale in 1933. Her mental scale largely includes test items termed *adaptive* or *language* by Gesell, with a few *personal-social* and *motor* items. The test was assembled to test the children in the Berkeley Growth Study. This cohort was selected from two hospitals. One accepted clinic cases and the other, only those who could pay for the delivery. All occupational classes were represented, but a disproportionate number of the fathers were students or in the professions. From 47 to 61 babies were tested each month. Preliminary standardization was based on these tests. Split-half reliabilities were only .63 and .51 for the first 2 months, but at subsequent ages, they ranged from .74 to .95.

The 1969 revision of the Bayley Mental and Motor Scales and Behavior Record included new items and changed the ordering of the test items. Standardization was done on a stratified United States sample of 1,262 children, with 83–94 tested at each age level. The split-half reliability coefficients for the 1969 scale were higher than those reported for the 1933 version of this test—from .81 to .93 (Werner & Bayley, 1966). Bayley (1969) wrote that the mental scale was designed to "assess sensory-perceptual acuities, discriminations, and the ability to respond to these; the early acquisition of 'object constancy,' memory, learning, and problem solving ability; vocalizations and the beginnings of verbal communication; and early evidence of the ability to form generalizations and classifications, which is the basis of abstract thinking" (p. 3).

This test was administered to approximately 50,000 8-month-old babies born in 12 different hospitals as a part of the Collaborative Perinatal Research Project (National Institute of Neurological Diseases and Stroke; Broman, Nichols, & Kennedy, 1975). These test results have been used in many investigations, which will be discussed later in this chapter.

Summary

These are the tests. Test–retest reliability where reported is adequate or good. What of the validity? All five tests have been widely used on research projects. It is only in the findings they have yielded that the value and thus the validity of the tests can be determined.

PREDICTION FROM INFANT TESTS

Gesell and Amatruda (1941) wrote, "diagnosis involves prognosis" (p. 349). Griffiths did not expect her test score to be predictive. Cattell (1940/1960) noted instances of marked variability in IQs of infants and young children on her test and concluded that "there is no age from birth to maturity at which it is safe to base an important decision on the results of intelligence tests alone" (p. 60). The purpose of her test, besides diagnosis, was "to add to existing knowledge as to the variability in the pattern of mental growth" (p. 5).

How well do these tests predict? As in most areas of human behavior, it depends! It depends on the integrity of the central nervous system, the genetic blueprint, in subtle and complex ways on experience, and, above all, on the interaction of all these factors. Investigators have begun to sort out the tangled skein and the picture is gradually becoming clearer.

The Gesell, Griffiths, Brunet– Lézine, Cattell, and Bayley tests have all been used extensively to predict later IQs from infant test scores. The method used is interage correlations. Agreement among investigators is high. Figure 1 shows interage correlations between Bayley's infant test scores and Stanford–Binet Form L IQs at 8 years for the Berkeley Growth Study sample (Bayley, 1949) and the Guidance Study sample (Honzik, Macfarlane, & Allen, 1948), born 1928–1929. These interage correlations are remarkably similar to those obtained for a Stockholm sample born between 1955 and 1961, tested in infancy on the Brunet– Lézine and at 8 years on Form L of the Stanford–Binet (Brucefors, 1972; Klackenberg-Larsson & Stensson, 1968). In neither of these longitudinal studies do IQs from infant mental test scores predict 8-year IQs, but both show increasing prediction during the age period of 1–3 years. This cross-validation of findings is impressive since the studies differ in three

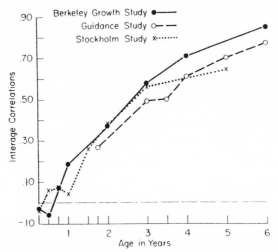

FIGURE 1.   Prediction of IQs on the Stanford–Binet, Form L or L-M, at 8 years from earlier scores on the California Infant and Preschool Scales in the Berkeley Growth Study and the Berkeley Guidance Study, and from earlier scores on the Brunet–Lézine in the Stockholm Study.

major respects: the infant tests used (Bayley versus Brunet–Lézine), cohort difference (Swedish versus United States samples), and a time period difference of 30 years.

These findings are further cross-validated by Cattell (1940/1960), who reported that scores on her test at 3 months correlated .10 with 3-year IQs on the Stanford–Binet, and by those of Hindley (1960), who found a correlation of −.13 between the 3- and 12-month test scores on the Griffiths for a London sample (see Table I). It should be noted that whereas the babies in the Stockholm and London studies were tested at ages 3, 6, 12, 18, and 24 months, the babies in Bayley's Berkeley Growth Study were tested each month and the scores were averaged for 3-month periods for purposes of determining the stability of test scores.

Table I shows that although the correlations over developmentally long time periods are negligible, adjacent ages yield moderately high *r*s. The findings of this table indicate what has become a truism in longitudinal studies of infants and children: The interage correlations are highly related to the age at testing and inversely related to the interval between tests. The negligible and even negative prediction from test scores obtained during the first months of life does not appear to be a chance phenomenon but rather a developmental fact. The moderate interage

TABLE I

Interage Correlations of Infant Mental Test Scores Found in Six Different Samples

Investigator	Study site	Test	n ($\cong$)	Interage Correlations (in months)					
				3 × 6 r	3 × 12 r	3 × 36 r	6 × 9 r	6 × 12 r	6 × 36[a] r
Bayley (1949)	Berkeley	Baley California Infant Test	61	.57	.28	−.09	.72	.52	.10
Klackenberg-Larsson & Stensson (1968)	Stockholm	Brunet–Lézine	140	.51	.36	−.08	.70	.59	−.05
Hindley (1960, 1965)	London	Griffiths	29–80	.53	−.13	—	—	.34	.40
Cattell (1940/1960)	Harvard	Cattell	35	—	—	.10	—	—	.34
Nelson & Richards (1939)	Fels	Gesell	48–80	—	—	—	—	.72	.46
Ramey & Haskins (1981a,b)	Chapel Hill	Bayley	26	—	—	—	.63	.66	.36

[a]The 36-month test was the California Preschool Scale in the Berkeley Study and forms of the Stanford–Binet in the other five investigations.

correlations between the 3- and 6-month scores of over .50 found in three different countries suggest that the changes in relative position take place gradually (Table I).

Brucefors, Johannesson, Karlberg, Klackenberg-Larsson, Lichen-stein, and Svenberg (1974), in an analysis of the test results for the Swedish sample, compared two groups of children who showed the greatest gains and losses over the age period 3 months to 8 years. The ascending group at 3 months had an average IQ of 84. The average IQ of this subsample at 8 years was 114 on the Terman–Merrill. The average IQ of the descending group was 111 at 3 months and 84 at 8 years. The rate of change in both of these extreme groups was gradual, with both groups earning the same score at about 18 months. Brucefors's interpretation of the marked changes, which also account in part for the negative correlations between the 3-month and childhood mental test scores, is in terms of *activity level*. Less active infants may be less responsive to test stimuli in the early months and earn lower scores, but they may be acquiring information and developing skills that serve them well at a later age.

Escalona (1968) wrote that "inactive babies show more sustained visual attention and do more tactile exploration of the immediate environment . . . up to at least 8 months than do active babies" (p. 241). In contrast, Escalona reported, "active babies, between the ages of 4 and 12 weeks, tend to develop responsiveness to relatively distant stimuli earlier, and acquire locomotion and the capacity for purpose manipulation of objects somewhat sooner" (p. 241). Escalona added that the claim that activity level demonstrably affects the course of development does nothing to define the process or mechanisms that account for variations. Her working hypothesis is that differences in activity level below the age of 8 months can be attributed to differences in the threshold for the release of movement. This hypothesis suggests a constitutional difference that predisposes the infant to react with activity or inactivity.

Support for the possibility that activity level may play a part in determining the age changes in mental test performance in infancy comes from Bayley and Schaefer (1964). They reported that the boys in the Berkeley Growth Study who had high IQs in later life were as infants calm, happy, and positive in their responses. They found that an abrupt shift in the nature of the correlations with the activity ratings at 18 months coincided with a drop in the boys' *r*s between mental and motor scores. They also found that activity in the child, his mental test scores, and hostility in the mother are correlated in the latter half of the first year but not thereafter. These authors concluded, "There is here a suggestion that hostile maternal behavior toward sons may goad them to activity and stimulate development until the boys begin to walk" (p. 38). They added, "If this is true, then we might postulate that the problems posed by active boys when they start running about and getting into things result in suppressive controls by mothers who are already hostile; at the same time, the accepting permissive mothers are encouraging activity in their relatively passive babies" (p. 38).

A similar trend was reported for the girls, but the impact of the mother's behavior on the girls' test score is much less clear. There is some evidence (Bing, 1963; Honzik, 1967) that the father's role is important in the development of the girls' cognitive skills, and the fathers were not rated in the Berkeley Growth Study. Taken together, these findings suggest that the road to high intellectual performance in later childhood is a complex one. Whether the baby is relatively active or relatively inactive in the first year may itself make a difference, but these individual differences also have an impact on the caretakers, whose responses may further affect cognitive development.

Effect of Including Mothers' Reports on Test Scores and Interage Prediction

Correlations between test scores earned at 6 months and 3 years (Table I) indicate that predictions are higher for the Griffiths and Gesell tests than for Bayley's Infant Mental Scale and the Brunet–Lézine. This difference may be due to the fact that the mothers' reports of test behaviors are not included in the scores on the Bayley test but play an important part in the scores on the Gesell and the Griffiths. It is possible that the more intelligent and more educated mother or caretaker gives a clearer picture of the child's behavior than the less able mother, thus adding her capabilities to the individual differences in the babies' test scores. In some instances the mothers' reports would contribute to more accurate scores. For example, Honzik, Hutchings, and Burnip (1965) found that infants who were "suspected" of having neurological impairment from their birth records vocalized with greater frequency during their mental tests at 8 months than was true of the normal control group. Honzik concluded that the 8-month babies in the control group were more inhibited by the strangeness of the test situation and vocalized less in the test situation, thus failing the vocalization items on the Bayley. On the Griffiths and the Gesell scales, the mothers' reports of vocalizing would have been credited. On future tests it would seem advisable to score a test with and without the caretakers' reports so that the advantages of both scores would be available. This is now possible on the Brunet–Lézine.

Vocalization Factor Score

Many researchers have concluded that early infant test scores are of little value because of the low negative interage correlations obtained when infant scores are compared with IQs obtained during childhood. Bayley's findings are among the most clear-cut in showing little or no prediction, and yet it was from a cluster analysis of the items on her test that were given each month to the children in the Berkeley Growth Study that a vocalization factor was obtained (Cameron, Livson, & Bayley, 1967). This factor score is moderately predictive to the age of 36 years, but for females only (Bayley, 1968). The age of first passing each test item was the score used in the correlation matrix from which the vocalization factor was derived. Vocalization definers include vocalizes eagerness, vocalizes displeasure, says "da da" or equivalent, and says two words.

In the same year that Cameron *et al.* (1967) published the results of the vocalization factor score, Moore (1967) in England reported for the London sample that a "speech quotient" derived from the hearing and speech section of the Griffiths Scale showed some constancy from 6 to 18 months in girls ($r = .51$, $p < .01$) but "virtually none in boys" ($r = .15$). This sex difference is of borderline statistical significance. Consistent with this finding is the failure of the boys' 6-month speech quotient to predict any later assessment, whereas that of the girls is significantly related to vocabulary at age 3 years. Moore concluded that "clearly, linguistic development runs a steadier course from an earlier age in female infants" (p. 95).

The sex difference in stability of vocalizations and a "speech quotient" was discovered independently from infant tests in London and in Berkeley. Further confirmation of this finding was reported by Kagan (1971), who found greater continuity and stability in the vocalizations of girls than of boys in a longitudinal study covering the first 3 years of life. He reported that the amount of vocalizing to facial stimuli was relatively stable in girls during the first year and that these vocalization scores predicted verbal behavior, but not overall IQ, at age $2\frac{1}{2}$ years.

A Sex Difference in Prediction from Infant Tests

The greater stability of infant vocalizations in girls than in boys suggests that there might be a sex difference in prediction of infant test scores.

McCall, Hogarty, and Hurlburt (1972) reported for the Fels longitudinal data that the 6-month Gesell DQs predicted $3\frac{1}{2}$-year Stanford–Binet IQs for girls ($r = .62$, $p < .01$), but not for boys ($r = -.01$). This difference is statistically significant at the .02 level. For this sample, the 12-month DQs yielded a fairly high correlation with the 6-year Stanford–Binet IQs for girls ($r = .57$, $p < .001$) but not for boys ($r = .22$, $p < .05$). Hindley (1965) also reported a sex difference in the stability of girls' DQs on the Griffiths as compared with that of boys (cf. Table II).

In contrast to these results, Klackenberg-Larsson and Stensson (1968) found no sex difference in DQ stability in the Stockholm sample, but they did find significant sex differences in mean DQ scores, favoring the girls at all age levels tested from 3 months to 5 years, and added that "the differences between the sexes are especially apparent in the language and personal-social scales" (p. 77).

TABLE II
Predictions of 5-Year Stanford–Binet IQs
from Griffiths DQs

	5-Year Stanford–Binet IQs	
	Boys ($n = 43$)	Girls ($n = 37$)
Griffiths DQs	r	r
6 months	.24	.41[a]
18 months	.35[b]	.48[a]

[a]$p < .01$ level.
[b]$p < .05$ level.

Goffeney, Henderson, and Butler (1971) reported the prediction of the Wechsler Intelligence Scale for Children (WISC) scores at age 7 years from the 8-month Bayley test scores for 626 children tested at the University of Oregon Medical School as a part of the Collaborative Perinatal Research Project. The mothers in this sample were below-average in socioeconomic status; 63% were black and 37% were white. The correlations between the 8-month Bayley test scores and the IQs on the WISC were .27, $p < .01$, for the females (.30, $p < .01$ for the blacks and .28, $p < .01$ for the white females) and .12, $p < .05$, for the males (.01 n.s. for the black and .16, $p < .05$ for the white males). Very similar correlations were obtained when the verbal IQs of the WISC were the predicted scores. The sex difference found in this study was minimally present for the total sample of 19,837 children in the Collaborative Project, which included cohorts born in 12 hospitals (Broman *et al.*, 1975). For this large

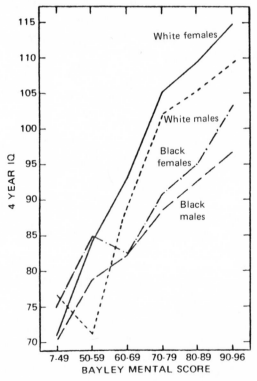

FIGURE 2.   Mean IQs on the Stanford–Binet, Form L-M, at 4 years according to scores on the Bayley Mental Scale at 8 months (Broman, Nichols, & Kennedy, 1975).

sample, prediction of 4-year Stanford–Binet IQs from Bayley test scores is shown in Figure 2. This figure suggests a much higher relationship between the 8-month and 4-year test scores than is indicated by the correlation coefficients, which are .21 for white boys (n = 4,569), .24 for black boys (n = 5507), .23 for white girls (n = 4,312), and .24 for black girls (n = 5507). These coefficients are all highly significant at better than the .01 level because of the large number of cases. These samples included a few children with handicapping conditions that would increase the correlations beyond those found for a normal, neurologically intact sample. In other words, in this large study there is positive but extremely low prediction, which is of negligible practical significance. The sex difference suggests greater mental test stability in girls but is not significant.

Socioeconomic Status and Prediction

The effect of the interaction of socioeconomic status and infant test scores on prediction was described by Willerman, Broman, and Fiedler (1970) for a Collaborative Project subsample of 3,037 babies born at the Boston Lying-In Hospital. This study shows that socioeconomic status has a significant effect on prediction for children earning low mental scores at 8 months. Children with low scores at 8 months who live in homes of low socioeconomic status will do poorly on the 4-year Stanford–Binet, while children with low scores at 8 months living in homes of high socioeconomic status will earn above-average IQs at 4 years. Willerman *et al.* concluded that "poverty amplifies the IQ deficit in poorly developed infants," or stated another way, "infants retarded at 8 months were seven times more likely to obtain low IQs at age 4 years if they come from the lower socioeconomic status than if they come from a higher socioeconomic level" (p. 69). This is a provocative paper but there may be alternative interpretations. Some of the children earning low scores at 8 months may be "late bloomers" who will eventually develop high verbal or other intellectual skills. The fact that these children live in homes of high socioeconomic status may not be the only determining factor. It may be that they are showing an increasing resemblance to parents who have similar skills. Support for this hypothesis comes from the studies of adopted children by Skodak and Skeels (1949) and Honzik (1957), who found that the IQs of adopted children who had never lived with their parents showed an increasing resemblance to the abilities of their biological parents but little increase in resemblance to the educational level of the adopting parents during the age period of 2–4 years.

Retardation, Prematurity, and Minimal Brain Impairment as Factors in Prediction

Relatively high correlations between infant test scores and childhood IQs appear from time to time in the literature. The inclusion of children with very low scores usually accounts for these findings. MacRae (1955) reported a correlation of .56 ($p < .01$) between ratings based on either the Cattell or the Gesell tests, administered to 40 children under 12 months of age, and WISC IQs obtained at 5 years or older. We note that some of the children in this sample were living in a home for the mentally deficient. The purpose of the testing was for adoption. MacRae found that the ratings based on the infant test scores were very helpful. More than half the ratings of the babies did not change at all between the first and sixth year, and ratings of only two children changed by more than one rating category. The rating categories were definitely deficient, somewhat below average, average, somewhat above average, and definitely superior.

Siegel (1981) reports relatively high interage correlations of test scores on the Bayley for a group composed of both preterm and full-term infants. These correlations have not been included in Table I because higher-than-average prediction is to be expected when children varying in maturity are included in the sample.

Knobloch and Pasamanick (1960) also found significant correlations between infant test scores and later IQs. For 147 prematures considered normal, Gesell DQs at 40 weeks correlated with 3-year Stanford–Binet IQs .43; but for 48 children who were "neurologically or intellectually abnormal," the correlation rose to .74. These authors stated that correlations of the same order of magnitude were obtained for both white and black children in the Baltimore sample, suggesting the validity of the findings.

The results indicating greater stability in the test scores of low-scoring babies are of importance to neurologists, pediatricians, psychologists, and others who depend on infant tests in the diagnosis of mental impairment and subnormality. However, test scores of infants and young children should always be adjusted for prematurity before correlations suggesting stability and thus prediction are computed (Hunt, 1977). This caveat applies equally to test scores of individual children.

DIAGNOSTIC VALUE OF INFANT TESTS

A major use of infant tests is in the diagnosis of mental defect. How accurate are these assessments? The fallibility of the tests themselves was suggested earlier in this chapter. On the other hand, the reliability of the tests is reasonably high and all tests are moderately predictive for low-scoring children. Before standardized tests were available, physicians, especially pediatricians, had to depend entirely on their background of experience in making a diagnosis of mental defect. To what extent does the information from an infant test help in the diagnosis? Illingworth (1960), a pediatrician, described in some detail the need for assistance in assessing the effects of such conditions as neonatal anoxia, head injury, virus encephalitis, or a subdural hematoma. He concluded from a review of the literature, "There is plenty of evidence that developmental tests, properly used, are of the utmost value" (p. 13). He went on to say, "No one, of course, should expect developmental tests when repeated to give a constant value" (p. 13). He stated that the objective test results should always be supplemented by a history and an evaluation of the quality of performance.

Studies designed to assess the usefulness of infant tests in aiding pediatricians' diagnoses of handicapping conditions were undertaken by Bierman, Connor, Vaage, and Honzik (1969) and Werner and associates (Werner, Honzik, & Smith, 1968; Werner, Bierman, & French, 1971), using the data of the Kauai Pregnancy Study. In these extensive investigations of the prenatal and postnatal development of all children ($n = 681$) born on the island of Kauai in 1954–1955, pediatricians rated the intelligence of the children at about 20 months of age as retarded, low normal, normal, or superior. A total of 93% were rated as normal; 5% as low normal; and .6% (1 girl and 3 boys) as retarded. These children were all tested on the Cattell at the same age, 20 months. The agreement between the pediatricians' appraisals and the Cattell IQs was not high ($r = .32$) but was statistically significant ($p < .01$). Of the children judged below normal by pediatricians at 20 months, 46% (16 out of

35) earned low IQs or were doing poorly in school at ages 5–9 years. The accuracy of these predictions of poor school performance was increased to 75% if the Cattell IQs were taken into account. Bierman *et al.* concluded that the test scores appear to be valuable to the pediatrician who does not want to err in the direction of giving a poor prognosis for a child who may later prove capable of adequate if not superior academic performance. Children who at 20 months had normal Cattell IQs but were misjudged as below normal in intelligence by the pediatricians were likely to have poor speech, show abnormal or slow motor development, and be of "poor physical status." Children who were judged *normal* by the pediatricians and whose Cattell IQs at 20 months were below 80 were not doing well in school at 6–9 years of age: "From one-half to two-thirds of these children were not capable of average academic work" (Bierman *et al.*, 1964, p. 688). In a follow-up at 10 years of all available children from the Kauai study, Werner *et al.* (1968, 1971) reported that the best single predictor of IQ and achievement at age 10 years was the Cattell IQ at age 20 months. For children with IQs below 80 at age 20 months, a combination of Cattell IQ and pediatricians' ratings of intelligence yielded a high positive correlation ($R = .80$) with the IQs on the Primary Mental Abilities tests at 10 years. The fact that the results of this study yielded positive results using 20-month Cattell IQs is remarkable. There is probably no age at which testing is more difficult than at the end of the second year, when children want to be autonomous and are therefore negativistic and resistant to adult suggestions, requiring great skill on the part of the examiner.

Infant tests are used to detect the effect of handicapping conditions known to have occurred at an earlier time, such as rubella during the pregnancy or asphyxia during the neonatal period. Honzik *et al.* (1965) compared the 8-month Bayley mental test scores of a group of infants suspected of having neurological handicaps, on the basis of hospital records, with the scores of a matched normal control group from the same hospitals. In this study the birth records of more than 10,000 babies born in seven different hospitals were scrutinized for evidence of complications of pregnancy or delivery and distress during the neonatal period. Of this group, approximately 2% were selected as possibly "suspect" of having brain impairment and were matched with a normal control group using the same records. The testers of the infants (128 males and 69 females) at 8 months were unaware of the neonatal appraisal as suspect or normal and rated most of the children "normal."

The Bayley mental test scores, however, differentiated the suspect from the control group at between the .05 and the .01 levels of significance. There were also statistically significant differences in the performance on individual test items. This investigation is noteworthy in showing that infant mental tests at 8 months can be differentiating of deficits that are not obvious even to experienced testers.

GENETIC STUDIES USING INFANT TESTS

The only direct evidence of the effect of the genetic structure on intelligence comes from chromosomal studies. The advent of hypotonic treatment of cells led to accurate counts of the number of chromosomes in the cells of normal human beings and to the possibility of correlating abnormal counts with tested abilities. One of the dramatic discoveries was that individuals with Down's syndrome have an extra 21st chromosome. Individuals with this genetic makeup seldom earn IQs of more than 50. The Bayley scales have been used to compare the abilities of Down's syndrome infants reared at home with those living in an institution (Bayley, Rhodes, Gooch, & Marcus, 1971). The test scores reflected not only the low scores of these children but also the beneficial effects of a home environment. The mean Bayley IQ of the children reared at home was 50; in the hospital, 35. This difference is statistically significant ($p < .005$).

Resemblance in the mental abilities of family members is the most frequently used method of assessing the relevance of heredity. Erlenmeyer-Kimling and Jarvik (1963) reported that for most relationship categories, the median of the empirical correlations closely approaches the theoretical value predicted on the basis of the genetic relationship alone. The median empirical value of the parent–child correlations is .50; the median r for fraternal twins, .53; and for identical twins reared together, .87, and reared apart, .75. These coefficients are based on samples of parents and children of school age. Are the findings similar for infant test scores? The answer is interesting. Parent–child resemblance is negligible in the first months of life, and the characteristic statistic of .50 is not reached until the end of the preschool period (Bayley, 1954; Hindley, 1962; Honzik, 1957, 1963.) Skodak and Skeels (1949) reported a similar finding for children who were adopted in the first months of life. The correlations between the adopted children's IQs and their natural parents' IQs (or years of schooling) increased from a near-zero value at 2 years to statistically significant correlations (p < .05–.01) at 4 years. The same increase in resemblance was not obtained between the children's IQs and their adoptive parents' years of schooling, although the children's above-average IQs reflected the good environments provided by the adopting parents. These findings suggest that changes in IQ during the early preschool years are not entirely due to environmental factors. If we can assume that the increasing resemblance between the scores of the natural parents and their children's IQs is due to heredity, the genetic factor still accounts for very little of the total variance in the children's mental test performance.

Twin studies are the ones most frequently cited as showing the effect of genetic similarity on intelligence. Freedman and Keller (1963) tested a group of same-sex twins every month during the first year on the Bayley scales. These investigators did not know the zygosity of the twins. They found that the intrapair difference on the combined mental and motor scales of the monozygotic twins was less in nearly all instances than that of the dizygotic twins (p < .01). Wilson (1972) and Wilson and Harpring (1972) reported within-pair correlations of monozygotic (MZ) and dizygotic (DZ) twins on the Bayley scales for a group of 261 pairs tested every three months during the first year (see Table III). These coefficients are similar to those found for older children and adults, except that the values for DZ twins are higher. The greater resemblance of fraternal twin babies than that of siblings is in all likelihood the result of the greater similarity in the experience of the twins than in the experiences of single-born children in the same family. There is also the possibility that the experiences of MZ twins are more similar than those of DZ twins. The greater similarity in the genetically similar MZ twins than in the less similar DZ twins suggests hereditary determination, but a crucial study would require that the twins be reared apart so that there is control of experiential factors.

A study by Nichols and Broman (1974), based on the 8-month Bayley tests given in the Collaborative Project, reports high intratwin correlations (.84 for 122 pairs of MZ twins and .55 for 227 pairs of DZ twins) but a relatively low correlation of .22 for siblings. The low resemblance between the 4,962 siblings tested at 8 months compared with the DZ intrapair r of .55 suggests that the experiences of infants living in the same family can be very different and that the large extent of the genetic similarity does not preclude great differences in test scores. There is always the possibility that the siblings were fathered by different men,

TABLE III
Within-pair Correlations of Twins
on Bayley Scales (261 Pairs)

Age in months	MZ[a] r	DZ[b] r
3	.84[c]	.67
6	.82	.74
9	.81[c]	.69
12	.82[c]	.61

[a]Monozygotic.
[b]Dizygotic.
[c]$p < .05$.

but this is less likely to have occurred in the Collaborative Project since the siblings were born and tested within a relatively limited time period. In fact, the authors refer to "full" siblings, suggesting that blood checks were done on the parents of these siblings. A significant finding in this study is that the high heritability estimate derived from the difference between the MZ and DZ correlations is due to the higher concordance of severe retardation in the MZ pairs. In the MZ sample, removal of all severely retarded children reduced the intrapair correlation of .84 to .55. Removal of severely retarded children had no effect on the DZ correlation of .55 or the sibling r of .22. These authors add that there is evidence that genetic factors are important in severe retardation, since the concordance ratio for such children on the Bayley scale is significantly higher in the MZ than the DZ twins.

EXPERIENCE AND MENTAL ABILITIES IN INFANCY

Interest in the effect of experience on the development of mental abilities has led to an upsurge of investigations of the possible effects of experience on the growth of intelligence in infancy.

The interaction of experience and heredity begins at conception, when the mother's nutrition and health begin to have their effect on the developing embryo. Infant tests are not needed to show the devastating effects of maternal rubella or Western encephalitis on the fetus, but they are being widely used to assess the effects of maternal and infant malnutrition, parent–child interaction, and intervention on the development of intelligence.

Malnutrition

Both the Gesell and the Bayley infant tests have been used extensively in assessments of the effects of malnutrition on intellectual development. Cravioto and Robles (1965) tested the hypothesis that the effect of severe malnutrition on mental development varies as a function of the period of life at which malnutrition is experienced. Twenty infants hospitalized for severe protein–calorie malnutrition were tested on the Gesell scale during hospital treatment and during the rehabilitation period. On the first examination, just after the acute electrolyte disturbance had been corrected, all infants scored considerably below the age norms. During rehabilitation, 14 of the children gained in DQs on the Gesell, but the 6 youngest did not, confirming the hypothesis that the younger the infant, the greater the adverse effect of poor nutrition. Pollitt and Granoff (1967) cross-validated these findings in Peruvian children with marasmus using the Bayley scales. The scores of this group of previously marasmic infants were compared with those of their siblings who had no unusual medical history and whose measurements were within normal limits. It was found that while the siblings were developing according to age expectation, 17 of the 19 children recovered from marasmus had severe mental and motor retardation. From these studies and those of Monckeberg (1968) and Chase and Martin (1969), Cravioto and Delicardie (1970) concluded that protein–calorie malnutrition occurring in the first year of life, if severe enough to retard physical growth markedly and to require hospitalization, may have adverse effects on mental development. These authors added that if the duration of the untreated episode is longer than four months, particularly during the first months of life, the effect may be so intense as to produce severe mental retardation, which is incompletely corrected by nutritional rehabilitation. With all their limitations, both mental and motor scores of infant tests helped describe the findings objectively and indicate their significance.

Maternal Deprivation

The impetus for studies of the effect of experience on intellectual development has come in part from Bowlby's monograph (1952) describing the ill effects of maternal deprivation. This monograph was the stimulus for research on the ill effects of institutional care, which was reviewed by Ainsworth (1962), Casler (1961), and Yarrow (1961). Casler concluded in his review that in institutional care, there is a lack or relative absence of tactile, vestibular, and other forms of stimulation, which accounts for some of the emotional, physical, and intellectual deficits. Casler (1965) later investigated the effects of extra tactile stimulation on infants living in an institution. He found that infants given specific tactile stimulation over a 10-week period performed significantly better on the Gesell Developmental Schedules than did a matched control group. This study, which used a rather small number of cases, should be replicated, but the findings are in line with those of comparative psychologists, such as Harlow (1958), as well as developmental psychologists (Bayley & Schaefer, 1964).

The next question is, how long-lasting are the ill effects of institutional care and the good effects of supplemental stimulation? A partial answer is provided by Dennis and Najarian (1957), who tested all the children in a Lebanese crèche where health needs were met but the infants were minimally stimulated. They found that the infants' Cattell IQs were normal at age 2 months but averaged only 63 during the age period 3–12 months. Children tested at ages 4 and 5 years obtained IQs of approximately 90, although they had spent all their lives in the crèche. These results suggest that the period of limited stimulation during infancy had little lasting effect on mental growth. A similar finding was reported by Kagan and Klein (1973) in a cross-cultural study of Guatemalan infants. The rural Guatemalan infants spend most of their lives in the small dark interior of a windowless hut. They are usually kept close to their mothers but are rarely spoken to or played with. Compared with American infants, they are extremely passive, fearful, unsmiling, and quiet. The Guatemalan infants studied were retarded with respect to "activation of hypotheses, alertness, onset of stranger anxiety, and object permanence" (p. 957). In marked contrast to the infants, the Guatemalan preadolescents were comparable to American middle-class norms on tests of perceptual analysis, inference, recall, and recognition memory. Kagan and Klein concluded that "infant retardation seems to be partially reversible, and cognitive development during the early years more resilient than had been supposed" (p. 957).

Apparently the effect of added stimulation may last no longer than that of relative deprivation. Rheingold and Bayley (1959) reported transitory gains in social responsiveness in institutionalized infants who were given "more attentive care" by one person from the sixth to the eighth month. These children performed no better on the Cattell test a year later than did a matched group who received no special attention.

Parent–Child Interaction and Infant Test Scores

Another approach to the problem of evaluating environmental influence is to relate infant test scores to socialization practices in the family. Baldwin, Kalhorn, and Breese (1945) were among the first to relate parental behavior to the test scores of infants and young children. He reported that for 94 children in the Fels longitudinal study tested on the Gesell, DQ changes were related to "freedom to explore," "emotional warmth," and "acceleratory methods." These variables, described in different ways, have been found relevant to gains in IQ in a number of subsequent studies. Bayley and Schaefer (1964) reported, for the small but intensively tested Berkeley Growth Study sample, that boy babies whose mothers evaluated them positively, granted them some autonomy, and expressed affection for them were "happy, positive, calm infants" (p. 67). These boys tended to earn below-average scores on the Bayley tests in the first year but made rapid gains in the next few years, when they were likely to earn high IQs. The authors reported that, conversely, boys who scored high in the first year were active, unhappy, and negative with mothers who were hostile and punitive. These boys tended to have low IQs after 4 years. It is difficult to sort out cause and effect in these relationships. Were the boys' Bayley test scores higher in infancy *because* they were stimulated by the punitive mothers? To what extent did the positive, affectionate mother affect the cognitive development of her son? For girls in this study, there was little correlation between the mothers' behavior in the first two or three years and their later intelligence. These results were to some extent cross-validated in the much larger Guidance Study sample (Honzik, 1967). Ratings of the *closeness of the mother–son relationship* at 21 months correlated with the mental test scores of the boys at this age ($r = .29$, $p < .05$). This correlation increased to .48, $p < .01$, at 9 years and was still significant at age 40 years (Honzik, 1972). *Closeness of mother to daughter* correlated significantly with the daughter's IQ at 2 years but not thereafter.

In the London longitudinal study, Moore (1967) reported a similar increasing correlation between the "emotional atmosphere of the

home," "toys, books, and experience," and "example and encouragement" and the children's IQs at 3 and 8 years. The interesting phenomenon here is that although the family variables were rated when the
children were aged 2½ years, the correlations, and thus the predictions,
of the 8-year IQs were all higher, and significantly so, than the 5-year
IQs. A major conclusion to be drawn from the four longitudinal studies
is that measurable experiences in the home in the first two years may
show an increasing correlation with IQs during childhood. It is understandable that evidences of ability of the parents, such as their education
or socioeconomic status, would show an increasing correlation with
their children's test scores. It is more difficult to comprehend why such
variables as *mother–son closeness,* which is not related to parental ability,
show an increasing relation to the son's IQ. Actually, *later* measures of
mother–son closeness do not show this correlation, which means that
the warmth and concern that the mother has for her son in the early
years has a greater effect on his later intellectual function than does
equivalent favoring and protectiveness in adolescence (Honzik, 1966).

Similar results are reported in an investigation by Bradley and Caldwell (1980). The home environments of 72 children (36 white and 36
black) was assessed when they were 6 and 12 months old. The children
were given the Bayley test at 1 year and the Stanford–Binet at 3 years.
Responsiveness of the mother and *play materials* at 12 months were significantly correlated with the Bayley test scores at 12 months and even more
highly with the Binet IQs at 3 years.

In a study of 41 black babies aged 5 to 6 months, Yarrow, Rubenstein, Pederson, and Jankowski (1972) differentiated the natural home
environment into (1) inanimate stimulation and (2) social stimulation.
These two types of environmental variables, obtained from time-sam-
pling observations, were not highly correlated but did correlate significantly with Bayley's mental test scores. It is of interest in this study that
the investigators considered the relation of other infant variables to *social*
and *inanimate stimulation.* The infant variable *vocalization to bell* correlated
significantly with social but not inanimate stimulation. *Goal-directed behaviors* correlated significantly ($p < .01$) with both social and inanimate
stimulation. Yarrow *et al.* concluded, "It is likely that the infant's orientation to objects and to people very early becomes part of a feedback
system with the environment. His smiling, vocalizing, and reaching out
to people; his visually attending to and manipulating objects tend to be
self-reinforcing and thus, to some extent self-perpetuating" (p. 217).

Two investigators have avoided the problem of inherited similarity of parents and children by studying the relation of the family milieu to infant test scores in adopted children. Beckwith (1971) correlated the Cattell IQs of 24 adopted infants with evaluations of their mothers' interactions with them in the home. Cattell IQs (at 8–10 months) were correlated with the infants' "social experience" ($p < .01$) and the extent to which he was "talked to," "touched," and "given an opportunity to explore the house." In this study, no relationship was found between the infants' IQs and the adoptive parents' socioeconomic status, but the IQs of the children did correlate with the natural mothers' socioeconomic class.

Another study of 40 adopted children at age 6 months was reported by Yarrow (1963). The maternal variables for the investigation were carefully chosen and covered three major maternal functions: (1) need-gratification and tension-reduction, (2) stimulation-learning conditions, and (3) affectional interchange. The maternal variables yielding the highest correlations with the 6-month Cattell IQs were *stimulus adaptation* ($r = .85$), *achievement stimulation* ($r = .72$), *social stimulation* ($r = .65$), and *physical contact* ($r = .57$). All these correlations were significant at the .01 level or better and were much higher than those reported for the natural mother and children. An important question here is how long-lasting the effects of the early infant experience are. These adopted children were tested again on the WISC at 10 years (Yarrow, Goodwin, Manheimer, & Milowe, 1973). Seven of the eight maternal variables assessed when these adopted children were aged 6 months correlated significantly with the WISC IQs at 10 years. However, when the correlations were computed for boys and girls separately, the findings were similar to those reported by Bayley and Schaefer (1964) and others: The relationships were negligible for the girls but highly significant for the boys. The range of *r*s for the girls was from .08 for *achievement stimulation* to .24 for *emotional involvement*. The range for the boys was from .43 ($p < .05$) for *achievement stimulation* to .68 ($p < .01$) for *physical contact*. These correlations, together with those of other investigators, clearly indicate the importance of early affective relationships and tactual stimulation to mental growth.

Cohen and Beckwith (1979) investigated the long-range effects on mental test scores of caregiver–infant interactions at 1, 3, 8, and 24 months in a group of 50 preterm babies. Gesell and Bayley test scores at 2 years (adjusted for prematurity) were predicted by social interactions as early as 1 month. The authors conclude that social transactions in the first months of life reflect some qualities of the relationship between caregiver and infant that are important to the child's mental growth. The implications of this study appear to be that the caretaking experienced by infants during the first two years of life has a pervasive effect on the development of mental abilities.

Intervention Effects on Tested Mental Abilities

The first intervention studies were targeted for preschool-aged children during the year or two before school entrance (Gray & Klaus, 1965). More recently, centers providing care for infants have made it possible to evaluate the effects of educational programs on mental development during infancy. Ramey and Haskins (1981a) studied the effect of an "educational day care program" by following groups of infants judged to be "at risk" for subnormal intellectual growth. A sample of infants with a High Risk Index (low income, low maternal IQ, little parent education, and other social factors) was divided into an Experimental and a Control Group. Both groups received diet supplements and social service, but the Experimental Group also participated in an educational program beginning in the second month of life. The day-care educational program was composed in part of activities designed to stimulate mental growth. Between 6 and 36 months, the Experimental children maintained normal intellectual growth with average Bayley IQs in the 95 to 105 range. Control children declined markedly in IQ beginning between 12 and 18 months and remained significantly below the Experimental group at 2 and 3 years. A second report (Ramey & Haskins, 1981b) adds that the group differences were maintained at ages 4 and 5 years.

SUMMARY

Studies cited suggest that depriving or stimulating experiences in infancy may have concurrent depressing or stimulating effects on infant test scores but that these effects may not be long-lasting if the later life experiences of the children are not consonant with the deprivation or stimulation experienced (Dennis & Najarian, 1957; Kagan & Klein, 1973;

Rheingold & Bayley, 1959). In contrast to these intervention studies, investigations of parent–child (or caretaker–child) interaction in relation to the mental test scores tend to show more lasting effects of attitudinal factors such as emotional atmosphere of the home, social interaction, being talked to, encouragement, acceleratory methods, freedom to explore, and stimulating toys and books (Baldwin *et al.*, 1945; Honzik, 1967, 1972). A tentative conclusion may be reached from these two sets of findings: (1) mental growth is positively affected by *continuing* interactions with the caretaker (parents or others) in infancy, and (2) moderate degrees of deprivation or stimulation may not have significant long-range effects on mental growth provided that these conditions do not persist. In other words, early experiences may have long-range consequences, but mental growth is not static, and concurrent experiences also affect mental development.

Reliability and Validity

The Bayley and Cattell infant tests are reliable and internally consistent as judged by the correlation of odd with even test items. Test–retest correlations are relatively high over short age periods but decline markedly as the time span between tests lengthens (for the Bayley, Brunet–Lézine, Cattell, Gesell, and Griffiths). The magnitude of the coefficients and the nature of the interage correlations have been cross-validated by studies in different countries using different tests at different periods, suggesting that a major determinant of these interrelationships is the rate of development of the human organism. Honzik (1938) noted that the magnitude of the *r*s varies with the age ratio of the first to the second test. Thus, the *r* between the 3- and 6-year test scores is roughly .50, as is the correlation between the 3- and 6- year scores. This age ratio underestimates the magnitude of the *r*s as the children grow older but is suggestive of the changing rate of development of mental abilities with age.

Another index of the validity of the tests is the agreement between scores on different tests. Erickson, Johnson, and Campbell (1970) reported a correlation of .97 ($p < .001$) between the Bayley and the Cattell scores of children who ranged in ability from profoundly retarded to normal, suggesting a high degree of validity of the scores of the children in this ability range.

Prediction from Infant Tests

Predictions based on the mental scores of infants depend not only on the growth processes but also on the effects of experience and on the nature of the tests used to measure the developing abilities. The following conclusions are reached from the studies discussed:

1. In neurologically intact infants, scores obtained on currently available tests during the first months of life are not predictive of later intelligence because of immaturity, rapidly changing behaviors, and the overriding significance of other behaviors such as the infant's relative *activity* during this age period (Bayley & Schaefer, 1964; Escalona, 1968).

2. Prediction of later intelligence test scores from infant tests begins to occur in the second half of the first year in girls only (Hindley, 1960, 1965), and more especially in certain specific abilities, such as vocalizations (Cameron *et al.*, 1967; Kagan, 1971).

3. Prediction of later intelligence does not accelerate until after the second birthday (Bayley, 1949; Brucefors, 1972; Hindley, 1965; Honzik *et al.*, 1948).

4. Prediction is markedly more accurate for low-scoring infants, regardless of whether the low score is due to chromosomal aberrations (e.g., trisomy 21), infection (rubella during the pregnancy), injury, perinatal anoxia, or generalized subnormality of unknown etiology.

5. The effect of experience on test constancy is suggested by a study showing that the interage correlations are noticeably higher for children living in the relatively constant environment of a day-care center (Ramey, Campbell, & Nicholson, 1973).

Diagnosis

Evidence from many investigations attests to the value of infant mental tests in the diagnosis of even minimal neurological lags or deficits (Bierman *et al.*, 1964; Honzik *et al.*, 1965). Infant test scores are diagnostic of deprivation experiences as well as of the effects of enriched environments. However, the value of these scores is greater in the assessment of deficits than of superiority, since high scores on infant tests are less stable than average or low scores. Precocity in infancy may reflect early maturing or the effects of a great deal of stimulation rather than higher potential for later above-average cognitive functioning.

What can be done to improve diagnoses? The use of mothers' reports as additional information may add to the value of the examination. It is often possible to determine the mother's estimate of the validity of the test by asking her if the baby responded to the test as she would expect or if she was surprised at what he or she could do.

The infant's cognitive style and reactions to the test should be recorded and evaluated. These evaluations may prove more useful than the test scores. Freedman and Keller (1963) found that monozygotic twins were significantly more alike that dizygotic twins on Bayley's behavior profile, which is a part of the Bayley tests. Actually the behavior profile was more differentiating than the mental and motor scales in this study or zygosity.

For more adequate diagnoses, a greater effort should be made to measure specific abilities and new groups of abilities. Also the findings of investigations of cognitive functioning should be considered as possible additions to the current tests. A premise of Fantz and Nevis's (1967) investigations was that "the early development of cognitive function is primarily through perception rather than action" (p. 351). They added that "later individual differences are more likely to be correlated with the early development of perception and attention than with action" (p. 351). Although it would not be feasible to duplicate Fantz's experiments in a testing situation, it would be possible to assess infants' attentiveness to schematic drawings as a part of the test. Lewis (1971) wrote that if one views *attention* and its distribution as an information-processing operation, attention can be viewed as a measure of cognitive functioning. Kagan (1971) reported social class differences in attentiveness in the first year, which further suggests its possible relevance to what is termed *intelligence* in the older child. *Attention* is easily assessed in the mental test situation, and if it proves diagnostic or predictive, it should be incorporated into the test score.

DISCUSSION

The value of infant tests is seriously questioned by Lewis (1973) and Lewis and McGurk (1972). This overview suggests why. Infant test scores are not stable over long periods of time, and their relationship to previous and concurrent experience is complex and only beginning to be understood. As Yarrow *et al.* (1973) wrote, "We are still in a rather primitive state with regard to concepts and methodology for handling the dynamic interplay among these sets of variables" (p. 1280). Perhaps the key word here is *dynamic*. Growth is seldom simple and seldom occurs at a constant rate; instead, it is rather highly interactive and occurs at a decelerating rate. A question that has not been raised but may prove highly relevant is whether experiences are more effective during periods of rapid or of relatively slower growth. Our hypothesis from what is known to date about intellectual development over the life span is that the effect of experiences, both those injurious and those beneficial to the organism's cognitive skills, is negatively related to growth and, thus, to age. In other words, the earlier the experience the greater the potential effect. This does not mean that the effects of experience are reflected immediately in the behavior of the infant. Some of the conditions, such as Western encephalitis or maternal care, are known to have more predictable effects on later cognitive development than on current cognitive skills. Another possibly confounding factor, suggested by the higher correlations between maternal care and infant scores in adopted than in own children, is that aspects of optimal maternal care may be negatively related to genetic potential. This was actually found in the Guidance Study, where the mother's education, which reflects her ability, was negatively related to the most relevant experiential variable, *closeness of mother and child* (Honzik, 1967).

Two later investigations suggest that aspects of "attentiveness to novelty in infancy" may be related to cognitive ability in childhood. Lewis and Brooks-Gunn (1981) report that a cognitive skill (visual attention to a novel stimulus) at 3 months predicts mental test scores on the Bayley at 2 years better than the 3-month Bayley test score. Lewis and Brooks-Gunn believe that information processing is central to cognitive functioning at this age and that this may account for the finding that "visual attention to a novel stimulus" is more predictive of later test scores than is the 3-month test score (p. 131). The second study, by Fagan and McGrath (1981), reports moderate correlations between "visual recognition memory" (preference for visual novelty) at 4–7 months and Peabody Vocabulary scores at 4 and 7 years. Critical evaluations of these two studies by McCall (1981) suggest that cautious optimism may be entertained as to the possibility that "infant visual recognition" may be a better predictor of later intelligence than scores on standardized tests and further, that "attitude toward and performance on novel tasks or parts of tasks may be a key aspect of intelligence from infancy on" (Sternberg, 1981, p. 154).

Attention and even attentiveness to novelty could possibly be assessed in the mental test situation. To the extent that they prove diagnostic or predictive, they could profitably be incorporated into the test scores of young infants.

The critiques of Lewis (1973) and Lewis and McGurk (1972) are valuable in asking some significant and cogent questions about attempts to measure "intelligence" in infancy. Infant tests obviously do not measure what is measured by the Stanford–Binet, the Wechsler, or the primary abilities tests; they measure abilities and skills that, to a large extent, are the bases and precursors of later mental development. It is clear that the tests could be improved by the elimination of items that are more motor than mental and possibly by the addition of new items suggested by recent research, for example, measures of attention, attentiveness to novelty, and preference for visual novelty. Infant tests, with all their limitations, have served us well. Possibly their main value has been in diagnosis, but they have also contributed substantially to our understanding of the many factors constituting the development of abilities in the first years of life.

Since mental growth over the life span (Eichorn, Hunt, & Honzik, 1981; Honzik, Hunt, & Eichorn, 1976) is dependent on the genetic blueprint—neurological integrity and experience (both earlier and concurrent)—investigators in the future may find that measures of the environment are more predictive of later test scores than are test scores obtained early in the first year. This is especially true for investigations attempting to predict later intelligence from tests given in infancy. Investigators will probably always want to use infant tests for the assessment of normalcy of responses, and the test situation may also serve as a relatively standard situation for observing the interaction of parent (or caretaker) and child, which has been found to have such significant and long-lasting effects.

CONCLUSION

The major question of this review is whether test scores accurately describe the growth of mental abilities and reflect the temporary and more permanent effects of experience. This we believe they do, and we believe that with amplification and careful use, their value can be enhanced. Standards and reference points are needed, and good tests can help serve this need. The purpose of infant testing is to determine the progress of an individual child or the mental development of all children. Prediction of later intellectual functioning is one worthy aim of infant tests but is secondary to the more important objective of adding to our understanding and knowledge of the course of development of mental abilities in infancy and early childhood.

REFERENCES

AINSWORTH, M. D. The effects of maternal deprivation: A review of findings and controversy in the context of research strategy. In *Deprivation of maternal care: A reassessment of its effects*. Geneva: World Health Organization, Switzerland, 1962.

BALDWIN, A. L., KALHORN, J., & BREESE, F. H. Patterns of parent behavior. *Psychological Monographs*, 1945, *58*(Whole No. 268).

BAYLEY, N. *The California first year mental scale*. Berkeley: University of California Press, 1933.

BAYLEY, N. Consistency and variability in the growth of intelligence from birth to eighteen years. *Journal of Genetic Psychology*, 1949, *75*, 165–196.

BAYLEY, N. Some increasing parent-child similarities during the growth of children. *Journal of Educational Psychology*, 1954, *45*, 1–21.

BAYLEY, N. Behavioral correlates of mental growth: Birth to 36 years. *American Psychologist*, 1968, *23*, 1–17.

BAYLEY, N. *Bayley scales of infant development*. New York: Psychological Corporation, 1969.

BAYLEY, N., & SCHAEFER, E. S. Correlations of maternal and child behaviors with the development of mental abilities: Data from the Berkeley Growth Study. *Monographs of the Society for Research in Child Development*, 1964, *29*(6, Whole No. 97).

BAYLEY, N., RHODES, L., GOOCH, B., & MARCUS, M. Environmental factors in the development of institutionalized children. In J. Hellmuth (Ed.), *Exceptional infant* (Vol. 2): *Studies in abnormalities*. New York: Brunner/Mazel, 1971.

BECKWITH, L. Relationship between attributes of mothers and their infants' IQ scores. *Child Development*, 1971, *42*, 1083–1097.

BERNSTEIN, B. Social class and linguistic development: A theory of social learning. In A. H. Halsey, J. Floud, & C. A. Anderson (Eds.), *Economy, education and society*. New York: Free Press, 1961.

BIERMAN, J. M., CONNOR, A., VAAGE, M., & HONZIK, M. P. Pediatricians' assessments of the intelligence of two-year-olds and their mental test scores. *Pediatrics*, 1964, *34*, 680–690.

BING, E. Effect of child-rearing practices on development of differential cognitive abilities. *Child Development*, 1963, *34*, 631–648.

BOWLBY, J. Maternal care and mental health. *Monograph Series* 1952, No. 2 (2nd ed.). Geneva: World Health Organization, 1952.

BRADLEY, R. H., & CALDWELL, B. M. The relation of home environment, cognitive competence and IQ among males and females. *Child Development*, 1980, *51*(4), 1140–1148.

BROMAN, S. H., NICHOLS, P. L., & KENNEDY, W. A. *Preschool IQ: Prenatal and early developmental correlates.* New York: Wiley, 1975.

BRUCEFORS, A. *Trends in development of abilities.* Paper given at Réunion de Coordination des Recherches sur la Croissance et le Développement de l'Enfant Normal, Institute of Child Health, London, 1972.

BRUCEFORS, A., JOHANNESSON, I., KARLBERG, P., KLACKENBERG-LARSSON, I., LICHENSTEIN, H., & SVENBERG, I. Trends in development of abilities related to somatic growth. *Human Development*, 1974, *17*, 152–159.

BRUNET, O., & LÉZINE, P. U. F. *Le développement psychologique de la première enfance.* Issy-les-Moulineaux: Éditions Scientifiques et Psychotechniques, 1951.

CAMERON, J., LIVSON, N., & BAYLEY, N. Infant vocalizations and their relationship to mature intelligence. *Science*, 1967, *157*, 331–333.

CASLER, L. Maternal deprivation: A critical review of the literature. *Monographs of the Society for Research in Child Development*, 1961, *26*(2, Whole No. 80).

CASLER, L. The effects of extra tactile stimulation on a group of institutionalized infants. *Genetic Psychology Monographs*, 1965, *71*, 137–175.

CATTELL, P. *The measurement of intelligence in infants and young children.* New York: Psychological Corporation, 1960. (Originally published, 1940.)

CHASE, H. P., & MARTIN, H. P. *Undernutrition and child development.* Paper read before the Conference on Neuropsychological Methods for the Assessment of Impaired Brain Functioning in the Malnourished Child, Palo Alto, Calif., 1969.

COHEN, S. E., & BECKWITH, L. Preterm infant interaction with the caregiver in the first year of life and competence at age two. *Child Development*, 1979, *50*, 767–776.

CRAVIOTO, J., & DELICARDIE, E. Mental performance in school age children. *American Journal of Diseases of Children*, 1970, *120*, 404–410.

CRAVIOTO, J., & ROBLES, B. Evolution of adaptive and motor behavior during rehabilitation from kwashiorkor. *American Journal of Orthopsychiatry*, 1965, *35*, 449–464.

DENNIS, W., & NAJARIAN, P. Infant development under environmental handicap. *Psychological Monographs*, 1957, *71*(7, Whole No. 436).

DODGSON, M. C. H. *The growing brain: An essay in developmental neurology.* Baltimore, Md.: Williams & Wilkins, 1962.

EICHORN, D. H., HUNT, J. V., & HONZIK, M. P. Experience, personality, and IQ: Adolescence to middle age. In D. H. Eichorn, J. A. Clausen, N. Haan, M. P. Honzik, & P. H. Mussen (Eds.), *Present and past in middle life.* New York: Academic Press, 1981.

ERICKSON, M. T., JOHNSON, N. M., & CAMPBELL, F. A. Relationships among scores on infant tests for children with developmental problems. *American Journal of Mental Deficiency*, 1970, *75*, 102–104.

ERLENMEYER-KIMLING, L., & JARVIK, L. F. Genetics and intelligence: A review. *Science*, 1963, *142*, 1477–1479.

ESCALONA, S. *The roots of individuality: Normal patterns of individuality.* Chicago: Aldine, 1968.

FAGAN, J. F., III, & McGRATH, S. K. Infant recognition memory and later intelligence. *Intelligence*, 1981, *5*(2), 121–130.

FANTZ, R. L., & NEVIS, S. The predictive value of changes in visual preferences in early infancy. In J. Hellmuth (Ed.), *Exceptional infant* (Vol. 1): *The normal infant.* New York: Brunner/Mazel, 1967.

FRANKENBURG, W. K., & DODDS, J. B. The Denver Developmental Screening Test. *Journal of Pediatrics*, 1967, *71*, 181–191.

FRANKENBURG, W. K., CAMP, B. W., & VAN NATTA, P. A. Validity of the Denver Developmental Screening Test. *Child Development*, 1971, *42*, 475–485.

FREEDMAN, D. G., & KELLER, B. Inheritance of behavior in infants. *Science*, 1963, *140*, 196–198.

GESELL, A., & AMATRUDA, C. *Developmental diagnosis*. New York: Paul B. Hoeber, 1941.

GOFFENEY, B., HENDERSON, N. B., & BUTLER, B. V. Negro-White, male-female 8-month developmental scores compared with 7-year WISC and Bender Test Scores. *Child Development*, 1971, *42*, 595–604.

GRAY, S. W., & KLAUS, R. A. An experimental preschool program for culturally deprived children. *Child Development*, 1965, *36*, 887–898.

GRIFFITHS, R. *The abilities of babies*, New York: McGraw-Hill, 1954.

HARLOW, H. F. The nature of love. *American Psychologist*, 1958, *13*, 673–685.

HINDLEY, C. B. The Griffiths Scale of Infant Development: Scores and predictions from 3 to 18 months. *Journal of Child Psychology and Psychiatry*, 1960, *1*, 99–112.

HINDLEY, C. B. Social class influences on the development of ability in the first five years. *Proceedings of the 14th International Congress of Applied Psychology*, 1962, *3*, 29–41.

HINDLEY, C. B. Stability and change in abilities up to five years: Group trends. *Journal of Child Psychology and Psychiatry*, 1965, *6*, 85–99.

HONZIK, M. P. Developmental studies of parent-child resemblance in intelligence. *Child Development*, 1957, *28*, 215–228.

HONZIK, M. P. A sex difference in the age of onset of the parent-child resemblance in intelligence. *Journal of Educational Psychology*, 1963, *54*, 231–237.

HONZIK, M. P. The environmental and mental growth from 21 months to 30 years. *XVIII International Congress of Psychology Proceedings*, 1966, *18*, 111–115.

HONZIK, M. P. Environmental correlates of mental growth: Prediction from the family setting at 21 months. *Child Development*, 1967, *38*, 337–364.

HONZIK, M. P. Intellectual abilities at age 40 in relation to the early family environment. In F. J. Monks, W. W. Hartup, & J. de Wit (Eds.). *Determinants of behavioral development*. New York: Academic Press, 1972.

HONZIK, M. P., MACFARLANE, J. W., & ALLEN, L. Stability of mental test performance between 2 and 18 years. *Journal of Experimental Education*, 1948, *17*, 309–324.

HONZIK, M. P., HUTCHINGS, J. J., & BURNIP, S. R. Birth record assessments and test performance at eight months. *American Journal of Diseases of Children*, 1965, *109*, 416–426.

HONZIK, M. P., HUNT, J. V., & EICHORN, D. H. Mental growth after age 18 years. *Compte-Rendu de la XIIIᵉ Réunion des Equipes Chargées des Études sur la Croissance et le Développement de l'Enfant Normal*. Rennes, France: Centre International de l'Enfance Press, 1976.

HUNT, J. V. Mental development of preterm infants during the first year. *Child Development*, 1977, *48*, 204–210.

ILLINGWORTH, R. S. *The development of the infant and young child: Normal and abnormal*. Edinburgh & London: E. & S. Livingstone, 1960.

KAGAN, J. *Change and continuity in infancy*. New York: Wiley, 1971.

KAGAN, J., & KLEIN, R. E. Cross-cultural perspectives in early development. *American Psychologist*, 1973, *28*, 947–961.

KLACKENBERG-LARSSON, I., & STENSSON, J. Data on the mental development during the first five years. In *The development of children in a Swedish urban community: A prospective longitudinal study*. (*Acta Paediatrica Scandinavica*, Supplement 187, IV.) Stockholm: Almqvist & Wiksell, 1968.

KNOBLOCH, H., & PASAMANICK, B. *An evaluation of the consistency and predictive value of the 40-week Gesell Developmental Schedule*. Paper presented at the Regional Research Meeting of the American Psychiatric Association, Iowa City, Ia., 1960.

KNOBLOCH, H., & PASAMANICK, B. *Gesell and Amatruda's developmental diagnosis* (3rd rev. and enlarged ed.). New York: Harper & Row, 1974.

KNOBLOCH, H., STEVENS, F., & MALONE, A. *The manual of developmental diagnosis.* New York: Harper & Row, 1980.

LEWIS, M. Individual differences in the measurement of early cognitive growth. In J. Hellmuth (Ed.), *Exceptional infant* (Vol. 2): *Studies in abnormalities.* New York: Brunner/Mazel, 1971.

LEWIS, M. Intelligence tests: Their use and misuse. *Human Development,* 1973, *16,* 108–118.

LEWIS, M., & BROOKS-GUNN, J. Visual attention at three months as a predictor of cognitive functioning at two years of age. *Intelligence,* 1981, *5*(2), 131–140.

LEWIS, M., & McGURK, H. Evaluation of infant intelligence. *Science,* 1972, *178,* 1174–1177.

MACRAE, J. M. Retests of children given mental tests as infants. *Journal of Genetic Psychology,* 1955, *87,* 111.

McCALL, R. B. Early predictors of later IQ: The search continues. *Intelligence,* 1981, *5,* 141–147.

McCALL, R. B., HOGARTY, P. S., & HURLBURT, N. Transitions in infant sensorimotor development and the prediction of childhood IQ. *American Psychologist,* 1972, *27,* 728–748.

MONCKEBERG, F. Effect of early marasmic malnutrition on subsequent physical and psychological development. In N. E. Scrimshaw & J. E. Gordon (Eds.), *Malnutrition, learning and behavior.* Cambridge, Mass.: MIT Press, 1968.

MOORE, T. Language and intelligence: A longitudinal study of the first eight years. Part I: Patterns of development in boys and girls. *Human Development,* 1967, *10,* 88–106.

NELSON, V. L., & RICHARDS, T. W. Studies in mental development. III: Performance of twelve-month-old children on the Gesell Schedule and its predictive value for mental status at two and three years. *Journal of Genetic Psychology,* 1939, *54,* 181–191.

NICHOLS, P. L., & BROMAN, S. H. Familial resemblance in infant mental development. *Developmental Psychology,* 1974, *10,* 442–446.

POLLITT, E., & GRANOFF, D. Mental and motor development of Peruvian children treated for severe malnutrition. *Revista Interamericana de Psicologia,* 1967, *1,* 93–102.

RAMEY, C. T., & HASKINS, R. The modification of intelligence through early experience. *Intelligence,* 1981, *5,* 5–19. (a)

RAMEY, C. T., & HASKINS, R. Early education, intellectual development and school performance: A reply to Arthur Jensen and J. McVicker Hunt. *Intelligence,* 1981, *5,* 41–48. (b)

RAMEY, C. T., CAMPBELL, F. A., & NICHOLSON, J. E. The predictive power of the Bayley Scales of Infant Development and the Stanford-Binet Intelligence Test in a relatively constant environment. *Child Development,* 1973, *44,* 790–795.

RHEINGOLD, H. L., & BAYLEY, N. The later effects of an experimental modification of mothering. *Child Development,* 1959, *31,* 363–372.

SIEGEL, L. S. Infant tests as predictors of cognitive and language development at two years. *Child Development,* 1981, *52,* 545–557.

SKODAK, M., & SKEELS, H. M. A final follow-up study of 100 adopted children. *Journal of Genetic Psychology,* 1949, *75,* 85–125.

STERNBERG, R. J. Novelty-seeking, novelty-finding, and the developmental continuity of intelligence. *Intellgence,* 1981, *5,* 149–155.

UŽGIRIS, I. C., & HUNT, J. McV. *An instrument for assessing infant psychological development.* Mimeographed paper, Psychological Development Laboratories, University of Illinois, 1966.

WERNER, E. E., & BAYLEY, N. The reliability of Bayley's revised scale of mental and motor development during the first year of life. *Child Development,* 1966, *37,* 39–50.

WERNER, E. E., HONZIK, M. P., & SMITH, R. S. Prediction of intelligence and achievement at 10 years from 20-month pediatric and psychologic examinations. *Chid Development*, 1968, *39*, 1063–1075.

WERNER, E. E., BIERMAN, J. M., & FRENCH, F. E. *The children of Kauai: A longitudinal study from the prenatal period to age 10.* Honolulu: University of Hawaii Press, 1971.

WILLERMAN, L., BROMAN, S. H., & FIEDLER, M. Infant development, preschool IQ, and social class. *Child Development*, 1970, *41*, 69–77.

WILSON, R. Twins: Early mental development. *Science*, 1972, *175*, 914–917.

WILSON, R. S., & HARPRING, E. B. Mental and motor development in infant twins. *Developmental Psychology*, 1972, *7*, 277–287.

WINICK, M. Fetal malnutrition and growth processes. *Hospital Practice*, 1970, *5*, 33.

YARROW, L. J. Maternal deprivation: Toward an empirical and conceptual reevaluation. *Psychological Bulletin*, 1961, *58*, 459–490.

YARROW, L. J. Research in dimensions of early maternal care. *Merrill-Palmer Quarterly*, 1963, *9*, 101–114.

YARROW, L. J., RUBENSTEIN, J. L., PEDERSON, F. A., & JANKOWSKI, J. J. Dimensions of early stimulation and their differential effects on infant development. *Merrill-Palmer Quarterly*, 1972, *18*, 205–218.

YARROW, L. J., GOODWIN, M. S., MANHEIMER, H., & MILOWE, I. D. Infancy experiences and cognitive and personality development at ten years. In L. J. Stone, H. T. Smith, and L. B. Murphy (Eds.), *The competent infant: Research and commentary.* New York: Basic Books, 1973.

4 A Conceptual Approach to Early Mental Development

ROBERT B. MCCALL

INTRODUCTION

One of the major preoccupations of our discipline is the study of mental development. Our thinking has been focused on several broad issues, namely, (1) the relative contributions of heredity and environment, (2) individual differences and the species-general developmental stages of mental development, and (3) consistency and change.

But scholars have tended to confine themselves to one aspect or another of these themes rather than attempting to view mental development simultaneously from each of these perspectives. That is, behavior geneticists tend to look for genetic contributors to individual differences and environmentalists look for environmental correlates of individual differences, Piaget focused on the species-general stages of development and was quite unconcerned about individual differences, and the entire discipline has been guided by a search for consistency rather than change in mentality across age. I have argued elsewhere (McCall, 1981) that our understanding of mental development is hampered by these allegiances, so it is valuable to look at each of them in more detail.

The Two Realms of Development

Suppose Figure 1 is a plot over age of some behavior X, for example, vocabulary, intelligence, or height. The heavy solid line represents the

ROBERT B. MCCALL • Communications and Public Service, The Boys Town Center, Boys Town, Nebraska 68010.

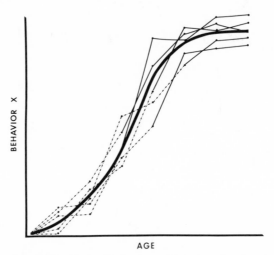

FIGURE 1.   A hypothetical plot of the developmental function of a given behavior for five individuals (thin lines) and the developmental function of the group (heavy line). Dashed portions of lines represent periods of instability in individual differences. (From "Transitions in Early Mental Development" by R. B. McCall, D. H. Eichorn, and P. S. Hogarty, *Monographs of the Society for Research in Child Development*, 1977, 42[171]. Copyright 1977 by the Society for Research in Child Development. Reprinted by permission.)

average of a group of subjects, while the thin solid and dashed lines represent age plots for individual subjects.

Statistically, we know that whether the average group performance rises, falls, or stays the same, individual differences within that sample may be consistent or inconsistent across age. For example, in Figure 1, individual differences are relatively inconsistent during the early years (represented by the dashed thin lines) but more consistent across age in the later years (represented by the thin solid lines). Within both early and late segments, the group average shows both slow and rapid change over age. Therefore, consistency over age in individual differences and consistency in average group performance are potentially independent.

Developmental Function

The *developmental function* of a specific characteristic is the average or typical value for that attribute plotted across age. A developmental function may pertain to an entire species, as in the case of Piaget's theory of the stages of early mental development; it can apply to identifiable subgroups, as in the case of different plots for males and females; or it may characterize an individual, as in the case of the solid or dashed thin lines in Figure 1.

Developmental functions are either *continuous* or *discontinuous*. A developmental function is continuous when the fundamental nature of the attribute remains the same over age. But when changes are qualitative, the developmental function is said to be discontinuous. Height in inches is continuous because the nature of the attribute remains fundamentally the same across age. Sensorimotor intelligence as defined by Piaget is discontinuous, because the fundamental qualitative nature of mentality is different from stage to stage.

Of course, whether an empirical function is continuous or discontinuous will depend not only on the nature of the attribute but also on how it is measured. For example, if Piaget's sensorimotor intelligence is measured by standard scores on instruments that change their qualitative character from age to age, the plot of the developmental function may be empirically continuous but conceptually discontinuous. Notice that a rapid shift in value from one age to a closely adjacent age is not a discontinuity by this definition. It may signal an underlying conceptual discontinuity, but whether a developmental function is continuous or discontinuous depends on its qualitative nature, not on its quantitative nature.

Individual Differences

The term *individual differences* refers to the relative within-sample position of individual subjects measured on a given attribute. Typically, we are interested in correlates of individual differences, involving either qualitatively different variables or the same variable measured at another time.

Individual differences are either *stable* or *not stable*. Stability implies that the relative rank ordering of subjects within a group on the same variable is somewhat consistent from one assessment occasion to another. If the relative rank ordering of subjects changes from one age to the next, individual differences are not stable.

Confusing the Two Realms

As discussed above and shown in Figure 1, the continuity or discontinuity of a developmental function is potentially independent of the stability or lack of stability in individual differences. We have not always kept this distinction in mind.

For example, when it was observed that the correlation between IQ at age 4 and IQ at age 17 was approximately .71, the square of which is .50, some interpreted this fact to indicate that half of adult intelligence was determined by age 4. While it is correct that approximately half the ordering of individual differences at age 17 could be seen in the distribution of subjects at age 4, it was not the case that 4-year-olds had acquired half of the mental capability of adults. It may be that differences in the heights of seedling sequoia trees correlate .71 with the relative rank ordering of the heights of those trees at maturity, but the 3-foot seedlings have not achieved half their ultimate stature, which may be well over 250 feet. One must be careful not to generalize from one realm of development to the other.

But having noted that individual differences and developmental functions are *potentially* independent, one must also bear in mind that the independence of these two realms is an empirical question. For example, it is quite possible that the stability of individual differences in relative performance on mental tests may show a drop in value at precisely the age at which major qualitative changes occur in the fundamental definition of mentality and the composition of the test. In fact, this does appear to occur during infancy (McCall, Eichorn, & Hogarty, 1977), but it did not have to be the case.

Consistency versus Change

I have argued elsewhere (McCall, 1977, 1981) that our discipline has a bias toward consistency with respect to developmental function and especially with respect to individual differences.

For many years we preferred to think of intelligence as essentially a unitary, constant characteristic of the organism that merely changed the nature of its manifestation over age. The brilliance or dullness of adults was there from birth, if not before; it was just more difficult to see or measure during infancy (Hunt, 1961). The developmental model for intelligence was something like a balloon being blown up. The balloon is essentially collapsed at birth but it has a basic shape to it even then. It becomes inflated during development, but its fundamental shape remains the same throughout: The principal change is in its size.

More recently, of course, Piaget proposed a discontinuous model of mental development. Piaget's orientation supposes a series of metamorphoses similar to those in the transformation of a tadpole into a frog. That is, the fundamental nature of mentality changes from one stage of development to the next. Although Piaget clearly championed a discontinuous developmental function for mental development, American researchers have not always attacked the problem from the same angle. Rather, we prefer to look for antecedents, consequences, and threads of communality from one stage to the next, rather than describing in detail the nature of the qualitative change.

The bias toward consistency is even stronger within the realm of individual differences. Who would correlate mental performance at one age with performance at another age and then be ecstatic at finding no correlation at all? Indeed, even our methodology favors the search for consistency. If one finds a significant correlation, then one has evidence for stability of one degree or another. And if the correlation is statistically significant but of modest size, we celebrate the finding of stability, although the amount of stability discovered may be substantially less than the amount of change unaccounted for. Should we find no significant correlation, the rules of scientific inference permit only the conclusion that "no evidence for consistency was observed," which each of us is well taught does *not* imply that we have found "evidence for change."

From one standpoint, the emphasis on consistency in either realm of development is antithetical to the definition of our discipline. "Development" means "change" (Wohlwill, 1973), and at the very least, we should be as vigorous about searching for and describing change as we have been in the search for and description of consistency.

Heredity and Environment

Perhaps no issue has been as pervasive, as long-lived, or as heated as the question of the relative contribution of heredity and environment. In principle, the question is worth pursuing, although years ago we decided that the answer for mental development made a simple dichotomous question absurd. The proper way to address the issue was not to ask "which" or even "how much" of a trait was determined by heredity and environment, but "how" that trait developed under the influence of both (Anastasi, 1958). But the question of "how" failed to generate the enthusiasm and attention that its simpler predecessors produced.

Confusing Realms

Another problem is that the nature–nurture battle has been waged largely with respect to individual differences, not the development function. Fundamentally, the strategy for determining the relative contribution of heredity and environment to mental performance is to determine what portion of individual differences is associated with genetic background and what proportion with environmental variations. Of course, the actual calculations are substantially more complex than this simplistic characterization implies, but the fact that heritability pertains only to individual differences is inescapable.

This one-sided study of the nature–nurture issue has gotten us into trouble on occasion. The Skodak and Skeels study (1949; Honzik, 1957) is well known for the fact that the IQs of the adopted children correlated .38 with those of their biological parents but essentially zero with an estimated index for their rearing parents. What seems less well known is that the average IQ of the children was 21 points higher than the average of their biological parents and nearly identical to the estimated average of their rearing parents. Obviously, hereditarians want to emphasize the individual difference result, while environmentalists will concentrate on the mean difference (i.e., developmental function). Both pieces of information are useful. They are not contradictory (Jensen, 1973), but they are distinct, and we rarely view them as two pieces of the same puzzle.

Failure to consider heredity and environment in both realms of development has also produced some apparent anomalies in the study of early mental growth. For example, although the debate continues, it appears to me that the genetic contribution to individual differences in mental performance is quite modest during the first 2 years of life. At the same time, although hard evidence is lacking, it seems obvious that nature is controlling the general ontogeny of mentality quite closely during this same period. Presumably, the explanation is that heredity is high for developmental function, but low for individual differences. This possible distinction has eluded us, because there are no methods for determining the heritability of a species-general developmental function (except to compare one species with another). Then too, just when it appears that nature is loosening its grip on the development of its children, not only do correlations with genetics increase, but correlations with environmental circumstances also increase. How is this mélange possible?

Conclusions

What we should have in broad strokes is a 2 × 2 × 2 design in which we look at developmental function and individual differences, consistency and change, and heredity and environment. We must respect the independence of factors and levels within factors on the one hand, but appreciate that the phenomenon we are studying will be understood only when we vigorously collect data and interpret them across all levels and all factors. Today we have theories and approaches within one or possibly two factors but not all three, and we are lacking data and methods in some cells altogether. At least we need to approach the study more comprehensively.

A Scheme for Thinking about Mental Development

I recently proposed an approach for thinking about early mental development (McCall, 1981). I deliberately avoided calling it a "model," because the approach does not make profound and detailed predictions about mental development, and because its importance, if any, at this stage of our thinking is to stand as a broad conceptualization of mental development that embodies the three factors described above, respecting their independence, but integrating them into a cohesive scheme.

The Scheme

My orientation rests heavily on the concept of canalization.

Canalization

The notion of canalization, once proposed by Waddington (1957) and more recently applied to mental development by Scarr-Salapatek (1976), postulates a species-typical path, called a *creod*, along which nearly all members of the species tend to develop. However, a given characteristic follows the creod only as long as the organism is exposed to species-typical, appropriate environments. In the presence of such environments, development proceeds "normally"; when the environment deviates markedly from the typical, development may stray from the creod.

An important corollary of canalization is the notion of a *self-righting tendency*. That is, when development is highly canalized, there is a marked tendency in the presence of species-typical environments for an organism to follow the species-general developmental path or creod. More important, this self-righting tendency is invoked following momentary deflections caused by biological or environmental circumstances. Thus, when a behavior is highly canalized, organisms tend to follow the creod under a wide diversity of environments and to return to the creod—that is, to self-right—even after highly atypical circumstances have occurred and are then removed. When development is less canalized, the self-righting tendency is weaker, individuals differ from one another to a greater extent, and individual differences are more likely to persist through time because there is less pull by nature to keep her children in single file.

The Scoop

The fundamental theoretical principle of my approach can be stated quite simply: Mental development is highly canalized at the beginning but becomes less so between approximately 2 and 6 years of age.

To understand the implications of this simple proposition, it helps to look at a diagram. Much as I have eschewed describing this scheme as a "model," it does help to have something visually concrete as an intellectual crutch. A "scoop" is pictured in Figure 2.

The scoop itself represents the creod or species-general developmental function for early mental development. Age is conceived to run from left to right. Notice that in the first 2 years of life, the creod is a deep trough that widens and becomes flatter after age 2. Inside the scoop, the different designs represent different qualitative stages of mental development. Notice these are not clearly demarked, but the transitions fade from one stage to the next. During childhood, grooves in the scoop can be seen that represent different biological emphases or paths that individuals might follow. The different patterns represent

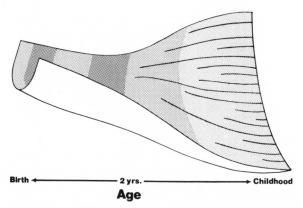

Birth ← ——————— **2 yrs.** ——————————→ **Childhood**

Age

FIGURE 2. The "scoop" approach to thinking about mental development. (From "Nature-Nurture and the Two Realms of Development: A Proposed Integration with Respect to Mental Development" by R. B. McCall, *Child Development*, 1981, 52, 1–12. Copyright 1981 by Society for Research in Child Development. Reprinted by permission.)

qualitative discontinuities in the species-general developmental function, while the grooves represent different individual differences in developmental paths.

Now imagine a lightweight ball representing an individual who begins life at the leftmost edge of the scoop. The incline in the scoop allows the ball to roll down the scoop over developmental time. This incline represents necessary and vital environmental contributors to the species-general function. In addition, environmental winds blow over the scoop. If they blow from left to right, they produce individual differences in the rate of development. If they blow in other directions, they can cause individual differences in the extent or nature of mental performance, depending on their strength and where the individual is on the creod.

The Dynamics

Now notice some of the dynamics of this system. A ball beginning life's course at the left is tightly contained by the steep walls of the scoop. Very strong winds are required to blow the individual completely off course, and small deflections from the main course are quickly "righted" and the ball forced back onto nature's narrow creod. The result is that individuals pass through pretty much the same stages, in the same sequence, with minor deviations in timing. Individual differences observed at one point in time are not likely to be stable over long developmental periods because of the strong self-righting tendency. Therefore, while the developmental function is highly canalized and tightly controlled by biology, individual differences are much more modestly related to both genetic and pervasive environmental circumstances and are usually temporary. It is not that individual differences are simply "error," but that they are temporary in the face of the steep sides of the creod, which produce the self-righting function.

I further speculate that the nature of the qualitative advances in developmental function at each of the early stages consists of relatively basic mental processes that, for the most part, are all-or-none. That is, all children acquire figure–ground and object-permanency, for example. Basically, we do not talk about individual differences in these characteristics, except in their rate of acquisition. Does anyone measure levels of proficiency in figure–ground or object-permanency among adults? Therefore, sensorimotor stages represent the acquisition of fundamental skills that all members of the species attain to essentially the same degree of proficiency.

At approximately 2 years, although the timing is more gradual than this implies, the sides of the scoop progressively fall away, the self-righting tendency diminishes, and environmental winds, especially crosswinds, are given greater access and potential to influence the developmental course of an individual. Of course, at the same time, the ball representing an individual is picking up speed, which represents accumulated experience and self-determination. The result is that there is an impressionable period, roughly between 18–24 months and perhaps 6 years, when the combination of the ball's slow speed and the decline of the self-righting tendency make the child most susceptible to environmental influence.

However, it should be added that everything is not fixed by age 3 or by age 6. The sides of the scoop remain low, the grooves in the scoop are not deep, and it is still possible for a sufficiently strong wind to deflect the ball in one direction or another. Therefore, while environmental and genetic influences are more likely to have long-term consequences after age 2 than before, the organism is still potentially plastic in the face of strong environmental circumstances.

IMPLICATIONS OF THIS APPROACH

Although the conceptual scheme offered here is quite simple, its implications are somewhat more complex. However, the utility of this approach is its ability to encompass diverse general themes in the data with a few conceptual principles.

Developmental Function—The Stages of Mental Development

Several people, including Piaget (1954/1966), Užgiris (1976), McCall *et al.* (1977), and Fischer (1980) have speculated on the nature of the developmental stages represented by the contrasting designs on the inside of the scoop.

Although important differences distinguish these theoretical orientations, the similarities across positions are striking. While there is less agreement on the fundamental theme underlying each stage, there is much agreement about the sequence of behavioral characteristics from one stage to the next and, surprisingly, about the age at which fundamental transitions occur. While some theorists postulate more transitions than others, there is at least some agreement that major stage boundaries occur at approximately 2, 7–8, 13, and 21 months of age during infancy (McCall *et al.*, 1977) and at 5–7 years during childhood (Gruen & Doherty, 1977; McCall, Appelbaum, & Hogarty, 1973; White, 1965). The particular nature of these stages has been described by the theorists cited above.

Individual Differences

Cross-Age Correlations

One implication of the scoop approach is that the steep sides and strong self-righting tendency during the first 2 years create modest and short-lived stability of individual differences in assessments of mental performance. Evidence for this point is presented in Tables I and II.

Table I presents the median correlations across several studies between test scores obtained at various ages during the first 2 years of life.

TABLE I

Median Correlations across Studies between Infant Test Scores
at Various Ages during the First 2 Years of Life[a]

Age in months	1–3	4–6	7–12	13–18
4–6	0.52 (8/6)	—	—	—
7–12	0.29 (14/6)	0.40 (18/10)	—	—
13–18	0.08 (3/3)	0.39 (6/6)	0.46 (9/6)	—
19–24	−0.04 (3/3)	0.32 (6/6)	0.31 (9/6)	0.47 (7/6)

[a]See McCall (1979) for particular studies represented in this table. From "The Development of Intellectual Functioning in Infancy and the Prediction of Later IQ" by R. B. McCall, in J. D. Osofsky (Ed.), Handbook of Infant Development, 1979. Copyright 1979 by John Wiley & Sons. Reprinted by permission.

These data are remarkably consistent across specific intervals of the same length (in approximate log units) during the first 2 years of life. Correlations along the main diagonal are roughly .45 for intervals of approximately 6 months (shorter at the younger ages, longer at the older ages) and approximately .33 for the first minor diagonal representing intervals of approximately 1 year (shorter at the younger ages, longer at the older ages). There is essentially no correlation to assessments after 1 year from initial assessments given in the first 3 months. A reasonable interpretation is that change across age during the first 2 years of life is the rule.

Table II presents analogous data from two of America's major longitudinal studies, the Fels Longitudinal Study and the Berkeley Growth

TABLE II

Age-to-Age Correlations for Childhood IQ
from the Fels Longitudinal Study and the
Berkeley Growth Study

Age in years	9	10	11	12
9	—	.90[a]	.82[a]	.81[a]
10	.88[b]	—	.90[a]	.88[a]
11	.90[b]	.92[b]	—	.90[a]
12	.82[b]	.90[b]	.93[b]	—

[a]Fels Study data.
[b]Berkeley Study data.

Study. Compare these correlations calculated across yearly intervals during midchildhood with those given in Table I for shorter intervals during infancy. It is clear that cross-age stability is much lower during infancy than during childhood, despite the shorter intervals used in the infancy table. This fact presumably testifies to the greater canalization of mental performance during infancy than during late childhood.

Prediction to Later IQ

Given the modest cross-age stabilities of individual differences within the infancy period, it is not surprising that correlations to child and adult IQs from the infancy period are quite modest. Table III presents a summary of these data for infant tests given between 1 and 30 months of age predicting childhood IQ tests given between 3 and 18 years of age. A close look at this table reveals two independent trends: Correlations increase with the age at which the test is given during infancy, and predictions are better the younger the age at which the childhood IQ test is administered. Together, these independent trends lead to the fact that the shorter the time between testings, the higher the correlation.

It should also be observed that the value of these correlations is not high in an absolute sense. Certainly, any long-term predictions from assessments made prior to 18 months, while not absolutely zero, are modest at best and of essentially no clinical utility. But notice that predictions from 19–30 months to ages throughout childhood are essentially the same value regardless of the prediction interval. That is,

TABLE III
Median Correlations across Studies between Infant Test Scores and Childhood IQ[a]

Age of childhood test (years)	Age of infant test (months)				
	1–6	7–12	13–18	19–30	Median
8–18	0.06 (6/4)	0.25 (3/3)	0.32 (4/3)	0.49 (34/6)	0.28
5–7	0.09 (6/4)	0.20 (5/4)	0.34 (5/4)	0.39 (13/5)	0.25
3–4	0.21 (16/11)	0.32 (14/12)	0.50 (9/7)	0.59 (15/6)	0.40
Median	0.12	0.26	0.39	0.49	

[a]See McCall (1979) for particular studies represented in this table. From "The Development of Intellectual Functioning in Infancy and the Prediction of Later IQ" by R. B. McCall, in J. D. Osofsky (Ed.), *Handbook of Infant Development*, 1979. Copyright 1979 by John Wiley & Sons. Reprinted by permission.

in Table I the correlation between 13–18 months and 19–24 months was .47, which is not very different from the values of .49, .39, and .59 found for predictions to childhood from 19–30 months (Table III). The implication is that whatever is holding back the predictive accuracy resides in infancy, not in childhood. That is, change must predominate before 2–3 years of life, with a shift in stabilities at or shortly after this point.

A similar phenomenon occurs in the Berkeley Growth Study data (McCall *et al.*, 1977) in which the correlations between the first principal components of assessments rise precipitously between 21 and 30 months of age (earlier for boys) and predict later components at the same level regardless of the age of the childhood assessment. These observations are consistent with the notion that canalization weakens after age 2 and that whatever abilities emerge at this point in development are relatively stable thereafter.

Dips in Cross-Age Stabilities of Individual Differences at Points of Discontinuity in the Developmental Function

I have argued above that stability of individual differences is *potentially* independent from continuities or discontinuities in developmental function. But in nature, it is quite possible that individual differences become rearranged at the advent of new skills, a phenomenon that would be reflected in reduced cross-age stabilities at points in development where major discontinuities in the developmental function occur. Consequently, cross-age stabilities should be higher for periods that do not span a discontinuity, and lower for comparable periods that do embrace a discontinuity in the developmental function.

As indicated above, some agreement exists for postulating discontinuities in the developmental function at 2, 7–8, 13, and 21 months of age. I have attempted to evaluate the cross-age correlations from Bayley (1933) and Wilson and Harpring (1972) from this standpoint, and the results are presented in Table IV. Admittedly, the differences of interest are not large, but the pattern revealed is provocative in its relative consistency.

Look first at the top of the table, which summarizes correlations from Bayley (1933) covering age spans of 1, 2, 3, and 4–9 months of age, either residing totally within a hypothesized stage or crossing one stage boundary. The data presented are averages of two to six correlations,

TABLE IV

Average Bayley Test-Score Correlations across Various Age Spans as a Function of Whether or Not Those Spans Are within a Stage or Cut across a Stage Boundary at 8, 13, or 21 Months of Age[a]

		Age span in months			
		1	2	3	4–9
Bayley (1933)[b]					
Within a stage		.84	.75	.82	.82
Across 1 stage boundary		.80	.74	.77	.68
	3	6	9	12	15
Wilson and Harpring (1972)[c]					
Within a stage	.52				
Across 1 stage boundary	.45	.39	.37		
Across 2 stage boundaries			.29	.28	.20

[a]From "Transitions in early mental development" by R. B. McCall, D. H. Eichorn, and P. S. Hogarty, *Monographs of the Society for Research in Child Development*, 1977 (Serial Number 171). Copyright 1977 Society for Research in Child Development. Reprinted by permission.
[b]Averages of 2–6 correlations. Data from 1 to 4 months have been omitted because reliability and cross–age *r*'s are known to be low for this period for reasons irrelevant to this comparison.
[c]Averages of 1–4 correlations for members of twin pairs.

and age span is not perfectly balanced with the particular ages involved in its definition.

Notice first that the correlations within a stage are remarkably similar in value regardless of the length of the age span involved: Three of the four correlations are within .02 of each other. This observation is in contrast to the well-known phenomenon of decreasing correlation with increasing age span within the infancy period, discussed above. In contrast, when the correlations span a stage boundary, they reveal the more typical pattern of declining value with increasing age span (notice that the irregularity for an age span of 2 months occurs both within and across stage boundaries and probably reflects some idiosyncratic characteristic of these particular data). Further, the correlations are ever so slightly lower across the stage boundary than within a stage for comparable age spans.

At the bottom of the table, the Wilson and Harpring (1972) data reveal progressive declines in correlations with increasing age span, and in the two comparisons possible, the correlations decrease when the age span includes one or two stage boundaries.

The Heritability of Individual Differences in Infant Test Performance

Since estimates of the genetic contribution to a characteristic rest totally with analyses of individual differences, heritabilities for infant test performance should be relatively low because of the instability of individual differences and the strong self-righting tendency that accompanies a highly canalized system. The low heritabilities should exist despite the assumption that nature tightly controls the developmental function.

Methods exist for evaluating the heritability of developmental functions for individuals, but they rely on individual differences in developmental function. No methods exist for evaluating the heritability of the species-general developmental function (except by comparing one species to another, which will produce a result directly proportional to the dissimilarity of the two species selected for study).

The heritability of infant test performance has been reviewed several times (e.g., Honzik, 1976; McCall, 1979; Scarr-Salapatek, 1976; Wilson, 1973). Readers interested in the details of this topic should consult these references carefully; other readers should be aware that considerable debate surrounds this field and that the brief summary presented here is heavily laden with the author's interpretation.

Parents and their children, siblings, and dizygotic twins all share 50% of their genes on the average. Therefore, the theoretical genetic correlation for each of these kinship pairs is .50. Systematic deviations from this theoretical genetic value presumably reflect the influence of nongenetic factors.

At the risk of glossing over important variations, consider some kinship correlations for infant tests within the first 2 years. Parent–child correlations for the Stanford–Binet performance of parents at age 17 and the Bayley infant test scores obtained on their children in the first 2 years of life are essentially zero (Eichorn, 1969). A study of infants separated from their biological mothers and reared in an orphanage (Casler, 1976) showed a similar result, except that there were some significant correlations between natural mothers' IQs and infants' Gesell performances at 2 months of age that did not persist at older ages. This suggests that prenatal and birth circumstances associated with maternal IQ may have an early but declining influence on infant test performance.

Correlations between parent and child when both are assessed on the same instrument (the Bayley) and at the same ages (i.e., parent and child both assessed as infants) are essentially zero during the first 2 years of life (Eichorn, 1969), although the correlations may begin to increase at 2 years, especially for like-sexed pairs (Eichorn, 1969; McCall *et al.*, 1973). In contrast, parent–child IQ correlations are approximately .50–.60 when the parent is assessed in adulthood, but .04–.38 when the parent is assessed at the same age and with the same tests as the child (McCall *et al.*, 1973).

Therefore, parent–child correlations are substantially higher than parent–infant correlations, especially when the parent is assessed as an adult. Moreover, the higher parent–child correlations for parent as an adult relative to parent as a child may reflect the environmental influence the parent has on the child's performance or the lack of consolidation of childhood IQ performance.

When infant siblings raised in the same home by the same parents are correlated for assessments made on the same instrument assessed at the same age (McCall *et al.*, 1973), the correlations are approximately .20 for a single assessment (although the similarity can increase if tests are averaged over several assessment periods). The correlation for siblings assessed as children or adults but not at the same age is approximately .55 (Jensen, 1969). Once again, sibling correlations are considerably smaller for infant tests than for child or adult IQs, presumably reflecting nongenetic circumstances during the infant years that influence individual differences in test performance.

Wilson and his colleagues at the Louisville Twin Study (Wilson 1974; Wilson & Harpring, 1972) have presented the most extensive data on monozygotic (MZ) and dizygotic (DZ) twin similarity for scores on the Bayley infant scale at different ages between 3 and 72 months. A summary of their correlations (McCall, 1979) shows that the within-pair similarity for DZ twins is remarkably consistent between 3 and 72 months of age and averages approximately .68. The within-pair similarity for monozygotic twins, which share all of their genes, was .83 for the same age span.

Notice two things: The DZ correlation of .68 is substantially higher than the correlation for siblings, despite the fact that both DZ twins and ordinary siblings share 50% of their genes. Presumably, the substantially higher concordance for twins is associated with nongenetic circumstances common to twins living together and tested at the same age and time but not shared by ordinary siblings, who are tested at the same age but at different times. This interpretation is quite consistent with the presumed ephemeral and nongenetic individual differences that occur in a highly canalized system. However, the fact that the similarity for DZs is remarkably consistent from 3 to 72 months (the oldest age considered) seems to be at odds with the hypothesized changing degree of canalization, unless the nongenetic circumstances diminish in perfect compensation with the rise in genetic influence.

The second thing to notice about these data is that the difference between MZ and DZ within-pair concordances is not as great as for older children and adults. For example, in Wilson's data, the maximum heritability (two times the difference in concordance for MZ versus DZ) varies unsystematically from age to age between 3 months and 6 years but averages approximately .30. This value assumes that all difference between the within-pair similarity of MZ versus DZ pairs is associated with genetic circumstances. Estimates of maximum heritabilities for MZ versus DZ twins based on assessments taken during later childhood and adulthood are substantially higher, from .50 to .70 (Jensen, 1969). Therefore, the heritability of individual differences during infancy and early childhood appears to be less than for older twins.

Thus far we have considered correlations and heritabilities for a single assessment during the early years of life. It is also possible, within pairs of related individuals, to assess the similarity of pattern of test performance over age apart from the general level of that performance. This has been done for ordinary siblings (McCall, 1970, 1972a,b; McCall et al., 1973) and for MZ and DZ twins (Wilson, 1974; Wilson & Harpring, 1972). The analysis of ordinary siblings provided no evidence at all for similarity in pattern of IQ change between 6 and 24 months, 42 to 132 months, and smaller time spans within these boundaries. On the other hand, Wilson's data reveal maximum heritabilities (as computed by McCall, 1979) between .10 and .82 for developmental pattern for various age spans between 9 and 72 months. In general, the heritabilities increase with age, but two unusually low heritabilities span the hypothesized 13-month and 5- to 7-year transition points.

It is difficult to interpret the difference between sibling and twin similarities in developmental function. There is ample reason to believe that twins will be more similar within pairs because of short-term nongenetic circumstances that are revealed when both individuals are assessed at the same time. However, just why such similarities should be greater for MZ than DZ twins is not at all clear. One problem in using twins for heritability estimates is that genetic circumstances producing serious mental deficiency are more common among MZ than DZ twins. When such subjects were eliminated in one study, all evidence for heritability of single assessments was eliminated (Nichols & Broman, 1974). However, it is only by inference that such a factor would apply to developmental profile as well as single-age assessments, and in any case this factor does not appear to operate as strongly in Wilson's data (Wilson & Matheny, 1976) as in the Collaborative Perinatal Study sample.

On balance, it appears that individual differences in infant test behavior are less strongly associated with genetic circumstances than are child and adult mental assessments. Moreover, within-pair similarities for DZ twins, for example, are substantially higher than for ordinary siblings, despite the fact that these kinship pairs share equal amounts of genetic material. Presumably, the higher DZ similarity derives from shared nongenetic circumstances that may be short lived but that contribute to twin similarity but not sibling similarity because the twins are measured at nearly the same time. Both of these observations are quite consistent with the notion of a highly canalized system early in life. These same data do not always show a gradual increase in the size of relationship between genetic circumstance and individual differences between 2 and 6 years as hypothesized, but other data do (Honzik, 1957; Jensen, 1969; McCall, 1979).

Correlations with Socioeconomic Status

In a highly canalized system, such as that hypothesized for mental development during the first 2 years of life, relationships between infant test performance and general indexes of environmental quality should be quite low. In contrast, contemporaneous correlations between test performance and certain environmental factors could be more substantial, but predictive relationships between the same variables would be expected to be short lived.

Recent reviews of the literature on the correlation between parental socioeconomic status (SES) and contemporary mental test performance (Golden & Birns, 1976; McCall, 1979) concur in concluding that there is no global relationship until 18–24 months of age. Across a diverse literature, the correlations prior to 18 months are essentially zero, and modest correlations (e.g., .20–.30) emerge between 18 and 24 months, followed by a steep rise to asymptotic levels (e.g., approximately .50) shortly thereafter. Some variation has been found with respect to the age at which the relationship increases, and some studies report this rise to occur earlier for girls than for boys. Notice that the increase starts at approximately 2 years of age. Also, although parental SES is frequently used to index environmental variables, it undoubtedly relates to genetic factors as well.

SES predicts later IQ better than do infant test scores obtained during the first 18 months of life. Thereafter, the infant test begins to predict later IQ, and it apparently reflects variability not directly associated with SES because the multiple correlations of infant tests and SES exceed the coefficients for either variable alone (Hindley & Munro, 1970; Ireton, Thwing, & Gravem, 1970; McCall, 1979). Therefore, from a practical standpoint, knowing the education of the infant's parents is a better predictor of the infant's childhood mental status than anything we have yet measured on the infant during the first 18 months.

It should also be noted that there is some tendency for low-scoring infants to have a greater likelihood of recovery during infancy or childhood if they come from upper-middle class homes than if they come from lower socioeconomic strata (Drillien, 1964; Knobloch & Pasamanick, 1967; Sameroff & Chandler, 1975; Scarr-Salapatek & Williams, 1973; Werner, Simonian, Bierman, & French, 1967; Willerman, Broman, & Fiedler, 1970). This is an important observation for the canalization approach, because it testifies to the self-righting tendency and suggests that it is more likely to operate under advantageous environments than under less advantageous circumstances.

Correlations with More Specific Environmental Circumstances

If individual differences tend to be produced by specific environmental conditions, then correlations with infant test performance should be higher for more specific measures of environmental supports. Further, since the organism is changing its fundamental characteristic from one stage to the next, the stability of these correlations will depend on the availability and appropriateness of the environmental factor across different stages.

McCall (1979) recently reviewed this literature and concluded that in the first 3 years of life, the presence of a variety of play objects (especially responsive toys) and the freedom to explore such an environment correlate with infant test performance. However, there needs to be a moderate signal-to-noise ratio, because excessive or unpatterned stimulation in the form of intense auditory and social stimulation as well as lack of orderliness in the home is related to lower performance.

High maternal verbal stimulation and verbal responsivity tend to correlate with test performance during the second year but not earlier, and there is some suggestion that this relationship is at least partly mediated by genetic factors (Beckwith, 1971). This observation is consistent with the proposition that the developmental function changes its primary characteristic from stage to stage, and environmental stimulation that supports the fundamental characteristic will correlate with test performance during that stage.

A diverse literature (see McCall, 1979) suggests that social, emotional, and personality factors are related to mental test performance, especially between 1 and 3 years of age. While maternal involvement, affection, and encouragement toward intellectual pursuits are associated with high scores for males, these maternal characteristics are meshed with the child's own affability, extroversion, or socially attentive disposition in predicting contemporary and future mental test performance for girls.

Perhaps the strongest results in this regard were reported by Yarrow (1963), who found correlations at 6 months of age for adopted children to be .85 for ratings of stimulus adaptation, .72 for achievement stimulation, and .65 for social stimulation. Moreover, several of these 6-month ratings correlated with 10-year WISC scores, but mainly for males (Yarrow, Goodwin, Manheimer, & Milowe, 1973).

Since it might be assumed that much intellectual stimulation is provided through a social medium and that parents adapt their social interactions to the skill level of the child, it may not be surprising that social factors are related to mental test performance and may evidence such relationships over a substantial age period.

Implications for Early Experience

The canalization approach and these data suggest that because the fundamental characteristic of mentality changes from one stage to the next during infancy, environmental stimulation promoting performance must also change concomitantly with the changing nature of the infant. Therefore, it is unlikely that there will be any kind of environmental experience at any single age during infancy that will ensure superior mental performance at substantially distant times in the future. There

will be no environmental inoculation in infancy against the mental rav-
ages of disadvantageous environments in childhood and beyond. In-
fant-stimulation programs cannot expect to have substantial long-term
consequences if they are terminated at 2 or 3 years of age. Either the
enrichment programs need to be continued and changed as the infant
grows older, or parents need to be trained to provide continuing and
changing enrichment.

While this prescription appears more complex than one might
hope, the other side of the coin is more joyful. Infants who are phys-
iologically or environmentally damaged early in life have a superb
chance of recovery if placed in advantageous home environments or
given supplementary stimulation in organized programs. Presumably,
however, such stimulation must match the particular stage and skill
level of the infant.

Beyond 18–24 Months

The decline of canalization beginning at approximately 2 years of
age implies that stages in the developmental function will be spaced out,
that there will be a greater diversity of behavior at any single age, that
individuals will tend to become channeled into separate paths and spe-
cializations, and that skills will emerge that are not all-or-none.

The child and adult literature are quite clear in showing substantial
genetic and environmental correlates of IQ test performance consistent
with the decline of canalization. That both genetic and environmental
correlates increase after approximately 2 years of age appears to be an
anomaly, since we usually think of these as opposing factors. However,
the relationships are not so high for one factor that they obviate relationships with the other. Moreover, genetic and environmental circumstances frequently covary themselves, implying that a substantial portion of the correlation of one factor with mental test performance is
shared by the correlation of the other factor with test performance.

Longitudinal stability in individual differences in IQ also increases
between 2 and 6 years, attaining asymptotic levels at about 6 years of
age. Although year-to-year correlations in late childhood and adulthood
are often .90 or higher, one should not assume that meaningful change
in relative mental performance cannot occur after age 6. McCall *et al.*
(1973) demonstrated that while approximately 40% of the Fels sample

did not change substantially in IQ between 2½ and 17 years of age, the remainder did. In fact, the average range of IQ performance was 28.5 IQ points, one in seven subjects changed 40 points or more, and one individual displayed a shift of 74 IQ points. It also appears that these are not random fluctuations about a common value, because most of the trends were simple linear, quadratic, and cubic functions spread over substantial periods of time. Since no similarity was found in the pattern of IQ change for siblings in the same sample, one might suppose that whatever produced these changes was nongenetic in character.

But what is it that might cause substantial shifts in mental performance after early childhood? Since siblings, who share parents and the same general climate of stimulation in the home, do not show similar patterns of IQ change, the environmental factors must be more specific than global indexes of characteristics that remain relatively constant.

McCall et al. (1973) suggested that it was the combination of environmental circumstances that matched the particular intellectual skills and interests of a child at a given age that tended to produce changes. The opportunity to visit the Kennedy Space Center and to be exposed to airplanes might stimulate one child in a family and leave another quite unaffected. Moreover, such events may influence a ninth grader but leave a fourth grader quite unimpressed. The "match" between environmental stimulation and the skills and interests of the child must be achieved, and some (McCall et al., 1973) have argued that a substantial portion of the environmental correlation with IQ is associated with this kind of "matched" environmental support rather than the general intellectual climate of the home.

Abnormality

What happens when an individual is pushed over the sides of the scoop?

It is possible for severe biological insult or extreme environmental circumstances to affect the mental performance of a child severely. The canalization approach suggests that during infancy there is a wide tolerance for adverse circumstances if the child is replaced into species-typical or advantageous environments. Evidence that recovery can occur has already been cited.

It has been widely assumed that advantageous circumstances must be instituted early in life, before 6 or 3 years of age, for recovery to occur. In view of case studies of children locked in unstimulating rooms for several years, we can say that it is possible for children to recover from the most deleterious circumstances even after age 6 (Clarke & Clarke, 1976; but see Sroufe, 1977). Of course there are limits, both in terms of the nature of the injury or circumstances and in terms of the age at which species-typical environments are instituted. But the human organism appears to be more flexible for longer periods of time than many have previously supposed.

It has sometimes been assumed that predictions to later IQ are higher for samples of infants designated to be at-risk for developmental abnormalities. This is partly true. A low score on an infant test, even in the absence of other symptoms, is somewhat more predictive of later abnormality than a high score is predictive of later superiority on mental tests. It is also the case that correlations increase somewhat earlier and

TABLE V

Correlations between Infant Test Scores and Childhood IQ for At-Risk Samples[a]

Childhood age of test (years)	Age of infant test (months)				Marginal values
	1–6	7–12	13–18	19–36	
8–18				.71[g,k] .46[g,j,k] .52[g,j,k] .39 (females) .33 (males)[h,k]	.48
5–7	.54[b]	.57[b]			.56
3–4		.37[c,j]	−.11[c] .48[d]		.40
2	.29[f] .42[e]	.44[c,j] .07[e], .39[e] .48[f]	.30	.63[e] .86[c]	.36
	.42	.31	.63	.64	

[a] From "Predicting Later Mental Performance for Normal, At-Risk, and Handicapped Infants" by C. B. Kopp and R. B. McCall, in P. B. Baltes and O. G. Brim, Jr. (Eds.), *Life-Span Development and Behavior* (Vol. 4), 1982. Copyright 1982 by Academic Press, Inc. Reprinted by permission.
[b] Drillien (1961).
[c] Hunt (1979, 1981).
[d] Knobloch and Pasamanick (1960).
[e] Siegal, Saigal, Rosenbaum, Young, Berenbaum, and Stoskopj (1979).
[f] Sigman (personal communication).
[g] Werner *et al.* (1968).
[h] Werner and Smith (1982).
[i] Marginal values are the average of the correlations presented in that row/column.
[j] Sample restricted to subjects scoring below a cut-off at both ages.
[k] Overlapping samples.

reach somewhat higher levels for samples of infants at-risk for developmental anomalies than for normal infants (Kopp & McCall, 1982).

Although the data are not as complete as for normal samples, Table V provides a survey of results for at-risk samples comparable to that presented in Table III for normals. While the general themes mentioned above can be seen in the comparison of these tables, it is also the case that the correlations for the at-risk groups are not substantial in an absolute sense. Predictions from the first 2 years of life are not so high as to make the infant test clinically useful by itself, and attempts to combine the infant test with a variety of pediatric factors have yielded mixed results.

This state of affairs is explained by the fact that many infants who suffer from a variety of disadvantageous birth circumstances recover. Also, many disabilities do not become manifest until 6, 12, or even 18 months of age. Once again, nature's flexibility is a bane to prediction but sometimes a boon to humanity. Screening and detection of abnormalities are difficult, but hope for recovery from early distress is often substantial.

REFERENCES

ANASTASI, A. Heredity, environment, and the question "how?" *Psychological Review*, 1958, *65*, 197–208.

BAYLEY, N. Mental growth during the first three years: A developmental study of 61 children by repeated tests. *Genetic Psychology Monographs*, 1933, *14*, 1–92.

BECKWITH, L. Relationships between attributes of mothers and their infants' IQ scores. *Child Development*, 1971, *42*, 1083–1097.

CASLER, L. Maternal intelligence and institutionalized children's developmental quotients: A correlational study. *Developmental Psychology*, 1976, *12*, 64–67.

CLARKE, A. M., & CLARKE, A. D. B. *Early experience: Myth and evidence.* London: Open Books, 1976.

DRILLIEN, C. M. Longitudinal study of growth and development of prematurely and maturely born children: VII. Mental development, 2–5 years. *Archives of Diseases of Childhood*, 1961, *36*, 233–240.

DRILLIEN, C. M. *The growth and development of the prematurely born infant.* Baltimore, Md.: Williams & Wilkins, 1964.

EICHORN, D. H. *Developmental parallels in the growth of parents and their children.* Presidential Address, Division 7, APA meeting, September 1969.

FISCHER, K. W. A theory of cognitive development: The control and construction of heirarchies of skills. *Psychological Review*, 1980, *87*, 477–531.

GOLDEN, M., & BIRNS, B. Social class and infant intelligence. In M. Lewis (Ed.), *Origins of intelligence.* New York: Plenum Press, 1976.

GRUEN, G. E., & DOHERTY, J. A. A constructivist view of a major developmental shift in early childhood. In I. Č. Užgiris & F. Weizmann (Eds.), *The structuring of experience.* New York: Plenum Press, 1977.

Hindley, C. G., & Munro, J. A. *A factor analytic study of the abilities of infants, and predictions of later ability.* Paper presented at the Symposium on Behavioral Testing of Neonates and Longitudinal Correlations, London, April 1970.

Honzik, M. P. Developmental studies of parent–child resemblance in intelligence. *Child Development*, 1957, *28*, 215–228.

Honzik, M. P. Value and limitations of infant tests: An overview. In M. Lewis (Ed.), *Origins of intelligence.* New York: Plenum Press, 1976.

Hunt, J. McV. *Intelligence and experience.* New York: Ronald, 1961.

Hunt, J. V. Longitudinal research: A method for studying the intellectual development of high-risk preterm infants. In T. M. Field (Ed.), *Infants born at risk: Behavior and development.* Jamaica, N. Y.: Spectrum, 1979.

Hunt, J. V. Predicting intellectual disorders in childhood for high-risk pre-term infants. In S. L. Friedman & M. Sigman (Eds.), *Pre-term and post-term birth: Relevance to optimal psychological development.* New York: Academic Press, 1981.

Ireton, H., Thwing, E., & Gravem, H. Infant mental development and neurological status, family socioeconomic status, and intelligence at age four. *Child Development*, 1970, *41*, 937–946.

Jensen, A. R. How much can we boost IQ and scholastic achievement? *Harvard Educational Review*, 1969, *39*, 1.

Jensen, A. R. Let's understand Skodak and Skeels, finally. *Educational Psychologist*, 1973, *10*, 30–35.

Knobloch, H., & Pasamanick, B. An evaluation of the consistency and predictive value of the 40-week Gesell developmental schedule. In C. Shagass & B. Pasamanick (Eds.), *Child development and child psychiatry.* Washington, D. C.: American Psychiatric Association, 1960.

Knobloch, H., & Pasamanick, B. Prediction from the assessment of neuromotor and intellectual status in infancy. In J. Zubin & G. A. Jervis (Eds.), *Psychopathology of mental development.* New York: Grune & Stratton, 1967.

Kopp, C. B., & McCall, R. B. Stability and instability in mental performance among normal, at-risk, and handicapped infants and children. In P. B. Baltes & O. G. Brim, Jr. (Eds.), *Life-span development and behavior* (Vol. 4). New York: Academic Press, 1982.

McCall, R. B. IQ pattern over age: Comparisons among siblings and parent-child pairs. *Science*, 1970, *170*, 644–648.

McCall, R. B. Similarity in developmental profile among related pairs of human infants. *Science*, 1972, *178*, 1004–1005. (a)

McCall, R. B. *Similarity in IQ profile among related pairs: Infancy and childhood.* Proceedings of the American Psychological Association meeting, Honolulu, 1972, 79–80. (b)

McCall, R. B. Challenges to a science of developmental psychology. *Child Development*, 1977, *48*, 333–344.

McCall, R. B. The development of intellectual functioning in infancy and the prediction of later IQ. In J. D. Osofsky (Ed.), *Handbook of infant development.* Hillsdale, N.J.: Lawrence Erlbaum, 1979.

McCall, R. B. Nature-nurture and the two realms of development: A proposed integration with respect to mental development. *Child Development*, 1981, *52*, 1–12.

McCall, R. B., Appelbaum, M., & Hogarty, P. S. Developmental changes in mental performance. *Monographs of the Society for Research in Child Development*, 1973, *38*(150).

McCall, R. B., Eichorn, D. H., & Hogarty, P. S. Transitions in early mental development. *Monographs of the Society for Research in Child Development*, 1977, *42*(171).

NICHOLS, P. L., & BROMAN, S. H. Familial resemblances in infant mental development. *Developmental Psychology*, 1974, *10*, 442–446.

PIAGET, J. *The origins of intelligence in children.* New York: International Universities Press, 1966. (Originally published in 1954.)

SAMEROFF, A. J., & CHANDLER, M. J. Reproductive risk and the continuum of caretaking causality. In F. D. Horowitz (Ed.), *Review of child development research* (Vol. 4). Chicago: University of Chicago Press, 1975.

SCARR-SALAPATEK, S. An evolutionary perspective on infant intelligence: Species patterns and individual variations. In M. Lewis (Ed.), *Origins of intelligence.* New York: Plenum Press, 1976.

SCARR-SALAPATEK, S., & WILLIAMS, M. L. The effects of early stimulation on low-birth-weight infants. *Child Development*, 1973, *44*, 94.

SIEGEL, L., SAIGAL, S., ROSENBAUM, P., YOUNG, A., BERENBAUM, S., & STOSKOPJ, B. *Correlates and predictors of cognitive and language development of very low birth weight infants.* Hamilton, Ontario, Canada: Department of Psychiatry and Pediatrics, McMaster University Medical Centre, 1979.

SKODAK, M., & SKEELS, H. M. A final follow-up study of 100 adopted children. *Journal of Genetic Psychology*, 1949, *75*, 85–125.

SROUFE, A. L. Early experience: Evidence and myth. *Contemporary Psychology*, 1977, *22*, 878–880.

UŽGIRIS, I. Č. Organization of sensorimotor intelligence. In M. Lewis (Ed.), *Origins of intelligence.* New York: Plenum Press, 1976.

WADDINGTON, C. H. *The strategy of the genes.* London: Allen & Son, 1957.

WERNER, E., SIMONIAN, K., BIERMAN, J. M., & FRENCH, F. E. Cumulative effect of perinatal complications and deprived environment on physical, intellectual, and social development of preschool children. *Pediatrics*, 1967, *39*, 480–505.

WERNER, E. E., & SMITH, R. S. *Vulnerable, but invincible: A longitudinal study of resilient children and youth.* New York: McGraw-Hill, 1982.

WERNER, E. E., HONZIK, M. P., & SMITH, R. S. Prediction of intelligence and achievement at 10 years from 20 months pediatric and psychologic examinations. *Child Development*, 1968, *39*, 1063–1075.

WHITE, S. H. Evidence for a hierarchical arrangement of learning processes. In L. P. Lipsitt & C. C. Spiker (Eds.), *Advances in child development and behavior* (Vol. 2). New York: Academic Press, 1965.

WILLERMAN, L., BROMAN, S. H., & FIEDLER, M. Infant development, preschool IQ, and social class. *Child Development*, 1970, *41*, 69–77.

WILSON, R. S. Testing infant intelligence. *Science*, 1973, *182*, 734–737.

WILSON, R. S. Twins: Mental development in the preschool years. *Developmental Psychology*, 1974, *10*, 580–588.

WILSON, R. S., & HARPRING, E. G. Mental and motor development in infant twins. *Developmental Psychology*, 1972, *7*, 277–287.

WILSON, R. S., & MATHENY, A. P., JR. Retardation and twin concordance in infant mental development: A reassessment. *Behavior Genetics*, 1976, *6*, 353–358.

WOHLWILL, J. R. *The study of behavioral development.* New York: Academic Press, 1973.

YARROW, L. J. Research in dimensions of early maternal care. *Merrill-Palmer Quarterly*, 1963, *9*, 101–114.

YARROW, L. J., GOODWIN, M. S., MANHEIMER, H., & MILOWE, I. D. Infancy experiences and cognitive and personality development at ten years. In L. J. Stone, H. T. Smith, and L. B. Murphy (Eds.), *The competent infant: Research and commentary.* New York: Basic Books, 1973.

5 Organization of Sensorimotor Intelligence

Ina Č. Užgiris

Introduction

Infancy is traditionally recognized as a distinct period in the course of human life. Even those who do not view ontogenesis in terms of qualitative transformations in psychological functioning seem to recognize a gap between functioning in infancy and functioning in subsequent age periods. The apparent limitations on self-initiated activity, on physical mobility, and on communication with others during infancy have impressed numerous observers and have led to the conjecture that the infant's world may be quite unlike the world known by adults. Nevertheless, infant functioning has been more often characterized in terms of lack or deficiency with respect to adult abilities than in terms of a coherent mode of organizing action in the world as it exists for the infant. Studies of infant intelligence have been concerned more with charting those infant behaviors that seem to indicate progressive approximation to adult patterns of action or those that seem to document acquisition of concrete information about reality than with any system manifest in the diverse activities of infants.

Research carried out in the past 10 or 15 years has demonstrated that infants possess a remarkable range of abilities, so much so that it has become almost obligatory to comment on the increase in our estimate of infant competence at the beginning of every paper on infancy. This accumulation of evidence on infant abilities for dealing with persons and things in an integrated fashion has encouraged theorizing about the mode of functioning during infancy in its own right, as a

Ina Č. Užgiris • Department of Psychology, Clark University, Worcester, Massachusetts 01610.

distinct form of human functioning, but one that is inherently related to later forms of functioning. Piaget's work has served as a starting point for much of this theorizing.

The model for the development of intelligence proposed by Piaget (1950, 1971a) recognizes the distinctiveness of intellectual activity during infancy but views this intellectual activity as an integral link in the evolution of human intelligence. While noting that the infant's intelligence is "practical"—directed at obtaining results in the environment—Piaget insists that it is characterized by an organization that reveals its filiation with higher forms of intellect. Since the infant does not possess language or a capacity for representation, intellectual constructions during infancy have to be based on perceptions and movements, and hence the characterization of infant intelligence as sensorimotor. Nevertheless, Piaget (1970a) claims that there is a "logic of actions" manifest in the coordinations between schemes (i.e., the repeatable and generalizable aspect of actions) that serves as the foundation for logico-mathematical structures. In particular, a form of conservation revealed in the construction of the object concept and a form of reversibility implicit in the coordination of displacements and positions in space are taken by Piaget to indicate the sensorimotor beginnings of these essential characteristics of operational thought. Thus, among the four periods that Piaget recognizes in the development of intelligence, the sensorimotor period represents the first structuration of intelligence, which subsequently becomes integrated into the higher-level structurations.

The three books dealing with functioning in the sensorimotor period stem from the middle years of Piaget's career (1952, 1954, 1962), yet the general conclusions reached in these works have been reiterated by Piaget in more recent expositions of his theory (e.g., Piaget & Inhelder, 1969). He has presented both an overall view of the changes in intellectual activity during infancy and a more specific account of progress in the construction of such categories of reality as object, space, causality, and time as well as of development in imitation and the capacity for representation. The sequence of the six sensorimotor stages describes the gradual differentiation of the assimilatory and accommodatory processes and their eventual coordination within a relatively stable structure. A critical outcome for the infant is the progressive objectification of reality and, as a complement, the evolution of the awareness of self as an agent in the world.

Piaget's proposal that understanding evolves from an initial state in which objective reality and the activity of the self are undifferentiated has been recently challenged from two positions. Evidence that even very young infants seem to perceive most aspects of the world in the same way as adults do (e.g., discriminating distance, location, and movement) and to relate inputs from different sensory modalities in a systematic manner has lent support to the view that infants do not need to construct an understanding of reality, only to learn to implement it (Bower, 1974). The inflow of sensory stimulation has been said to hold the information necessary for perceiving the world objectively; the task for the infant is only to discover what the world happens to be like (Butterworth, 1981). The studies providing evidence for this view will not be reviewed in the present chapter, since their implications for conceptualizing development in understanding reality are just beginning to be worked out. Much more about the perceptual abilities of infants is known at present than at the time of Piaget's writing, and a number of his statements concerning perceptual functioning need to be revised. Nevertheless, the genesis of understanding remains a separate problem, whatever the perceptual abilities of young infants. Piaget (1977b) asserted that understanding may be prefigured in perception, but that it is not derived from perception. Action in the sense of transformation of perceived reality is held to be essential for development in understanding. The role of the currently better known perceptual abilities of infants in the advancement of intellectual functioning requires consideration from within the framework of Piaget's theory.

The second position from which Piaget's stance has been challenged is represented by work on the social orientation of even very young infants (e.g., Schaffer, 1977). It has been suggested that the infant must have some understanding of the self in order to interact with others in ways that are adjusted to the mutual experience of both partners in an ongoing interaction (Trevarthen, 1980). This precocious ability to engage in well-regulated interactions with other persons suggests a cleavage between functioning in relation to persons and in relation to the rest of the world at the very outset of intellectual development. Whether this disparity in functioning also indicates separate courses in the construction of understanding within the social and nonsocial spheres and whether progress in one sphere leans on progress in the other remain unresolved issues (cf. Lewis & Brooks-Gunn, 1979). The essential question for understanding development in intellectual functioning is whether the singularity of actions in relation to persons contributes to the structuration of intelligence in a distinct way. The evidence that has accumulated in the past few years concerning the abilities of infants to participate in social exchanges will not be reviewed here, but the question of the manner in which the special character of social actions enters the formation of human intelligence must be underlined as central for future work.

Piaget's account of development during the sensorimotor period has been presented in terms of a sequence of six stages. To the extent that the notion of *stage* implies a distinct form of organization in intellectual activity, the issue of congruence in the form or organization evidenced across different domains of functioning and within an interval of time arises with respect to this period. The fact that progress in the various domains has been depicted in terms of the same number of stages suggests a coordination among the transformations in each domain. On the other hand, the claim that intercoordination between schemes constructed in different domains is an achievement of sensorimotor development suggests that congruence among functioning in various domains should be expected only at the culmination of the sensorimotor period.

The issue is complicated by Piaget's use of the term *stage* to mean several things. Most generally, it is applied to the major structurations in the development of intelligence, namely, the sensorimotor, the preoperational, the concrete-operational, and the formal-operational: These will be called *periods* in the present chapter. Reference to these periods as *stages* in the development of intelligence most clearly meets the criteria of hierarchical organization of qualitatively distinct structurations, each of which represents a level of equilibrium closer to a fully equilibrated totality (cf. Pinard & Laurendeau, 1969). Yet the numerous findings that the appearance of achievements in different domains of functioning is not synchronous has stimulated considerable debate regarding the theoretical status of the stage notion (e.g., Brainerd, 1978). In addition, the term *stage* has been frequently applied to three levels in the development of cognitive structures within a given period; this yields the stage of entrance into a period, the stage of transition, and the stage of consolidation. From this viewpoint, fluctuation in the application of newly forming structures is to be expected and the issue of congruence of achievements across different domains of functioning is important only with respect to the stage of consolidation. The disparities usually covered by the term *horizontal décalage* would be expected to disappear once the structuration is fully elaborated. This usage of the term *stage* puts a greater emphasis on the criterion of hierarchization than on wholistic structuration. With respect to the sensorimotor period, it leads to an expectation of congruence in the level of functioning across domains only in the last one or two subdivisions of this period. Finally, Piaget has used the term *stage* to refer to the six subdivisions of the sensorimotor period.

The status to be given to the stages of the sensorimotor period is further complicated by the fact that Piaget has not attempted to characterize their organization in a formal way. He has provided numerous examples of behaviors that he considered indicative of a given stage in sensorimotor development, but he did not provide abstract models for the links between general intellectual functioning (described largely in terms of levels of circular reactions) and functioning in other domains (described in terms of domain-specific phenomena). Consequently, the sensorimotor stages tend to be discussed in the research literature not with respect to structures or levels of organization, but in relation to typical behaviors. The statement by Piaget (1971b) claiming that intellectual activity in the sensorimotor period, as in subsequent periods, reflects overall structures—that is, organized wholes having the properties of self-regulation and transformation—would seem to fit the culmination of sensorimotor development better than it does each of the six stages. Consequently, the congruence of achievement in different domains of functioning will be one question that will guide this examination of research on intellectual functioning in the sensorimotor period. A second question that will be addressed pertains to the criterion of invariant sequence, that is, the attainment of behaviors exemplifying the sensorimotor stages in the expected order. Manifestation of the same organization of achievements across domains and in the same sequence in different populations will be also considered.

A very brief characterization of the six stages in the development of sensorimotor intelligence proposed by Piaget will be presented to facilitate discussion in subsequent sections of this chapter. More complete expositions of his view are readily accessible (e.g., Hunt, 1961). The age ranges cited by Piaget for each of the stages have become widely accepted, although they were not meant to be normative except in the most general way.

Stage I, the use of reflexes, spans the period of birth to about 1 month of age. Although using the term *reflex*, Piaget seems to be interested in those activities of the infant that embody the characteristics of schemes, that is, possess generalizable and repeatable aspects. These reflexes or schemes are thought to be sustained through exercise, thus calling the processes of assimilation and accommodation into play. The infant is seen as active in making contact with the environment, yet at the same time dependent on such contacts for coming to know the world in which he or she exists. Stage II, the stage of primary circular reactions, spans the period of 2 to 4 months of life and is marked by the appearance of new adaptations termed *habits*. The use of schemes is said to lead eventually to somewhat stable modifications in the schemes. These modified schemes, which also come to be repeatedly used, indicate the beginnings of differentiation in the infant's activities. Although some of the behavioral examples resemble instances of conditioning, Piaget insists that the behavioral changes reflect an integration of new elements into existing schemes and not an association between an arbitrary stimulus and some behavioral response. Stage III, the stage of secondary circular reactions, spans the period between around 4 to 8 months of life. It is characterized by Piaget as a transition period between activity based on simple habits and intelligent activity. The assimilation of accidentally produced outcomes engenders repetition of schemes so as to reproduce those outcomes, which entails an accommodation to the circumstances at hand. This is said to lead to new coordinations between schemes as well as to a gradual dissociation between goals and means, since the same means come to be used in more and more varied contexts.

The achievement of coordination between means and end marks the beginning of intelligent activity for Piaget. While the magnitude of the transformation involved in becoming able to intend goals prior to embarking on the means for their attainment was stressed more in Piaget's earlier writings, the reorganization taking place between Stages III and IV, nevertheless, is clearly a major landmark in sensorimotor development. Stage IV, the coordination of secondary schemes and their application to new situations, spans the period between 8 and 12 months. The various schemes constructed at the previous stage come to be coordinated with each other in goal-directed sequences. The important novelty stressed by Piaget is that the coordinations between schemes involved in the goal-directed sequences are freely constructed in each situation, albeit from well-practiced schemes, indicating the differentiation of means from goal. Stage V, the tertiary circular reaction stage, spans the first half of the second year, between about 12 and 18 months of age. This stage is marked for Piaget by an increase in the importance of the accommodatory process, so that schemes are repeated less in order to preserve them unchanged and more so as to incorporate new elements of reality into them. The experimentation observed in infants' activities is described by Piaget both as an interest in novelty and as systematic groping that leads to invention. In goal-directed activities, the infant appears capable of trial-and-error learning, since the scheme serving as means is flexible enough to be varied systematically and, thus, to be adjusted to the goal. Although Piaget insists on interpreting trial-and-error learning as a process directed by both the goal of the action and the scheme serving as the initial means and not as a blind process, he nevertheless assigns considerable importance to such activity within intelligent action.

Although differentiation between means and end achieved in Stage IV is an important mark of intelligent activity, Piaget suggests that from the vantage point of Stage VI, the three preceding stages might be grouped together in that they all reflect adaptation through adjustment to a concrete situation by overt activity, by taking into account the constraints of the situation gradually and only subsequent to acting in it. Stage VI, the invention of new means through mental combinations, is distinguished for Piaget by the fact that new coordinations appear as if through insight, sudden understanding, or some similar process. Rather than viewing it as the sudden appearance of true intelligence, Piaget prefers to consider it a direct outgrowth of development in the previous stages. Beginning around 16 to 18 months of age, the infant is said to become capable of internalized groping—of covert activity—so that the overt action finally engaged in appears already adjusted to the requirements of the situation. One of the most interesting questions concerning sensorimotor development clearly pertains to this process of internalization, to the mechanisms that make possible such abbreviation and condensation, so that previously overt acts come to serve as signifiers of reality and can be acted on covertly. Acquisition of language is thought to facilitate this internal activity greatly, but not to be responsible for its initial appearance. In this way, also, Piaget insists that Stage VI sensorimotor intelligence is but a completion of the accomplishments of previous stages.

Research on intellectual functioning in the sensorimotor period has moved in two directions. One line of studies has taken the phenomena described by Piaget and has subjected them to more rigorous scrutiny in order to determine the variables affecting infant performance as well as to test some implications drawn from Piaget's theory with respect to these phenomena. The great majority of these studies pertain to object permanence and spatial understanding. Their specific findings will not be systematically presented in this chapter, since they have somewhat limited bearing on the issue of organization of sensorimotor intelligence and they are reviewed in readily available sources (e.g., Bower, 1975; Gratch, 1975, 1977; Haith & Campos, 1977). A second line of studies has used systematic assessment procedures to examine a broader range of functioning in more diverse populations. The results from these studies will be reviewed with respect to the questions of sequence, congruence, and universality discussed previously. Then, several explicit proposals for conceptualizing the organization of sensorimotor intelligence, founded at least in part on Piaget's ideas, will be presented together with data that are available.

Assessment of Sensorimotor Intelligence

Scales for Assessment

Albert Einstein Scales of Sensorimotor Development

Following systematic observation of a large sample of infants in New York City, three scales spanning the period of sensorimotor intelligence were constructed by Escalona and Corman (undated), with the help of a number of co-workers. These scales are designed to establish the stage of functioning according to Piaget's theory. The Prehension Scale, Object Permanence Scale, and the Space Scale have been distributed in mimeographed form (no date) and have been used in several studies.

The Prehension Scale spans the first three stages in sensorimotor development, although the items are scored in terms of evidence for Stage II or Stage III functioning. It contains 16 items, which were administered to 51 infants between 5 and 35 weeks of age and to 14 infants studied longitudinally. Corman and Escalona (1969) reported that Green's Index of Consistency (I) is .66 for this scale and commented that none of the infants studied longitudinally manifested behaviors at a higher level before those at a lower level on this scale.

The Object Permanence Scale includes 18 items and spans Stages III through VI. It was administered to 113 infants between 5 and 26 months of age and to 15 infants followed longitudinally. Scored in terms of stages, this scale is reported to be perfectly ordinal ($I = 1.00$). A more interesting finding obtained by Corman and Escalona (1969) pertains to the degree of completion of one stage prior to entrance into the following one. They found progression from one stage to another to be much more gradual than might be expected from the view of stages as sudden transformations. For example, infants entering Stage IV were found to pass, on the average, 82% of the items at the previous stage, but the range was 40%–100%. Similarly, infants entering Stage VI were found to pass, on the average, 74% of the items at Stage V, but the range was 20%–100%. From these data it appears that there is considerable variability in infant functioning across specific situations, if all the items used as indexes of a particular stage are in fact equally appropriate for its assessment.

The Space Scale also spans Stages III through VI and includes 21 items. It was administered to 83 infants between 5 and 26 months of age and to 16 infants studied longitudinally. An invariant order in the attainment of stages was found ($I = .98$), but completion of items at one stage prior to entry into the next was again variable. On the average, infants achieved 76% of items at Stage IV prior to entry into Stage V, but the range was 25%–100%; similarly, infants achieved 56% of the items at Stage V prior to entry into Stage VI, with a range of 20%–100% (Corman & Escalona, 1969). The findings are somewhat complicated by the fact that there is an unequal number of opportunities to show stage-specific behaviors for each of the stages. For example, for the Object Permanence Scale, there are 4 opportunities at Stage III and 11 at Stage VI. Nevertheless, these findings suggest a need for additional studies concerning the correlates and implications of variability in functioning at any one stage even within a single domain. If such variations are meaningful, then scoring infant development by stages glosses over important differences. Until the meaning of these variations is understood, however, scoring in terms of number of items passed, which has been adopted by some investigators, does not seem to be justified. In the latter case, the assumption is made that the greater the number of actions performed at one stage level, the greater the advancement within that stage and (implicitly) the sooner the entrance into the subsequent stage. Such quantitative relationships do not follow clearly from Piaget's theory of sensorimotor development or from Corman and Escalona's study.

These scales do not seem to have been administered to a single sample of infants, and comparisons of stage level congruence across domains have not been reported.

Casati–Lézine Scale

This scale was constructed in France and is described by Casati and Lézine (1968) in a monograph that provides instructions for its administration. Information concerning the performance of 305 infants tested between 6 months and 2 years of age with this scale and two infants tested longitudinally is presented by Lézine, Stambak, and Casati (1969). The scale is composed of four tests, each containing a variable number of items.

The first test concerns object search. It spans Stages III through VI and contains seven items. The items appear to be selected for the purpose of diagnosing each stage and, therefore, are usually assigned two per stage, one to indicate entry into a stage and the second to indicate its completion. Their findings are presented in terms of the percentage of infants passing each item at each month of age. With very few exceptions, a greater percentage of infants was found to pass each item at each successive month of age (the exceptions indicate a reduction by one or two infants). Similarly, the item indicating entry into a stage was passed by a greater percentage of infants than the item indicating completion of that same stage at every month of age, with very few exceptions. Although formal scaling analyses were not performed, it is reported that there were only two individual instances of success on a higher-level item following failure on a lower item.

The second test concerns the use of intermediaries, such as a string, supports (a cloth and a pivoting board), and a rake or stick in order to obtain an object placed out of reach. The four string items span Stages IV and V, while the support and stick items span Stages IV through VI, containing seven and six items, respectively. Again, with few exceptions, a greater percentage of infants was found to pass each item at each successive month of age; the exceptions seem small enough to be due to sample variations. For all of the items used, a greater percentage of infants passed the item marking the entry into each stage than the item marking completion of that stage. Examination of individual protocols revealed four instances of an infant's succeeding on a higher-level item following failure on a lower item.

The six items in the test for exploration of objects span Stages IV through V. The objects used in the test are a mirror and a box. The group results indicate a regular increase in the percentage of infants passing each item with age and a larger percentage of infants passing the lower of the two items marking each stage. In individual records, only one instance of inversion in the sequence was reported.

The fourth test, concerned with combination of objects, is applicable at Stages V and VI. The eight items require the combination of a tube and rake or a tube and a chain. For this test, the increase in percentage of infants passing each item at each successive month of age is less regular, although the higher-level items are passed by fewer infants than the lower-level items at most ages. The assignment of items as indexing Stage V versus Stage VI seems to be the least clear for this test compared to the others.

Each infant in the sample was administered the various tests, so that examination of congruence in stage level across tests was possible. Findings pertinent to this issue have been reported by Lézine *et al.* (1969), but their discussion will be deferred to a subsequent section of this chapter. These tests have been used by other investigators, whose findings provide information concerning the performance of different groups of infants on them.

Scales of Psychological Development in Infancy

A set of seven scales has been constructed by Užgiris and Hunt that is also derived from Piaget's original observations, but the scales are not scored in terms of sensorimotor stage level. These scales had been available in mimeographed form (Užgiris & Hunt, 1966, 1972) before they were published (Užgiris & Hunt, 1975) together with results on the 84 Illinois infants observed in situations contained in these scales. The infants ranged from 1 to 24 months in age and each was examined with all the scales.

The first scale pertains to Visual Pursuit and Permanence of Objects and has 14 steps. It attempts to tap the same domain of functioning as that described previously by the Object Permanence and Object Search scales. Given the central importance attributed by Piaget to construction of the notion of object in the sensorimotor period, it is not surprising that all scales include an assessment of this domain. The second scale pertains to the Development of Means for Obtaining Desired Events and has 13 steps. This scale contains situations similar to those grouped by Casati and Lézine as the Use of Intermediaries and, at the higher levels, is directed to the same domain as their Combination of Objects test. The two scales pertaining to Imitation (Vocal and Gestural) have nine steps each. Although recently there has been increasing interest in imitation during infancy (e.g., Jacobson, 1979; Kaye & Marcus, 1981; Meltzoff & Moore, 1977; Užgiris, 1981), no other imitation scales have been constructed.

The scale pertaining to Development of Operational Causality has seven steps. It is concerned with progress in objectification of causality, which is considered by Piaget to be importantly involved in the construction of reality. Similarly, the scale pertaining to the Construction of Object Relations in Space deals with objectification of space. It has 11 steps and is concerned with the same domain as Escalona and Corman's Space Scale. The scale pertaining to the Development of Schemes for Relating to Objects has 10 steps and is directed at a more general assessment of the differentiation and coordination of schemes. It seems to be related to the Exploration of Objects test included by Casati and Lézine.

These scales were administered twice within a 3-day period to each infant in the sample in order to obtain an indication of stability of infant performance. Reliability was established by having the performance of each infant scored by two observers. The steps in each scale were analyzed for ordinality, and correlations of performance with chronological age were also determined. These results are summarized in Table I.

Other Scales

Two scales grounded in Piaget's theory were constructed by Décarie (1965, 1972). The first scale is concerned with the development of the object concept and has eight items, scored in terms of stage level. It was the first systematic scale to be derived from Piaget's observations. In Décarie's sample of 90 infants between 3 and 20 months of age, no infant was found to pass an item indexing a higher stage after failing on an item indexing a lower stage. The correlation between stage level and chronological age was found to be .86. A scale pertaining to the development of causality has also been reported (Décarie, 1972). This scale is

TABLE I

Summary of Results for the Scales of Psychological Development in Infancy

Scale	Observer reliability (% agreement)	Infant performance stability (% agreement)	Scalability (Green's *I*)	Correlation with age (*r*)
Visual pursuit and permanence of objects	96.7	83.8	.97	.94
Development of means	96.2	75.5	.81	.94
Imitation, vocal	91.8	72.6	.89	.88
Imitation, gestural	95.7	70.0	.95	.91
Development of operational causality	93.7	71.2	.99	.86
Construction of object relations in space	96.9	84.6	.91	.91
Schemes for relating to objects	93.0	79.0	.80	.89

also scored in terms of stage level, and it contains 15 items. Décarie appears to have constructed these scales to meet the needs of her research and not to serve general assessment purposes. An interesting feature of both these scales is that the infant's level of progress may be assessed with respect to persons and with respect to inanimate objects. More recently, Simoneau and Décarie (1979) reported a Test of Object Permanence designed specifically for assessing the early stages in the construction of the object, which are rather sparsely represented in other object permanence scales. In addition, a number of investigators have presented data, most often pertaining to the domain of object concept development, using a series of items expected to be of increasing difficulty without subjecting them to a formal scaling procedure (e.g., Bell, 1970; Corrigan, 1978; Ingram, 1978; Rogers, 1977). Thus, although the various scales are similar, their proliferation makes generalization across studies a difficult task.

A Cognitive Development Scale based on Piaget's work has been described by Mehrabian and Williams (1971). It contains 28 items concerned with observing responses, denotation and representation, object stability, causality, and imitation. These items are organized into a single scale and are scored mostly in terms of pass/fail (a few items have a larger range of scores), but without consideration for stage level. The scale has been administered to a sample of 196 infants from the Los Angeles area, ranging between 4 and 20 months in age. As a result of multidimensional homogeneity analysis, the authors concluded that the entire scale measures a unitary ability, which contrasts with the conclusion of most studies using the previously described scales. Many of the items were found to show a low correlation with age, but a test–retest correlation of .72 was reported for 43 infants retested after an interval of 4 months. This scale is directed toward assessing those cognitive achievements considered particularly relevant for development in representational ability.

With increasing interest in the relation of language to progress in symbolization, a need appeared to systematize ways for determining an infant's level of functioning in the realm of symbol use. Nicolich (1977; McCune-Nicolich, 1981) has proposed a system for scoring play that is closely tied to Piaget's theory and has reported findings indicating the concurrent appearance of symbol use in several domains of functioning. Taking an even broader perspective, Seibert (1980) has devised seven scales (grouped into three clusters) to assess development in social-communication skills during the sensorimotor period. These scales are based on more recent explications of organization in sensorimotor functioning, but they still maintain considerable links to Piaget's theory.

Stage Sequence and Congruence

Invariant Sequence

The expectation of an invariant progression through the sequence of stages in the sensorimotor period has been examined in normal and in retarded populations. Relevant data are evidence of an orderly advance in terms of stages with increasing chronological age and the results of scaling analyses. In general, studies of normal infants support the expectation of an invariant sequence more clearly than those of retarded populations.

In addition to the results already mentioned in describing the assessment scales, several longitudinal studies are relevant to this question. Using the Casati–Lézine Scale, Kopp, Sigman, and Parmelee (1974) tested a sample of 24 infants at monthly intervals between the ages of 7 and 18 months. Although an increase in the percentage of infants achieving each stage level was generally obtained, the progression was considerably less regular than that reported by Lézine et al. (1969) for their cross-sectional sample. In order to examine the variability in performance, Kopp et al. (1974) calculated a percentage of decline for each infant, representing the percentage of tests on which the infant scored lower than in the previous testing. For all infants combined, the percentages of decline were not very high, but they did occur, reaching 20% at 15 months of age and ranging between 14% and 16% for tests occurring between 13 and 18 months of age. The authors state that declines across stage boundaries were less frequent but were observed as well.

Similarly, longitudinal observations carried out on Baoulé infants from the Ivory Coast have revealed regressions in the stage of functioning determined through the Casati–Lézine Scale (Dasen, Inhelder, Lavallée, & Retschitzki, 1978). The children were tested every 3 months during the second year of life; about 30 children contributed to the analyses. Regressions were most frequent between the completion of Stage IV and entrance into Stage V, when the children were between 13 and 15 months of age. The authors comment that the greatest number of children refused to participate in the examinations when they were 15 months of age, suggesting that some of the regressions might be due to inadequate assessment. Regressions occurred most frequently in the Use of Intermediaries—Support test and least frequently on the Exploration of Objects test, although no single pattern was dominant in the data, with some children showing regression on only one of the tests and others on several tests, both in the same and in different examinations.

A comparable pattern of results has been obtained in a longitudinal sample of 12 United States infants assessed with the Užgiris–Hunt Scales (Užgiris, 1973a). Regression at the time of transition was quite common: An infant would perform in a situation at a higher level than previously, would drop back in the next session, and would again perform at the higher level at a subsequent session and continue at the higher level from there on. There were few exceptions from the postulated order of attainment for the sample as a whole. The order for a single pair of two adjacent steps appeared to be inverted for the whole sample on each of the scales, with the exception of Object Permanence and Causality. These inverted pairs of steps were not concentrated at any one age period. In addition, individual infants were observed to attain the higher of two adjacent steps first on almost every scale, although these irregularities appeared to be due to idiosyncratic factors in that no more than one or two infants showed any particular inversion. Since these scales are not scored according to stage level, each irregularity should not be interpreted as indicating variability in stage progression. It must be remembered that Corman and Escalona (1969) found little consistency among infants as to which items indexing a particular stage they achieved prior to entry into the next stage. To the extent that adjacent steps on the Užgiris–Hunt Scales may be considered to index the same sensorimotor stage, variability in order of achievement need not be troublesome.

In a different report on apparently the same sample of infants mentioned previously, Kopp, Sigman, and Parmelee (1973) presented the results of scalability analyses for three of the Casati–Lézine tests. With respect to the Object Search test, the reproducibility index I was found to be acceptably high at each of five ages (9, 12, 15, 18, and 20 months), although at 12 months of age it reached only .75. Similarly, for the Use of Intermediaries—Strings, $I = 1.00$, except at 15 months of age, when it reached only .65. However, for the Use of Intermediaries—Support, I implied a high level of scalability only at 9 and again at 20 months of age, being particularly low at 18 months. Dasen *et al.* (1978) also found the greatest frequency of regressions on the Support items, suggesting that this test from the Casati–Lézine Scale may have specific problems. In general, the findings from the longitudinal studies indicate that although there is a definite order in sensorimotor attainments, individual infants do backslide in concrete test situations more at some ages than at others and more in some domains of functioning than in others.

A study conducted by Miller, Cohen, and Hill (1970) investigated the effect of order of item presentation on infant performance using the Užgiris–Hunt Object Permanence scale. Three age groups of infants were presented with appropriate items from this scale either in the order of difficulty (the usual administration procedure) or in a random order. Each age group was administered a predetermined subset of the items. With the constricted range of test situations, the scalability analysis did not yield an acceptable I coefficient for any of the age groups in either of the two administration conditions. Examination of their graphs suggests that the largest deviations from the postulated sequence occurred on two tasks, which might be due to a different interpretation of those tasks by these investigators.

In a replication study carried out by Kramer, Hill, and Cohen (1975), a single subset of test situations was presented to infants tested three times over a period of 6 months. Green's I coefficient indicated acceptable scalability for this subset of items and was considerably higher for the ordering of the items given in the Užgiris–Hunt Scale than for the ordering proposed by Miller et al. (1970). Furthermore, 80% of the infants whose level of performance changed across the three sessions progressed in the expected sequence.

The question of the invariance of the sequence of progression in retarded populations has been addressed almost exclusively through scalability analyses. When severely retarded groups are examined, including older children and adults, the appropriateness of the test situations that have been designed to suit young infants is questionable. Moreover, to the extent that actions characteristic of a given level of functioning disappear when a new level of understanding comes to guide action, a simple scoring for the presence or absence of actions characteristic of each stage or level in sensorimotor functioning will not produce an ordinal scale. Different means have to be adopted to demonstrate the presence of the achievement aspect of the earlier levels of functioning. Studies of retarded populations to date have not dealt with this problem by devising different test situations; usually, behaviors indicative of each level of functioning have been scored as present or absent using the situations designed for infants. Sometimes specific scales have been omitted from analyses because they were thought to include achievements that are merged into others at higher levels of functioning.

Kahn (1976) assessed 63 severely retarded children between $3\frac{1}{2}$ and $10\frac{1}{2}$ years of age on the Užgiris–Hunt Scales and performed scalogram analyses on the data for all scales with the exception of the Schemes scale. Green's I coefficients ranged between .81 for the Spatial Relations scale and 1.0 for the Imitation scales, indicating a sequence of progression highly similar to that observed for normal infants. In contrast, a study by Silverstein, Brownlee, Hubbell, and McLain (1975) on 64 institutionalized retarded children with a mean chronological age of 14 years obtained much lower ordinality. The Object Permanence and the Space scales constructed by Escalona and Corman and the corresponding scales from the Užgiris–Hunt instrument were used. Green's I coefficients were acceptably high for the scales evaluating understanding of object permanence (.58 and .70, respectively) but were much lower for the two scales concerned with spatial understanding (.46 and .30, respectively), suggesting either a different sequence of attainment for this population or a lack of systematic progression. A more complete review of sensorimotor functioning in retarded populations has been prepared by Kahn (in press).

It must be emphasized once more that an invariant order of achievement is claimed only for the sensorimotor stages and not for samples of tasks that might be selected to index one or more of these stages. In fact, examples of backsliding in the behavior of individual infants on specific tasks may be found among Piaget's observations as well. Variation in performance from one day to another and from one task to another is to be expected; what seems to be needed is better explication of criteria for determining the consolidation of a stage, subsequent to which backsliding within the same domain ought not to occur except in very unusual circumstances. However, as long as particular performances in a single situation are treated as equivalent to stage achievement, findings of regressions and inversions for individual infants are bound to be reported.

Stage Congruence across Domains

The classification of sensorimotor functioning into domains has been somewhat arbitrary. Theoretically, each domain should pertain to activities based on a specifiable cognitive structure. The domains individually discussed by Piaget were chosen on theoretical grounds; however, their structuration and types of interdependence were suggested rather than explicitly analyzed by him. Those who have grounded their work in Piaget's observations have either accepted Piaget's classification of domains or have arrived at more delimited groupings of competencies on some intuitive basis. Consequently, the theoretical meaningfulness of comparisons across domains so variously formed may be questioned, since findings of lack of congruence would have different implications depending on whether the two domains might be said to rely on a single structure. A more formal characterization of structures and levels of organization in sensorimotor intelligence may facilitate both the evaluation of tasks presented to infants and of findings regarding congruence in performance on such tasks. With these limitations in mind, the available research on stage congruence will be examined.

One of the first attempts to examine congruence in sensorimotor development across domains was carried out by Woodward (1959), not on infants, but on severely mentally retarded children. A total of 147 children ranging up to 16 years in age were administered a variety of tasks described by Piaget, which were scored in terms of stage level. Overall, Woodward reported that success on two tasks classified as indicative of the same stage was significantly related for this sample. In addition, when the children were assigned to a stage of sensorimotor development on the basis of their performance on various problem-solving tasks (items similar to coordination between schemes and to the use of means or intermediaries in other scales) and also to a stage in object concept development, considerable consistency in stage level was obtained (87% of the children were classified at the same stage in both). Less consistency was obtained when stage in problem-solving was compared to stage based on type of circular reactions shown (43% were classified in the same stage in both). Given the considerable variability in level of development in different domains that has been observed in children attaining concrete operations, Woodward's findings are impressive. One may be tempted to attribute these results to the nature of her population, which might be viewed as no longer advancing, having been somehow arrested in intellectual development.

A similarly impressive degree of stage consistency has been reported for Down's syndrome infants. Cicchetti and Mans (in press) assessed 43 infants longitudinally at 5, 9, 13, 16, 19, and 24 months of age using the Užgiris–Hunt Scales, from which specific items were converted into stage designations. The degree of stage congruence across the four domains represented by the scales of Object Permanence, Means-Ends, Causality, and Relations in Space varied with the age of the infants. At 5 months, the percentage of concordance was 71%; it dropped to 58% at 9 months and 50% at 13 months. At 16 months, the percentage of concordance was 54%, but rose to 73% and 79% for the 19-month and 24-month assessments, respectively. In view of the fact that congruence was measured across four domains, the extent of agreement obtained is quite high.

A lower degree of congruence across domains was reported by Dunst, Brassell, and Rheingover (1981) in their study of 143 retarded youngsters falling into three mental age groups: 3–8 months, 8–12 months, and 12–18 months. The performance of these children on the Užgiris–Hunt Scales was converted to a stage designation following the procedure described by Dunst (1980). The mean percentage of stage agreement for all 21 pairwise comparisons of the seven scales was 39% for the lowest mental age children, 28% for the middle group, and 31% for the highest mental age children. For individual pairs of scales, percentage of stage congruence ranged between 2% and 64%. A comparable degree of congruence was obtained by Rogers (1977) for a sample of institutionalized retarded children assessed for sensorimotor Stages II through VI by means of items devised for that purpose or adopted from other scales, representing the domains of object permanence, causality, space, and imitation. An overall average of 29.6% of stage agreement was reported. Since the method for determination of stage level from observed performance has varied considerably across studies, the inconsistency in results may be due to differences in stage determination as well as to differences in populations studied.

A substantial degree of stage consistency across domains has been reported also by Lézine et al. (1969) for their sample of French infants. Comparing stage congruence across their tests taken in pairs, they found an overall percentage of concordance to be 71%, ranging from 58% to 87% for individual pairings of tests. For example, the stage on Object Search was equivalent to that obtained in the Use of Intermediaries—Strings for 74% of the infants, equivalent to that obtained in Object Exploration for 72% of the infants, and equivalent to that obtained in the Use of Intermediaries—Rake-Stick for 59% of the infants. Among the discrepancies found, over 80% amounted to a discrepancy by a single stage. It may be of interest to note also that infants tended to be advanced most frequently on the Object Search test. In addition, the researchers carried out such analyses for various age groups separately; it turned out that stage congruence was more prevalent at some ages than at others. Overall, there was highest congruence at 9 and 10 months of age and lowest congruence at 14 and 15 months of age, with the percentage of concordance generally lower in the second year of life. This observation complements the finding of Cicchetti and Mans (in press) as well as that of Kopp et al. (1974) that the greatest instability in performance on longitudinal testing occurs at around 15 months of age. It is intriguing to consider this instability as evidence for a period of transition and reorganization prior to the consolidation achieved at Stage VI.

Findings reported by Dasen *et al.* (1978) indicate an even higher degree of stage congruence on the Casati–Lézine Scale for the Baoulé infants. Their sample of 63 infants ranged between 4 and 19 months at the start of the study. The infants were examined more than once, generally between three and five times, at 3-month intervals; consequently, individual children contributed unevenly to the data. For all pairwise comparisons of tests, the average percentage of stage congruence was 79%, with a range between 59% and 96%. The results are not given separately for different ages, making it impossible to determine whether congruence was higher toward the end of the sensorimotor period.

As the preceding account indicates, stage congruence has been empirically tested by looking at the incidence of functioning at the same stage level in different domains at some moment in time. Such congruence might be expected when stage consolidation has taken place, but would not necessarily hold during the course of development. The pattern of relations in stage achievement across domains has not been considered in the research literature. Wohlwill (1973) pointed out some time ago that in the course of development, one domain might consistently lead another, they may reciprocally influence each other, or they may gradually diverge. Knowledge of the pattern of relations between various domains would be of no less value than knowledge of the degree of strict concordance for understanding the organization of sensorimotor functioning. The available data have been considered by Curcio and Houlihan (in press) in relation to the different types of across-domain relations described by Wohlwill.

A few studies have looked at stage VI consolidation across different domains. To look at the relationship of language development to level in sensorimotor functioning, Zachry (1978) used the Užgiris–Hunt Scales, but changed the scoring to a stage level system. For his sample of 24 infants, ranging in age between 13 and 23 months of age, he found considerable variation in stage level across the six domains tapped by these scales. An interesting finding reported by Zachry pertains to the scale combinations on which infants scored at Stage VI. All 22 infants who reached Stage VI were at Stage VI on Object Permanence; thus, the four infants who scored at Stage VI on only one scale did so on Object Permanence. All infants who scored at Stage VI on one additional scale added Relations in Space. Thereafter, however, the combinations of scales were more varied. Nevertheless, a Guttman scalar analysis suggested an orderly sequence in Stage VI development as follows: Object Permanence, Relations in Space, Imitation, Causality, and then Means-Ends. These results imply that there may be orderly consolidation at the Stage VI level. Some confirmation for this order is provided by data on retarded children. Kahn (1975) found that the greatest number of children in his sample had attained Stage VI in Object Permanence, slightly fewer had attained Stage VI in Means-Ends and Causality, but considerably fewer had done so in Imitation. For Down's syndrome children at 24 months, Cicchetti and Mans (in press) reported that most had attained Stage VI in Object Permanence, Relations in Space, and Means-Ends, but few had done so in Causality or in Vocal and Gestural Imitation.

The approach to the issue of common structuration across domains by the method of comparing agreement in stage level presents some difficulties. If the theoretical point that stage level is to be inferred from overall performance and not equated with passing of a test item is taken seriously, it becomes clear that an independent determination of stage for each domain is difficult with the assessment instruments available. Some investigators have turned to correlational approaches in order to study consistency of achievement across different domains without having to make stage determinations. Correlations, however, by focusing on the relative rank of individual children in a group entail a fundamentally different approach to this issue (for a discussion, see Curcio & Houlihan, in press).

A correlational analysis of scores on the Užgiris–Hunt Scales (Užgiris, 1973a) attained by infants studied longitudinally indicated few substantial correlations for pairs of scales at any single month of age between 2 and 24 months. The only notable clusters of correlations were obtained between scores on Object Permanence and Relations in Space scales for the period of 5 to 8 months of age, and between scores on Relations in Space and the Means-Ends scales for the periods of 4 to 6 months and, also, 14 to 19 months. Even taking into account the small size of the sample, these results are at variance with expectation. One problem with this type of correlational analysis is that chronological age is used to group infants. If it is assumed that infants progress at different rates and that congruence across domains may be expected only at times of stage consolidation, then carrying out such analysis by age would tend to obscure any meaningful pattern in the results. In fact, examination of infants grouped in terms of their Object Permanence scores suggested more interesting patterns for the same data.

In a study by King and Seegmiller (1973), infants living in Harlem were administered the Užgiris–Hunt Scales at 14, 18, and 22 months of age. The performance of these infants was compared across domains at these three age levels. The results are again subject to the limitations of analysis by chronological age. All correlations were rather low; the greatest number of significant correlations was obtained at 14 months of age. The Object Permanence scale scores were significantly related to those on Relations in Space, Means-Ends, Causality, and Gestural Imitation scales. In addition, scores on the Relations in Space scale were significantly correlated with scores on Schemes, Causality, Means-Ends, and Gestural Imitation. Similarly, scores on the Causality scale were significantly correlated with scores on the Schemes scale, and scores on the Means-Ends scale with those on Gestural Imitation. The one scale that did not correlate with any of the others pertains to Vocal Imitation. On the other hand, at 18 and 22 months of age there were practically no significant intercorrelations among the scales. The almost single exception was the correlation between scores on the Schemes scale and Vocal Imitation at 18 months and again at 22 months.

The suggestion that grouping infants by chronological age may mask developmental relationships is supported by the correlational results obtained by Kopp et al. (1974) for the Casati–Lézine Scale. Although considerable stage level congruence has been reported for these tests, Kopp et al. obtained only eight significant correlations for all the age levels analyzed, and no more than two at any one age level. Most of the significant correlations involved items from the Use of Intermediaries tests. In addition, even where significant correlations have been obtained, the magnitude of the correlations has been quite low.

In a longitudinal study in which 25 infants were tested at 14, 18, and 22 months of age with the Užgiris–Hunt Scales and a Foresight Scale, Wachs and Hubert (1981) found very low intercorrelations among scale scores at each of the ages used and little change in the magnitude of the correlations with age. On the basis of these studies with normal infants, it seems that the strategy of correlational analysis is not successful in demonstrating even those interdependencies between achievements in different domains that seem to be indicated by the stage congruence analysis.

In contrast, correlational analyses carried out on samples of retarded infants provide somewhat stronger support for interdependence
between functioning in different domains. Kahn (1976) calculated intercorrelations among scores on the seven Užgiris–Hunt Scales for his
sample of retarded children and found all 21 correlations to be significant, ranging between .43 and .93. In the sample studied by Dunst *et al.*
(1981), mentioned previously, not all of the intercorrelations were significant: About two-thirds of the 21 correlations computed were significant
for the children in the 3- to 8-month mental age group and only about
half were significant for the children in the two higher mental age
groups. The magnitude of the correlations was also lower than in Kahn's
study, averaging .41 for the first group, .30 for the 8- to 12-month mental age group, and .28 for the 12- to 18-month mental age group. An
interesting pattern of increasing interdependence between domains
over the sensorimotor period was obtained by Cicchetti and Mans (in
press) for Down's syndrome infants. Since the Schemes scale was not
given, 15 intercorrelations were computed for every age of testing. At 5
months, only 3 intercorrelations were significant, but for the testings at
9, 13, and 16 months, around 10 intercorrelations were significant, and
at 19 months, 14 of the 15 intercorrelations were significant. The magnitude of the correlations was also moderately high at this testing. It is
feasible to view this pattern as evidence for consolidation of sensorimotor intelligence at around this age. At the 24-month testing, intercorrelations were again lower, reaching significance in 7 of the 15 comparisons,
but it is accounted for by the authors as a ceiling effect producing reduced variability. If a slower rate of progression characterizes retarded
populations, consistencies in their functioning may be more readily revealed through correlational analyses.

Interrelationships between sensorimotor domains have been also
studied by the techniques of factor analysis and cluster analysis on measures of interdependence obtained for the Užgiris–Hunt Scales. Silverstein, McLain, Brownlee, and Hubbell (1976) used the intercorrelation
matrix with age partialed out for the Užgiris and Hunt (1975) sample as
well as the intercorrelation matrix for the 14-month testing of the King
and Seegmiller (1973) sample, and subjected them to hierarchical cluster
analysis and then to principal components factor analysis. Three very
similar factors were obtained for the two samples by both methods. The
Causality, Relations in Space, and Schemes scales formed one cluster;
Object Permanence, Means-Ends, and Gestural Imitation scales formed

the second cluster, and Vocal Imitation stood apart from the other scales. Two other studies using similar techniques noted that the clustering of the scales changes with the age of the children contributing to the intercorrelation matrix.

Wachs and Hubert (1981) performed a principal components factor analysis on their 14-month, 18-month, and 22-month data separately. At the youngest age, Schemes, Causality, and Relations in Space scales defined the first component, matching one of the clusters obtained by Silverstein *et al.* (1976). Means-Ends and Gestural Imitation scales loaded positively and the Vocal Imitation scale loaded negatively on the second component. The third component was defined by the Object Permanence scale, loading positively, and their Foresight scale, loading negatively. Except for the loading of Object Permanence scale on a separate component, the grouping of the scales is relatively similar to that obtained in the analysis by Silverstein *et al.* At 18 months, however, a different grouping of the scales was obtained. The first component was defined by Object Permanence, Means-Ends, Relations in Space, and Vocal Imitation scales. The second component was defined by the Causality and Foresight scales loading positively, with the Means-Ends and Schemes scales loading negatively. The Gestural Imitation scale defined the third component. At 22 months, four components were extracted. The Object Permanence scale and their Foresight scale loaded highest on the first component, Means-Ends and Relations in Space scales defined the second component, Schemes and Vocal Imitation scales defined the third component, while Causality and Gestural Imitation scales loaded on the fourth component. The grouping of Vocal Imitation with Schemes at 22 months corroborated King and Seegmiller's (1973) finding of a significant correlation between these two scales for their 22-month group.

Hierarchical cluster analyses were also performed by Dunst *et al.* (1981) on their sample of retarded children grouped by mental age. Different groupings of the scales were obtained for the three levels of mental age. At the 3- to 8-month level, Object Permanence and Means-Ends scales formed one cluster, the Vocal Imitation scale stood alone, and the remaining four scales were joined in a third cluster. At the 8- to 12-month level, Object Permanence, Means-Ends, Schemes, and Relations in Space scales formed one cluster; Gestural Imitation and Causality scales were joined in a second cluster; and the Vocal Imitation scale again stood apart. At the 12- to 18-month level, with the exception of the continued independence of the Vocal Imitation scale, the grouping of the other scales was quite changed. Gestural Imitation and Relations in Space scales formed one cluster, while the Causality, Schemes, Object Permanence, and Means-Ends scales formed another cluster.

Because of the numerous differences in samples and procedures, comparison across studies is not feasible. The only moderately consistent finding is that the Vocal Imitation scale is most independent of the other scales. Otherwise, these analyses indicate that the Užgiris–Hunt Scales are neither uniformly interrelated nor completely distinct, but the nature of the interdependencies at various levels in development needs further study. In this regard, the outcome is similar to findings for other measures of infant intelligence. A scalogram analysis of the Bayley Mental Scale test (Kohen-Raz, 1967) also produced several ordinal sequences rather than a unitary scale. Moreover, McCall and his collaborators (McCall, Hogarty, & Hurlburt, 1972; McCall, Eichorn, & Hogarty, 1977) demonstrated a component structure in psychometric test data for infants that changes with the age of the infants. A central issue that clearly cannot be resolved solely through empirical analyses is the identification of the domains of functioning that should be distinguished within the sensorimotor period.

Progression with Age

Age Sequence

The clearest result that has been obtained with all the different assessment scales is the regular relationship of progress along the scales to chronological age. All who constructed the various sets of scales to assess sensorimotor development have reported high and significant correlations of scores with age (Casati & Lézine, 1968; Corman & Escalona, 1969; Décarie, 1965; Užgiris & Hunt, 1975). Subsequent studies have also generally obtained regular increments in scores with age, at least for samples of infants taken as a group (e.g., Dasen et al., 1978; King & Seegmiller, 1973; Lewis & McGurk, 1972; Paraskevopoulos & Hunt, 1971; Užgiris, 1973a). Some exceptions have been found only in protocols of individual infants tested repeatedly. Nevertheless, since these results have been obtained in studies using different tests, different populations of infants, and even scoring procedures such that items presented as alternative indicators of a stage were added up to give a quantitative score, the results should be considered robust. They satisfy the minimal validity requirement for these scales.

Rate of Progression with Age

Another question frequently addressed in the literature pertains to the correlation of individual scores over age. The expectation of high correlations presupposes either uniform rates of progress or stable rates for individuals. This question is foreign to a Piagetian view of development. Nevertheless, it seems to arise from the attempt to transpose the scales that have been devised to assess the level of sensorimotor functioning into psychometric devices, which are grounded in a different set of assumptions (see, e.g., Furth, 1973; Užgiris, 1973b; Užgiris & Hunt, 1975). A number of studies have examined correlations between scores at one age level and scores at some subsequent age level, with the expectation that those infants who attain the highest scores at the earlier age should be the ones to attain the highest scores at a later age. If, however, developmental progress is viewed as discontinuous, there is little reason to expect such correlations, especially if achievement is scored in terms of stage level—that is, the initially more advanced infant might be extending the range of functioning at some given stage level during the intertest interval while the initially less advanced infant is reaching that stage level, so that at the second testing, both may be at the same stage, reducing any interage correlation. Similarly, unless it is proposed that coordination of each stage is equally rapid, there seems little reason to expect similarities in rates of progress through them. In fact, the evidence appears to indicate a lack of consistency in rate of progress over age.

King and Seegmiller (1973), in the study already referred to, correlated the scores obtained by infants at 14 months with their scores at 18 and 22 months on the Užgiris–Hunt Scales. By and large, they found nonsignificant correlations between scores on the various scales over the intervals used. Aside from the questions that may be raised about seeking such correlations on conceptual grounds, it should be pointed out that the age range chosen in this study was not optimal, since coordination of sensorimotor intelligence is thought to be accomplished somewhere between 16 and 24 months. Since the scales are designed to assess presence of various coordinations in sensorimotor functioning and not the ease or the breadth of their application, they would not be likely to detect many individual differences between 18 and 22 months. For the interval between 14 and 22 months, there was little predictability for individual scales, but correlation for a total score based on all the scales was moderate ($r = .56$) and significant.

Negligible interage correlations were also obtained by Lewis and McGurk (1972) for scores on the Escalona and Corman's Object Permanence scale. For infants tested at 3, 6, 9, 12, 18, and 24 months of age, they obtained significant correlations only between scores at 3 months and scores at 12 and 18 months. The authors concluded that relatively more advanced functioning at an early age was not predictive of functioning at a later age. Using the Užgiris–Hunt Scales, Wachs (1979) also found negligible predictability of scores on individual scales over intervals varying between 3 and 12 months during the second year of life. Object Permanence scores for male infants were an exception, correlating significantly across all age intervals. The general picture is, however, of little consistency in the rank order of individual infants over several-month intervals during the first 2 years of life.

The longitudinal study of Down's syndrome infants carried out by Cicchetti and Mans (in press) revealed a closer relationship between relative advancement at one age and that at a later age, particularly during the second year of life. Scores on the Užgiris–Hunt Scales at 5 months of age in most cases did not correlate with scores at later ages, while scores at 13 months correlated significantly with scores at subsequent testings in 30 of 36 comparisons. Nevertheless, the magnitude of the correlations was only moderate. In general, however, the possibility that there might be regularities in progression through developmental cycles has not been adequately investigated, since the available studies have not differentiated between items indexing entrance into a stage and items marking extension of functioning within the same stage when calculating interage correlations. It would seem most plausible to expect a relationship between ages for achieving entrance into successive stages and possibly between ages for consolidating successive stages.

Influence of Environmental Conditions

Although rapidity of progress through the stages of sensorimotor intelligence is not a question of central concern for a stage theory of development, the correlation of specific environmental conditions with achievement of given levels of functioning is of interest, since such correlations provide leads concerning the prerequisites for particular achievements. The effects of environmental conditions on intellectual functioning in infancy have been considered in several recent papers (e.g. Super, 1981; Užgiris, 1977; Wachs, in press), so that only studies using Piagetian assessment instruments will be mentioned here.

In her study of object concept development, Décarie (1965) included in her sample infants from three different backgrounds: their natural homes, foster homes, and institutions. However, since the effect of the infant's circumstances on development was not a major concern of the study, only minimal characterization of the conditions in the foster homes or the institution was provided. Whatever the actual differences between these environments, they did not account for much of the variance in object concept ranks (over 3%), even though their contribution was statistically significant.

Golden and Birns (1968) were more directly interested in the influence of environmental conditions and compared home-reared infants from three socioeconomic groups on Escalona and Corman's Object Permanence scale. No significant differences in achievement were obtained for any of the age groups tested (12, 18, and 24 months). Since the range of scores was similar for each of the socioeconomic groups, it does not seem that a ceiling effect was responsible for these findings. Similarly, Cobos, Latham, and Stare (1972) reported that 292 infants tested with the same Object Permanence scale in Bogota showed no significant differences in performance when contrasted by social class. Also, these infants did not differ significantly from the New York sample. On the other hand, malnourished infants had significantly lower scores after the age of 10 months when compared with well-nourished youngsters from the same population. The specific scores at different ages have not been reported for this study.

A related study by Wachs, Užgiris, and Hunt (1971) used an early version of the Užgiris–Hunt Scales to examine groups of infants at 7, 11, 15, 18, and 22 months of age from two socioeconomic levels. On the Object Permanence scale, a significant difference in favor of the higher-socioeconomic-level infants was obtained only at 11 months of age. Thus, this finding does not contradict the report of Golden and Birns (1968). However, on the Means-Ends scale, a significant difference was obtained in favor of the higher-socioeconomic-level infants for every age group except the 15-month group. The Vocal Imitation scale was administered only to the three oldest groups, but significant differences were obtained on it for every age group tested. However, on the Schemes scale, there were no significant differences by socioeconomic level for any of the age groups. However, the most interesting aspect of this study pertains to correlations obtained between a variety of specific characteristics of the home environments and the achievements of this sample of infants as a whole. The authors concluded that intensity of stimulation, variety of changes in the environment, and exposure to spoken language were the most significant dimensions of the environment with regard to infant achievement at this period.

Wachs (1976, 1979) has also conducted a longitudinal study of 39 infants living at home. Detailed observations of the home environments were made, starting when the infants were 11 months of age and continued at regular intervals during the second year. The Užgiris–Hunt Scales were administered at 12, 15, 18, 21, and 24 months of age. Correlations between achievement level on the various scales and the home environment rating during the preceding 3 months revealed a number of significant relationships. Wachs specified four categories of environmental characteristics that appeared to be particularly relevant to early intellectual functioning: (1) the predictability and regularity of the environment of the child; (2) adequacy of stimulation offered the child; (3) intensity of stimulation (related negatively), when accompanied by lack of possibilities for withdrawal from it; and (4) verbal stimulation. The importance of these findings lies not in the demonstration of the influence of environmental conditions, but in the demonstration that different characteristics of the infant's environment are related to specific achievements at different periods of the infant's life. Unless one assumes that all the important characteristics of environments are positively interrelated, such findings support the notion that rates of progress shown by individual infants should vary from one age interval to the next.

A comparison of the rates of progress in the sensorimotor period for infants from different cultural environments indicates only coarsely the effects of circumstances, without specifying the experiential factors linked to those effects. Nevertheless, such data are relevant to the issue of modifiability of sensorimotor development. The Casati–Lézine Scale has been administered to infants of comparable age in France (Lézine et al., 1969), the Ivory Coast (Dasen et al., 1978), the United States, and India (Kopp et al., 1974; Kopp, Khoka, & Sigman 1977). The outstanding result is the similarity of functioning across these cultural groups. In comparing the performance of the 83 Indian infants with that of 71 United States infants, Kopp *et al.* (1977) obtained no significant differences for infants between 7 and 8 months of age; between 9 and 10 months, the United States infants scored higher on Object Search and both Intermediaries tests (String and Support), but these differences were attributed by the authors to the infants' response to the test situation. Among the 11- to 12-month infants, United States infants again scored higher on the Object Search and Intermediaries—Support tests. When the range of achievement in terms of stage level was considered, there was greater variability among the Indian than the United States infants.

Dasen *et al.* compared their results for the Baoulé children directly with Lézine *et al.*'s results for French children. With one exception, all the significant differences were in favor of the Baoulé children. These differences were particularly evident on the Combination of Objects test and the Intermediaries test, excluding the Support items. The one exception was that the French infants were able to deal with multiple strings earlier, a task that involves discrimination in the face of potentially misleading cues. Clearer differences were found in a sample of 23 matched pairs of Baoulé children among the well-nourished and the moderately malnourished members. The adequately nourished children scored higher overall, particularly on the Object Search and Intermediaries tests, and more so in the second year of life at Stage V and Stage VI levels. These indications that not all aspects of sensorimotor functioning are uniformly affected by a set of environmental conditions reinforce the conclusion drawn by Wachs (in press) regarding the specificity of environmental influences.

The studies described so far suggest that a broad range of environmental conditions provide sufficient support for development in the sensorimotor period that differences between cultural groups are minimal. In contrast, the research by Hunt and his collaborators (Hunt, 1980; Hunt, Mohandessi, Ghodssi, & Akiyama, 1976; Paraskevopoulos & Hunt, 1971) on infants reared in institutional settings demonstrate both the marked lag in various achievements that can result from certain types of rearing and, through their modification of the rearing conditions, some of the types of experiences that may be important for early development. A particularly interesting implication of this work is that early cognitive development is highly dependent on experiences gained in interpersonal interactions with others, a position also supported by

studies linking caregiver sensitivity and responsivity to intellectual functioning (e.g., Clarke-Stewart, 1973; Donovan & Leavitt, 1978).

Correlations with Other Measures

Many considerations complicate a direct comparison of measures of
sensorimotor development with other measures of infant intellectual
functioning. Even when the comparisons are done contemporaneously,
the conceptual grounds for expecting a parallel ranking of individuals on
different measures are not clear. Assuming that both measures provide
valid assessments, either a biological inclination for rate of achievement
of competencies or some overlap in tasks has to be postulated in order to
predict a high correspondence between various types of measures. To
the extent that different infant competencies are differentially dependent on the presence of environmental supports, specific opportunities,
and specific previous learnings as well as individual proclivities, and to
the extent that such specific supports and opportunities vary freely,
high correspondence between achievements need not occur. In addition, if the measures of sensorimotor development in fact index the level
of reorganization of actions, while the psychometric devices assess degrees of competence along dimensions that have been found to detect
individual differences, they may produce largely unrelated rankings of
infants. It has been pointed out in earlier sections of this chapter that a
conception of sensorimotor development in terms of transformations in
overall structures does not demand complete congruence of functioning
within different domains, particularly during the earlier stages. There
seems to be even less reason to expect congruence between assessment
methods that use different criteria to determine advancement. The complications are further increased when the comparisons between assessments are made over time, since either the importance of intervening
experiences must be discounted or they must be assumed to be comparable for every infant in the group.

Given these considerations, the general pattern of low correlations
between different assessment measures is unremarkable. Lewis and
McGurk (1972) correlated scores on Escalona and Corman's Object Permanence scale and Bayley's Mental Development Index for six age levels
between 3 and 24 months of age. At 6 months, the correlation between
these two measures was significant ($r = .60$), but there was essentially no

relationship at all subsequent testings. Similarly, Golden and Birns (1968) obtained a significant correlation between performance on the Cattell and on the same Object Permanence scale at 12 months of age ($r = .24$), but the correlations between these same measures were not significant for infants at 18 and 24 months of age. King and Seegmiller (1973) correlated infants' scores on individual Užgiris–Hunt Scales with their scores on Bayley's Mental Scale and on Bayley's Psychomotor Scale separately. At 14 months of age, Bayley's Mental Scale correlated significantly with most of the measures, except for Object Permanence and the two Imitation scales (rs between .32 and .42). The significant correlations with Bayley's Psychomotor Scale were fewer and somewhat lower. On the other hand, the different measures were practically unrelated to each other at 18 months. At 24 months, scores on Means-Ends, Schemes, and Vocal Imitation scales correlated significantly with Bayley's Mental Scale scores.

An interesting aspect of such studies is that correlations between sensorimotor functioning level and intellectual functioning measured by psychometric tests between 2 and 3 years of age seem to be stronger than concurrent correlations, although the magnitude of the correlations is still moderate. Wachs (1975) related the performance of 39 normal infants assessed on the Užgiris–Hunt Scales at 12, 15, 18, 21, and 24 months of age to their Stanford–Binet scores at 31 months. He found that at each successive age level, a greater number of the scales were significantly related to the child's score on the Stanford–Binet, with Object Permanence showing the most consistent relationship. The specific scales providing the best prediction varied with the age of assessment.

A similar study was conducted on a sample of 80 preterm and 68 full-term infants (Siegel, 1979). These infants were assessed repeatedly at 4, 8, 12, and 18 months with the Užgiris–Hunt Scales and at 30 and 36 months on the Stanford–Binet. A number of significant relationships were found. The total score on the Užgiris–Hunt Scales for each assessment was significantly related to the 30-month and 36-month Stanford–Binet score, but the magnitude of the correlations ranged between .29 and .45 for the various time intervals. Of the individual scales, Object Permanence, Means-Ends, Relations in Space, and Vocal Imitation correlated most consistently with scores on the Stanford–Binet, especially for the testings during the first year. Siegel (1981) has also reported significant correlations between scores on the Užgiris–Hunt

Scales and scores on the Bayley Scales administered at 2 years of age for the same sample of infants. It seems reasonable to speculate that the greater variability in performance produced by the inclusion of preterm infants in the sample and the size of the sample were responsible for these results.

A number of studies recently have searched for relationships between sensorimotor attainments and various aspects of language development. A discerning discussion of the literature has been prepared by Bates and Snyder (in press). Two broad lines of endeavor can be distinguished: Some investigators seem interested in determining the sensorimotor achievements prerequisite for communication and speech, while others are using relative standing on sensorimotor scales as a general predictor of language competence at some later age. A relationship between Stage VI on at least some of the Užgiris– Hunt Scales and language use has been reported in studies of retarded populations (Kahn, 1975) as well as between the achievement of Stage V level, particularly on the Means-Ends scale, and the beginnings of intentional communication in studies of both normal and impaired infants (Bates, Camaioni, & Volterra, 1975; Curcio, 1978; Harding & Golinkoff, 1979; Steckol & Leonard, 1981). Significant correlations between the Užgiris–Hunt Scales given as early as 4 months of age and the Receptive and Expressive Emergent Language (REEL) scale scores have been also found (Mahoney, Glover, & Finger, 1981; Siegel, 1981). The question of the relationship between sensorimotor intelligence and development in symbol use is clearly important, but in order to clarify it, studies that include measures of pretend play, imitation, and communicative abilities in addition to the more traditional sensorimotor assessments are required.

FORMALIZATIONS OF THE ORGANIZATION OF SENSORIMOTOR INTELLIGENCE

Introduction

It has been stated already that a more formal characterization of the levels of sensorimotor intelligence than has been explicitly presented by Piaget would be desirable for studying sensorimotor functioning in infancy. Evaluation of the adequacy of various tasks as indexes of some one level of functioning would be facilitated, as would evaluation of functioning in different domains, which are usually assessed by diverse tasks. More importantly, such characterization would focus attention on the process of change from one level to another in terms of specific mechanisms rather than general variables of growth and experience. Beginning steps in this direction have recently been taken.

For conceptualizing levels in sensorimotor functioning, it seems useful to bear in mind a distinction made by Piaget (1970b) between experience that leads to knowledge of physical reality and "logicomathematical" experience—that is, experience that is derived from the subject's own actions on objects and that abstracts the relations imposed on objects. Although these two types of experience are fused at the outset, they become more distinct during the sensorimotor period. In describing infant actions, the word *scheme* has been used to refer to the pattern inherent in actions without distinction as to the type of experience resulting from them. Thus, *scheme* sometimes refers to activities such as sucking, shaking, or throwing, by means of which the infant may be expected to discover the actual properties of various objects, and sometimes to activities such as joining, ordering, or exchanging, from which the infant may be expected to abstract relations present in coordinations of actions irrespective of the objects employed. The recent work by Langer (1980, 1981) demonstrates that painstaking observation of infant activities can reveal examples of schemes directed toward physical knowledge as well as of schemes directed toward operational constructions, both of which undergo systematic change in the direction of greater complexity during this period. A facet of this distinction has been elaborated by Piaget (1977a) in his discussion of the types of compensations that seem possible at different stages in sensorimotor development.

The characterizations of levels in sensorimotor functioning that attempt to go beyond Piaget's original description have not progressed to the point of tying the formalizations of structural features to specific assessment procedures and to collection of data bearing directly on these formalizations. For purposes of illustration, my characterization of four levels in sensorimotor functioning will be presented together with some observations that suggest the types of activities expected to be seen at each level. The data come from a longitudinal study of 12 infants, who have been observed at regular intervals between the ages of 1 month and 2 years using the Užgiris–Hunt Scales (Užgiris, 1973a, 1977).

In a subsequent section, other recently proposed formalizations of development in sensorimotor intelligence will be considered.

Four Levels in Organization of Actions

My description does not address the type of actions shown by the neonate. Piaget's account includes a level (Stage I) characterized by actions that partake of external reality as global totalities but are not deliberately directed toward it, consistent with his position on the initial nondifferentiation between subjective and objective reality. Since my observations did not cover the neonatal period, the question of the type of actions characterizing this period is left open.

Level of Simple Unitary Actions

At this level, roughly analogous to Piaget's Stages II and III, infant actions have the character of single-unit behaviors, lacking evidence of internal differentiation. Although they are directed toward external objects (in the broadest sense of the term), these objects have importance only insofar as they permit or hinder the flow of activity. In terms of the equilibration process (Piaget, 1977a), when disturbances occur, they are dealt with primarily by cancellations rather than by integration of modifications into the system of actions. Knowledge of object properties appears to be limited and the infant's repertoire of schemes is applied to objects indiscriminately. Shifts from one scheme to another occur in rhythmic alternation without coordination between successive schemes. Among the infants observed in the longitudinal study, the predominant actions on objects during this period consisted of mouthing, looking, and some variety of shaking or banging; however, these activities were not combined with each other in a regulated way. When they followed each other, the shifts seemed to be due to incidental factors. In addition, these schemes were applied to one object at a time, that is, the infant did not construct relations between objects in such a way that there would be the opportunity to abstract relations from his or her own actions.

The transition to the next higher level seems to be linked to three parallel developments. First, application of different schemes to the same object in close succession or simultaneously, as the infant begins to monitor visually his or her own behavior, may be expected to substantiate objects as reality elements at the intersection of diverse schemes, each object providing particular resistences to assimilation by the different schemes. This development may be observed in the appearance of the "examining" scheme (Užgiris, 1969). All 12 infants observed longitudinally engaged in examination of objects prior to showing behaviors that would place them at the next level in sensorimotor development. Second, interaction with responsive objects, whether inanimate objects designed to provide regular but varied inputs (such as musical toys) or persons who regularly respond to the infant's vocal and facial overtures, may be expected to promote a primordial differentiation of self as agent. Again, all the infants in the longitudinal sample were observed to make attempts to reproduce interesting spectacles and to evolve "procedures," that is, actions that are transplanted from one concrete situation to another when the infant is confronted with interesting events during this period. In analyzing traditional infant tests, McCall *et al.* (1972) also isolated a main component at 6 months of age that was heavily defined by items concerned with exploration of test materials or with production of perceptual consequences through action. Third, execution of different schemes in close succession even on different objects may be expected to lead to the coordination of actions through the simple relation of "joining." Once such joining of schemes appears, the construction of two-unit actions and means–end relations becomes feasible. Piaget (1977a) has also emphasized the conjunction of schemes achieved through secondary circular reactions in promoting compensatory regulations.

Level of Differentiated Actions

The entrance into this level may be equated with the transition to Stage IV in Piaget's theory. Piaget (1977a) suggests that true compensating regulations first appear at the Stage IV level due to the increased differentiation between actions and objects. This level is characterized by the appearance of subcomponents in actions, achieved through coordination of the simple schemes in means–end relations. The infant's repertoire of schemes at first simply expands and then is more discriminately applied to objects in accord with their diverse characteristics, as success or failure to attain goals provides feedback to differentiate the actions. Specifically, the repertoire of schemes composed of mouthing, grasping, looking, and hitting expands to include a variety of manipulations such as shaking, striking, dropping, stretching, tearing, crumpling, and so forth. Similarly, on expansion of the repertoire of schemes, the initial combination of schemes through joining gives way to more diverse interrelations between schemes. Means–end differentiation in behavior fosters selective joining of schemes and the implicit coordination in terms of "ordering" or "inclusion"; the combination of reciprocal schemes allows the discernment of the "inverse" relation. Such sequences of actions may be observed in attempts to remove obstacles, to employ intermediaries in the attainment of goals, to explore properties of objects, and also, to explore spatial and causal relationships between objects. Coordinated sequences of actions appear first with respect to single objects (e.g., letting fall and picking up, crumpling and straightening) and then with multiple objects (e.g., putting one object into another and taking it out, building a structure up and knocking it down). The repetitive execution of many infant activities may be considered practice of these coordinations.

A number of observations from my longitudinal sample of infants illustrate functioning at this level. With respect to object concept development, while correlations in the age of achievement between one step in the sequence and other steps were not high overall, indicating little individual consistency in the rate of progress, correlation between the age of beginning search for an object completely covered by a screen and the extension of search to a number of separate screens was fairly high. For my sample, the beginning of search for a partially covered object (which does not require a differentiated action) correlated slightly (r = .31) with the beginning of search for a completely covered object; however, the beginning of search for a completely covered object correlated substantially ($r = .76$) with search for an object under one of two or one of three screens, and moderately ($r = .53$) with search for an object following a series of visible displacements.

A number of associated achievements appeared in other domains during the interval spanned by the extension of object search from one to a number of locations (for my sample, an interval of four months, on the average). In terms of schemes shown in relation to objects, 10 of the 12 infants were not actively dropping objects and observing their fall at the beginning of this interval, but 11 of the 12 were engaging in this activity at the time they searched for objects following a series of visible displacements. The dropping activity seems to involve a two-component action, since the object is released by the infant to move through space and is then visually relocated. Similarly, while none of the 12 infants showed differentiated schemes at the beginning of this interval, 11 of the 12 engaged in such activities at the end. Moreover, adults who interact with the infant may have an important role in facilitating the construction of more complex actions. Many of the games played with infants are such that the adult partner performs one link in a sequence of actions by the infant, thereby completing the organization of the activity. For example, in the early dropping of objects and exploration of their movements through space, infants depend on someone else for the retrieval of those objects. Later, infants construct sequences of dumping and retrieval on their own, taking over the segment previously carried out by the partner. It may be important to study the extent to which sensitivity to the structurations being worked on by infants and willingness to fill in appropriate gaps during interactions with them forms a part of "good mothering," reflected in the correlation between rate of infant development and maternal "interest" or "responsiveness to baby" typically found in studies concerned with environmental influences on early development.

In the domain of imitation, 10 of the 12 infants in my sample did not imitate sound patterns familiar to them at the beginning of this interval, but only 2 did not do so by the end. Also, none of the 12 infants imitated complex actions composed of familiar schemes at the beginning, but only 5 did not do so by the end. The development in imitation at this level seems to involve the taking of the modeled action as a goal, with an attempt to construct appropriate actions to reproduce it; hence, it may be treated as requiring the capacity for differentiated actions. Instances of an infant reproducing only some aspect of the modeled action are quite instructive. In another domain, while 11 of the infants did not, at the beginning of this interval, make use of relationships between objects (such as that of support) all did so by the end. It seems that a number of achievements in different domains requiring a differentiation of action from goal in order to commence at least a two-component action appear in parallel and, thus, indicate the beginning of the structuration of actions. In studying the taking into possession of multiple objects, Bruner (1970) observed the appearance of the strategy of setting one object in reserve while picking up another in infants of roughly the same age as the age of infants at this level in my sample. Many of the games played with infants of around this age require the ability to coordinate sequences of actions (Gustafson, Green, & West, 1979).

What does not occur at this second level is an immediate modification of actions as a result of their outcome. The schemes appear to function still as elementary units; the coordinations observed take place between available schemes. If a constructed sequence of actions is unsuccessful, the infant's ability to modify the sequence seems limited to a fresh construction of another sequence from schemes already in the repertoire, without an immediate modification of any single scheme in the sequence. The compensations, therefore, are not yet of the type that integrate the disturbing element into the scheme of action (Piaget, 1977a). Thus, while outcomes do regulate actions, the regulation is global, leading to a reconstruction of the whole action sequence rather than to a specific modification of a component in the sequence. In novel problem situations, attainment of solutions through gradual approximation or from observation of demonstrated solutions does not seem to take place.

Level of Actions Regulated by Differentiated Feedback

This level in sensorimotor development is characterized by the beginning of a specific regulation of actions by their outcomes and, thus, by an adjustment of actions (involving more than one component) so as to attain various goals. In terms of the equilibration process, clear instances of beta-type compensations occur (Piaget, 1977a).

Even global responsivity to feedback from outcomes evident at the previous level may be expected to promote objectification of reality and an increasing selectivity in actions. Toward the beginning of the second year, several changes in infants' behavior toward objects is evident; there is a reduction in the number of schemes applied to a particular object, and there is a conventionalization of the schemes applied. Probably due to awareness of the social outcomes of actions (i.e., differential adult responses to hitting and hugging of dolls or to mouthing and rolling of toy cars) as well as to progress in imitation of novel actions, the activities of infants begin to reflect culturally favored modes of interacting with objects, particularly those objects that are miniaturizations of things important in adult life. The responsivity to outcomes in the social domain may be expected to contribute to self–other differentiation, especially with respect to self and others as agents capable of instigating action. This differentiation may be observed in "showing-sharing" schemes, through which the infant manifests both an expectation for responsiveness from another and a recognition that such responsiveness has to be instigated by the other person. Increase in the variety of relations through which schemes come to be coordinated contributes to construction of sequences of action composed of multiple-component schemes. This in turn seems to distance the schemes from particular contexts of application and to increase the flexibility of coordinations into which they enter. In short, individual schemes may be said to become more context-free as well as more mobile action components.

One of the important coordinations established at this level involves the relation of compensating adjustment, which might be considered an outgrowth of the inverse relation between two schemes established at the previous level. The regulation of actions by outcome in a differentiated way involves making compensating adjustments. Likewise, for the regulation to be differentiated rather than global, characteristics of reality that are independent of the infant's actions have to be understood. Once feedback is differentiated, even in novel problem-solving situations, available schemes can be adjusted in a graded manner in coordination with other schemes. Thus, action sequences seem to be no longer constructed and reconstructed as totalities, but tend to be modified in their components to suit the demands of the situation. The trial-and-error behavior, or "groping" (to use Piaget's term), exhibited by infants in problem-solving situations may be taken as evidence of such compensating regulation. The attainment of behavioral regulation by differentiated feedback may be expected to promote imitation of novel events and improvement in skilled actions.

The most prominent feature of various achievements at this level is the alteration of actions through gradual approximation, the taking into account of correspondence between outcome and goal. The appearance of regulation by outcome is most clearly evident in the infant's activities in relation to objects, in imitation, and in means behavior. For example, at the time when infants begin to search for an object in several locations or under a number of screens, almost none show activities with objects that indicate a social influence on their actions. However, very soon afterwards the social influence on their actions becomes evident, so that instead of examining, banging, dropping, and stretching, they begin to push toy cars around, to hug dolls or cuddly animal toys, to build with blocks and to put necklaces around their necks—the actions specific to each object. Furthermore, at about this time varied relationships between objects—for example, one serving as an extension of another—begin to be exploited in obtaining desired goals. The use of these relationships is often achieved through a gradual modification of initially ineffective actions.

In regard to imitation, the infant's behavior indicates not only an attempt to reconstruct the model by means of known schemes but also the ability to recognize failure to reproduce the model accurately. Infants generally imitate unfamiliar gestures visible to them more readily when they are performed in relation to an object than when presented as a gesture. For instance, sliding a piece of paper back and forth is imitated more readily than the sliding action alone. This suggests that an act with a definite result may facilitate reproduction by providing an outcome for the infant to work toward. The successive modifications that the infants show in gradually approximating the model are also illuminating. This level corresponds roughly to the appearance in the domain of object construction of search for objects hidden by means of an invisible displacement in one or a number of locations—for my sample, between 14 and $15\frac{1}{2}$ months, on the average. At the appearance of search for an object hidden by an invisible displacement under one of three screens, 10 of the 12 infants regularly showed socially influenced activities with objects. Similarly, at the same point the majority of infants (7 of the 12) began to imitate novel sounds presented by the examiner directly, after a period of responding to them by varied vocal responses. All infants attempted to imitate novel visible gestures by means of gradual approximation at the time that they began to search for an object hidden by an invisible displacement, although only 5 of the 12 imitated novel gestures directly. In addition, most infants at this point seemed to recognize centers of causality in objects and, therefore, learned from demonstrations how to activate objects or produce spectacles, frequently through gradual approximation.

Thus, while a diversity of new achievements may be observed to occur in parallel at this level, including a great deal of specialization in action, the appearance of regulating relationships between outcomes and actions (with compensating modification of successive actions) seems to be the main advance in structural organization.

Level of Anticipatory Regulation of Actions

This level is assumed to manifest the culmination of sensorimotor intelligence and to bridge the practical intelligence of infancy and the representational intelligence of early childhood. Since the equilibration process is not only compensatory but also formative (Piaget, 1977a), the adjustments constructed at the previous level would be assumed to be retained in the intercoordinations of schemes, in other words, through schemes for relating actions.

The beginnings of language may be seen as both an outgrowth of and a further instrument for the coordinated intellectual functioning of this period. The appearance of make-believe play may be taken as evidence for decontextualization of schemes in that their application to objects is no longer governed exclusively by the evident properties of the object. Anticipatory coordination of schemes would account for both inference and evocation of absent aspects of reality. The extensive network of coordinations between schemes constructed at the previous level of functioning may be thought to allow for the regulation of sequences of actions, in many instances prior to overt execution of those sequences, and prior to feedback from actual outcomes. Thus, sequences of actions appear to be coordinated not through after-the-fact compensations, but by means of anticipatory compensations, resulting in already adapted overt behavior. This change may be seen as analogous to the change that takes place during skill acquisition in a task such as target tracking from compensatory to anticipatory regulation, so that the perfected skill seems to be governed by foresight.

The notion of object attained at this level of functioning may be thought to stand for the invariance inherent in a group of spatial displacements and a coordinated network of schemes applicable to objects. Accordingly, the infant may be expected not only to conserve the object through a series of inferred displacements, but also to be sensitive to the identity/nonidentity of objects. The ability to coordinate actions by means of anticipatory regulations may be expected to promote more articulated self–other distinctions and to lead to increased appreciation of others as spontaneous and less predictable (in contrast to inanimate objects) centers of action. The self-consciousness and wariness sometimes observed in children at this level of development may be taken as a manifestation of such abilities.

If the achievement of the highest level in object construction, that is, the beginning of search for objects following a series of invisible displacements with the ability to reconstruct the path of an object in reverse, is taken as an index for this level of development (21–23 months of age on the average, in my sample), a number of parallel achievements in other domains may be noted. By the time the infants in my sample began to search for objects following a series of invisible displacements, all were engaging in activities showing a social influence, all were imitating novel sounds directly, and all were imitating novel visible gestures directly. They were beginning to imitate novel invisible gestures, that is, facial gestures that one cannot see oneself perform and, thus, cannot obtain direct visual feedback about the outcome of the attempt. However, by the time these infants began to reconstruct the path of the object in reverse, 9 of the 12 were imitating even invisible gestures directly. Similarly, at the same time, 9 of the 12 infants were imitating new words regularly, something they were just starting to do at the beginning of this level.

In regard to activities with objects, the naming of objects in recognition appeared, an activity that should be distinguished from the use of verbal labels, since it involves examination of an object and use of the name to express the possibilities suggested by the object, often in the context of interacting with another person and communicating this understanding to the other person rather than making a request. Of the 12 infants, 10 were naming objects in recognition by the time they constructed a reverse path in their search for objects. In addition, their activities with objects frequently revealed evocation of objects or events not within their perceptual field at the moment (e.g., shown in symbolic play). Furthermore, if they found direct access to a desired object blocked, the majority of infants by this time constructed detours through space in order to reach their goals. Evidence for the construction of solutions to problem situations without overt groping was also obtained at this level: All infants began to show foresighted behavior in one problem situation (necklace and container) and 8 of the 12 did so in another (the solid ring). It seems that the multiplicity of correspondences established between multicomponent action sequences and differentiated outcome evaluations come to make possible at this level the nonovert adaptation of actions.

These four levels in sensorimotor intelligence have been described in terms of the types of action sequence that are constructed by the infant and in terms of the adjustments that seem to be made in such sequences as a result of their outcome. These two aspects imply sensitivity to characteristics of reality and to characteristics of sequences of action. The interweaving of understanding regarding the properties of objects and understanding derived from coordination of actions seems to be involved in transition from one level to another. Working with an independent set of data and a different methodology, McCall *et al.* (1977) have described development in mental functioning during infancy in terms of a series of five stages, finding considerable parallels between their Stages II–V and the four levels described for my longitudinal sample.

An attempt to obtain direct support for these four levels was made by Wachs and Hubert (1981) in a study referred to previously. In analyzing the performance of their sample of 25 infants on the Užgiris–Hunt Scales by means of principal components factor analysis, they related the steps of the scales loading on the first component at each age of assessment to the organization of actions characteristic of one of the levels in my formalization. At 14 months, the first component was taken to be defined by actions fitting Level 2: Differentiated actions. At 18 months the first component seemed to be defined by actions matching Level 3: Regulation by feedback. At 22 months, loadings on the first component were related to Level 4: Anticipatory regulation. A more complete description of the type of actions shown in the situations presented that were taken as indicative of each level would have been desirable as well as inclusion of a broader range of situations for observing infant actions. The set of scales for studying development in communication competence that have been constructed by Seibert (1980) in terms of these four levels will eventually provide another source of data regarding this formalization.

Other Formalizations of Levels in Sensorimotor Development

Two other formalizations of levels of functioning during the sensorimotor period have recently been proposed in the context of more comprehensive theories of cognitive development (Case, 1978; Fischer, 1980). Both theories acknowledge Piaget's influence, but in fundamental respects they present a different conception of development. They retain the notion of hierarchically organized and qualitatively different periods in development (linked to those delineated by Piaget) and propose a number of distinct stages within each period. However, the stages are presented as structurally very similar within each period, differing mostly in the elements that are being structured and, thereby, in the type of cognitive functioning that is possible. A fundamental difference from Piaget's conception seems to lie in these theories' weighting the elements being structured more than the organizational totality in determining the possible functioning. Piaget's notion of *vertical décalage* contained the idea of recycling through similar structurations at each period in development, but the stages within each period were never placed in exact correspondence. The theories also differ from Piaget in their account of transformation from one stage to another.

Fischer (1980) presents cognitive development as taking place in recurring cycles (or tiers) of four levels each. The first tier is the sensorimotor; the next, the representational; and then, the abstract tier. The levels are defined in terms of the type of skill that can be controlled. Skills are conceived as units of behavior that involve a set of acts and the relations between those acts. In the sensorimotor tier, the skills consist of sensorimotor actions.

Level 1, that of simple sets, has a structure of single elements (sets). The sets consist of acts that are adjusted to some aspect of the environment; primary-circular-reaction-type activities typify these sets. At the first level, the infant is said to be unable to control relations between sets. Level 2 is characterized by mappings between simple sets, for instance, means–ends mappings, and thus, involves unidirectional relations between two sets. Here the infant can control one action in order to bring about another. Both secondary-circular-reaction-type activities and Stage IV actions, in which means are differentiated from ends, seem to typify this level. Although Piaget and most investigators who have studied development in infancy mark a clear transition at the point of differentiation of means from end, Fischer seems to deemphasize this transition. The third level is described as a sensorimotor system, involving a reversible relation between two sets. The activities characterizing this level resemble Stage V tertiary circular reactions described in Piaget's theory and include sensorimotor classification activities with bidirectional relations. Level 4 is both a culmination of sensorimotor development and the beginning of the representational tier. From the perspective of the earlier levels, it is characterized as a system of sensorimotor systems, achieved by the construction of a reversible relation between two systems of Level 3. From the perspective of the representational tier, it constitutes a simple representational set. At Level 4, the child is said to be able to represent the properties of objects, events, or people independently of his or her own actions. The same structural relations (sets, mappings, systems, and systems of systems) characterize the four levels in the other tiers.

Skills are claimed by Fischer to be specific to domains of functioning. Consequently, progress through the hierarchical levels is said to take place fairly independently in various domains. The issue of congruence across domains in level of functioning, therefore, does not arise. However, the definition of what constitutes a domain remains unresolved (as Fischer acknowledges), and therefore, there is a danger of accounting for every discrepancy in performance found on tasks indexing the same level as a domain difference. Empirical evidence with regard to this theory as applied to the sensorimotor period is also lacking. Once empirical work demonstrates how the types of relations between sets are determined from observations of infants' activities, it will be possible to gauge the productivity of this theory better.

Case (1978) has presented his theory as neo-Piagetian, but in substance it seems to differ more drastically from Piaget's approach than does Fischer's theory. Although four major periods corresponding to Piaget's are identified, the changes within each period are conceived in terms of information processing that can be carried out and the load that such processing places on memory space. The structures are conceived as strategies that increase in complexity and power. As in Fischer's theory, the sequence of substages maintains a formal similarity across the four periods but differs in content. In the sensorimotor period, the strategies are motoric, while in the preoperational period they are representational, and at the next period they are logical. The substages within each period are defined by the child's ability to handle one additional feature at each successive substage.

There are five substages in the sensorimotor period. Substage 0 is labeled "Reactive Exercise" and corresponds to Piaget's Stage I. Substage 1: "Isolated Centration" corresponds to Piaget's stage of primary circular reactions and seems to match Fischer's Level 1. It is defined by the ability to store one item of information, usually a satisfying action or an interesting input. Substage 2: "Relational Centration" is characterized by the ability to center not only on the infant's action, but also on the relationship between the action and some consequence in the external world. Thus, two items are being stored. Case indicates that secondary-circular-type activities belong at this substage. Substage 3: "Birelational Centration" is characterized by actions that produce interesting results indirectly. The removal of obstacles and the activation of agents are cited as examples; hence, this substage seems to correspond to Piaget's Stage IV. The three items stored are the infant's action, the relation of this action to a second action, and the consequence that obtains. Substage 4: "Birelational Centration with Particularization" is defined by the apprehension of the specific relations in a sequence of actions. The use of various intermediaries characterizes this substage. The child's action on the intermediary, its consequence, the action on the goal, and the consequence of that action are the four items that are stored. Representational ability evident at Stage VI in Piaget's theory is discussed by Case as the first substage of the preoperational period.

The preliminary nature of this formulation is acknowledged by Case. It evidently represents a downward extension of his formalization designed for the concrete operational period and rests on a translation of observations on infants made by investigators holding other perspectives to fit the formal categories of his theory. He suggests that memory sets limits on the complexity of the strategy that can be assembled by the infant, so that memory growth in addition to specific experience is an important factor in the transition from one substage to another. What seems to be most clearly lacking to make this theory useful for studying development in the sensorimotor period is a set of independent procedures to determine the infant's memory capacity and a way to compute the memory load of specific tasks.

These formalizations require additional elaboration to permit a full appreciation of their differences and similarities, their applications in research contexts, and their implications.

Summary and Conclusions

The available literature on infant functioning during the sensorimotor period demonstrates that an orderly sequence of achievements is manifested in the formation of a number of competencies. The regularity of these sequences and their high correlation with chronological age are among the most consistently reported findings. However, with respect to a number of other questions, the evidence is far from definitive.

The lack of clarity about the notion of stage with respect to sensorimotor intelligence in the empirical literature makes it difficult to evaluate the findings. If a stage progression is posited, then questions about transition from one stage to the next must be separated from questions pertaining to modification in functioning within a given stage. However, since determination of stage level is often based on performance in a particular task, the separation of stage transition from change in performance on a task becomes difficult. Moreover, investigators vary in their choice of tasks for indexing stage level, and most do not try to present their rationale for choosing a particular task. The greatest uniformity obtains in tasks used to assess stages in the construction of the object. In addition, the transition to a higher stage is neither sudden nor fully realized in an instance, at least according to the Piagetian conception of stages. A determination of stage level by means of a single task presented once is, therefore, suspect. Attention to whether a child's performance indicates initial accession to a stage level, inconsistent functioning at that level, or consolidated achievement would be helpful in evaluating stage-related findings across different studies.

There seems to be considerable consensus that development in sensorimotor functioning is multifaceted and must be assessed in more than one domain. However, the delineation of domains, the choice of domains to be used in assessment, and the choice of tasks within those domains are rarely explicated clearly. The domains represented in the different scales that have been constructed to assess progress in sensorimotor functioning vary from scale to scale. Even when the same domain appears to be represented within two assessment scales, the tasks included to tap functioning within that domain differ. Thus, it is well nigh impossible to deal with the question of congruence in the level of cognitive functioning across domains. The one apparent consistency concerns the reported relative advance in object concept development as compared with other domains of functioning (e.g., Cicchetti & Mans, in press; Dasen *et al.*, 1978; Lézine *et al.*, 1969; Užgiris, 1973a). Since a multifaceted conception of sensorimotor intelligence contrasts with the unitary view of intellectual progress adopted by most psychometric tests for infants, the problem of domain definition seems to deserve further effort.

The more explicit formalizations of the organization of sensorimotor intelligence that have been recently proposed are all preliminary in nature, but they should facilitate a more systematic analysis of various tasks used to assess sensorimotor functioning. Unless the requirements of the tasks are understood in a formal way, linkage of the assessment procedures to the theories of intellectual development is dubious. Without such a tie, the status of infant performances in these assessment measures becomes no different from that of the various items contained in the standardized infant intelligence tests. Even more important is to continue linking assessment to the ways that are used by infants to deal with the situation presented; dichotomization in terms of success or failure is insufficient. The manner of compensating for disturbances, of dealing with contradictions, and of handling memory demands is highly relevant to determining an infant's level of functioning.

The issue of ways to conceptualize individual differences in functioning within Piaget's theory is not unique to the study of the sensorimotor period; it is critical only because much of our interest in infant functioning is related to interest in individual variation. The most frequent method for examining individual differences is by means of rates of progress. However, if sensorimotor intelligence is the foundation for subsequent intellectual development, it will be coordinated and then reorganized by most children at an early age. Are differences of a few months in achieving some level of functioning of real interest? On the other hand, an expectation that the rate of progress shown at one level will be continued at the next cannot be sustained. There are no theoretical reasons for expecting stability in rate of progress, and all the empirical evidence points to the contrary. A different way of thinking about individual differences would be to determine the domains or even specific contents in which an individual excels, as well as the solidity of such excellence, that is, the breadth of excellence over diverse tasks. Similarly, individuals may differ in their readiness to reorganize their level of functioning on exposure to special situations providing supports for higher-level functioning, that is, there may be differences in openness to developmental change in the sense of Vygotsky's (1978) proximal zone of development. It is important to search for ways to think about individual differences that would be meaningful in the context of Piaget's theory, since the problems of the normative approach have been sufficiently revealed with other types of assessment procedures.

The newly evident appreciation that intellectual functioning during infancy has a distinct form that needs to be understood in its own right augurs well for future work in this area. Knowledge of the genuine abilities of infants is important as a first step in conceptualizing the structure of a system that would allow for those abilities and their orderly change. Yet there is a danger in extrapolating too readily from models constructed on the basis of knowledge about intellectual functioning at higher levels of development, since sensorimotor functioning has some truly distinguishing aspects. Although the linkage of functioning at the sensorimotor level with functioning at higher levels of development is highly significant, ties between the character of sensorimotor functioning and the constraints of contextual reality are no less significant. Understanding the sensorimotor mode of functioning presents the challenge of explicating relations between ways of operating on contents and the contents themselves.

REFERENCES

BATES, E., & SNYDER, L. S. The cognitive hypothesis in language development. In I. Č. Užgiris & J. McV. Hunt (Eds.), *Research with scales of psychological development in infancy.* Urbana, Ill.: University of Illinois Press, in press.

BATES, E., CAMAIONI, L., & VOLTERRA, V. The acquisition of performatives prior to speech. *Merrill-Palmer Quarterly,* 1975, 21, 205–226.

BELL, S. M. The development of the concept of object as related to infant attachment. *Child Development,* 1970, 41, 291–311.

BOWER, T. G. R. *Development in infancy.* San Francisco: Freeman, 1974.

BOWER, T. G. R. Infant perception of the third dimension and object development. In L. B. Cohen & P. Salapatek (Eds.), *Infant perception* (Vol. 2). New York: Academic Press, 1975.

BRAINERD, C. J. The stage question in cognitive developmental theory (with open peer commentary). *The Behavioral and Brain Sciences,* 1978, 2, 173–213.

BRUNER, J. The growth and structure of skill. In K. Connolly (Ed.), *Mechanisms of motor skill development.* New York: Academic Press, 1970.

BUTTERWORTH, G. *Structure of the mind in human infancy.* Paper presented at the meetings of the International Society for the Study of Behavioral Development, Toronto, August 1981.

CASATI, I., & LÉZINE, I. *Les étapes de l'intelligence sensori-motrice.* Paris: Les Editions du Centre de Psychologie Appliquée, 1968.

CASE, R. Intellectual development from birth to adulthood: A neo-Piagetian interpretation. In R. Siegler (Ed.), *Children's thinking: What develops?* Hillsdale, N.J.: Lawrence Erlbaum, 1978.

CICCHETTI, D., & MANS, L. Sequences, stages, and structures in the organization of cognitive development in Down's syndrome infants. In I. Č. Užgiris & J. McV. Hunt (Eds.), *Research with scales of psychological development in infancy.* Urbana, Ill.: University of Illinois Press, in press.

CLARKE-STEWART, K. A. Interactions between mothers and their young children: Characteristics and consequences. *Monographs of the Society for Research in Child Development,* 1973, 38, No. 6–7 (Serial No. 153).

COBOS, L. F., LATHAM, M. C., & STARE, F. J. Will improved nutrition help to prevent mental retardation? *Preventive Medicine,* 1972, 1, 185–194.

CORMAN, H., & ESCALONA, S. Stages of sensorimotor development: A replication study. *Merrill-Palmer Quarterly,* 1969, 15, 351–361.

CORRIGAN, R. Language development as related to stage 6 object permanence development. *Journal of Child Language,* 1978, 5, 173–189.

CURCIO, F. Sensorimotor functioning and communication in mute autistic children. *Journal of Autism and Childhood Schizophrenia,* 1978, 8, 281–292.

CURCIO, F., & HOULIHAN, J. Varieties of stage organization in the sensorimotor functioning of normal and atypical populations. In I. Č. Užgiris & J. McV. Hunt (Eds.), *Research with scales of psychological development in infancy.* Urbana, Ill.: University of Illinois Press, in press.

DASEN, P., INHELDER, B., LAVALLÉE, M., & RETSCHITZKI, J. *Naissance de l'intelligence chez l'enfant Baoulé de Côte d' Ivoire.* Berne: Hans Huber, 1978.

DÉCARIE, TH. G. *Intelligence and affectivity in early childhood.* New York: International Universities Press, 1965.

DÉCARIE, TH. G. *La réaction du jeune enfant a la personne étrangère.* Montreal: Les Presses de l'Université de Montréal, 1972.

DONOVAN, W. L., & LEAVITT, L. A. Early cognitive development and its relation to maternal physiologic and behavioral responsiveness. *Child Development,* 1978, *49,* 1251–1254.

DUNST, C. J. *A clinical and educational manual for use with the Užgiris and Hunt Scales of Infant Psychological Development.* Baltimore, Md.: University Park Press, 1980.

DUNST, C. J., BRASSELL, W. R., & RHEINGOVER, R. M. Structural and organizational features of sensorimotor intelligence among retarded infants and toddlers. *British Journal of Educational Psychology,* 1981, *51,* 133–143.

ESCALONA, S., & CORMAN, H. *Albert Einstein scales of sensorimotor development.* Unpublished manuscript, Department of Psychiatry, Albert Einstein College of Medicine, no date.

FISCHER, K. W. A theory of cognitive development: The control and construction of hierarchies of skills. *Psychological Review,* 1980, *87,* 477–531.

FURTH, H. G. Piaget, IQ and the nature-nurture controversy. *Human Development,* 1973, *16,* 61–73.

GOLDEN, M., & BIRNS, B. Social class and cognitive development in infancy. *Merrill-Palmer Quarterly,* 1968, *14,* 139–149.

GRATCH, G. Recent studies based on Piaget's view of object concept development. In L. B. Cohen & P. Salapatek (Eds.), *Infant perception* (Vol. 2). New York: Academic Press, 1975.

GRATCH, G. Review of Piagetian infancy research. In W. F. Overton & J. McC. Gallagher (Eds.), *Knowledge and development* (Vol. 1). New York: Plenum Press, 1977.

GUSTAFSON, G. E., GREEN, J. A., & WEST, M. J. The infant's changing role in mother-infant games: The growth of social skills. *Infant Behavior and Development,* 1979, *2,* 301–308.

HAITH, M., & CAMPOS, J. J. Human infancy. In *Annual review of psychology.* Palo Alto, Ca.: Annual Reviews, 1977.

HARDING, C., & GOLINKOFF, R. The origins of intentional vocalizations in prelinguistic infants. *Child Development,* 1979, *50,* 33–40.

HUNT, J. McV. *Intelligence and experience.* New York: Ronald, 1961.

HUNT, J. McV. *Early psychological development and experience.* Worcester, Ma.: Clark University Press, 1980.

HUNT, J. McV., MOHANDESSI, K., GHODSSI, M., & AKIYAMA, M. The psychological development of orphanage-reared infants: Interventions with outcomes (Tehran). *Genetic Psychology Monographs,* 1976, *94,* 177–226.

INGRAM, D. Sensori-motor intelligence and language development. In A. Lock (Ed.), *Action, gesture and symbol.* New York: Academic Press, 1978.

JACOBSON, S. W. Matching behavior in the young infant. *Child Development,* 1979, *50,* 425–430.

KAHN, J. V. Relationship of Piaget's sensorimotor period to language acquisition of profoundly retarded children. *American Journal of Mental Deficiency,* 1975, *79,* 640–643.

KAHN, J. V. Utility of the Užgiris and Hunt Scales of sensorimotor development with severely and profoundly retarded children. *American Journal of Mental Deficiency*, 1976, *80*, 663–665.

KAHN, J. V. Uses of the scales of psychological development with mentally retarded populations. In I. Č. Užgiris & J. McV. Hunt (Eds.), *Research with scales of psychological development in infancy*. Urbana, Ill.: University of Illinois Press, in press.

KAYE, K., & MARCUS, J. Infant imitation: The sensory-motor agenda. *Developmental Psychology*, 1981, *17*, 258–265.

KING, W. L., & SEEGMILLER, B. Performance of 14 to 22-month old black, firstborn male infants on two tests of cognitive development. *Developmental Psychology*, 1973, *8*, 317–326.

KOHEN-RAZ, R. Scalogram analysis of some developmental sequences of infant behavior as measured by the Bayley infant scale of mental development. *Genetic Psychology Monographs*, 1967, *76*, 3–21.

KOPP, C. B., SIGMAN, M., & PARMELEE, A. H. Ordinality and sensory-motor series. *Child Development*, 1973, *44*, 821–823.

KOPP, C. B., SIGMAN, M., & PARMELEE, A. H. Longitudinal study of sensorimotor development. *Developmental Psychology*, 1974, *10*, 687–695.

KOPP, C. B., KHOKA, E. W., & SIGMAN, M. A comparison of sensorimotor development among infants in India and the United States. *Journal of Cross-cultural Psychology*, 1977, *8*, 435–451.

KRAMER, J. A., HILL, K. T., & COHEN, L. B. Infants' development of object permanence: A refined methodology and new evidence for Piaget's hypothesized ordinality. *Child Development*, 1975, *46*, 149–155.

LANGER, J. *The origins of logic: Six to twelve months*. New York: Academic Press, 1980.

LANGER, J. Logic in infancy. *Cognition*, 1981, *10*, 181–186.

LEWIS, M., & BROOKS-GUNN, J. *Social cognition and the acquisition of self*. New York: Plenum Press, 1979.

LEWIS, M., & McGURK, H. Infant intelligence. *Science*, 1972, *178*, 1174–1177.

LÉZINE, I., STAMBAK, M., & CASATI, I. *Les étapes de l'intelligence sensorimotrice*. Paris: Les Editions du Centre de Psychologie Appliquée, 1969.

MAHONEY, G., GLOVER, A., & FINGER, I. Relationship between language and sensorimotor development of Down Syndrome and nonretarded children. *American Journal of Mental Deficiency*, 1981, *86*, 21–27.

McCALL, R. B., HOGARTY, P., & HURLBURT, N. Transitions in infant sensorimotor development and the prediction of childhood IQ. *American Psychologist*, 1972, *27*, 728–748.

McCALL, R. B., EICHORN, D. H., & HOGARTY, P. S. Transitions in early mental development. *Monographs of the Society for Research in Child Development*, 1977, *42*, No. 3 (Serial No. 171).

McCUNE-NICOLICH, L. Toward symbolic functioning: Structure of early pretend games and potential parallels with language. *Child Development*, 1981, *52*, 785–797.

MEHRABIAN, A., & WILLIAMS, M. Piagetian measures of cognitive development up to age two. *Journal of Psycholinguistic Research*, 1971, *1*, 113–126.

MELTZOFF, A. N., & MOORE, M. K. Imitation of facial and manual gestures by human neonates. *Science*, 1977, *198*, 75–78.

MILLER, D. J., COHEN, L. B., & HILL, K. T. A methodological investigation of Piaget's theory of object concept development in the sensory-motor period. *Journal of Experimental Child Psychology*, 1970, *9*, 59–85.

NICOLICH, L. McC. Beyond sensorimotor intelligence: Assessment of symbolic maturity through analysis of pretend play. *Merrill-Palmer Quarterly*, 1977, *23*, 89–101.

Paraskevopoulos, J., & Hunt, J. McV. Object construction and imitation under differing conditions of rearing. *Journal of Genetic Psychology*, 1971, *119*, 301–321.

Piaget, J. *Psychology of intelligence*. London: Routledge & Kegan Paul, 1950.

Piaget, J. *The origins of intelligence in children*. New York: Norton, 1952.

Piaget, J. *The construction of reality in the child*. New York: Basic Books, 1954.

Piaget, J. *Play, dreams, and imitation in childhood*. New York: Norton, 1962.

Piaget, J. *Genetic epistemology*. New York: Columbia University Press, 1970. (a)

Piaget, J. Piaget's theory. In P. H. Mussen (Ed.), *Carmichael's manual of child psychology*. New York: Wiley, 1970. (b)

Piaget, J. *Biology and knowledge*. Chicago: University of Chicago Press, 1971. (a)

Piaget, J. The theory of stages in cognitive development. In D. R. Green, H. P. Ford, & G. B. Flamer (Eds.), *Measurement and Piaget*. New York: McGraw-Hill, 1971. (b)

Piaget, J. *The development of thought: Equilibration of cognitive structures*. New York: Viking Press, 1977. (a)

Piaget, J. The role of action in the development of thinking. In W. F. Overton & J. McC. Gallagher (Eds.), *Knowledge and development* (Vol. 1). New York: Plenum Press, 1977. (b)

Piaget, J., & Inhelder, B. *The psychology of the child*. New York: Basic Books, 1969.

Pinard, A., & Laurendeau, M. Stage in Piaget's cognitive developmental theory: Exegesis of a concept. In D. Elkind & J. H. Flavell (Eds.), *Studies in cognitive development*. New York: Oxford University Press, 1969.

Rogers, S. J. Characteristics of the cognitive development of profoundly retarded children. *Child Development*, 1977, *48*, 837–843.

Schaffer, H. R. (Ed.). *Studies in mother–infant interaction*. New York: Academic Press, 1977.

Seibert, J. *Developmental assessment for early intervention: Testing a cognitive stage model*. Technical report from the Mailman Center for Child Development, University of Miami, 1980.

Siegel, L. Infant perceptual, cognitive, and motor behaviors as predictors of subsequent cognitive and language development. *Canadian Journal of Psychology*, 1979, *33*, 382–395.

Siegel, L. Infant tests as predictors of cognitive and language development at two years. *Child Development*, 1981, *52*, 545–557.

Silverstein, A. B., Brownlee, L., Hubbell, M., & McLain, R. E. Comparison of two sets of Piagetian scales with severely and profoundly retarded children. *American Journal of Mental Deficiency*, 1975, *80*, 292–297.

Silverstein, A. B., McLain, R. E., Brownlee, L., & Hubbell, M. Structure of ordinal scales of psychological development in infancy. *Educational and Psychological Measurement*, 1976, *36*, 355–359.

Simoneau, K., & Décarie, Th. G. Cognition and perception in the object concept. *Canadian Journal of Psychology*, 1979, *33*, 396–407.

Steckol, K., & Leonard, L. Sensorimotor development and the use of prelinguistic performatives. *Journal of Speech and Hearing Research*, 1981, *24*, 262–268.

Super, C. M. Behavioral development in infancy. In R. H. Munroe, R. L. Munroe, & B. B. Whiting (Eds.), *Handbook of cross-cultural human development*. New York: Garland STPM Press, 1981.

Trevarthen, C. The foundations of intersubjectivity: Development of interpersonal and cooperative understanding in infants. In D. R. Olson (Ed.), *The social foundations of language and thought*. New York: Norton, 1980.

Užgiris, I. Č. *Some antecedents of the object concept*. Paper presented at a symposium on the object concept at the meetings of EPA, Philadelphia, April 1969.

UŽGIRIS, I. Č. Patterns of cognitive development in infancy. *Merrill-Palmer Quarterly*, 1973, *19*, 181–204. (a)

UŽGIRIS, I. Č. *Infant development from a Piagetian approach: Introduction to a symposium*. Paper presented at the American Psychological Association Convention, Montreal, 1973. (b)

UŽGIRIS, I. Č. Plasticity and structure. In I. Č. Užgiris & F. Weizmann (Eds.), *The structuring of experience*. New York: Plenum Press, 1977.

UŽGIRIS, I. Č. Two functions of imitation during infancy. *International Journal of Behavioral Development*, 1981, *4*, 1–12.

UŽGIRIS, I. Č., & HUNT, J. McV. *An instrument for assessing infant psychological development*. Unpublished manuscript, 1966.

UŽGIRIS, I. Č., & HUNT, J. McV. *Toward ordinal scales of infant psychological development*. Unpublished manuscript, 1972.

UŽGIRIS, I. Č., & HUNT, J. McV. *Assessment in infancy*. Urbana, Ill.: University of Illinois Press, 1975.

VYGOTSKY, L. S. *Mind in society*. Cambridge, Ma.: Harvard University Press, 1978.

WACHS, T. D. Relation of infants' performance on Piaget scales between twelve and twenty-four months and their Stanford-Binet performance at thirty-one months. *Child Development*, 1975, *46*, 929–935.

WACHS, T. D. Utilization of a Piagetian approach in the investigation of early experience effects. *Merrill-Palmer Quarterly*, 1976, *22*, 11–30.

WACHS, T. D. Proximal experience and early cognitive-intellectual development: The physical environment. *Merrill-Palmer Quarterly*, 1979, *25*, 3–41.

WACHS, T. D. Early experience and early cognitive development: The search for specificity. In I. Č. Užgiris & J. McV. Hunt (Eds.), *Research with scales of psychological development in infancy*. Urbana, Ill.: University of Illinois Press, in press.

WACHS, T. D., & HUBERT, N. C. Changes in the structure of cognitive-intellectual performance during the second year of life. *Infant Behavior and Development*, 1981, *4*, 151–161.

WACHS, T. D., UŽGIRIS, I. Č., & HUNT, J. McV. Cognitive development in infants of different age levels and from different environmental backgrounds. *Merrill-Palmer Quarterly*, 1971, *17*, 283–317.

WOHLWILL, J. F. *The study of behavioral development*. New York: Academic Press, 1973.

WOODWARD, M. The behavior of idiots interpreted by Piaget's theory of sensorimotor development. *British Journal of Educational Psychology*, 1959, *29*, 60–71.

ZACHRY, W. Ordinality and interdependence of representation and language development in infancy. *Child Development*, 1978, *49*, 681–687.

6   An Evolutionary Perspective on Infant Intelligence

Species Patterns and Individual Variations

SANDRA SCARR

> Since selection can and did occur in terms of developments at all ontogenetic points, the entire life span is a product of evolutionary adaptation, and a psychologist interested in causes of behavior must simultaneously consider phylogeny and ontogeny, difficult as it may seem. (Freedman, 1967, p. 489)

Any attempt to construct an evolutionary view of infant intelligence should raise a certain skepticism in the reader's mind. What, after all, is the nature of intelligence in infancy? And how shall the validity of an evolutionary account by judged? Not, certainly, by its predictive power for the future evolution of infant behavior! On the first question I shall defer largely to Piaget (1952), whose descriptions and explanations of infant intelligence I find consistent with an evolutionary view. On the second question, a few words about evolutionary theory may be helpful.

The central tenet of evolutionary theory is natural selection, an exceedingly simple idea. Organisms differ from one another. They produce more young than the available resources can sustain. Those best adapted survive to pass on their genetic characteristics to their offspring, while others perish with fewer or no offspring. Subsequent generations therefore are more like their better adapted ancestors. The result is evolutionary change (Ghiselin, 1969, p. 46). Elaborations of the idea of

---

SANDRA SCARR • Department of Psychology, Yale University, New Haven, Connecticut 06520.

natural selection, as it applies to period in the life span, learned characteristics, and speciation, appear throughout this chapter.

An evolutionary account of any human behavior is by definition a
historical reconstruction. We cannot observe our behavioral past. There
are limits, however, to the fancifulness of a useful evolutionary construction: the known facts must fit and contrary facts must be few and
isolated. Most important, the hypothetical account must be open to
falsification; it cannot contain statements that could explain every possible outcome—and thus be unfalsifiable. These criteria are especially
important for *ad hoc* theories, since predictions about human evolution
cannot be tested within the life span of any investigator. Some testable
hypotheses can be generated, however, about phenomena not directly
used to construct the account. The implications of the theoretical construction will, I hope, extend beyond the immediate boundaries of its
most central facts. In these ways evolutionary views can be scientifically
tested.

Within an evolutionary framework, I want to make a radical argument about the natural history of human, infant intelligence. The argument revolves around the primary nature of early intelligence—a nonverbal, practical kind of adaptation. Sensorimotor behaviors must, I
think, have emerged very early in primate evolution, certainly before
man split off from the great apes. There is simply too high a degree of
parallelism in the early intelligence of apes and man to suggest independent, convergent evolution. The phylogeny of infant intelligence seems
to be very ancient history.

The ontogeny of infant intelligence has a distinctive pattern and
timing. The species pattern, I would argue, is not an unfolding of some
genetic program but a dynamic interplay of genetic preadaptations and
developmental adaptations to features of the caretaking environment.
Individual variation is limited by canalization on the one hand and by
common human environments on the other. From the common behavioral elements to be seen among individuals, one can abstract a species
pattern to describe and contrast with the patterns of other species. One
must be ever mindful, however, that what exists are individuals, each
different from the other; a species-typical pattern is an abstraction from
reality. The development of infant intelligence has both a species-typical
pattern and individual variation. How and why the species theme and
individual differences exist is the subject of this chapter.

Four hypotheses about the nature and evolution of human infant intelligence are basic to my argument:

1. That infant intelligence evolved earlier in our primate past than ontogenetically later forms of intelligent behavior and remains virtually unchanged from the time that hominids emerged.
2. That selection pressures resulting in the present pattern of sensorimotor intelligence acted both on the infant himself and on the caretaking behaviors of his parents.
3. That infant intelligence is phenotypically less variable than later intelligent behavior because it has been subjected to longer and stronger natural selection.
4. That the phenotypic development of infant intelligence is governed both by genetic preadaptation (canalization) and by developmental adaptation to human physical and caretaking environments.

AN EVOLUTIONARY VIEW OF INFANT INTELLIGENCE

The Nature of the Sensorimotor Period

The primary tasks of infant primates are to survive the first two years and to learn to operate effectively in the physical and social environment. The attachment system is of critical importance to survival and to learning species-appropriate social interactions. Sensorimotor skills are critical to survival and to adaptation in the physical and social worlds. As several authors have noted (e.g., Bell, 1970; Bowlby, 1969, 1973), the development of social attachments is intertwined with increasing cognitive skills, such as object or person permanence. I divide the cognitive and affective domains here more for convenience of discussion than for any good conceptual reasons. Infant primates' survival depends on the protection of their caretakers while they become competent to explore and learn. The increasing distance permitted between infant and mother is correlated with increasing sensorimotor skills. Both serve survival and adaptation.

Infant primates are remarkably curious and open to learning how to be practical experimenters. The presymbolic skills of human infancy that

Piaget has so richly described also characterize our nearest primate relatives. The great apes and even Old World monkeys master sensorimotor skills that are very like those of human infants.* Later in the life span, human and nonhuman primates show different forms of adaptation. Different selective pressures, particularly those that led to man's cultural revolution, have produced forms of childhood and adult intelligence quite dissimilar to those of nonhuman primates.

Man's gradual accumulation of culture has great relevance to his evolution past infancy. Culture provided new environments to which childhood and adult adaptations could occur. As McClearn (1972) said:

> First steps toward culture provided a new environment in which some individuals were more fit, in the Darwinian sense, than others; their offspring were better adapted to culture and capable of further innovations; and so on. The argument can be made that, far from removing mankind from the process of evolution, culture had provided the most salient natural selection pressure to which man has been subject in his recent evolutionary past. (p. 57)

The pressures of culture on intelligence are self-evident. The greater the ability of some individuals to learn and to innovate, the more likely they were to survive to reproduce and the more likely it was in the long run that their progeny would have even greater fitness in the new environment. But I would argue that the symbolic cultural revolution had practically no effect on the evolution of infant intelligence.

The distinctly different nature of infant intelligence was recognized by Florence Goodenough, who noted:

> The unsettled question as to whether or not true intelligence may be said to have emerged before symbolic processes exemplified in speech may have become established. Attempting to measure infantile intelligence may be like trying to measure a boy's beard at the age of three. (Cited in Elkind, 1967)

Sensorimotor intelligence is qualitatively different from later symbolic operations, whose evolution may have quite a different history. I do not propose a common primate history for formal operations or even for concrete operations, although some symbolic and conceptual skills are shared by apes and man (e.g., Premack, 1971). I do propose that the natural history of sensorimotor intelligence is independent of skills that

---

*I do not claim that other mammals are not capable of some aspects of sensorimotor intelligence, such as object permanence. The manipulative, tool-using skills, however, are largely limited to species with good prehension.

evolved later and that there is no logically necessary connection between them.

Indeed, the empirical connection between sensorimotor skills and later intellectual development is very tenuous (Stott & Ball, 1965). Children with severe motor impairments, whose sensorimotor practice has been extremely limited, have been shown to develop normal symbolic function (Kopp & Shaperman, 1973). The purported dependence of symbolic activity on sensorimotor action has not been demonstrated. One reason for the lack of correlation may be different sources of individual variation. If sensorimotor and symbolic skills have different genetic bases, they could well be uncorrelated. Sensorimotor skills are best seen as a criterion achievement—that is, individual differences are found in the *rate* but not the final *level* of sensorimotor development. Symbolic intelligence has individual differences in both rate and level of achievement, and the rate of development is correlated with the final level (witness the substantial correlations between IQ at ages 5 and 15). Infant intelligence is characterized by universal attainment by all nondefective species members. Its evolution has a more ancient history than does symbolic reasoning, and individual differences do not have the predictive significance of variations in later intelligence.

Infant Learning

The fact that human infants learn is of paramount importance to understanding the evolution of infancy and infant development. All normal babies interact with their social and physical worlds, structure and interpret their experiences, and modify their subsequent interactions. As Piaget has described, human infants set about learning in a graded sequence of intellectual stages that reflect their growing awareness of the effects of their actions and of the properties of the physical and social worlds around them.

A critical feature of human learning is its flexibility. In infancy we see the major transitions from reflex organization to a flexible, experimental approach to the world. By 1–1½ years babies have become impressive, practical experimenters. The rapid development of practical intelligence leaves the rest of the preadolescent period for mental adaptations. While formal operational thought may not develop in all normal species members, sensorimotor intelligence does.

In a brilliant and provocative paper Bruner (1972) outlined the nature and uses of immaturity for human development. He identified the "tutor-proneness" of the young, their readiness to learn through observation and instruction. Infants are ever ready to respond to novelties provided by the adult world. Further, they use play, according to Bruner, as an opportunity to work out their knowledge in safety—without the consequences that would befall adults who were in the initial stages of learning both sensorimotor skills and how to be a responsible social animal. The distinctive pattern of immaturity lends itself to more flexible adaptation for the species. The usefulness of opportunities for learning depends on the behavioral flexibility of the infant to acquire by learning what has not been "built into" the genome.

Two facts of human evolutionary history are particularly salient for infancy: the necessity of infant–mother dyads and the consistent availability of a larger human group into which the dyad is integrated. No surviving infant was without a social context throughout human history.† The evolution of infant development has occurred, therefore, in the context of normal infant environments. This context has, I think, profound implications for the lack of developmental fixity (Lehrman, 1970) in infant behavior. Foremost, it has been unnecessary for selection to build into the genotype those behaviors that all infants would develop experientially in their human groups. All normal infants would have close contact with mothers and other conspecifics and with tools and material culture, thus giving them opportunities to learn object manipulation, social bonds, and a human language. What has evolved genotypically is a bias toward acquiring these forms of behavior, a bias that Dobzhansky (1967) calls human educability.

The Evolution of Infancy

Infancy is a mammalian theme. A period of suckling the dependent young evolved as an efficient way to increase the survival chances of fewer and fewer offspring. Although extended care of dependent young is a burden and a risk for their parents, it is of greatest evolutionary importance to the mammalian pattern of reproduction and parental be-

---

†The few reported cases of feral children, even if they are believed, have contributed little to the human gene pool and the subsequent evolution of infant behaviors.

havior. The more an organism is protected from the vicissitudes of the environment, the greater the role of intraspecific competition. What one offspring requires of its parents are energy and resources not available to another offspring of those same parents. It became advantageous to have fewer and better-equipped offspring and to have long life spans. Both competition for females and demands for long parental care put a premium on long life span, and this again decreased the number of offspring still further (Mayr, 1970, pp. 338–340).

Primate infancy is an elaboration (exaggeration?) of the mammalian pattern: a single infant born not more than once a year and requiring years of parental care. What advantages can such a pattern confer? Highly developed parental care allows a fundamental change in the genetics of behavioral development. Primate infants have a more "open program" for learning than other mammals. Such an open program requires a far larger brain in the adults who provide the care and in the infants who must learn what information is needed. Primate intelligence is a coadapted product of evolutionary changes in the duration and the intensity of infant dependence and parental care. No one product could have evolved independently of the others.

I would argue, however, that the pattern of development for human infants in the sensorimotor period was basically established in common with other closely related primates. The later evolutionary history of apes and man led to species differences in the degree of immaturity at birth, the degree of flexibility in learning, and the length of the socialization period. In considering infancy alone, however, I am struck by incredible similarities in the sensorimotor period, similarities that should be considered apart from the later, more obvious differences. Prolonged infancy evolved as a primate variation on the mammalian theme. Human infancy is a further evolution of the primate pattern. Contemporary apes have evolved patterns of infant development that still share much with the human species. These similarities originated in our common primate past.

Every period of the human life span is a product of selection (Mayr, 1970, p. 84). Multiple pressures, about which we can speculate only *post hoc*, must have played interacting roles in the evolution of prolonged infancy. LaBarre (1954) argued for an increasing specialization of human infants in *brains*. One-seventh of the newborn's weight is brain. With limitations to the female pelvic girth, infants were born less and less mature to assure the safe passage of the big brained fetus into the world. Changes in adult behaviors must have accompanied the increasingly long dependence of a less mature infant:

> Curiously enough, as human females became better mammals (through sexual availability and permanent breasts) and as human males increased in constancy of sexual drive, the human infant seems simultaneously to be specializing in mammalian infancy. In helplessness and dependency, human babies and children are about as infantile as mammalian infants come. (La-Barre, 1973, p. 29)

LaBarre's account of the coordinated adaptations in adult male, adult female, and infant includes the structure of the family, which, he says, depends on the sexual availability of the female to keep the father home, on the father's strong sexual drive, and on the infant's attachment relation with his mother (LaBarre, 1954). LaBarre's account of the evolution of human immaturity is highly speculative. Mayr (1970, p. 407) argued that brain size could have increased still further if (1) the female pelvic size increased, (2) pregnancy were shortened, or (3) more brain growth was postnatal. Any of these adaptations would permit further evolution of brain size (although no increase in brain size has occurred in the last 30,000 years of man's evolution, presumably because there is no longer a selective premium on it). Omenn and Motulsky (1972) noted that human newborns are delivered at a less advanced stage of development than newborn apes and monkeys, a fact that they attribute to two adaptational differences. First, the female pelvis narrowed with the adaptation to bipedal locomotion, and the restriction in the bony birth canal required earlier birth of fetuses. Second, the slow maturation of human infants is ideally adapted to the molding of species-specific behaviors by social input.

It is impossible at present to decide which set of factors in evolutionary history accounted for the correlated shifts in infant intelligence, immaturity, and parental behaviors. They are coadapted. The total phenotype is, after all, a compromise of all selection pressures, some of which are opposed to each other (Mayr, 1970, p. 112). The evolution of neoteny and infant intelligence most likely represents a compromise solution among pressures on adults to provide increased infant care (a liability), pressures for increasing brain size and flexible learning ability (a benefit, we presume), reproductive economy, and other factors about which we can only guess.

Restrictions on Phenotypic Variability

In the case of infant intelligence, the flexibility in learning that is typical of humans must have some bounds. Species adaptation depends on a rather limited range of behavioral phenotypes. Some characteristically human patterns need to emerge in every individual. There are two principal mechanisms for limiting the possible number of phenotypes that develop: *canalization* by genetic preadaptation and *developmental adaptation*.

Canalization is a genetic predisposition for the development of a certain form of adaptation, guided along internally regulated lines. Environmental features are necessary for complete development or for the full expression of the adaptation, but the direction of the development is difficult to deflect. Environmental inputs that are necessary for canalized development to occur must be universally available to the species, or this form of adaptation would not work.

Embryologists, particularly Waddington (1957, 1962, 1971), have long recognized the "self-righting" tendencies of many aspects of growth. The difficulty of deflecting an organism from its growth path (which Waddington calls a *creod*) is expressed in the idea of canalization. Canalization restricts phenotypic diversity to a limited species range while maintaining desirable genetic diversity. If all genetic diversity were phenotypically expressed, there would be such enormous behavioral differences among people that it is difficult to see how any population could reproduce and survive (Vale & Vale, 1969). There are obviously functional equivalences in many genotypes (they produce similar phenotypes) for the most basic human characteristics.

Canalization is a very conservative force in evolutionary history. A well-knit system of canalization tends to restrict evolutionary potential quite severaly. It accounts for the maintenance of particular phenotypes throughout a family of related species for no obvious reason, since a different phenotype seems to serve another taxon equally well in the same environment (Mayr, 1970, p. 174). In the case of infant intelligence, the similarities among primate species suggest a relative immunity to recent evolutionary pressures.

A major reason for the perseverence of particular phenotypes is that new characters or traits are produced not by isolated mutations but by a reorganization of the genotype. It requires a genetic revolution to break up a well-buffered developmental pattern. Second, most genetic variability can be hidden by canalized development and therefore by immune to selective pressures:

> A tight system of developmental homeostasis helps to shield the organism against environmental fluctuations. However much genetic variation there in a gene pool the less of it penetrates into the phenotype, the smaller the point of attack it offers to selection. (Mayr, 1970, p. 39)

The total genome is a "physiological team." No genes are soloists; they must play harmoniously with others to achieve selective advantage because selection works on the whole person and on whole coadapted gene complexes in the population. As Dobzhansky (1955) has said, evolution favors genes that are "good mixers," ones that make the most positive contributions to fitness against the greatest number of genetic backgrounds.

Selection is always for coadapted gene complexes that fit a developmental pattern. The sheer number of gene differences between individuals or species is not a good measure of overall difference. To express individual or population differences as differences in the number of nucleotide pairs of the DNA is like trying to express the difference between the Bible and Dante's *Divine Comedy* in terms of the frequency of letters used in the two works (Mayr, 1970, p. 322). The developmental pattern of infant intelligence is, I would argue, a strongly buffered epigenotype that is shared by our closest primate relatives. To break it up would require multiple rewritings of the primate manuscript.

Compared with canalization, *developmental adaptation* is a more flexible arrangement to ensure survival in various possible environments. The genetic program does not specify a particular *response* to any environment, but it specifies a generalized *responsiveness* to the distinctive features of environments within a permissible range of variation. In practice it is very difficult to distinguish between *developmental adaptation* and *genetic preadaptation* (through selection), because they serve the same goal—to limit the possible behavioral phenotypes that develop.

The contrast between canalization and developmental adaptation is not a distinction between genetic and environmental determinants of development. Every human characteristic is genetically based (because the entire organism is), but a useful distinction can be made between genetic differences and nongenetic differences. *Nongenetic* means simply that the differences between two phenotypes are not caused by genetic differences. The capacity of a single genotype to produce two or more phenotypes is itself genetically controlled, of course (Mayr, 1970). The notion of a genetic blueprint for ontogeny means that each genotype has its own canalized course of development, from which it can be deflected only with difficulty. In the case of strong genetic canalization, individual phenotypic differences are presumably genetic because one genotype cannot produce a variety of phenotypes. In the case of weak canalization, one genotype can and does produce multiple phenotypes among which the differences are not genetic.

Two puzzling examples of human adaptation illustrate the difference between genetic adaptation as a result of natural selection and developmental adaptation as a result of genetic flexibility (strong versus weak canalization). Milk "intolerance" normally develops in most humans after the preschool years. The ability to digest large quantities of milk in adulthood is the result of prolonged lactase activity in some populations that have practiced dairying for the past several thousand years. Is the continued secretion of lactase in adulthood a developmental adaptation to continued milk drinking past weaning? Or is it a result of natural selection for lactase activity in those peoples for whom some selective advantage was derived from milk in their adult diets?

The second example is adaptation to life at high altitudes. One feature of high altitudes is reduced oxygen concentrations in the air. Peoples in Ethiopia and in the Andes at elevations above 10,000 feet typically have large lung capacities and deep "barrel chests." Peoples who live at lower altitudes have smaller chests and lung capacities. Is this primarily a developmental adaptation or a result of natural selection for adaptation to a high-altitude niche?

In both cases, either a developmental or a selective adaptation would accomplish the same goal of better utilization of the available resources—in one case nutrition, in the other case oxygen. For reasons beyond the comprehension of this author, the case of milk "intolerance" seems to be primarily the result of natural selection acting on the gene frequencies for lactase activity past childhood (Gottesman & Heston, 1972). The second case—adaptation at high altitudes—is primarily a developmental phenomenon. We know these explanations to be the primary ones because in the case of lactase activity, continued milk-drinking into later childhood does not maintain lactase activity in intolerant people at levels adequate for comfortable absorption of a significant portion of their nutrition through milk, and discontinued milk drinking does not terminate lactase activity in people who are genetically tolerant of milk. In the lactose-intolerant group, loading the stomach with milk at any time results in renewed lactase activity. In the lactose-tolerant case, lactase activity declines despite continued stimulation through milk consumption.

The high-altitude example could well have represented genetic selection for life under unusual oxygen tension (Baker, 1969). After 15,000 years in the high Andes, however, Peruvian Indians who descend to the lowlands have children with little evidence of barrel-chestedness, and Indians who migrate from lowland to highland areas have children who exhibit the phenomenon. Harrison (1967) reported that Amharic Ethiopians who migrate from 5000- to 10,000-foot altitudes develop some chest enlargement even in adulthood.

What kinds of human behavioral characteristics are likely to show developmental adaptation more than genetic preadaptation? Omenn and Motulsky (1972) proposed that older (in an evolutionary sense) forms of adaptation are more likely to have limited genetic variability and a higher degree of canalization. Specifically, the brain stem, the midbrain, and the limbic structures that evolved earlier are less polymorphic than cortical areas of the brain. Behavioral characteristics associated with higher cortical centers are newer evolutionary phenomena and are likely to develop more variable phenotypes. Behaviors associated with older areas of the brain, those we share with other primates, are genotypically and phenotypically less variable. Their development is more highly canalized. This hypothesis has clear implications for infant intelligence, as contrasted with later forms of intelligence.

EVIDENCE ON CANALIZATION AT SPECIES, POPULATION, AND INDIVIDUAL LEVELS OF ANALYSIS

To evaluate the research evidence on the canalization of infant intelligence, we must coordinate the data gathered with several methodological approaches. Ethological and comparative studies of primates speak to the canalization of infant intelligence at a species level. Behavior genetic studies of variation analyze sources of individual differences within populations, and cross-cultural studies deal with population differences in development. Four operational definitions (or primitive models) are proposed to integrate comparative and ethological descriptions of species patterns with analytical studies of variation, including population and individual levels of analyses. Predictions can be made from any of the four:

1. Functional equivalencies in both genotypes and environments are interpreted as strong canalization at a species level. If neither genotypic nor environmental differences contribute much to phenotypic diversity, there will be a restricted range of individual differences, moderate heritability, and a distinctive species pattern.

2. Functional differences in genotypes but equivalencies in environments are interpreted as strong canalization at an individual, not a species, level. If genetic differences are the primary contributors to phenotypic differences, then heritability will be high within a population and between populations, if the distribution of genotypes is different.

3. Functional equivalencies of genotypes but not environments are interpreted as weak canalization at individual and population levels, with low heritabilities and a weak species pattern.

4. Functional equivalencies of neither genotypes nor environments will yield extreme individual phenotypic variation and moderate heritabilities within and between populations, if genotypes are differently distributed.

The implications of an evolutionary account for varied data on infant intelligence can now be tested. If infant intelligence indeed evolved early in primate history, if its development is to some extent canalized, and if both genotypes and environments are largely equivalent functionally, then contemporary primates should share a similar pattern of infant intelligence, individual diversity within the human species should be restricted, and the heritability of sensorimotor intelligence should be moderate, not high.

Infant Intelligence as Species-Specific Behavior

The notion of species-specific behavior is an abstraction from the reality of individual variation. Some behavioral geneticists deny the concept of "species-typical" any heuristic value (Bruell, 1970); others would support its usefulness as a statement about the highly leptokurtic shape of the distribution of individual differences within a species, measured on a species-comparative scale. Genetically conditioned homogeneity within a species is seen as a species-specific characteristic; genetically conditioned heterogeneity is seen as individual variation within a species (Gottesman & Heston, 1972).

There is confusion inherent in the contrast between genetically conditioned homogeneity and heterogeneity in behavioral characteristics because (1) the notion of species-specific behavior is always an abstraction, (2) complex behaviors are always polygenic and to some degree phenotypically heterogeneous, and (3) the degree of phenotypic homogeneity is always relative to the scale on which the phenotype is measured. For example, take linear height. In the human population, adult heights vary between, say 3 feet and 7 feet, with the median height being about 5 feet 6 inches. From a within-species vantage point, the distribution is somewhat leptokurtic, with perhaps 95% of the world population distributed between 5 feet and 6 feet 2 inches. If we scale human heights on a species-comparative scale from .01 inches to 240 inches (from protozoans to giraffes), the human distribution appears strongly leptokurtic. A "species-typical" height of about 5 ½ feet represents a useful value in relation to other species. Actually, of course, the human variation is quite large if one's perspective is intraspecific. And so it is with nearly all human behaviors.

Robin Fox (1970) has argued for the usefulness of the species-specific concept. Language capacity is one obvious example, but kinship, courtship and marriage arrangements, political behaviors, and male groupings that exclude females appear to be other species-specific human traits. There are limits, he argues, to what the human species can do and to what we can understand in another's behavior. There must be "wired-in" ranges for the information-processing capacity that responds only to certain kinds of inputs. Our abilities to process information and to respond to the inputs of another's behavior are strongly tied to our phylogeny and to timing in the life cycle.

We are faced with an apparent paradox: that species-specific behaviors do not exist but are an abstraction from the reality of individual variation, yet the concept of species-typical does have heuristic value on a species-comparative scale. We can better approach the problem of variation and the species-typical concept, I believe, by looking at what limitations there are on variability within species and by what mechanisms variation is limited.

Biases in Learning

Though it hardly needs saying, human infants tend to learn some things rather than others. One example is language acquisition, for which underlying sensitivities to speech sounds, both comprehension and production, combine with the stimulation of a language environment to produce a speaking human child. Another example is hand–eye coordination. At around 3 months, normal infants gaze extendedly at their hands as though they were detached objects. One might think that visually guided reaching followed from such accidental experiences. In fact, blind infants "gaze" at their hands in prolonged fashion at about the same age as seeing infants (Freedman, 1974). The canalization of arm–hand motor development seems to bring all infants' hands within their visual range at that point in development. Experience with hand regard doubtless plays a role in subsequent coordinations, but the opportunity for hand–eye coordination to develop has not been left to experiential chance.

Seligman (1970) has shown that mammals come to a learning situation with a good deal of built-in bias to learn particular things. It is simply not the case that any stimulus can be equally well associated with any response or reinforcement. I would argue that human infants have built-in biases to acquire certain kinds of intelligent behaviors that are consonant with primate evolutionary history, that these biases are programmed by the epigenotype, and that human environments guarantee the development of these behaviors through the provision of material objects that are assimilated to them.

We seldom emphasize the role of common human environments in development, being attuned as we are to looking at distinctive features. The environments for highly canalized behaviors such as walking are seldom even studied. Lipsitt (1971, p. 499) gave a charming description of an infant who is "ready" to walk being propped up on his legs and flopped back and forth between adults. The acquisition of walking undoubtedly has experiential components that can be studied (Zelazo, 1974). On the other hand, all human environments seem to provide the necessary and sufficient conditions for walking to begin between 10 and 15 months. Only physically infirm infants (handicapped or malnourished) and those deprived of firm support (Dennis, 1960) fail to walk during infancy.

A similar point can be made about language acquisition. All normal, hearing infants have a human language environment, regardless of which language is spoken, that provides the necessary and sufficient conditions for acquisition. Infant intellectual development has some of the same properties in that it follows a species pattern of sensorimotor skills that assimilate whatever material objects the culture offers. The overall species patterns for motor, language, and cognitive development seem to be well ordered by the chromosomes and the common human environment. While experimental interventions may accelerate the acquisition of these behaviors, all normal infants acquire them in due time, and it is not clear that acceleration has any lasting impact on subsequent development.

Deprivation Effects

If infant intelligence is highly canalized at a species level, one would predict that environmentally caused retardations of sensorimotor development would be overcome once the environmental causes were eliminated. Canalization implies such an outcome. Recently Kagan and Klein (1973) published a cross-sectional study of infant and childhood development in Guatemala. Their assessment of infant development in an Indian village suggested to them that the children were behaviorally quite retarded at the end of the first year. Older children in the same setting, however, approached the performance levels of United States children on a variety of learning and perceptual tasks. From the observation of "retarded" infants and intellectually "normal" older children, Kagan and Klein concluded that human development is inherently resilient, that is, highly canalized at the species level:

> This corpus of data implies that absolute retardation in the time of emergence of universal cognitive competences during infancy is not predictive of comparable deficits for memory, perceptual analysis, and inference during preadolescence. Although the rural Guatemalan infants were retarded with respect to activation of hypotheses, alertness, and onset of stranger anxiety and object permanence, the preadolescents' performance . . . were comparable to American middle class norms. Infant retardation seems to be partially reversible and cognitive development during the early years more resilient than had been supposed. (p. 957)

What Kagan and Klein (1973) suggested about canalization is that the caretaking practices of rural Guatemalans significantly retard the rate of infant development but that this deflection is only temporary because later child-rearing practices compensate for the early deprivation. In Waddington's terms, the Guatemalan infants' mental development is asserted to have been temporarily deflected from its canalized course by environmental deprivations but to have exhibited the same kind of "catch-up" phenomenon claimed for physical growth among children who have been ill or malnourished for brief periods of time. Unfortunately, serious ceiling effects on the later tests make it difficult to judge whether the older Guatemalan children have intellectual skills typical of United States white children. Thus, arguments for the canalization of infant intelligence at a species level are not well supported by this study.

The Guatemalan data do suggest that environmental deprivation can retard sensorimotor development. Studies of institutionalized infants (Dennis, 1960; White, 1971) also support the conclusion that social and physical deprivation retard infant intelligence. One can question, however, whether sensorimotor skills fail to emerge eventually in even moderately deprived infants. While there is no question that the rate of acquisition is affected, is there any evidence that infants who have any contact with physical and social objects fail to develop criterion-level sensorimotor skills by 2–3 years of age?

Clearly, one could design a featureless, contactless environment that would turn any infant into a human vegetable. Extreme deprivation will prevent the emergence of the species-typical pattern. But the more interesting questions are how much input is necessary for adequate sensorimotor development and how many naturally occurring environments fail to provide the necessary conditions for criterion level development.

The proposal that sensorimotor intelligence is to some degree a canalized form of development does not require that the behaviors emerge in an environmental vacuum. *Canalization does not imply that species-typical development will occur under conditions that are atypical of those under which their evolution occurred.* It does imply that within the range of natural human environments most genotypes will develop similarly in most environments.

The Guatemalan data suggest that in at least one naturally occurring human environment the rate of sensorimotor development is slower than in some other conditions. An alternative explanation is also available, however: that the differences observed are due to genetic differences between groups in the rate of sensorimotor development. Whether the differences between Guatemalan and United States infants are genetic and/or environmental, the data provide some evidence against an extreme canalization position. There must be some developmental adaptation to enriched or impoverished environments and/or some group differences in genotypic responsiveness to sensorimotor environments that affect the rate of infant intellectual development. There is no evidence, however, that nondefective genotypes and naturally occurring environments are not equivalent in producing, eventually, the species-typical pattern.

Other Primates

The ethological, comparative evidence suggests that we share with at least the great apes a primate form of infant intelligence. The homologous, intelligent behaviors of infant apes and humans strongly suggest common origins in our primate past. During the first 18 months of human life there are few intellectual accomplishments that are not paralleled in nonhuman primates, particularly the apes. Both develop object concept, imitation, spatial concepts, cause–effect relations, and means—ends reasoning. In brief, both young apes and young humans become skillful, practical experimenters.

Our knowledge of chimp intellectual development comes primarily from home-reared animals, whose progress on form-board problems and the like exceeds that of their human infant companions in the first year of life (Hayes & Nissen, 1971). Even at the age of 3, Viki, the Hayeses' chimp, closely resembled a human child of 3 on those items of the Gesell, Merrill–Palmer, and Kuhlmann tests that do not require language:

> Viki's formal education began at 21 weeks with string-pulling problems. At 1 year she learned her first size, form, and color discriminations. By 2 ½ years of age she could match with an accuracy of 90% even when a 10-second delay was imposed. (Hayes & Nissen, 1971, p. 61)

Viki was reared in a human child's environment, and her non-linguistic attainments are impressive. Certainly her sensorimotor intelligence was as adequate as that of a human infant. In the wild, Van Lawick-Goodall's (1971) observations confirm the excellent sensorimotor intelligence of chimps at later ages, but few data are available on their intellectual development in the first year of life.

Hamburg (1969) noted the many similarities between man and chimpanzees in the number and form of chromosomes, in blood proteins, in immune responses, in brain structure, and in behavior. The more we see of their behavior, he said, the more impressed we are by their resemblance to man: "This is not to imply that we inherit fixed action patterns. The chimpanzee's adaptation depends heavily on learning, and ours does even more so!" (p. 143).

Hamburg further suggested that there are probably important biases in what chimps and humans learn: "Our question is: Has natural selection operated on early interests and preferences so that the attention of the developing organism is drawn more to some kinds of experiences than others?" (p. 144). Both chimp and human infants attend to physical problem-solving tasks and to relational problems in their environments.

The nature of learning processes in chimp and human infants is virtually the same. Both profit particularly from observational learning, a skill that is a forte of primate adaptation. From observing the behavior of conspecifics, primates imitate and then practice the observed sequences of behavior over and over again:

> The chief mode of learning for the non-human primate is a sequence that goes from observation to imitation, then to practice. They have full access to virtually the whole repertoire of adult behavior with respect to aggression, sex, feeding, and all other activities. The young observe intently, and then imitate, cautiously at first, all the sequences they see. Then they may be seen practicing these sequences minutes or hours after they have occurred. This observational learning in a social context becomes extremely important for the young primates. It takes the place of active instruction on the part of adults, which never seems to occur. (Hamburg, 1969, p. 146)

The active instruction of human infants by adults probably exceeds that provided by other primate parents. In most parts of the world, however, infants are not instructed on the development or use of sensorimotor schemes. Although both home-reared chimps and human infants may profit from active instruction, it is not clear that the normal development of sensorimotor intelligence requires more than *opportunities* for exploration and learning.

The Gardners' chimp, Washoe, exhibited observational learning of even the most "unnatural" behaviors, such as signs, although most of the signs were deliberately taught to her. She learned the sign for "sweet" from the Gardners' use of it in connection with her baby-food desserts. Later reinforcement of her use of the sign increased the reliability of her use of "sweet," but she acquired it from observation (Gardner & Gardner, 1971). She freely combined signs in novel utterances, reflecting her primate ability to make flexible combinations.

What differences, then, exist between the chimp and the human infant in sensorimotor intelligence? I would argue that the differences are in degree, not in kind. As Bruner (1972) has said, the difference between nonhuman primates and humans is in the *flexible use* and *combinatorial quality* of schemes, not in the schemes themselves. This is especially true in infancy, in which the greater cortical development of the human species has only barely begun to show its eventual effects. Human infants may exceed chimps in the combinatorial quality of their schemes, but the evidence is not so striking that observers of chimpanzee infants have noticed any great differences from human infants.

There is no question that after the age of 3, chimps and human children are intellectually different. Despite extensive tutoring in sign language and conceptual skills, Washoe's and Viki's problem-solving skills at 4 years were hardly a match for those of an ordinary 4-year-old child. In infancy, however, their skills were entirely comparable to those of a normal human infant.

The commonalities between apes and man in sensorimotor intelligence suggest that within each species, most genotypes and environments are functionally equivalent in producing the recognizable species (perhaps, panprimate) form of development. The commonalities also suggest that this ancient phylogenetic adaptation has been highly resistant to evolutionary change—a characteristic of canalized behaviors.

Interspecies similarity is always greater between early forms of development than between later, more differentiated forms. The most extreme statement of this point of view is that ontogeny recapitulates phylogeny. Although we have all been taught to reject this rigid view, there is a perfectly good observation that has been thrown out in the process. Embryologists can tell the difference between a human embryo and a fish embryo even though both have gill slits, but the embryonic forms have more in common than do adult forms of the two taxons. It is not too great a leap, I hope, to note that early behavioral forms among primates have more in common than later behavioral forms. This is not to say that chimps and human infants have identical forms of behavior, only that they have more in common in the first 18 months than they do in later life.

An elaboration of this view, suggested by John Flavell, proposes that early human behavior has qualities that are panmammalian (e.g., sucking); later in the sensorimotor period, we can no longer refer to panmammalian but only to panprimate forms of behavior. By adolescence, human intelligence is uniquely human and other primate intelligence is unique to those species. The progressive divergence of intellectual development is analogous to the progressive differentiation of embryos. At no point are different species' forms indistinguishable, but early forms have more in common than later ones.

The restricted range of individual variation is another characteristic of canalization at a species level. Such individual differences as exist arise in the *rate* of sensorimotor development, not in the level eventually attained. Differences in the rate of sensorimotor development are small, relative to later intellectual differences. The overall pattern of sensorimotor intelligence is quite homogeneous for the species, since criterion performance is accomplished in 15–20 months for the vast majority of human infants. When one compares this restricted range of phenotypic variation with the range of intellectual skills of children between 11 and 12 years, for example, it is readily apparent that sensorimotor skills are a remarkably uniform behavioral phenomenon.

The hypothesis that infant intelligence is a more highly canalized form of development than later intelligence does not mean that environmental influences are inconsequential, either for development or for individual differences. Even strongly canalized behaviors respond to experience. Learning strongly affects the subsequent sexual behavior of castrated male cats, whose normal sexual development requires only opportunities to perform. Male cats castrated after copulatory experience are vastly superior in sexual performance to inexperienced castrates. Nest building in rabbits improves steadily over the first three litters, even though the differences among strains of rabbits in nest-building skills are largely due to genetic differences (Petit, 1972). Rather, I would argue that infant intelligence shows some signs of canalization in the timing and the general outline of its program but clearly develops in response to the sensorimotor environment. Later intellectual development, particularly around adolescence, seems to have a far less definite form and timing for all members of the species.

All nondefective infants reared in natural human environments achieve all of the sensorimotor skills that Piaget has described. (Do you know anyone who didn't make it to preoperational thought?) This is not a trivial observation, or at least no more trivial than the observations that all nondefective human beings learn a language, are attached to at least one caretaker, achieve sexual maturity, and die in old age, if not before. One cannot say that all nondefective human beings develop formal operational logic, learn a second language, are attracted to the opposite sex, or have musical talent. There is a fundamental difference between these two sets of observations: in the first case, everyone does it; in the second case, only some do.

Uniformity of achievement may be due to limited genetic variability, to canalized development that hides genetic variability, to uniform environments, or to some combination of the three causes. The evidence suggests to me that there is less genetic variability in infant intelligence than in later intelligence, that much of the genetic variability that exists is hidden in a well-buffered, epigenetic system, and that many environments are indeed functionally equivalent for the development of sensorimotor skills. I would argue that the genetic preadaptation in sensorimotor intelligence is a strong bias toward learning the typical schemes of infancy and toward combining them in innovative, flexible ways. What human environments do is to provide the materials and the opportunities to learn. For the development of sensorimotor skills, nearly any natural, human environment will suffice to produce criterion-level performance.

Canalization at the Individual Level

Wilson (1972a,b) has argued, on the basis of his data on twins' development, that infant mental development is highly canalized at the individual level, difficult to deflect from its genotypic course, and unaffected by differences in an average range of home environments. If Wilson is correct, the heritability of infant intelligence scores should be very high, phenotype variation should be fairly large, and the data should fit canalization model 2 (p. 203 of this chapter):

> Therefore, the hypothesis is proposed that these socioeconomic and maternal care variables serve to modulate the primary determinant of developmental capability, namely, the genetic blueprint supplied by the parents. On this view, the differences between twin pairs and the similarities within twin pairs in the course of infant mental development are primarily a function of the shared genetic blueprint.
>
> Further, while there is a continuing interaction between the genetically determined gradient of development and the life circumstances under which each pair of twins is born and raised, it requires unusual conditions to impose a major deflection upon the gradient of infant development. (Wilson, 1972b, p. 917)

The primacy of "genetic blueprints" for development is a view shared by Sperry (1971). With respect to the importance of infancy and early childhood, Sperry said:

> The commonly drawn inference in this connection is that the experiences to which an infant is subjected during these years are primary. I would like again to suggest that there might be another interpretation here, namely,

that it is the developmental and maturational processes primarily that make these years so determinative.

During the first few years, the maturational program is unraveling at great speed. A lot of this determination seems to be inbuilt in nature; this is becoming increasingly clear from infant studies. I think we ought to keep our minds open to the possibility that the impression these first years are so critical is based to a considerable extent on the rapid unraveling of the individual's innate character. (p. 527)

Two lines of evidence have been used to support a strong canalization position on individual differences in infant mental development: family correlations and studies of individual consistency over time.

Family Studies

Table I shows the results of four family studies of twins and siblings, using infant mental tests.

Wilson's conclusion about the "genetic blueprint" for development is based on the very high monozygotic (MZ) correlations obtained on the same day by co-twins (Wilson & Harpring, 1972). The co-twin correlations at the same point were much higher, in fact, than the month-to-month correlations for the same infant.

Nichols and Broman's (1974) data from the Collaborative Study support Wilson's findings of high MZ correlations. Monozygotic twins could hardly have been more similar. The two studies differ, however, in their results for dizygotic (DZ) pairs. The genetic correlation between DZ co-twins is estimated to be between .50 and .55, the larger figure based on parental assortative mating. But note that Wilson's DZ pairs were considerably more similar than expected. Wachs (1972) replied that "This degree of correlation indicates the operation of nongenetic factors in the dizygotic twins' mental test performance" (p. 1005). Indeed, Nichols and Broman's dizygotic twins displayed the level of similarity predicted by a genetic model. Both same- and opposite-sexed twins have correlations of .50 ± .09, which are well within the 95% confidence interval around .5 in this study.

Now look at the siblings. Although they share the same percentage of genes, on the average, as dizygotic twins, the Fels study and the Collaborative Study found them to be far less similar in mental development during infancy. With sample sizes between 656 and 939 pairs, Nichols and Broman reported average correlations of about .20 for siblings; McCall reported .24. There is no question that sibs are less similar

TABLE I

Infant Mental Scale Correlations for Related Pairs in the First Year of Life[a]

Author	Date	Test	Age (months)	Twins						Siblings				Estimates of genetic variance	
														Twins	Sibs
				MZ	(n)	SSDZ	(n)	OSDZ	(n)	SS	(n)	OS	(n)	2(*ri*MZ *ri*DZ)	2(*ri*)
Wilson	(1972b)	Bayley	3	.84		.67								.34	
			6	.82		.74								.16	
			9	.81	(~82)	.69	(~101)[b]							.24	
			12	.82		.61								.42	
Nichols & Broman	(1974)	Bayley													
		Whites	8	.83	(48)	.51	(41)	.56	(62)	.17	(887)	.22	(939)	.64	.39
		Blacks	8	.85	(74)	.43	(47)	.57	(78)	.22	(656)	.16	(745)	.84	.38
		Total	8	.84	(122)	.46	(88)	.58	(140)	.21	(1,543)	.20	(1,684)	.76	.41
McCall	(1972a)	Gesell	6 & 12	Variance within MZ pairs significantly lower than variance within DZ pairs (n = 20)											
Freeman & Keller	(1963)	Bayley	2–12							−.24 (142)—					.48

[a]Abbreviations: MZ, monozygotic; DZ, dizygotic; SS, same sex; OS, opposite sex; *ri*, intraclass correlation coefficient.
[b]There were a few opposite-sex pairs included.

than DZ twins and that the explanation must be based on the greater environmental similarity of twins, both pre- and postnatally.

The comparison of sibling and DZ twin results is puzzling. The maximum heritability that can be obtained for any characteristic is twice the sibling correlation (Falconer, 1960). This calculation assumes that *all* of the variance between sibs is genetic and that no environmental variance is present. For behavioral traits this is an absurd assumption, and the heritability should most often be less than twice the sib correlation. A comparison of the McCall and the Nichols and Broman sibling data with the latter's twin results quickly shows a substantial difference in calculated heritability. Twice the sibling correlation varies around .40; heritabilities based on the twin results are much higher, around .75.

Since twins are nearly always tested on the same day and thus are the same age at testing, while sibs are most likely slightly different ages when tested, Nichols and Broman (1974) examined their data for age differences between sibs at testing, which were inconsequential. Then they tested for uniform correlations across the range of scores to assess the influence of extremely low scores. Extreme scores, which are much more frequent for twins in general, also showed greater concordance than higher scores among MZ twins. After eliminating the twin pairs in which one or both scored less than 50, Nichols and Broman found that the MZ correlation was reduced to .63, while the DZ correlation increased slightly to .57. Low scores had inflated the heritability estimate by a factor of 6! Although the best estimate of heritability for a population should include some low scores, the distribution of scores in a twin sample should represent the population distribution. Nichols and Broman concluded:

> These results suggest that the influence of genetics (differences) on scores on the Bayley Mental exam is greatest at the low end of the distribution, and underline the need for caution when interpreting twin correlations. (p. 5)

The hypothesis that a "genetic blueprint" programs individual infant mental development does not stand up as well as the high MZ correlations would lead us to believe.

Canalization of Patterns of Infant Mental Development

There is an additional hypothesis that deserves mention: that patterns of change in infant mental development are programmed by the individual genotype. Waddington (1971) proposed that the degree of canalization can vary depending on the alleles present at relevant loci, which would suggest that some genotypes are better buffered than others. Wilson (1972a) found that the profiles of scores obtained from the MZ twins over the first 2 years were significantly more similar than those obtained from DZ pairs, that is, that MZ co-twins show more similar responses to their common environments. McCall (1970) found no similarity in sibling profiles of intellectual development. Apart from the methodological arguments, which I will not detail here (see McCall, 1970, 1972b; McCall, Appelbaum, & Hogarty, 1973; Wilson, 1972b; Wilson & Harpring, 1972), there is a substantive question again about the interpretation of twin data. Co-twins must share very common rearing environments as well as genotypes. In infancy, the effects of shared prenatal environments may be more important than they are at later ages. Sibling data provide a crucial check on the generalization of twin results.

Continuity in Development

Continuity in developmental levels and profiles has been used as evidence for canalization. In longitudinal studies of singletons, less continuity of intellectual level has been found in infancy than in later years (Bayley, 1965). Although one recent study with a small sample failed to find any continuity (Lewis & McGurk, 1972), there are most often correlations of 0.2–0.6 in mental levels across the first 2 years. Wilson (1972a; Wilson & Harpring, 1972) has attributed the lower correlations among ages under 2 to the genetic blueprint, which has genotypically different spurts and lags in its course. Others have argued for discontinuities in the skills being tested at various ages (McCall et al., 1973; Stott & Ball, 1965).

Continuity from infant to later development can be observed for some infants who score poorly on infant mental scales. They remain retarded more often than others who are not impaired in early life. But the prediction from the first year to later childhood is greatly enhanced by consideration of the caretaking environment, which if poor, increases the risks for poor development of "retarded" infants (Sameroff & Chandler, 1975; Scarr-Salapatek & Williams, 1973; Willerman, Broman, & Fiedler, 1970). Infants who perform poorly in the first year but who have middle-class families are rarely retarded by school age. Infants at risk for retardation whose families are lower class show greater continuity in poor development (Scarr-Salapatek & Williams, 1973; Willerman *et al.*, 1970).

The reasons for later retardation may vary between middle-class and lower-class groups, but the continuous caretaking environment is at least one apparent difference. Sameroff and Chandler (1975) presented a transactional model that ascribes consistency both to organismic variables and to caretaking environments that support and maintain responses in the system. For example, infants with "difficult" temperaments are more likely to evoke assaultative behavior from their caretakers, whose battering increases the probability of more maladaptive behavior by the infants, and so forth. It is not clear that continuity in infant mental development can be attributed primarily to individual genetic blueprints.

Canalization at the Population Level: Group Differences

If infant development is highly canalized at a species level, one might expect to find universal patterns and rates of infant behavioral development, regardless of differences in child-rearing practices. No one has recently argued that the *sequences* of infant behavioral acquisitions are different across cultures. Piaget's descriptions of the important sensorimotor stages seem to apply to all normal infants. Differences in *rates* of development, however, have been noted for infants and older children of various cultural groups.

There are at least three problems with the cross-cultural paradigm in studies of canalization. First, genetic differences in rate of development may exist between populations. Relatively isolated gene pools may have evolved somewhat different patterns of infant development. Second, cross-cultural studies are fraught with methodological problems (Pick, 1975; Warren, 1972) that may apply less to infant studies than to studies of older children but that cannot be ruled out entirely. Third, the cultural practices that may, in fact, affect rates of infant development may not be identified by investigators, who may be at a loss to know what comparisons to make. These three problems—possible genetic differences, methodological problems, and identification of relevant environmental contingencies—make the interpretation of cross-cultural research on infant development difficult. Nevertheless, what has been observed?

Compared with United States white infants, those reared in other groups have been observed to be accelerated or retarded in sensorimotor development. African infants have often been found to be precocious (Freedman, 1974; Warren, 1972), particularly in the early appearance of such major motor milestones as sitting, standing alone, and walking. Although some investigators have related the precocity of African infants to child-rearing practices (Geber, 1958), United States black infants have also been found to be precocious in the same ways (Bayley, 1965; Knobloch & Pasamanick, 1953; Nichols & Broman, 1975). The similar patterns of precocity of urban United States black infants and rural African infants would seem to reduce the efficacy of a cultural argument to explain the phenomenon.

Navaho infants have been reported to be somewhat retarded in motor development, an observation that has been attributed to he cradle board but that may reflect gene pool differences. The latter explanation is particularly interesting in light of Freedman's (1974) report of the flaccid muscle tone and paucity of lower limb reflexes in Navaho newborns.

Several other reports of behavioral differences among newborns from different populations are suggestive of gene pool differences (Brazelton, Robey, & Collier, 1969; Freedman, 1974), although prenatal differences are not easily ruled out. In a particularly well-designed study, Freedman and Freedman (1969) did show differences between small samples of Chinese-American and Caucasian-American newborns whose mothers were members of the same Kaiser-Permanente hospital group. Presumably, many possible differences in prenatal life could be ruled out as competitive hypotheses.

There are few comparable studies of infant mental or language development cross culturally. We do not know when object permanence or first words appear in various groups; a first step toward studies of canalization at a population level should certainly include the simple description of the existing group variation.

The evidence from cross-cultural studies suggests that there are variations among groups in the rates of infant development. The origins of these differences are possibly cultural in part and probably genetic in part. Further studies at a descriptive level would clarify the degree of variation among groups in developmental patterns. Studies of infants from two gene pools—some of whom were reared by members of their own culture, compared with others adopted into families of a different group—would clarify the roles of genetic and environmental differences among groups. If canalization is strong for infant development in both groups, then rearing conditions should affect neither the differences among infants from different gene pools nor the similarities among infants from the same gene pool. Opportunities for such studies exist, as in the cases of black and Asian infants adopted into United States Caucasian families. Is their rate of infant development similar to Caucasian infants in the same families or to infants from the same gene pool reared by members of their own groups?

Whither Studies of Canalization?

Hypotheses about the strong canalization of infant development at species, population, or individual levels have not yet been thoroughly investigated. Studies of canalization at an individual level can benefit from several research strategies. Adoptive studies also provide a useful technique for examining the influence of shared genotypes and shared environments. Comparisons of infants with their biological relatives can be made for groups reared by their own parents and others reared by adoptive families. Further family studies of siblings and half-siblings, reared together and apart, would enhance our knowledge of genotypic differences in development. An ingenious natural experiment can be found in the families of adult monozygotic twins. In the family constellations are MZ twins, siblings, parents and their children, half-sibs, and separated "parent"–child pairs (composed of the MZ twins with the co-twin's children). A beautiful part of the design is the intactness and normality of the families who are related in all of those varied ways.

High heritabilities of infant development within a population would suggest that the environments sampled are functionally equivalent and that genotypic differences are important sources of variation. This would be evidence for canalization of that development within the context of average infant environments. Current evidence from twin and sibling studies of mental development leaves this model in doubt, however, even for the one population studied. There is even less evidence available for the canalization of mental development at a population level.

At a species level, an argument can be made for considerable restriction in phenotypic variation and for a recognizable species pattern, a pattern shared with our closest primate relatives. Whatever the sources of variation, there is a typical form of sensorimotor intelligence that develops over the first 18 months of human life. This pattern, I would argue, depends on the functional equivalence of most genotypes and environments within the species. Canalization of infant sensorimotor intelligence is not a genetic blueprint for the emergence of particular responses. It is, rather, a preadapted responsiveness to certain learning opportunities. The full development of the sensorimotor skills depends on the infants' encountering the appropriate learning opportunities, but most human environments are rich in the physical and social stimuli that infant intelligence requires. Differences in rate of sensorimotor development are not yet assignable to genetic or environmental causes, but they are relatively unimportant variations on a strong primate theme.

ACKNOWLEDGMENTS

I want to express my gratitude to Professors William Charlesworth and John Flavell for their careful, critical reviews of the manuscript. Their challenging ideas have been sometimes incorporated in the chapter, but they are not responsible for any errors of presentation. The research and review were supported by the Grant Foundation and the National Institute of Child Health and Human Development (HD-06502 and HD-08016).

REFERENCES

BAKER, P. Human adaptation to high altitude. *Science,* 1969, *163,* 1149.
BAYLEY, N. Comparisons of mental and motor test scores for ages 1–15 by sex, birth order, race, geographical location, and education of parents. *Child Development,* 1965, *36,* 379.
BELL, S. M. The development of the concept of the object and its relationship to infant-mother attachment. *Child Development,* 1970, *41,* 291.
BOWLBY, J. *Attachment and loss* (Vol. 1): *Attachment.* New York: Basic Books, 1969.

BOWLBY, J. *Attachment and loss* (Vol. 2): *Separation*. New York: Basic Books, 1973.

BRAZELTON, T. B., ROBEY, J. S., & COLLIER, G. A. Infant development in the Zinacanteco Indians of southern Mexico. *Pediatrics*, 1969, 44, 274.

BRUELL, J. Behavioral population genetics and wild *Mus musculus*. In G. Lindzey & D. D. Thiessen (Eds.), *Contributions to behavior genetic analysis: The mouse as a prototype*. New York: Appleton, 1970.

BRUNER, J. S. The nature and uses of immaturity. *American Psychologist*, 1972, 27, 687.

DENNIS, W. Causes of retardation among institutional children: Iran. *Journal of Genetic Psychology*, 1960, 96, 47.

DOBZHANSKY, T. A review of some fundamental concepts and problems of population genetics. *Cold Spring Harbor Symposium on Quantitative Biology*, 1955, 20, 1.

DOBZHANSKY, T. On types, genotypes, and the genetic diversity in populations. In J. Spuhler (Ed.), *Genetic diversity and human behavior*. Chicago: Aldine, 1967.

ELKIND, D. Cognitive development. In Y. Brackbill (Ed.), *Infancy and early childhood*. New York: Free Press, 1967.

FALCONER, D. S. *Introduction to quantitative genetics*. New York: Ronald Press, 1960.

FOX, R. The cultural animal. *Encounter*, 1970, 42, 31.

FREEDMAN, D. G. A biological approach to personality development. In Y. Brackbill (Ed.), *Infancy and early childhood*. New York: Free Press, 1967.

FREEDMAN, D. G. *An ethological perspective on human infancy*. Hillsdale, N.J.: Lawrence Erlbaum Associates, 1974.

FREEDMAN, D. G., & FREEDMAN, N. C. Behavioral differences between Chinese-American and European-American newborns. *Nature*, 1969, 24, 1227.

FREEDMAN, D. G., & KELLER, B. Inheritance of behavior in infants. *Science*, 1963, 140, 196–198.

GARDNER, B. T., & GARDNER, R. A. Two-way communication with an infant chimpanzee. In A. M. Schrier & F. Stollnitz (Eds.), *Behavior of nonhuman primates*. New York: Academic Press, 1971.

GEBER, M. The psychomotor development of African children in the first year, and the influence of maternal behavior. *Journal of Social Psychology*, 1958, 47, 185.

GHISELIN, M. *The triumph of the Darwinian method*. Berkeley, Ca.: University of California Press, 1969.

GOTTESMAN, I. I., & HESTON, L. I. Human behavior adaptations: Speculations of their genesis. In L. Ehrman, G. S. Omenn, & E. Caspari (Eds.), *Genetics, environment, and behavior*. New York: Academic Press, 1972.

HAMBURG, D. A. Sexual differentiation and the evolution of aggressive behavior in primates. In N. Kretchmer & D. N. Walcher (Eds.), *Environmental influences on genetic expression*. Washington, D.C.: United States Government Printing Office, 1969.

HARRISON, G. A. Human evolution and ecology. In *Proceedings of the Third International Congress of Human Genetics*. Baltimore, Md.: Johns Hopkins University Press, 1967.

HAYES, K. J., & NISSEN, D. H. Higher mental functions of a home-raised chimpanzee. In A. M. Schrier & F. Stollnitz (Eds.), *Behavior of nonhuman primates*. New York: Academic Press, 1971.

KAGAN, J., & KLEIN, R. E. Cross-cultural perspectives on early development. *American Psychologist*, 1973, 28, 947.

KNOBLOCH, H., & PASAMANICK, B. Further observations on the behavioral development of Negro children. *Journal of Genetic Psychology*, 1953, 83, 137.

KNOPP, C. B., & SHAPERMAN, J. Cognitive development in the absence of object manipulation during infancy. *Developmental Psychology*, 1973, 9, 430.

LaBarre, W. *The human animal.* Chicago: University of Chicago Press, 1954.

LaBarre, W. The development of mind in man in primitive cultures. In F. Richardson (Ed.), *Brain and intelligence: The ecology of child development.* Hyattsville, Md.: National Educational Press, 1973.

Lehrman, D. Semantic and conceptual issues in the nature-nurture problem. In L. R. Aronson, E. Tobach, & E. Shaw (Eds.), *Development and evolution of behavior.* San Francisco: Freeman, 1970.

Lewis, M., & McGurk, H. Evaluation of infant intelligence. *Science,* 1972, *178,* 1174.

Lipsitt, L. P. Discussion of paper by Harris. In E. Tobach, L. R. Aronson, & E. Shaw (Eds.), *The biopsychology of development.* New York: Academic Press, 1971.

Mayr, E. *Populations, species, and evolution.* Cambridge, Mass.: Harvard. University Press, 1970.

McCall, R. B. IQ pattern over age: Comparisons among siblings and parent-child pairs. *Science,* 1970, *170,* 644.

McCall, R. B. Paper presented at the meeting of the American Psychological Association, Honolulu, 1972, (a)

McCall, R. B. Similarity in developmental profile among related pairs. *Science,* 1972, *178,* 1004. (b)

McCall, R. B., Appelbaum, M. I., & Hogarty, P. S. Developmental changes in mental performance. *Monographs of the Society for Research in Child Development,* 1973, *38*(3, Whole No. 150).

McClearn, G. E. Genetic determination of behavior (animal). In L. Ehrman, G. S. Omenn, & E. Caspari (Eds.), *Genetics, environment, and behavior.* New York: Academic Press, 1972.

Nichols, P. L., & Broman, S. H. Familial resemblance in infant mental development. *Developmental Psychology,* 1974, *10,* 442.

Nichols, P. L., & Broman, S. H. *Preschool IQ: Prenatal and early developmental correlates.* New York: Wiley, 1975.

Omenn, G. S., & Motulsky, A. G. Biochemical genetics and the evolution of human behavior. In L. Ehrman, G. S. Omenn, & E. Caspari (Eds.), *Genetics, environment, and behavior.* New York: Academic Press, 1972.

Petit, C. Qualitative aspects of genetics and environment in the determination of behavior. In L. Ehrman, G. S. Omenn, & E. Caspari (Eds.), *Genetics, environment, and behavior.* New York: Academic Press, 1972.

Piaget, J. *The origins of intelligence in children.* New York: International Universities Press, 1952.

Pick, A. D. The games experimenters play: A review of methods and concepts of cross-cultural studies of cognition and development. In E. C. Carterette & M. P. Friedman (Eds.), *Handbook of perception.* New York: Academic Press, 1975.

Premack, D. On the assessment of language competence in the chmipanzee. In A. M. Schrier & F. Stollnitz (Eds.), *Behavior of nonhuman primates.* New York: Academic Press, 1971.

Sameroff, A. J., & Chandler, M. J. Reproductive risk and the continuum of caretaking casualty. In F. D. Horowitz, M. Hetherington, S. Scarr-Salapatek, & G. Siegel (Eds.), *Review of child development research* (Vol. 4). Chicago: University of Chicago Press, 1975.

Scarr-Salapatek, S., & Williams, M. L. The effects of early stimulation on low-birth-weight infants. *Child Development,* 1973, *44,* 94.

Seligman, M. E. P. On the generality of laws of learning. *Psychological Review,* 1970, *77,* 406.

Sperry, R. W. How a developing brain gets itself properly wired for adaptive function. In

E. Tobach, L. R. Aronson, & E. Shaw (Eds.), *The biopsychology of development.* New York: Academic Press, 1971.

STOTT, L. H., & BALL, R. S. Infant and preschool mental tests: Review and evaluation. *Monographs of the Society for Research in Child Development,* 1965, *30*(Whole No. 101).

VALE, J. R., & VALE, C. A. Individual differences and general laws in psychology. *American Psychologist,* 1969, *24,* 1093.

VAN LAWICK-GOODALL, J. *In the shadow of man.* New York: Dell, 1971.

WACHS, T. Technical comment. *Science,* 1972, *178,* 1005.

WADDINGTON, C. H. *The strategy of the genes.* London: Allen & Son, 1957.

WADDINGTON, C. H. *New patterns in genetics and development.* New York: Columbia University Press, 1962.

WADDINGTON, C. H. Concepts of development. In E. Tobach, L. R. Aronson, & E. Shaw (Eds.), *The biopsychology of development.* New York: Academic Press, 1971.

WARREN, N. African infant precocity. *Psychological Bulletin,* 1972, *78,* 353.

WHITE, B. L. *Human infants: Experience and psychological development.* Englewood Cliffs, N.J.: Prentice-Hall, 1971.

WILLERMAN, L., BROMAN, S. H., & FIEDLER, M. Infant development, preschool IQ, and social class. *Child Development,* 1970, *41,* 69.

WILSON, R. S. Twins: Early mental Development. *Science,* 1972, *175,* 914. (a)

WILSON, R. S. Similarity in developmental profile among related pairs of human infants. *Science,* 1972, *178,* 1005. (b)

WILSON, R. S., & HARPRING, E. B. Mental and motor development in infant twins. *Developmental Psychology,* 1972, *7,* 277.

ZELAZO, P. R. *Newborn walking: From reflexive to instrumental behavior.* Paper presented at the annual meetings of the AAAS, Symposium on Psychobiology: The Significance of Infancy, San Francisco, 1974.

7 Early Learning and Intelligence

JOHN S. WATSON AND
RICHARD D. EWY

INTRODUCTION

There are three important ways that learning in infancy might be related to what could be termed an infant's intelligence. For one, the infant's ability to have and/or benefit from early learning experiences may be a function of the infant's level of intelligence. Differences among infants in intelligence would be reflected in concurrent differences in infants' performances in learning situations—the amount of learning, the form of learning, or the type of experience that results in learning. Observation of an infant's learning could then be used to reveal something about the infant's level of intelligence.

A second possible relationship between intelligence and early learning has the opposite causal linkage. The infant's level of intelligence may change as an effect of learning experiences. This relationship might take the form of either a "deficit" model, wherein the effects of experiences are to be found in the buildup of negative effects for infants who do not receive certain types of learning opportunities (see, for instance, Jensen, 1974; Seltzer, 1973), or a "cumulative environmental effects" model, wherein different amounts of gain in intelligence correspond to the amount and/or form of early learning opportunities engaged by the infant. Either view suggests a relationship between the amount and/or form of early learning and subsequent level of intelligence.

The third view stands quite separate from notions of cause-and-effect relations. Early learning may be unrelated to intelligence in terms

JOHN S. WATSON AND RICHARD D. EWY • Department of Psychology, University of California, Berkeley, California 94720.

of being either a result of level of intelligence or as a determiner of level of intelligence, but it may still bear a useful predictive relationship to later expressions of intelligence. In this chapter, each of these three potential relations between infant learning and intelligence is considered with respect to past and future relevant research. Before we turn to that discussion, however, it is necessary to say a few words about what is meant by *learning* and *intelligence* as these terms are used in this chapter.

DEFINITION OF TERMS

Learning is used here to refer to the process in which an individual's behavior or disposition to behave is changed by experience—with the exception of changes caused by the direct biological effects of fatigue, physical injury, or drugs. This definition is meant to include such traditional learning phenomena as operant and classical conditioning, sensitization and habituation, and imitation and observational learning, as well as such things as the adaptations of sensory motor schema described by Piaget (1952). This broad definition is consistent with the nonexclusive stance taken by most reviews of infant learning (e.g., Brackbill & Koltsova, 1967; Brackbill, Fitzgerald, & Lintz, 1967; Fitzgerald & Brackbill, 1976; Fitzgerald & Porges, 1971; Hulsebus, 1973; Lipsitt, 1963; Marquis, 1931; Papousek, 1967; Rovee-Collier & Gekoski, 1979; Rovee-Collier & Lipsitt, 1980; Sameroff, 1971; Sameroff & Cavanaugh, 1979; Schaffer, 1973).

Intelligence is used here to refer to the dispositional property or properties of an individual that function in the determination of efficiency and quality of cognitive behavior and cognitive adaptation to the individual's experience. The reference to cognitive behavior and adaptation is purposely loose enough to remain compatible with most variants of professional definition (e.g., Anastasi, 1958; Bayley, 1970; Mussen, Conger, & Kagan, 1979; Stevenson, 1970; Vernon, 1979) and, one hopes' is consistent with common usage as well. About the only classes of behavior meant to be excluded are those of a purely emotional, motivational, or accidental form. The reference to "dispositional property" is a relatively technical philosophical point that deserves brief elaboration. . The reference is to the conceptual tradition of Carnap (1938) and Ryle (1949) as integrated with an assumption of materialism (e.g., Armstrong, 1968; Weissman, 1965). Briefly stated, this involves two assumptions about intelligence. One is that the property may be said to exist even when it is not being substantiated by criterial behavior; for instance, an individual may be said to be very intelligent even at times when his or her behavior is quite ordinary. The second assumption is that individual differences in intelligence refer to a real (i.e., material structural) variable as opposed to an intervening (i.e., mathematical relational) variable introduced to simplify behavioral laws (MacCorquodale & Meehl, 1948). From this perspective, intelligence is viewed as one of a variety of ability or capacity constructs that have frequent usage in both professional (e.g., Baldwin, 1958) and common-sense explanations of behavior (e.g., the "naive" theory of action as constructed by Heider, 1958).

The acceptance of intelligence as an ability construct introduces the opportunity to make a distinction that is almost inevitably raised in either lay or professional considerations of ability. The distinction concerns the extent of ability that exists at the present time ("actual ability" according to Baldwin, 1958) as contrasted with the maximal extent of ability to exist at some future time in the life of the individual ("potential ability" according to Baldwin). Clearly, this distinction has been a central concern of the intelligence-testing field from the time Binet accepted the task of sorting Parisian children into those who would and those who would not come to have the ability to succeed in formal schooling. We shall return to this distinction between actual and potential intelligence at a later point in the discussions of the three forms of relationship between early learning and intelligence.

One final point should be made concerning the use of the terms learning and intelligence. These terms will often be used as if they referred to single entities—that is, as if learning were a unitary process or intelligence a unitary trait. This will be done for the sake of simplicity in discussions that do not require maintaining a differentiated conception of the terms. However, it should be noted that the weight of existing evidence indicates that intelligence is more likely a set of traits than a single trait (Anastasi, 1958; McCall, Hogarty, & Hurlburt, 1972), and learning appears to proceed in a variety of rather independent processes (Stevenson, 1970). Our choice to simplify reality by using the terms learning and intelligence to refer to the general conceptual classes should not result in any misrepresentation of the relations to be discussed, and indeed it should be quite possible to substitute particular

learning processes or intellectual traits for the general references in almost any of the discussions. It is worth keeping in mind, however, that
any of the relationships may be stronger for certain processes and traits
than for others.

INTELLIGENCE AS A DETERMINANT OF EARLY LEARNING

Let us begin consideration of the three potential relationships by
focusing on the possibility that an infant's learning is in some degree
determined by his or her intelligence. To a certain extent this possibility
appears to flow directly from the general definitions of the two terms. If
one accepts that intelligence is in part embodied by the disposition (or
dispositions) for cognitive adaptation to experience and that learning is
the process of behavioral or dispositional change resulting from experience, then it follows logically that whenever learning involves adaptive
change in cognitive behavior, that learning should be influenced by—
and to that extent reveal—the individual's intelligence. In short, higher
intelligence should express itself in faster or better learning or in a wider
range of situations in which learning may take place. At least this should
be so when the learning involves adaptive changes in cognitive behavior.

Studies that have compared traditional measurements of intelligence with learning scores of various kinds have observed only modest
levels of correlation between the two variables. In reviewing this area,
Stevenson (1970) concluded that while the extent of the relation does not
appear to be as low as early studies implied (e.g., Garrett, 1928; Garrison, 1928; Husband, 1941; Simrall, 1947), the correlations observed in
more recent studies (e.g., Duncanson, 1964; Stake, 1961; Stevenson,
Hale, Klein, & Miller, 1968; Stevenson & Odom, 1965) have only been in
the moderate range of $r = .20-.40$. Moreover the higher correlations
appear to be limited to learning situations involving "verbal material,
tests of the ability to acquire associations, and material similar to that
found in intelligence tests" (Stevenson, 1970, p. 914). One is left to
wonder, therefore, whether the modest correlations that have been
found are reflections of truly general functional relationships between
intelligence and learning or whether they reflect relationships that depend on relatively specialized response capacities (e.g., verbal skills)
that are relevant to the shared content of these particular intelligence

and learning tasks. So then, while the general definitions of intelligence and learning lead to the expectation that higher intelligence will display itself in superior learning capacity, the available data do not forcefully substantiate this logical expectation.

It is clearly too early to conclude that early learning capacity and intelligence are independent, however. As has been observed by previous reviewers of children's learning (e.g., Brackbill & Koltsova, 1967; Stevenson, 1970), studies that have examined this question are surprisingly few in number. If the scarcity of such studies of children seems remarkable, then the void of such studies of infants (i.e., those under 2 years) should seem even more so. A variety of infant intelligence tests has been available for decades (Bayley, 1970), and techniques for studying learning and conditioning in infants have proliferated in the last 20 years (Brackbill & Koltsova, 1967; Lipsitt, 1963, 1969; Papousek, 1959; Rovee-Collier & Gekoski, 1979). So why does there appear to be no single study in which the mental age (MA) or IQ scores of infants have been related to some aspect of their performance in a standard learning situation (e.g., operant or classical conditioning)? Perhaps it is that researchers on infant behavior are less diversified in their experimental versus differential psychological training and/or interests than those researchers working with older subjects. On the other hand, perhaps infant researchers have been discouraged from initiating such a study because they are aware that the weight of existing evidence implies that (1) learning has at most a moderate relation to intelligence in older age levels and (2) that intelligence test performance in infancy has little relation to intelligence test performance at the older age levels (Bayley, 1970). The meagerness of the relation between learning and intelligence together with the instability across age of the IQ measures perhaps has suggested to many that research energies may be more fruitfully devoted to other questions of infant development.

The point remains that regardless of the difficulty in arriving at a data set that will allow some form of conclusive statement, the question of the relationship between intelligence and early learning is of basic concern in the study of the development of mental behavior. The failure to find a substantial relationship at older age levels should not be discouraging. The relationship could well change with age. As Gagné (1965) points out, there is a developmental extension in the complexity and forms of learning. Along with this transition in capacity, it seems reasonable that as the child matures, the most meaningful or focal learning experiences are occurring primarily at the frontier of expansion—at the most complex and advanced forms of learning. If one accepts intelligence as a determining factor of the efficiency and quality of cognitive adaptation to experience, then it should not be surprising to find that the form of learning that is associated with intelligence in infancy is different from the form associated with intelligence in older children or adults. Continuing this line of reasoning, one might sensibly search for potential relationships between simple instrumental learning and intelligence in early infancy even though there is a fair amount of evidence to imply that the efficiency of such nonverbal learning is minimally, perhaps even slightly negatively (Brackbill & Koltsova, 1967), associated with intelligence in older children.

McCall and colleagues (McCall, Eichorn, & Hogarty, 1977: McCall *et al.*, 1972) have made a related point concerning transitions of item structure in intelligence tests across the course of infant development. Basically, their point is that intelligence may be expressed in different behavioral functions at different ages, and that our attempt to understand and predict the developmental course of infant intelligence would fare better if we were to focus on the behaviors that are reflective of the infant's intellectual organization at the time of testing. If level of intelligence is a global variable, remaining stable across time, then the systematic variation it produced would be most clearly observable in those behavior patterns most closely related to the infant's current state of cognitive adaptation. Thus, as the thrust of development shifts—for example, from calibration and coordination of the sensory-perceptual systems (age 0–2 months) to active exploration of the environment (up to 7 months, see McCall *et al.*, 1977)—there could be an associated shift in the assessment value of behaviors related to the different areas of organization.

On the basis of a principal components analysis of intelligence test performance in a longitudinal sample of infants, it would seem that a transition in the behavioral expression of intelligence does occur in infancy (McCall *et al.*, 1977). The principal components structure in the test data showed periods of stability of individual differences and consistency of item structure, followed by unstable transition periods after which new periods of stability emerged with different item structures. On the basis of McCall's interpretation of the behavioral content of the major components in each epoch of stability, it would seem that the wave of intellectual advance passes from a behavioral area of exploration and instrumental learning ending at 7 months, to imitative acts (especially vocal) between 8 and 13 months, followed at 14–18 months by verbal recognition and labeling. Symbolic relations emerge at about 21 months and characterize the next half year. Moreover, the first principal components at each age displayed high interage correlations (i.e., with *other* principal components). The range of these correlations is greater than the range reported by Bayley (1949) on these same infants for the composite test scores. Bayley (1970) suggested partially on the basis of these low correlations for the composite scores that there was a lack of continuity between precocity on these tests and later higher intelligence. However, the correlations for principal components are in the range that suggests continuity, at least in the first 60 months of life. Although McCall *et al.* (1977) report that 6-year and older IQ assessments don't show high correlations with principal components below 21 months of age for males, such correlations for females extend as far down as 4 months and suggest the potential for future early measures of intelligence. One interesting facet of McCall's data is their implication that early expression of intelligence may be most appropriately assessed by an examination of the kind of simple nonverbal learning that appears to have so little relation to intelligence at later ages.

With the implication that instrumental learning may embody a central expression of intelligence at around 4–7 months of age, one may wonder whether this remains so at even earlier ages or whether another form of learning—classical conditioning, for example—might provide the central expression of intelligence in younger infants. This would seem a reasonable possibility, particularly if one tends to view classical conditioning as a simpler form of learning than instrumental or operant conditioning. On the other hand, Sameroff (1971) has made a strong case for viewing classical conditioning as a developmentally more advanced form of learning than operant conditioning on the bases of both empirical data and Piagetian theory. If Sameroff is right, then perhaps one should expect instrumental learning to provide an earlier expression of intelligence than classical conditioning. That may depend, of course, on whether intelligence is best displayed by a learning capacity at the time it is emerging or at the time it is reaching consolidation. For instance, Lewis (1969, 1971) has found that habituation rate appears to reach a developmental peak at about 12–13 months (see also Lewis & Brooks-Gunn, 1981). While this learning capacity surely emerges much earlier—some reports find it in the first week of life—Lewis found a significant association between relative speeds of habituation at 12–13 months and intelligence test performance at $3\frac{1}{2}$ years of age. If this correlation is accepted as indicating an expression of existing or actual intelligence in the 12-month-old (Lewis does propose viewing the habituation measure as an index of "cognitive efficiency"), then habituation would appear to be an example of a learning capacity that expresses intelligence when the capacity is reaching developmental consolidation.

In the void of empirical data about the direct relations between learning capacities and intelligence in infancy, there are many reasonable theoretical options for the developmental order of learning forms that best express intelligence. The purpose of the present discussion has not been to argue for any particular sequence in which intelligence might be seen to determine, and thus to be expressed in, various forms of early learning. The purpose has been to argue that while few if any researchers have been moved to invest energy in studying the relations between an infant's intelligence and his or her developing learning capacities, this area of study would seem to be a poteneially fruitful one for both the practical objective of finding early measures of intelligence and the theoretical objective of charting the early relations between intelligence and the capacity to learn.

Cognitive Processes within Learning

It was said earlier that the expectation of observing a relationship between early learning and intelligence followed from the possibility that early learning would be determined by intelligence, and it was said that that possibility appears to flow directly from the general definitions of the two terms. However, this logical expectancy flows from only one portion of the general definition of intelligence, the portion referring to the dispositional property (or properties) for cognitive adaptation to the individual's experience. The remaining portion of the general definition refers simply to the *existing* status, in terms of efficiency and quality, of cognitive behavior as opposed to the readiness to adapt or change that behavior as a function of experience. Both portions of the definition were included because they both have historical precedence. However, it is worth noting that Binet did not define intelligence in terms of the capacity for cognitive adaptation, and some researchers have been led to conclude firmly, if not convincingly, that intelligence cannot be defined as the ability to learn the kinds of behaviors for which intelligence tests test (e.g., Simrall, 1947). For the sake of discussion, let us assume that a valid conventional definition of intelligence does not include the portion referring to the capacity to adapt cognitive behavior to experience. Could one propose that early learning might yet reveal an infant's existing level of intelligence? The answer still would seem clearly to be yes, and it would seem worth considering how this might be so whether or not one seriously contemplates excluding the adaptation portion of the definition of intelligence.

Characteristically, studies of conditioning or learning represent the learning process with some single score indicating change in rate or relative probability of the focal behavior. If learning is a unitary process, than it makes good sense to look for the best single score to represent this process. However, if learning is itself the product of some set of component processes, then it is possible that rate or probability of behavior might change while one or more component processes remained unaltered by experience. This point relates to the question of whether intelligence is revealed by early learning. It is possible to conceive of learning being dependent on a component cognitive process that is in some way assessable in the learning situation but that is itself not necessarily altered by the learning. Some work by Watson on the component processes of instrumental learning in early infancy can serve to illustrate the form of cognitive-process assessment being considered here. The major point to be made is that while a relatively stable cognitive process may be centrally involved in an act of learning, it is possible that other factors may operate to mask its role and thus to limit the extent to which summary scores of the total learning function can provide a reliable estimate of an underlying cognitive process. As shall be seen, however, less contaminated measures of the component processes may be readily available.

Watson (1966, 1967) has described early instrumental learning as involving "contingency analysis," "contingency memory," and "response recovery." Further elaboration of the model (Watson, 1979) has added a construct of "focal time" as an attentional variable within contingency analysis. It is proposed that when an infant perceives the occurrence of an interesting event, he or she will seek to establish the relationship that may exist between that event and some aspect of his or her behavior. It is assumed that the event's occurrence will arouse the infant to search memory records of stimuli and responses under the guidances of a primitive "learning instruction" of the form "find and repeat the response that preceded the occurrence of the interesting stimulus." The effect of making these assumptions is that three separate processes arise as internal to instrumental learning. First, there is the process of an active short-term memory trace of recent stimuli and behavior that will fade in time. This process is termed *contingency memory*. Second, there is the process of seeking out the response trace as marked by its temporal relation to the stimulus trace—that is, the process termed *contingency analysis*. The later elaboration of the model proposes that this process includes developing information analogous to conditional probabilities from numerous occurrences of the stimulus and the response. And third, there is a role played in learning by the temporal distribution of the effective behavior—that is, the *response recovery* process. If the infant had no memory, then no learning could occur. If the time required for analysis of the contingency were longer than the trace storage time, then no learning could occur. And if response recovery did not occur before trace storage had faded, then even if the correct response had been selected from memory, the selection could not be confirmed if it were forgotten by the time the response occurred again.

It is not important that the reader accept this specific three-component model of early instrumental learning. For the sake of the model's illustrative function, it is important only that the reader note that of the three processes proposed in this model, only two resemble cognitive acts. The third, speed of response recovery, is a process that one might imagine to be reliably measurable, but neither common sense nor professional experience would lead one to view this motoric variable as representative of what might come under the category of a cognitive process. By contrast, both contingency analysis and contingency memory have the appearance of processes traditionally associated with the cognitive variables assessed by intelligence tests—analytic thinking and short-term memory. We might suspect that greater intellectual capacity would be composed of quicker and more accurate contingency analysis as well as a more retentive contingency memory. Yet, even if this were so, variation in response recovery time could effectively mask the expression of these cognitive variables in summary learning scores (e.g., rate and asymptote). What is needed in a case like this is some means of measuring the component processes in a manner independent of the criterial measurement of the overall learning.

In the work with the contingency awareness model of early infant learning, a methodological development holds some promise as a means for assessing individual learning records and providing direct measurement of speed of contingency analysis and temporal length of contingency memory. Watson (1974) analyzed the interresponse times of 8-week-old infants who were exposed to a mobile that turned contingent on head movement for some infants and feet movements for others. The distributions of interresponse times of rewarded responses were found to deviate from chance expectancy when analyzed in terms of the critical time demarcations of 1 and 7 seconds. The specific time demarcations were originally derived from analysis of grouped data on instrumental eye movements (fixations) of 8- and 12-week-old infants (Watson, 1967). The more recent data are notable because they were obtained with a method that was applied to individual response records and because the statistical assessment was focused on the shape of the distribution of interresponse times, which makes the assessment quite independent of indexes of learning as provided by general response rate. Although the work to data has not examined the question of the existence of reliable individual differences in contingency analysis and memory times for infants of a specific age, the method of interresponse time analysis would seem to offer a way to approach the question of individual differences in the two cognitive variables while bypassing the masking effects of the response recovery variable as the latter affects response rate.

It is not yet clear whether the interresponse time distribution analysis will make a definitive contribution to the examination of the potential link between early instrumental learning and the infant's existing level of intelligence. Nevertheless, the example illustrates a form of component analysis of learning. We would propose that such component analyses ought to be applied to infant learning before any final conclusions are drawn to the effect that early learning is not determined by an infant's intelligence.

EARLY LEARNING AS A DETERMINANT OF INTELLIGENCE

Whether or not an infant's intelligence affects his learning, one may yet ask whether his learning affects his subsequent intelligence. That is, is it the case that learning experiences in infancy play a role in determining an individual's intelligence at some point following the learning?

Surely an affirmative answer to this question was implicit within John B. Watson's claim that he could produce an adult befitting any walk of life if only given the opportunity to mainpulate the individual's learning experiences from birth (Watson, 1930). That claim caused a storm of professional debate and a flurry of popular interest that have never really subsided during the course of the subsequent half century. The concern for and credibility of teaching people to be more intelligent have risen sharply, however, since the publication of Hung's *Intelligence and Experience* (1961). Enrichment programs have proliferated both at the level of small experimental studies (Friedlander, Sterritt, & Kirk, 1975; Hess & Bear, 1968; Starr, 1971) and at the level of national demonstration projects (e.g., Head Start, Follow Through, "Sesame Street"). Yet it must be noted that this growth of investment in the proposed learning-determines-intelligence function is a growth that has been nurturd more by faith than by fact. Even though many books are appearing in the popular press on "how to raise your child's IQ," little more is known today than was known in Watson's day about how experience might be arranged to increase the intelligence of the average healthy child.

Of course one might remark that it should not be surprising that we have accumulated so little knowledge about affecting the intelligence of the average healthy child, since the average healthy child has rarely been the object of study in experiments or the object of concern in demonstration projects. Rather, effort has been concentrated on the less-than-average—or at least the expected-to-be-less-than-average (e.g., the disadvantaged and the institutionalized). While the social and moral virtue of this specialized effort is unquestionable, there is a danger at the level of theory. The problem with studying disadvantaged groups is that the power of the intervention to raise intelligence may be only at the level of eliminating deficits that otherwise accrue in the impoverished environments. Once the environment is brought into the "normal" range, differences in opportunities to learn may have no effect on the dispositional properties we label "intelligence." An analogy would be that soil conditions have no effect on plant growth over a wide range of types of soil until conditions are met that the plants are not able to handle, and deficits begin to accumulate. Such a model underlies the work of many researchers in the areas of IQ differences and compensatory education. It is discussed in terms of a "cumulative deprivation hypothesis" (Schulz & Aurbach, 1971) or a "cultural deprivation syndrome" (Ramey & Smith, 1976; Seltzer, 1973). Enrichment programs are said to halt the onset of these accumulating deficits.

This is contrasted with the model held by those researchers who have dealt with "normal range" samples (e.g., Hanson, 1975; Metzl, 1980). The suggestion here is that experiential effects do not decrease to zero once the deprived environment is brought up to par. That is, a meaningful amount of variability in intelligence within the assumed homogeneous middle class can be accounted for by differential amounts or types of experiences. Unfortunately, much work still needs to be done in this area before we can claim to know the exact nature of this formulation of early learning as a determinant of intelligence.

Another issue that arises from the predominant use of "at risk" samples is that the choice of techniques and materials incorporated into most enrichment studies has been guided by the environmental model provided by the surroundings of children judged to be developing in a normal and healthy way, that is, by the middle-class home. There is an understandable appeal to this modeling procedure when we consider the pragmatic goal of raising the less-than-average up to par. Yet it must be recognized that the procedure leads to replications of rather than experiments on environmental effects. The problem with such replications is that whether or not they work, the results will probably make little difference to our understanding of the learning–intelligence relation. For instance, if it could be demonstrated that introducing disadvantaged children to a middle-class environment leads to a normal (i.e., middle-class) distribution of intelligence, we would know little more than we knew to start with concerning the basis of the environmental effect. We would have learned that there is apparently no difference between the disadvantaged and the middle-class child as regards the effect of the middle-class environment on intelligence. We also would have learned that the two environments have different effects on the development of intelligence in disadvantaged children. Yet we already knew that the middle-class environment supported the growth of normal intelligence in the middle-class child. So the only real gain in information about the effect of learning on intelligence would be that something in the way the two environments differ makes a significant difference to the growth of intelligence in the disadvantaged child.

We may engage in *ex post facto* speculation about the dimensions of observed contrast between the two environments that we would guess are the determinative variables for intelligence, but until we can demonstrate that manipulation of these variables can either raise intelligence beyond the middle-class average or depress it below the disadvantaged average (the latter being a study we could not morally engage in), then our grasp of the environmental effect on intelligence would be far from firm. We would be little better than the simple farmer who notices that his corn grows better in field A than in field B. He is wise enough to put all his corn in field A, but he is unable to improve productivity of that field and he has no clear idea why it is better for corn than is field B. It seems fair to say that for all our effort to affect the growth of intelligence over the past 50 years, we have not done much better than the simple farmer has with his corn. When a developmental psychologist is confronted by a young mother who earnestly wants to aid, if possible, the development of her child's mind, even the most ardent of environmentalists can do little more than repeat middle-class child-rearing ideology and thus direct her to our field A.

If our lack of progress in obtaining clear and conclusive evidence about the learning-determines-intelligence function is a surprise to anyone, then it might be an even greater surprise to learn how rarely the subjects of enrichment studies have been infants (less than 2 years old). Our efforts with school-age and preschool-age children have been bountiful, particularly in the past decade. Infants, however—because of their inaccessibility, their need for individual treatment, or their enigmatic form of humanity—have received only a fraction of the attention given to older children. From either a historical or a theoretical vantage point, this lack of attention to infants is rather surprising. Both J. B. Watson and Hunt accentuated the importance of development during infancy. Both carried out empirical studies of infant behavior. However, neither they nor the many researchers influenced by them were moved to invest much energy in the specific task of discovering what (if any) learning experiences in infancy will affect the growth of intelligence.

Considering the pervasive, almost pernicious, interest in intelligence-rating in our society and the similar commitment to control in American behavioral psychology, one might well expect to find many experiments on the infant learning-determines-intelligence function. After all, what would it take to carry out a basic study of this type? Randomly assign infants to two groups. Give the infants in one group a special learning experience. Then give a posttest of intelligence, immediate and/or delayed, to both groups. If the experimental group tested statistically significantly higher than the control group, that would be rousing evidence for the effectiveness of the learning experience. Replication of the effect and refinement of controls would be pursued, of course. Yet the basic study would not be insubstantial on its own. So where are these studies? Given the interest in the topic and the simplicity of the basic design of the study, we might expect to uncover a profusion of them in the literature. For some reason, however, they are few and far between. And for the most part, those that do exist have either introduced a relatively short (two weeks or less) learning experience (e.g., Grubman, 1978; Schaffer & Emerson, 1968) or, like the majority of studies with older children, they have involved the presentation of middle-class experiences to disadvantaged or institutionalized infants (e.g., Dennis & Sayegh, 1965; Rheingold, 1956).

There have been two impressive experiments with non-middle-class infants that cannot be classified as merely replication studies for the inquiry into the learning-determines-intelligence relation. Both White's work with institutionalized infants (White, 1968, 1969, 1971) and Heber's work with disadvantaged infants (Garber & Heber, 1973; Heber & Garber, 1970, 1975; Heber, Garber, Harrington, Hoffman, & Falender, 1972) reported experimental effects that placed their average experimental subjects beyond the level of cognitive capacity anticipated. In a series of experiments, White provided special handling and altered the immediate (crib) environment of infants in an effort to accelerate the development of visual and prehensile skills. The results of this work make it clear that the introduction of various modifications within the first 6 months of life can appreciably alter the growth of sensorimotor capacity, and it seems that the development of visually directed reaching can be accelerated so that it becomes established at 5 months rather than at 7 months. White argued that visually directed reaching is very possibly the major developmental accomplishment of an infant's first half year, and therefore to advance its arrival by 2 months is probably a very meaningful advance in cognitive-perceptual life for the infant. White's study provides an important contribution to the investigation of the learning-determines-intelligence relation by giving us a good example of how the *form* of learning experiences can be structured to affect at least certain aspects of early cognitive-perceptual skill. Although it is not clear how extensively intelligence might be affected by a limited number of relatively specific learning experiences, White's results of specific environmental manipulations provide the welcome beginnings of an articulation in our meager knowledge about the learning-determines-intelligence relation.

By contrast to the rather specific behavioral acceleration in White's work, Heber reported more general enrichment effects, here showing how the intensity rather than form can be advantageously manipulated (Garber & Haber, 1973; Heber & Garber, 1970, 1975; Heber et al., 1972). Infants in a high-risk category for mental retardation (mother's IQ below 70, residing in the most disadvantaged neighborhood in Milwaukee) were given massive amounts of individual attention in an effort to displace all the presumed negative factors in the infant's social environment. At 18 and 22 months, the 20 experimental subjects were performing substantially above age norms, and 20 control infants were at age norms. This initial effect of enrichment was maintained if not slightly improved upon with later assessments of intelligence. At 42 months, experimental infants had an average IQ 33 points above the controls' average, and at 5 years of age the average IQ for the experimental group was 118, while the control groups average was 92.

Close inspection of the data shows that the difference seems to be largely due to a decline in the control group, those that remained in the impoverished environment. At 12 months of age, the experimental group's average IQ was 115, the control group's average was 113. The question relevant to our earlier discussion is, "Did the enrichment program help in the positive sense, or did it merely stave off the detrimental effects of an impoverished environment?" Part of the answer to this question involves looking at middle-class samples to see if intervention programs can record gains where the environment is not impoverished. Recently, Metzl (1980) devised an intervention program containing specific instructions that she applied to samples of normal infants (at least 5.5 lbs. birthweight: Apgar scores at least 7; two-parent, self-supporting families with at least high-school education). The program started at 6 weeks of age and continued until 6 months of age with pre- and post-administrations of the Bayley Scales of Infant Development. Only the mother was involved for one sample, both parents participated in another sample, and a control group received only the pre- and posttest assessments. The control group showed no change, the group with one parent involved increased significantly on the mental subscales to a score of 104.1, and the group with both parents involved increased even more, to a score of 111.5. By contrast, there were no increases or differences between groups on the motor subscales. Henderson (1977) reported similar short-term improvements in intelligence after a 6-month enrichment program. The point stressed by both Metzl and Henderson is that the intervention programs, although simple to teach the parents, deal with specific types of experiences—parents in Metzl's study were instructed (among other things) to imitate and respond to their infants' vocalizations, while Henderson devised a personalized program for each child involving specific learning activities. It is possible that these programs achieved their success simply by increasing environmental complexity. Learning effects do not necessarily depend on contingency experience. However, a recent dissertation by Marguerite Stevenson found that complexity of the caretaking environment as measured by Caldwell's Home Observation for Measurement of the Environment (HOME) rating scale (Caldwell, Heider & Kaplan, 1966) did not correlate with cognitive competence at 12 months of age. This finding strengthens the implication that Metzl's and Henderson's positive effects were probably dependent on the contingencies incorporated in their enrichment procedures. If future research supports this interpretation, then our earlier comments about the lack of knowledge of the specific nature of the environmental effect may have to be revised, and happily so. Such success, if it were to hold up as a change in intelligence and not just temporary learning effects, would suggest that such effects as those achieved by Heber's general enrichment program can be achieved by a program as specific and as easily instituted as White's. What is needed, of course, is follow-up assessment to see whether these effects remain after a span of time. Very few such follow-ups exist. Heber followed his subjects to age 5, and Fowler (1972; Fowler & Khan, 1974) reports a similar follow-up. Fowler's intervention program produced gains in both advantaged and disadvantaged samples at least twice as high as the 7-point IQ gain for a control cohort. A follow-up after 2 years, when subjects were between 4 and 6 years, unfortunately found that the middle-class samples—including the control group—were hovering around 130. The disadvantaged group was the only sample to decline although it still displayed an unusually high average IQ of 114. This mixed pattern of results clearly leaves much to be learned about early enrichment effects.

Moreover, if we wish to examine the causal connection between early learning and later intelligence, we have to separate the criteria of the cause from those of the effect. As it is, the definitions of intelligence and learning allow the possibility that the particular dispositional property whose change serves as the criterion of learning is a cognitive capacity whose change is simultaneously serving as the criterion of a change in intelligence. If learning is a cause of change in intelligence, then the amount of learning should be related to the amount of change in intelligence. Yet if the same disposition is taken to serve as evidence of both the learning and the change in intelligence, then a test of the relation between the two is impossible because there is no freedom for the relation to be less than perfect. What is required, of course, is separation of criteria either by content or by time. Without that we have, at best, evidence that change in intelligence is related to the opportunity for learning. Most existing studies—including White's, Heber's, and Fowler's—suffer from this limitation. The problem is that a situation offering an opportunity for learning may offer other things as well, for instance, emotional arousal. While there are many nonexperimental (i.e., correlational) studies implying that early learning is a positive influence on the development of intelligence (e.g., see the reviews by Hunt, 1961, and White, 1971), others, particularly the recent work of Kagan and his colleagues in Guatemala (Kagan & Klein, 1973; Kagen, Klein, Finley, Rogoff, & Nolan, 1979), have been interpreted as providing evidence that a great reduction in early learning opportunity has negligible effects on later display of intelligence. It seems obvious that we need more experimental investigation of the learning–intelligence relation. However, if such experiments are to take full advantage of their potential to sort out cause and effect, then they will have to incorporate separate measures of learning and intelligence. When and if that is accomplished, we would very likely possess a far more satisfactory picture of the learning-determines-intelligence function.

Early Learning as a Means for Predicting Intelligence

It is conceivable that infant learning is neither a cause nor an effect of intelligence but that it yet may provide a useful predictive relationship to future individual differences in intelligence. Learning might simply "expose" individual differences in intelligence that were not visible in its absence. This distinction between learning-exposes-intelligence and the previously discussed learning-determines-intelligence or intelligence-determines-learning is not immediately obvious. In an effort to draw this distinction most simply and clearly, it is helpful to consider a hypothetical analogy from the domain of motor development. There is no intention of implying any meaningful relationship between the development of motor and intellective capacities. The point to be made is strictly limited to the conceptual level as regards the development of *any* capacity (i.e., dispositional property). The analogy will be followed quickly by a consideration of some real and potentially relevant data.

Let us assume that we are interested in the development of the capacity for throwing a ball—specifically, the capacity to throw it fast. Further assume that we and our friends in professional baseball are aware of the following hypothetical facts: (1) Neither periodic nor prolonged training in childhood has any appreciable effect on the average speed with which adults are capable of throwing a baseball; (2) when tested in a situation in which they were enticed to throw a baseball, adults throw it with an average speed of 50 mph (SD = 10 mph); (3) when tested in a situation in which they are enticed to throw a golf ball, 18-month-old infants throw it with an average speed of 15 mph (SD = 5 mph); and (4) the correlation between throwing capacity at 18 months and adulthood is r = .10. We can imagine that this very modest association of capacity at the two ages might have some theoretical interest to physical anthropologists, but it would surely be of little interest to our practical-minded friends in professional baseball.

Now assume that we decide to introduce a short-term training program prior to testing. The training consists of practice trials interspersed with exposure to a model who illustrates efficient foot and leg movements and oppositional arm action. When the subject reaches asymptotic performance, the test situation is presented. It is conceivable that the following facts might emerge: (1) it is still true that neither periodic nor prolonged training in childhood has any effect on the average speed with which adults are capable of throwing a baseball; but immediately following a training program, (2) 18-month-olds throw the golf ball with an average speed of 20 mph (SD = 5 mph) and (3) adults throw the baseball with an average speed of 55 mph (SD = 10 mph); (4) the correlation between the capacity to throw the ball before versus immediately following the training program is r = .20 for 18-month-olds and r = .90 for adults; and most notably (5) the correlation between throwing capacity following training at 18 months and throwing capacity in adulthood, with or without training, is r = .70. These new facts would be of additional theoretical interest for researchers focusing on the development of physical capacities. And it is easy to imagine that those people who are interested in improving the quality of professional baseball would be very excited about the new-found predictive relationship between throwing capacity at 18 months and adult capacity. Even though childhood training would not be expected to affect later throwing capacity, the predictive power of 18-month-old capacity would yet be a fact having significant practical value. It could be used to select those children to whom special attention might be most fruitful to provide special training in other capacities that are required by the sport—fielding and batting, for example.

One might explain the predictive power of the posttraining capacity at 18 months by proposing that the test score following training reflects the capacity of the individual to learn to throw a ball. The ability to throw a ball at any age could be assumed to depend on past learning, and so the posttraining score at 18 months predicts the adult capacity to throw (with or without training) because it distinguishes between individuals who are and those who are not likely to learn to throw well over the course of experience between 18 months and adulthood. However, in the present hypothetical example there is strong evidence that adult capacity to throw is not appreciably affected by learning, inasmuch as neither periodic nor prolonged training in childhood affects adult capacity.

There is an alternative explanation of the predictive power of the posttraining capacity at 18 months that is more consistent with all the hypothetical facts at hand. One might propose that the capacity to throw a ball at any age is dependent on both experience in throwing and physical characteristics (e.g., skeletal, neurological, and muscular) that are largely governed by an individual's genetic and maturational status. With a modest and yet critical amount of experience, the differences between individual levels of capacity can be almost totally accounted for by differences in genetic and maturational status. Yet with less than the critical level of experience, differences in capacity might be predominantly accounted for by the amounts of experience individuals did have.

One may further assume that if the probability of certain experiences is relatively constant in an individual's environment, then his likelihood of having been exposed to some minimal level of these experiences will increase as time passes (i.e., as he or she ages). Now if early physical status were correlated with adult physical status, then the assumptions made so far would lead to an expectancy that (1) capacity in ball-throwing would be more closely associated with physical status in adulthood than in early childhood because of the adult's greater likelihood of having been exposed to the critical level of experience, and (2) the provision of training prior to the assessing of capacity in childhood should increase the predictive relation between childhood capacity and adult capacity, since it would increase the extent to which physical status would be a major factor in accounting for individual differences in childhood capacity. This explanation would be consistent with the facts that average capacity changes by a large amount with age and yet is little affected by periodic or prolonged training in childhood. The large age differences are attributed to maturation. It should also be noted that there would be nothing mysterious about the fact that training 18-month-olds has a large effect on the predictive correlation while it has only a small effect on the average level of capacity at that age. The situation would be one in which the training affects the distribution around the mean by affecting the performance of some individuals very much. For this hypothetical example, then, the distribution of capacity following training at 18 months is more closely aligned with the distribution of adult capacity than is capacity prior to training at 18 months.

One way of labeling this distinction between capacity prior to and following short-term training is to call the former *manifest capacity* and the latter *latent capacity*. Each is a dispositional property statement about the existing status of an individual, and thus each is a form of what Baldwin (1958) termed "actual capacity." Furthermore, each is to be distinguished from future states of capacity such as Baldwin's "potential capacity," which refers to the maximal capacity that the individual has the possibility of attaining in the future.

Figure 1 presents a hypothetical graph illustrating a simple and parallel age change in the mean values of manifest and latent capacity. Figure 2 presents a more complicated and nonparallel age change in the two forms of existing capacity. One can imagine that the differences in the relations illustrated in Figures 1 and 2 might exemplify the contrast in growth functions of separate types of behavioral capacity (e.g., ball-throwing versus memory for digits), or the difference between Figures 1 and 2 might exemplify the contrast in growth functions arising for a single type of behavioral capacity as this develops in two quite different environments (e.g., middle class versus lower socioeconomic class). Figure 3 illustrates the contrast that could conceivably exist between the predictive validities of the correlations between manifest capacity in

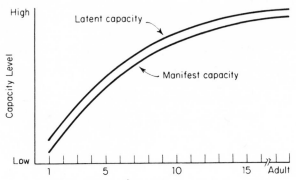

FIGURE 1. Hypothetical growth curves of manifest and latent capacity illustrating cases of parallel growth.

adulthood (defined as 18 years or older) and the two measures of capacity at earlier ages.

Returning to the example of ball-throwing, it should be clear that the learning experience involved in the assessment of latent capacity would not be viewed as either a cause or an effect of the future level of manifest capacity, nor would it be viewed as a cause or an effect of the association between the present state of latent capacity and the future state of manifest capacity. The role of the short-term training seems best described as one of exposing the association between present and future capacity. Surely the difference in performance observed prior to, versus following, training is attributable to learning during the training procedure, but one would not contend that the difference in predictive

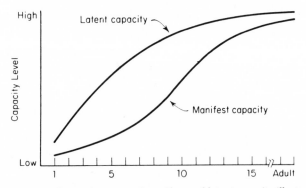

FIGURE 2. Hypothetical growth curves of manifest and latent capacity illustrating cases of nonparallel growth.

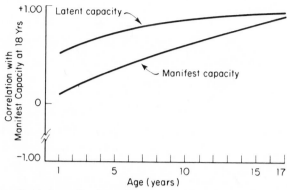

FIGURE 3. Hypothetical graph of correlations of both manifest and latent capacity at successive ages with manifest capacity at 18 years.

validity of manifest versus latent capacity is itself an effect of learning—at least not in the sense that we normally use the term *learning*. We would say a subject learned his new level of behavior (e.g., he learned to throw a ball at 26 mph), but we would not say that he learned the new level of predictive validity (e.g., he learned to display a closer association between his present and his future capacity to throw a ball relative to the capacities of other children). Indeed we have no way to conceive of how such a thing could be learned!

How does this hypothetical example help us understand real-world relationships between learning and intelligence? If it helps, it does so by illustrating a potential relationship between infant learning and future intelligence, which is neither of the form intelligence-determines-learning nor of the form learning-determines-intelligence. In an analogy with the motor capacity example, early learning may simply provide a means of exposing underlying (i.e., latent) differences in intellectual capacity that happen to have a greater developmental stability than those differences in intellectual capacity that are visible without the provision of a special learning experience. It is interesting to note that while the developers of intelligence tests have long been concerned with the possibility of contaminating effects that might be introduced by experience in the taking of an intelligence test (Anastasi, 1958), the fact that individuals will improve with experience has not been closely examined for its potential use in improving the predictive validity of early test scores.

The potential for such improvement exists, as is illustrated by data supplied by Broman, Nichols, and Kennedy (1975) concerning racial differences in preschool IQ. Blacks averaged 13 points behind whites, similar to differences found in most such studies (and sample sizes were quite large—over 12,000 white children and over 14,000 black children). However, the gap may reflect differences between blacks and whites in manifest capacity and may not reflect true differences in latent capacities of the two groups. This is a reasonable possibility because predictor variables in the multiple regression analyses were found to account for 25%–28% of the variance in IQ for white children, but only 15%–17% among black children. If this were the effect of greater variation between manifest and latent intelligence among black children, then one might find special advantage for the assessment of black children if early learning measures were used.

We could find no specific application of this principle of learning as an "exposing agent" in the attempt to predict later intelligence. However, two reports on preschool children come relatively close to this goal. Jacobson, Berger, Bergman, Millham, and Greeson (1971) gave a conceptual learning program to 3- and 4-year-old children from poverty backgrounds. After 20 hours of participation in the program, all subjects had acquired learning sets. Those subjects with low (46–83) starting IQs gained an average of about 20 points, and children starting with higher IQs gained about 9 points. The missing piece here is how performance on the learning set acquisition (number of trials to criterion) relates to the posttest IQ (presumably a better measure of latent intelligence than the pretest IQ). Similarly, Kinnie and Sternlof (1971) trained several groups of middle-class whites and lower-class whites and blacks either for familiarization with the language and materials used on the test (similar to a learning set program) or for familiarization with middle-class adults similar to the traditional IQ tester. Again, no data are presented on how individual performance on the familiarization task relates to individual differences in assessed IQ. We cannot fault these studies for that failure since they were designed with other goals in mind. But they are nonetheless illustrative of the kind of relationship we are suggesting for learning as an "exposing agent" of intelligence.

A related concept is Vygotsky's "zone of proximal development" (Vygotsky, 1978). This zone is the distance between a child's actual developmental level and the level of competence he or she could achieve with aid and guidance. The difference between Vygotsky's concept and our concept of latent capacity is that Vygotsky suggests aiding, prompting, and guiding the child during the assessment, while we suggest that training take place prior to assessment. This removes the potential problem that achievement with assistance may be more a function of the one giving the test than the child who is being assessed. However, it is apparent that research inspired in part by Vygotsky often makes operational the concept of "achievement with aid" as training prior to a standard form of assessment. For instance, Field (1981) reports a number of studies involving pretest, training, and posttest on Piagetian conservation tasks. She concludes that such training "appears to be a particularly good tool to distinguish between true retardates and children whose learning disabilities have other sources" (p. 496). We believe that such findings should encourage others to apply latent capacity analysis to concerns about predictive validity in intelligence testing of normal subjects.

Nearly 2 decades ago, in an attempt to improve the predictive relationship between IQ tests and later scholastic progress, MacKay and Vernon (1963) began work on intelligence tests that involve learning in a controlled situation. They concluded that the effort was somewhat disappointing, that the predictive value of their various learning tests varied widely, and that the gain scores held no real advantage. Rather, the score with highest predictive validity was the final score, achieved after instruction and practice. It is this final score, however, that fits our definition of latent capacity. The finding is therefore consistent with our claim that this measure could maintain a better predictive relationship to later achievement than could either pretraining scores or learning scores within training.

The previously mentioned study by Simrall (1947) provides some additional relevant data, although it is neither a study of infants nor a study of predictive relations between early and adult intelligence measures. Recall that Simrall provided high-school children with training on intelligence test behavior and found that their gains in performance were unrelated to their pretraining scores. She concluded from this fact that the performance gain (i.e., the learning) was unrelated to intelligence (i.e., the original performance scores). But we may conclude something quite different. We may conclude that the predictive value of the new performance (original performance plus gain) can be very different from the original performance. Had the gains been perfectly correlated with the original scores, then addition of the gains to the original scores would not appreciably change the predictive value of the sum score from that of the original. If intelligence test performance gains are as independent of original scores in infancy as they are in older children (and we have no reason to believe that they would not be), then the stage is clearly set for looking at the potential interplay of latent intelligence in infancy and intelligence test performance in later years.

Whether we are concerned with the practical problems of prediction or with the theoretical implications of dispositional continuity, one thing seems clear at the start: Any attempt to assess latent intelligence in infancy should be encouraged by the fact that it has little chance of uncovering lower associations between intellectual capacity in infancy and adulthood than have to date been obtained with existing tests of manifest infant intelligence (e.g., Bayley, 1970). If one considers the proposals here, together with the emergence of interest in Vygotsky's zone of proximal development, and the progression of work from MacKay and Vernon (1963) through Kinnie and Sternlof (1971) and Jacobson *et al.* (1971) to Field (1981), one cannot help but feel that assessment of infant intelligence is a field that has room for improvement—or to put it another way, the latent capacity of intelligence testing in infancy exceeds the manifest capacity the field currently enjoys.

SUMMARY AND CONCLUSIONS

In this chapter we have considered the potential relation between early learning and intelligence in terms of three specific forms that the relation might take: intelligence-determines-learning, learning-determines-intelligence, and learning-exposes-intelligence. We have considered these three forms separately, but of course it is quite possible that two or more forms may coexist and even interact with one another. For example, intelligence may determine learning, which in turn determines subsequent intelligence. However, before much energy is spent on working through the potential interweavings of these forms, it should be clear that we should first invest sufficient energy in establishing the existence of a substantial fabric for any of these forms. Earlier, more general reviews have noted with some astonishment just how little we know about the relation of learning and intelligence (Stevenson, 1970; White, 1971). When one limits one's sight to infancy, the extent of our knowledge is little indeed. Thus, to a large extent this chapter has been an exercise in dividing a small pie. If there is any merit to such an exercise, it is perhaps that in dividing a small morsel, there is the potential of increasing the number of appetites aroused.

REFERENCES

ANASTASI, A. *Differential psychology* (3rd ed.). New York: Macmillan, 1958.

ARMSTRONG, D. A. *A materialist theory of the mind.* New York: Humanities Press, 1968.

BALDWIN, A. L. The role of an "ability" construct in a theory of behavior. In D. C. McClelland, A. L. Baldwin, U. Bronfenbrenner, & F. L. Strodtbeck (Eds.), *Talent and society.* Princeton, N.J.: Van Nostrand, 1958.

BAYLEY, N. Consistency and variability in the growth of intelligence from birth to eighteen years. *Journal of Genetic Psychology,* 1949, *75,* 165–196.

BAYLEY, N. Development of mental abilities. In P. H. Mussen (Ed.), *Carmichael's manual of child psychology* (3rd ed., Vol. 1). New York: Wiley, 1970.

BRACKBILL, Y., & KOLTSOVA, M. M. Conditioning and learning. In Y. Brackbill (Ed.), *Infancy and early childhood.* New York: Free Press, 1967.

BRACKBILL, Y., FITZGERALD, H. E., & LINTZ, L. M. A developmental study of classical conditioning. *Monographs of the Society for Research in Child Development,* 1967, *32*(Whole No. 8).

BROMAN, S. H., NICHOLS, P. L., & KENNEDY, W. A. *Preschool IQ: Prenatal and early developmental correlates.* New York: Wiley, 1975.

CALDWELL, B. M., HEIDER, J., & KAPLAN, B. *The inventory of home stimulation.* Paper presented at the meetings of the American Psychological Association, New York, September, 1966. (Available from the Center for Early Development and Education, University of Arkansas at Little Rock, Little Rock, Arkansas 72204.)

CARNAP, R. Logical foundations of the unity of science. In *International encyclopedia of unified science* (Vol. 1, Part 1). Chicago: University of Chicago Press, 1938.

DENNIS, W., & SAYEGH, Y. The effect of supplementary experiences upon the behavioral development of infants in institutions. *Child Development,* 1965, *36,* 81.

DUNCANSON, J. P. *Intelligence and the ability to learn.* Princeton, N.J.: Educational Testing Service, 1964.

FIELD, D. Comparison of the conservation acquisition of mentally retarded and non-retarded children. In M. P. Friedman, J. P. Das, & N. O'Connor (Eds.), *Intelligence and learning.* New York: Plenum Press, 1981.

FITZGERALD, H. E., & BRACKBILL, Y. Classical conditioning in infancy: Development and constraints. *Psychological Bulletin,* 1976, *83,* 353–376.

FITZGERALD, H. E., & PORGES, S. W. A decade of infant conditioning and learning research. *Merrill-Palmer Quarterly,* 1971, *17,* 79–117.

FOWLER, W. A developmental learning approach to infant care in a group setting. *Merrill-Palmer Quarterly* 1972, *18,* 145.

FOWLER, W., & KHAN, N. *A follow-up investigation of the later development of infants in enriched group care.* Paper presented at the Annual Meeting of the American Educational Research Association, Chicago, April 1974. (ERIC Document Reproduction Service No. ED 093 506)

FRIEDLANDER, B. Z., STERRITT, G. M., & KIRK, G. E. (Eds.). *Exceptional infant* (Vol. 3): *Assessment and intervention.* New York: Brunner/Mazel, 1975.

GAGNÉ R. M. *The conditions of learning.* New York: Holt, Rinehart & Winston, 1965.

GARBER, H., & HEBER, R. *The Milwaukee Project: Early intervention as a technique to prevent mental retardation.* Storrs, Ct.: University of Connecticut Technical Paper, 1973. RICE Document Reproduction Service No. ED 080 162)

GARRETT, H. E. The relation of tests of memory and learning to each other and to general intelligence in a highly selected adult group. *Journal of Educational Psychology,* 1928, *19,* 601.

GARRISON, K. C. The correlation between intelligence test scores and success in certain rational organization problems. *Journal of Applied Psychology,* 1928, *12,* 621.

GRUBMAN, D. *Pretest stimulation effects upon infant developmental test performance.* Doctoral dissertation, University of California, Berkeley, 1978.

HANSON, R. Consistency and stability of home environmental measures related to IQ. *Child Development,* 1975, *46,* 470–480.

HEBER, R., & GARBER, H. *An experiment in the prevention of cultural-familial mental retardation.* Madison, Wisc.: Regional Rehabilitation Research and Training Center in Mental Retardation, 1970. (ERIC Document Reproduction Service No. ED 059 762)

HEBER, R. & GARBER, H. The Milwaukee Project: A study of the use of family intervention to prevent cultural-familial mental retardation. In B. Z. Friedlander, G. M. Sterritt, & G. E. Kirk (Eds.), *Exceptional infant* (Vol. 3). New York: Brunner/Mazel, 1975.

HEBER, R., GARBER, H., HARRINGTON, S., HOFFMAN, C., & FALENDER, C. *Rehabilitation of families at risk for mental retardation.* Madison, Wisc.: Regional Rehabilitation Research and Training Center in Mental Retardation, 1972. (ERIC Document Reproduction Service No. ED 087 142)

HEIDER, F. *The psychology of interpersonal relations.* New York: Wiley, 1958.

HENDERSON, B. A. *Infant enrichment: The effects of a prescribed curriculum on cognitive development* (Doctoral dissertation, University of Wisconsin, Madison). *Dissertation Abstracts International,* 1977, *38*(6–A), 3370.

HESS, R. D., & BEAR, R. M. (Eds.). *Early education.* Chicago: Aldine, 1968.

HULSEBUS, R. C. Operant conditioning of infant behavior: A review. In H. W. Reese (Ed.), *Advances in child development and behavior* (Vol. 8). New York: Academic Press, 1973.

HUNT, J. McV. *Intelligence and experience.* New York: Ronald, 1961.

HUSBAND, R. W. Intercorrelations among learning abilities: III. The effects of age and spread of intelligence upon relationships. *Journal of Genetic Psychology,* 1941, *58,* 431.

JACOBSON, L., BERGER, S., BERGMAN, R., MILLHAM, J., & GREESON, L. Effects of age, sex, systematic conceptual learning, acquisition of learning sets, and programmed social interaction on the intellectual and conceptual development of preschool children from poverty backgrounds. *Child Development,* 1971, *42,* 1399–1415.

JENSEN, A. R. Cumulative deficit: A testable hypothesis? *Developmental Psychology,* 1974, *10,* 996–1019.

KAGAN, J., & KLEIN, R. E. Cross-cultural perspectives on early development. *American Psychologist,* 1973, *28,* 947–961.

KAGAN, J., KLEIN, R. E., FINLEY, G. E., ROGOFF, B., & NOLAN, E. A cross-cultural study of cognitive development. *Monographs of the Society for Research in Child Development,* 1979, *44*(Whole No. 5).

KINNIE, E., & STERNLOF, R. The influence of nonintellective factors on the IQ scores of middle- and lower-class children. *Child Development,* 1971, *42,* 1989–1995.

LEWIS, M. A development study of information processing within the first three years of life: Response decrement to a redundant signal. *Monographs of the Society for Research in Child Development,* 1969, *34*(Whole No. 9).

LEWIS, M. Individual differences in the measurement of early cognitive growth. In J. Hellmuth (Ed.), *Exceptional infant* (Vol. 2). New York: Brunner/Mazel, 1971.

LEWIS, M. & BROOKS-GUNN, J. Visual attention at three months as a predictor of cognitive functioning at two years of age. *Intelligence*, 1981, 5, 131–140.

LIPSITT, L. P. Learning in the first year of life. In L. P. Lipsitt & C. C. Spiker (Eds.), *Advances in child development and behavior* (Vol. 1). New York: Academic Press, 1963.

LIPSITT, L. P. Learning capacities of the human infant. In R. J. Robinson (Ed.), *Brain and early behavior*. New York: Academic Press, 1969.

MacCORQUODALE, K., & MEEHL, P. E. Hypothetical constructs and intervening variables. *Psychological Review*, 1948, 55, 95.

MacKAY, G. W. S., & VERNON, P. E. The measurement of learning ability. *British Journal of Educational Psychology*, 1963, 33, 177–186.

MARQUIS, D. P. Can conditioned responses be established in the newborn infant? *Journal of Genetic Psychology*, 1931, 39, 479–492.

McCALL, R. B., HOGARTY, P. S., & HURLBURT, N. Transitions in infant sensorimotor development and the prediction of childhood IQ. *American Psychologist*, 1972, 27, 728–748.

McCALL, R. B., EICHORN, D. H., & HOGARTY, P. S. Transitions in early mental development. *Monographs of the Society for Research in Child Development*, 1977, 42(Whole No. 3).

METZL, M. Teaching parents a strategy for enhancing infant development. *Child Development*, 1980, 51, 583–586.

MUSSEN, P. H., CONGER, J. J., & KAGAN, J. *Child development and personality* (5th ed.). New York: Harper & Row, 1979.

PAPOUSEK, H. A method of studying conditioned food reflexes in young children up to the age of six months. *Pavlov Journal of Higher Nervous Activity*, 1959, 9, 136–140.

PAPOUSEK, H. Conditioning during early postnatal development. In Y. Brackbill & G. G. Thompson (Eds.), *Behavior in infancy and early childhood*. New York: Free Press, 1967.

PIAGET, J. *The origins of intelligence in children*. New York: International Universities Press, 1952.

RAMEY, C., & SMITH, B. Assessing the intellectual consequences of early intervention with high-risk infants. *American Journal of Mental Deficiency*, 1976, 81, 318–324.

RHEINGOLD, H. L. The modification of social responsiveness in institutionalized babies. *Monographs of the Society for Research in Child Development*, 1956, 21(2).

ROVEE-COLLIER, C. K., & GEKOSKI, M. J. The economics of infancy: A review of conjugate reinforcement. In H. W. Reese & L. P. Lipsitt (Eds.), *Advances in child development and behavior* (Vol. 13). New York: Academic Press, 1979.

ROVEE-COLLIER, C. K., & LIPSITT, L. P. Learning, adaptation, and memory. In P. M. Statton (Ed.), *Psychobiology of the human newborn*, New York: Wiley, 1980.

RYLE, G. *The concept of mind*. New York: Barnes & Noble, 1949.

SAMEROFF, A. J. Can conditioned responses be established in the new-born infant? *Developmental Psychology*, 1971, 5, 1.

SAMEROFF, A. J., & CAVANAUGH, P. J. Learning in infancy: A developmental perspective. In J. D. Osofsky (Ed.), *Handbook of infant development*. New York: Wiley, 1979.

SCHAFFER, H. R. The multivariate approach to early learning. In R. H. Hinde & J. Stevenson-Hinde (Eds.), *Constraints on learning: Limitations and predispositions*. New York: Academic Press, 1973.

SCHAFFER, H. R., & EMERSON, P. E. The effects of experimentally administered stimulation on developmental quotients of infants. *British Journal of Social and Clinical Psychology*, 1968, 7, 61.

SCHULZ, C. B., & AURBACH, H. A. The usefulness of cumulative deprivation as an explanation of educational deficiencies. *Merrill-Palmer Quarterly*, 1971, 17, 27–39.

SELTZER, R. The disadvantaged child and cognitive development in the early years. *Merrill-Palmer Quarterly*, 1973, *19*, 241–252.

SIMRALL, D. Intelligence and the ability to learn. *Journal of Psychology*, 1947, *23*, 27.

STAKE, R. E. Learning parameters, aptitudes, and achievements. *Psychometric Monographs*, 1961, *9*.

STARR, R. Cognitive development in infancy: Assessment, acceleration, actualization. *Merrill-Palmer Quarterly*, 1971, *17*, 153.

STEVENSON, H. W. Learning in children. In P. H. Mussen (Ed.), *Carmichael's manual of child psychology* (3rd ed., Vol. 1). New York: Wiley, 1970.

STEVENSON, H. W., & ODOM, R. D. Interrelationships in children's learning. *Child Development*, 1965, *36*, 7.

STEVENSON, H. W., HALE, G. A., KLEIN, R. E., & MILLER, L. K. Interrelations and correlates in children's learning and problem solving. *Monographs of the Society for Research in Child Development*, 1968, *33*(7).

VERNON, P. E. *Intelligence: Heredity and environment.* San Francisco: Freeman, 1979.

VYGOTSKY, L. S. *Mind in society.* M. Cole, V. John-Steiner, S. Scribner, & E. Souberman (Eds.). Cambridge, Mass.: Harvard University Press, 1978.

WATSON, J. B. *Behaviorism* (Rev. ed.). New York: Norton, 1930.

WATSON, J. S. The development and generalization of "contingency awareness" in early infancy: Some hypotheses. *Merrill-Palmer Quarterly*, 1966, *12*, 123.

WATSON, J. S. Memory and "contingency analysis" in infant learning. *Merrill-Palmer Quarterly*, 1967, *13*, 55.

WATSON, J. S. *Early infant learning: Some roles and measures of memory, thinking, and trying.* Paper presented at Annual Meeting of the British Psychological Society, Bangor, North Wales, 1974.

WATSON, J. S. Perception of contingency as a determinant of social responsiveness. In E. Thoman (Ed.), *The origins of social responsiveness.* Hillsdale, N.J.: Lawrence Erlbaum Associates, 1979.

WEISSMAN, D. *Dispositional properties.* Carbondale: Southern Illinois University Press, 1965.

WHITE, B. L. Informal education during the first months of life. In R. D. Hess & R. M. Bear (Eds.), *Early education.* Chicago: Aldine, 1968.

WHITE, B. L. Child development research: An edifice without a foundation. *Merrill-Palmer Quarterly*, 1969, *15*, 50.

WHITE, B. L. *Human infants: Experience and psychological development.* Englewood Cliffs, N.J.: Prentice-Hall, 1971.

8　Environmental Risks in Fetal and Neonatal Life as Biological Determinants of Infant Intelligence

JANE V. HUNT

INTRODUCTION

The infant who is at risk for intellectual impairment because of environmental events in fetal or neonatal life is the special concern of this chapter. Investigators who study the underlying causes of impairment approach the question from varied perspectives. Both medical and social science research will be reviewed here to provide the reader with some synthesis, examine what is known, and consider research methods that offer promise for investigating unresolved questions.

At the outset, some definition of terms may be helpful. *Environmental risk* is defined here, and differentiated from other events and conditions resulting in intellectual impairment, by including only those risks that are (1) culturally determined, (2) recognized as risk factors, and (3) currently modifiable. Both biological and social risks can meet these criteria. This review will emphasize early biological risks, but will include some consideration of interaction effects with factors from the social environment. Some aspects of medical care and treatment are included in this definition. Rapid advances in fetal and neonatal medicine have generated research questions about both traditional and new medical practices and there is a general, heightened public awareness of

JANE V. HUNT • Institute of Human Development, University of California, Berkeley, California 94720.

treatment alternatives. New technologies for monitoring the fetus and newborn are making some studies more feasible than in the recent past.

A major distinction between *fetal* and *neonatal life* is the kind of risks that can occur. Most environmental risks in fetal life are mediated through the infant's mother. Infants are their mothers' victims when a risk such as maternal drug abuse or malnutrition is at issue but, at the same time, they are insulated from many risks of extrauterine life. However, because fetuses are becoming more and more medical patients in their own right, they may be encountering new risks. For example, they may be exposed to drugs *in utero* prescribed specifically for them (e.g., steroids to hasten the maturation of lung function when preterm delivery is unavoidable). The line between fetus and newborn is becoming blurred in other ways. Intrusive procedures such as amniocentesis (sampling the fluid surrounding the fetus) and intrauterine blood transfusions are established special procedures. Fetal surgery to correct congenital problems is already a reality and will become more widespread in the future. We can observe fetal development and even such behaviors as thumb sucking by means of sonographic images and can monitor vital functions with increasing precision. Some new procedures may suggest new risks, but other innovations provide the means to define old risks (e.g., fetal hypoxia) with more precision.

Environmental risks in neonatal life are more diverse and, often, more readily specified and studied than are those encountered in fetal life. Most of the risks associated with the newborn are exemplified in studies of preterm infants, a population that has received considerable research emphasis in recent years. The rapid increase in neonatal care facilities has called attention to these infants and made them more available for study, for many now survive extremes of prematurity and neonatal complications that would have proved fatal before the advent of intensive care. Research questions are raised about the quality of the survivors and the long-term consequences of treatment procedures. The infant may spend weeks or months in the intensive care nursery, an environment very different from either intrauterine experience or the home environment, and his or her special environmental needs and risks are being investigated. Development during the preterm period is of interest because this is the "fetal-infant" described by Gesell almost 4 decades ago as being "a kind of open-sesame for defining the nature of [fetal] development, for he is not walled off from direct observation as is his unborn counterpart" (Gesell & Amatruda, 1945, p. 107). To this observation we add that such infants also are not walled off from a myriad of threatening environmental events.

In the context of risk studies, measurement of *infant intelligence* implies the determination of intellectual impairment associated with the risk. Impairment in infancy is defined here as more a construct than a reality, as is the generic concept of infant intelligence. With few exceptions, the increased likelihood of subsequent intellectual disabilities at older ages is inferred from deviant infant characteristics that are thought to be relevant. These include neurological abnormalities observed through clinical examinations or neurophysiological measurements, atypical behavioral responses obtained in a variety of situations, and delays noted in standard clinical assessments of development or rates of mental growth. The predictive significance of the atypical characteristics in infancy is not certain, and indeed, is one major research interest.

The rapid pace and predictable sequence of development during infancy permit a sensitive monitoring of both depressed function and delayed organization of behavior. Standard assessments of infant mental development—such as those of Bayley (1969), Cattell (1940), and Gesell and Amatruda (1941)—provide empirical evidence for a sequence of observable behaviors in infancy, with both the ordering and timing of these behaviors being predictable within limits. Such assessments provide a comprehensive evaluation of the infant's current status, but there is ample evidence that scores during the first year of life are not correlated with IQ at older ages. There is other evidence (see Hunt, 1979, 1981) that some intellectual disabilities in childhood are not predicted from infant scores. Consequently, research in recent years has explored alternatives to standard developmental tests when examining associations between risk conditions and infant behavior. Alternatives have included measurements of sensory perception and integration, operant learning, social interaction and specific components of standard tests. Measurements of neonatal behavior have also been of interest, including both specific abilities and more general assessments, such as the Brazelton Neonatal Behavioral Assessment Scale (Brazelton, 1973).

Valid long-term prediction for the individual from any specific measurement of infant behavior is unlikely. The demonstration of a direct association between infant status and later intellectual attainment is complicated by the mediating effects of environmental events on the continuing differentiation of brain structure and function beyond infancy. Such effects are well documented in research using animal models (cf. Rosenzweig & Bennett, 1978). Except in cases of devastating damage to brain structures, these mediating effects are expected to be potent and, often, controlling (cf. Sameroff & Chandler, 1975). Studies of infant status, whether or not they are predictive and precisely because they do not reflect the total long-term effects of all environmental differences, may be the most appropriate way of determining some direct associations between early environmental insults and behavior.

The term "high-risk infants," which now appears frequently in published research, is used to denote infants with several different risk conditions. Usually, one of two definitions of intellectual impairment is implicit in such studies. In the first case, early brain damage is suspected or confirmed and there is an assumption that the original intellectual potential of the individual has been reduced. In the second case, the underlying potential is not thought to be damaged, but rather it is assumed that mental development is being depressed or distorted. This distinction is reflected in the choice of variables studied. Research paradigms generally reflect an emphasis on either biological variables or social risk states. The distinction may be largely heuristic. Brain damage secondary to biological events often is inferred from behavioral or neurological data without any certain knowledge of structural or physiological changes. On the other hand, the more general environmental effects on infant development may ultimately be explained in physiological terms. In fact, interaction effects between the two risk categories are generally acknowledged, and either conceptual orientation leads to similar research issues. These issues include determining the immediate behavioral consequences of risk, discriminating transient from long-term effects, and tracing the consequences of risk through succedent stages of intellectual development.

OVERVIEW OF BIOLOGICAL RISKS

The fetus and neonate are at risk for a variety of environmental insults that are potentially damaging to the brain during the time that brain growth and differentiation are most rapid. Two general developmental periods of greatest vulnerability have been identified by Dobbing and Sands (1973). The first occurs during the period of neuronal multiplication between 10 and 18 weeks of gestation, and the second, associated with the period of most active dendritic branching and synaptic connections, occurs during the last 3 months of pregnancy and on through the first 18 months of postnatal life. Brain tissue can be damaged by exposure to toxic substances, deprivation of essential nutrients, disruption of mechanisms of oxygen and carbon dioxide exchange, mechanical damage, and combinations of these events. These problems are not mutually exclusive and may occur as chained or interactive events. For example, responses of cerebral blood vessels to abnormalities in blood oxygen and carbon dioxide levels can result in cerebral hemorrhage (Dykes, Lazzara, Ahmann, Blumenstein, Schwartz, & Brann, 1980); infants born to heroine addicts may respond less well to increased levels of carbon dioxide (Olsen & Lees, 1980).

Specific brain structures develop at different rates during gestation and infancy. Those systems undergoing active organization have the highest metabolic rates and are most vulnerable to damage (cf. Hutchings, 1978). Some cortical loci undergo cell differentiation relatively late in gestation and so are particularly vulnerable to events associated with preterm delivery and resuscitation. These maturational factors suggest not only that the effects of an insult may be more intense when it occurs early in life but also that the specific time of encounter may influence the precise nature of effects.

The complex interactions among risk variables, brain maturation, brain damage, and behavioral effects can be studied systematically in subhuman species to explore the correspondence between brain and behavior. Such direct associations are difficult to document for the surviving human infant. Although new techniques such as computerized tomography and sonography have enhanced our ability to visualize some morphologic features and events such as cerebral bleeding, direct evidence for more subtle changes in brain structure and function are generally relegated to research using animal models. For human infants, brain damage usually is inferred from selected behavioral observations. By using this research strategy, we can study the association between risk variables and behaviors for both specific risks (e.g., a drug) and also more general conditions (e.g., maternal malnutrition) even when the exact effects on the brain are uncertain. The strategy has some obvious drawbacks, namely, cause and effect are difficult to establish, and individual exceptions are often noted. However, controlled group comparisons may suggest an association that will justify further refinement and scrutiny.

The biological risks reviewed in this chapter have been selected for their general interest. These topics are considered representative of current research issues in the investigation of earliest environmental determinants of infant intelligence. Specific studies are presented in some detail to emphasize methodologic differences. The cluster of studies presented for each topic is intended to provide some information about the range and duration of behavioral effects. The review is not exhaustive for any topic but, where possible, the reader is referred to more detailed discussions in recent publications.

CULTURAL DRUGS

The consequences of exposure to new bioactive chemical substances in the environment have become an important public health issue, with mounting concern for the special hazards that such exposure may pose to the fetus and infant. Examples range from chemicals used in agriculture and industry to prescribed medicines. Fetal effects are typically investigated by using animal models to determine the incidence of mortality and structural abnormality in offspring following deliberate and controlled maternal exposure during gestation. Behavioral effects may also be studied, particularly when structural abnormalities in the central nervous system are documented. Public concern has not been limited to new environmental hazards. Some relatively common substances in the environment, such as lead, have proven to be neurotoxic, with serious behavioral consequences for exposed infants. Such findings have raised questions about other common substances in the environment, including some cultural drugs that have gained broad acceptance.

Deliberate drug ingestion during pregnancy has long been recognized as a potential source of fetal insult, and pregnant women under medical care may be cautioned against some commonly used substances such as alcohol, cigarettes, and nonprescription drugs. The most obvious drug effects are those that cause physical abnormalities, for example, the deformities caused by thalidomide. Such marked behavioral effects on the newborn infant as depression (from morphine) and heightened reactivity (from heroin) have long been noted, but systematic studies of the neonate and especially of more prolonged drug effects on behavioral development are relatively new. "Behavioral teratology" is now emerging as a defined research interest (Hutchings, 1978).

Cultural drugs are difficult to study with precision if, as is often the case, women are unable to give accurate retrospective estimates of timing or dosage throughout pregnancy. One research strategy is to identify a population of chronic, heavy users of the drug under study (e.g., chronic alcoholics or heroin addicts). However, less extreme fetal drug exposure is more difficult to evaluate. Moderate to heavy use of popular drugs such as cigarettes, coffee, and alcohol may be noted in varied combinations. Other confounding variables such as the adequacy of prenatal care, maternal health and nutrition, infant nutrition, and early socialization may be associated with the socioeconomic status of the mother which, in turn, may be related to drug use. Such considerations suggest that stringent controls are needed to assess the effects of a cultural drug on infant development. Either the study must be large enough to evaluate important variables simultaneously, or it must be specific and select enough to avoid some of the more important confounding conditions.

Fetal Exposure to Alcohol

Maternal alcohol use is an example that serves well as a model in considering studies of cultural drug effects on the fetus. Children born to alcoholic mothers sometimes have an abnormal appearance so distinctive that it has been termed "fetal alcohol syndrome" (Jones & Smith, 1973). The syndrome includes abnormal facies, joint and other deformities, growth deficiency at birth and in childhood, and mental deficiency in childhood. In a review of this syndrome, Streissguth (1976) cited evidence that the syndrome had been recognized in a number of independent clinical reports. Clarren and Smith (1978) subsequently published a review of 250 cases from the world literature. Although the physical stigmata can be observed at birth, the accompanying mental retardation is the most persistent common finding. In a recent review, Abel (1980) concludes, "The evidence of cognitive impairment is now so compelling that the FAS [fetal alcohol syndrome] is among the most commonly recognized congenital neurological disorders" (p. 32).

Data from the Collaborative Perinatal Project of the National Institute of Neurologic Disease and Stroke (the Collaborative Project) were examined for independent substantiation of the early clinical reports by invesitgators at the University of Washington (Jones, Smith, Streissguth, & Myriathopoulos, 1974). The Collaborative Project was a national prospective study of 53,000 pregnancies and the resulting offspring; children were assessed repeatedly through 7 years of age (Broman, Nichols, & Kennedy, 1975). Although alcohol consumption during pregnancy had not been included as a research variable, Jones, Smith, and colleagues found 23 cases clearly identified as alcoholics before and during pregnancy and matched each with controls on the basis of socioeconomic status, education, age, race, parity, marital status, and geographic region. They found that 43% of the infants of alcoholic mothers either died in the perinatal period (four infants) or evidence stigmata of the fetal alcoholic syndrome (six infants), compared with 2% (one neonatal death, no stigmata) in the control group. Group differences were not striking when mental test scores at 8 months and 4 years were examined, but at age 7, 44% of those tested who were born to alcoholic mothers had IQs below 79, compared with 9% of the matched controls.

The Collaborative Project data suggest the possibility that unexamined environmental variables may have contributed to the differences in intellectual status at 7 years. The data also indicate that not all infants of alcoholics were affected by physical or mental abnormalities, nor was there a complete correspondence between the two abnormalities. However, the case for the increased likelihood of mental deficiency in the offspring of alcoholics was clearly supported in these prospective data. Similar findings were reported for retrospective and prospective studies in Sweden (Olegard, Sabel, Aronsson, Sandin, Johansson, Carlsson, Kyllerman, Iversen, & Hrbek, 1979); the authors concluded that "damage to the fetus by alcohol is now the largest known health hazard by a noxious agent that is preventable" (p. 120).

The precise cause of the mental deficiency associated with maternal alcoholism during pregnancy is unclear. Structural anomalies have been found in the fetal brains of animals exposed to alcohol *in utero* and in human infants who died shortly after birth. In the latter case, structural defects were primarily associated with aberrant patterns of brain cell migration (Abel, 1980). Abel has suggested, however, that conditions secondary to alcohol consumption, such as altered nutrition, cannot be ruled out; casual studies using animal models have produced equivocal results (and see review by Streissguth, Landesman-Dwyer, Martin, & Smith, 1980).

Although maternal alcohol abuse may be recognized as detrimental to normal intellectual development of the offspring, more culturally approved alcohol use during pregnancy has not been so widely accepted as a risk to subsequent infant development. As is the case with many cultural drugs, the burden of proof is on the researcher. As noted, such proof may be difficult to establish, particularly when relatively long-term behavioral effects are at issue.

The most comprehensive study to date assessed the development of 462 infants in a prospective study of maternal alcohol, nicotine, and caffeine use during pregnancy (Streissguth, Barr, Martin, & Herman, 1980). Mothers were interviewed during the fifth month of pregnancy and scores were derived for the use of these substances. The sample was predominantly white, married, and middle class. Infant mental and motor development were assessed at 8 months with the Bayley Scales of Infant Development. The motor scores showed a linear decrease with increasing levels of alcohol use. Scores on the mental scale indicated a threshold effect for women who consumed the equivalent of four or more drinks per day. Neither nicotine nor caffeine use showed significant associations with either motor or mental scores, and substance interaction effects were not related to infant outcome. The incidence of reported use of street drugs was low and did not contribute to the overall findings of alcohol effects. The alcohol-related differences in infant scores remained significant when comparisons were adjusted for gestational age, maternal education level and parity; average scores for all comparison groups were within the normal range.

Possible intervening variables were examined in separate studies that have been summarized and considered together (Streissguth, Martin, Martin, & Barr, 1981). No group differences were found between heavier drinking mothers and controls for maternal illnesses, number of overnight separations from the infants, or social upheavals in families (e.g., divorce, loss of employment). The observed level of environmental stimulation provided by the mothers did not differ between the heavier-drinking and control groups. Of interest is an observation that women at the highest educational level (master's degree or higher) had the highest level of reported alcohol use. Despite the finding that lower developmental scores at 8 months were associated with moderate to heavy social drinking and despite the reported search for intervening variables, there remains the possibility that other environmental effects could account for the differences observed in these essentially normal infant scores. Alternatively, other differences suggesting underlying neurological abnormalities may become identified at older ages. This sample of infants is being followed to 7 years of age.

Behavioral differences in the newborn period have been reported in this and other studies that suggest some initial compromise of infant status. Factors reported for the newborns of social drinkers but not for those of controls in the Washington studies (see Streissguth *et al.*, 1981, for a review) included increased jitteriness, decreased sucking pressure, atypical behavior patterns, and decreased operant learning. Sleep disturbances and poorer state regulation have been reported by others (Rosett, Snyder, Sander, Lee, Cook, Weiner, & Gould, 1979). Some atypical newborn behaviors suggest withdrawal symptoms (Abel, 1980) and so may reflect functional disruption rather than neurological damage. The importance of such behavioral deviancies to later intellectual status remains to be determined.

Cigarette Smoking during Pregnancy

Cigarette smoking by pregnant women was recognized as a fetal risk when an association was established between smoking and low birthweight (Simpson, 1957). The growth retardation *in utero* is thought to be caused by fetal hypoxia, a risk also associated with neurological impairment. Many of the longitudinal studies of pregnancy outcome included data on maternal smoking. The large Collaborative Project (described previously) included data on years of smoking and number of cigarettes per day smoked during pregnancy: maternal smoking habits were ascertained at 5 months' gestation in the University of Washington study detailed previously; in Britain the National Child Development Study (a longitudinal study of all the children born in the same week of 1958 in England, Scotland, and Wales) included mothers' accounts of smoking during pregnancy; in Finland, data on maternal smoking were gathered for more than 12,000 women during pregnancy and again following delivery. The results of these studies can be summarized.

Smoking data gathered by the Collaborative Project were analyzed as predictors of IQ at 4 years of age (Broman *et al.*, 1975). Neither years of smoking nor estimates of cigarettes smoked per day during pregnancy made significant independent contributions to childhood IQ level when maternal age, race, and socioeconomic status were controlled. Significant negative correlations between number of cigarettes smoked and infant birth weight were in agreement with previous findings. Data from infant behavioral assessments (Bayley test scores at 8 months and neurological examinations at 12 months) were treated as predictor variables of childhood IQ in this analysis and so associations between infant status and smoking data were not reported.

Approximately 5,000 of the cases enrolled in the Collaborative Project were seen at Johns Hopkins Medical Institutions in Baltimore. The Baltimore data were analyzed independently by Hardy and Mellits (1972) to determine the effects of smoking during pregnancy on intellectual functioning. The maternal records of all children who had been tested at 7 years of age were examined; 88 cases were selected in which mothers reported smoking at least 10 cigarettes a day throughout pregnancy, and these were matched with 88 control cases (nonsmoking mothers) on the basis of race, date of delivery, maternal age and education, and sex of child. Another control group (55 cases) was additionally matched for birth weight because of the expected lower birth weight in the smoking group and the probably association between low birth weight and poorer intellectual performance. As expected, the unselected control group had significantly greater average weight and length at birth than did the smoking group. Neither comparison was significant beyond 1 year of age. Head circumference did not distinguish smoker from control groups at any age, and there were no differences in neurological status at 12 months (Bayley scores at 8 months were not reported). No differences in IQ or other measurements of intellectual performance were found at 4 or 7 years. The authors conclude that, for this inner-city population, "if the child survives the neonatal period, no significant differences beyond the first year are demonstrated in either physical or mental development to seven years of age" (Hardy & Mellits, 1972, p. 1336).

Fogelman (1980) investigated the associations between maternal smoking and development for nearly 6,000 children who had been assessed most recently at age 16 years for the British National Child Development Study. His findings were similar to those reported for the children in this study at 7 and 11 years. Children of mothers who reported smoking 10 or more cigarettes per day after the fourth month of pregnancy had significantly lower average scores on tests of reading and mathematics; those whose mothers smoked fewer cigarettes did less well on the tests than did children of nonsmoking mothers. Height for boys (but not for girls) was similarly associated with the three classifications of maternal smoking. In contrast to the analyses at 7 and 11 years, Fogelman controlled for some associated variables including social class, family size, and maternal height. No behavioral data from infancy were reported.

A study conducted in Finland examined the effects of maternal smoking on birth weight and health for 12,068 infants (Rantakallio, 1978a,b). When maternal age, parity, marital status, and place of residence were controlled, maternal smoking was found to reduce birth weight in a dose-related manner. Differences in health between children of smokers and matched controls were examined through age 5 years. Significant differences that favored the controls were noted between the groups for the incidence of diseases of the nervous system and sense organs (including strabismus). However, there was an even greater group distinction for respiratory and skin diseases. The author suggested that the latter problems may have been associated with continued maternal smoking in childhood, whereas the neurological problems suggested prenatal effects of smoking. However, the children of smokers were hospitalized more during the first year of life because of respiratory diseases, suggesting a possible secondary postnatal source of central nervous system problems.

The conflicting results obtained in the four long-term studies suggest that unspecified population differences influenced development into childhood. Such prenatal differences as the incidence of combined smoking and drinking may also have been a factor in outcome. These behaviors are highly correlated—women who drink heavily also tend to smoke heavily (see discussion by Abel, 1980). Both separate and interaction effects have been reported when heavy smoking and drinking were considered together. In a study of 204 women identified as alcohol abusers during pregnancy (Sokol, Miller, & Reed, 1980), the risk of intrauterine growth retardation was estimated to be increased 2.4-fold in association with alcohol abuse alone, 1.8-fold with smoking alone, and 3.9-fold when both risks were present. Both drug variables were considered in the Washington University study (Streissguth, Barr, Martin, & Herman, 1980) and, as noted, smoking was not related to Bayley scores at 8 months nor were the interaction effects between the two drugs. Some interaction effects were found for neonatal behaviors in this study population in separate investigations of atypical sleep positions, operant learning tasks, and strength of suck.

Neonatal behavior was examined for the effects of smoking in a small study by Saxton (1978). Fifteen infants of mothers who smoked more than 15 cigarettes per day throughout pregnancy were compared with 17 infants of nonsmokers, matched for maternal age, social class, parity, normal birth weight, sex, and obstetric factors. Performance was compared at 4 to 6 days of age on the Brazelton Neonatal Behavioral Assessment Scale, and no significant group differences were obtained for mean scores of any item. A trend for items with some auditory component to differentiate between the groups was emphasized by the author, with infants of smokers thought to demonstrate somewhat higher auditory thresholds.

Despite the great interest in the effects of maternal smoking on intellectual development, little evidence emerges from the studies reviewed here to indicate specific or long-lasting central nervous system deficits in the infants of smokers. In contrast, the finding of low birth weight and shorter body length in the newborns of smokers has been repeatedly noted in these and other studies.

Subcultural Psychotropic Drugs

Some drugs are not uncommon in a particular subculture of society but have not gained broad general acceptance. Heroin, for example, is predominantly a drug of the urban poor. Some other illicit street drugs are used by a broader socioeconomic cross-section of society but are viewed as unusual and dangerous by the majority. Unlike the more common cultural drugs previously considered, there may be little distinction made between use and abuse of these substances. Because they have obvious effects on the adult central nervous system, such drugs are considered potentially dangerous for the developing nervous system of the fetus exposed to them *in utero*. Heroin and its replacement drug, methadone, have been the focus of several studies of behavioral effects in human infants.

Heroin use by pregnant women can result in a recognizable pattern of withdrawal symptoms in the newborn. These symptoms, which include hyperactivity, tremors, low threshold to stimuli, and gastrointestinal problems, often require sedation. Because a history of maternal addiction is not often volunteered, most infants of heroin users are identified because of their withdrawal symptoms (Wilson, Desmond, & Verniaud, 1973).

Wilson and colleagues (1973) examined the neonatal records of 30 infants enrolled in a nursery follow-up study of high-risk infants whose mothers had a history of heroin addiction. Two were born to mothers who were in custody and did not take heroin during pregnancy; 12 had mothers who voluntarily altered or discontinued heroin use during pregnancy, often by drug substitution; 16 infants were born to chronic addicts who used heroin throughout pregnancy. Some withdrawal symptoms were noted in the records of 24 infants. As expected, the severity of withdrawal was associated with maternal drug use. Continued or recurrent symptoms that persisted for 3 to 6 months were noted in 82% of the affected infants. These symptoms were described as irritability, exaggerated rooting and oral activity, wide-amplitude tremors, spontaneous startles, vasomotor instability, exaggerated sensitivity to sounds, and poor socialization. There were no differences between infants being reared at home and those (53%) placed in court-appointed foster homes. Of 14 infants evaluated between 15 and 34 months of age, 9 were found to have some abnormalities such as hyperactivity, brief attention span, temper tantrums, sleep disturbances, irregular neurological findings, and excessive sweating. The behavioral problems were not thought to be associated with the social environment because all infants were by then in foster homes. Problems were associated with severe withdrawal symptoms in the newborn period, impaired somatic growth, and, to a lesser extent, with fetal growth retardation. The infants performed within the normal range on the Gesell scales of motor development, adaptive behavior, language, performance, and personal-social development. A consistent discrepancy was noted between fine and gross motor development, with fine motor being less advanced. The authors concluded that these clinical findings were suggestive of some persistent behavioral problems being directly associated with fetal exposure to heroin.

Wilson and colleagues (Wilson, McCreary, Kean, & Baxter, 1979) subsequently reported a controlled study of behavioral development for 77 preschool children of heroin-addicted mothers, based on comparisons with three control groups: 20 children whose mothers did not use heroin during pregnancy but either lived with narcotics addicts or subsequently used heroin themselves (the drug-environment group); 15 medically high-risk infants with neonatal conditions similar to those of heroin-exposed newborns, but whose mothers denied taking psychotropic drugs at any time (the high-risk group); and 20 children with no suspected drug exposure *in utero* and no medical complications of pregnancy, delivery, or neonatal life, who lived in the same geographic area (the socioeconomic comparison group).

The heroin-exposed and high-risk groups had significantly less prenatal care, lower average birth weights but comparable gestational ages, lower one-minute Apgar scores, and longer average postnatal hospitalization than did the other control groups. The four groups did not differ on measurements of parent education and occupation or participation in day-care programs, but children of heroin-addicted mothers often lived with a substitute mother (50%) and half of those infants were subsequently legally adopted. A factor analysis of the children's physical environment revealed no group differences, nor did factors derived from a rating of parental attitudes.

The heroin-exposed group was significantly below the other three groups at preschool age for measurements of height, weight, and head circumference. No differences were found in indexes of health and neurological function. Psychometric test scores were within the normal range for the heroin-exposed group but this group perfomed significantly lower on the General Cognitive Index of the McCarthy Scales of Children's Abilities (McCarthy, 1972) and on three of five subtests: perceptual performance, quantitative, and memory. No differences were found between groups in parents' perceptions of the children's intellectual functioning, but the heroin-exposed group differed significantly in having more parents' ratings of problems such as temper outbursts, impulsiveness, poor self-confidence, aggressiveness, and inability to make and keep friends. Compared with the socioeconomic control group (matched for age, sex, race, and socioeconomic status), the heroin exposed group showed some deficits in physical, intellectual, perceptual, and behavioral measures. The authors noted that heroin exposure could not clearly be identified as the causal agent because there often was a history of multiple drug abuse in the mothers. Although cognitive development was not greatly impaired in the heroin-exposed group, the authors concluded that "these children must be considered more vulnerable to suboptimal social and environmental conditions" (Wilson et al., 1979, p. 141).

Methadone has been used as a replacement drug in the treatment of heroin since 1965, and its effects on infants exposed prenatally have been investigated. Because methadone is prescribed, both dosage and onset of use are more certain than is usually the case for illicit drugs.

However, there is often uncertainty about the use of other drugs by pregnant women while on methadone maintenance. The extent of this problem is illustrated in a report of 35 infants whose mothers were enrolled in a methadone maintenance program during pregnancy (Ramer & Lodge, 1975). When neonatal withdrawal symptoms were compared with infant drug levels, six infants who showed the most severe symptoms also showed morphine (a heroin derivative) as well as methadone in their urine.

The same investigators (Lodge, Marcus, & Ramer, 1975) attempted to distinguish between heroin and methadone use by comparing both behavioral and electrophysiological data from the newborn period for four small subgroups: regular use of methadone during pregnancy (11 cases), heroin during pregnancy except for methadone during the last month (9 cases), regular and simultaneous use of heroin and methadone during pregnancy (7 cases), and no-drug controls (10 cases). The Brazelton Scale and items from the Bayley Scales were administered. Few differences were found among these small groups, but, in general, the methadone–heroin subgroup demonstrated behaviors that differed most from the normal controls. There were indications that addicted babies from all three subgroups were below the nondrug controls in measurements of visual attention and orientation and were relatively more responsive to auditory stimuli. Addicted babies also vocalized more and mouthed and fingered hands more.

Auditory and visual evoked-response data for the same infants complemented the behavioral findings. In general, visual evoked responses were poorly defined whereas auditory evoked responses were better integrated in the addicted group; responses of addicted infants were generally characterized as more irregular and unreliable than those of the control group. Similar behavioral differences were noted by Davis and Shanks (1975), who reported that methadone-exposed neonates reacted more to sound than to visual stimuli on measures from the Brazelton Scale. In both studies, the Bayley Mental Scale was administered to some infants during the first year, and scores were within the normal range.

Development during infancy was evaluated by other investigators (Kaltenbach, Graziani, & Finnegan, 1979) who reexamined some infants of methadone-treated mothers at either 1 or 2 years of age. In this study, comprehensive interdisciplinary care had been provided to drug-dependent pregnant women who had entered the program at all stages of pregnancy and were then maintained on methadone for the duration of pregnancy. When relevant antecedent factors were compared at follow-up for the 26 1-year-olds and 17 2-year-olds, significant differences were found in the daily maternal methadone dose but no differences were noted in the proportion of infants who had required medication for neonatal withdrawal symptoms (62% and 67%, respectively).

Both groups of drug-exposed infants were compared with control infants at the same age, randomly selected from a population with comparable socioeconomic, racial, and medical backgrounds. Infants were evaluated by neurological assessments and by the Bayley Mental Scales. Neurological findings were within normal limits for all children. Average Bayley scores were normal for both groups at each age, but the scores at 1 year were significantly lower for the drug-exposed group. No score differences were found between the 2-year-old groups, but some language items were failed more frequently by the drug-exposed infants. Average scores for all 2-year-olds were lower than those for 1-year-olds, a finding in accord with longitudinal data for children from comparably low socioeconomic environments. The authors concluded that infants with methadone-treated mothers functioned within normal limits at 1 and 2 years of age.

In another study, both the mental and motor scales of the Bayley Scales of Infant Development were given to infants of methadone-treated women at 1 year of age (Strauss, Starr, Ostrea, Chavez, & Stryker, 1976). Unlike the findings of the previous study, no differences were noted between the average mental scale scores of these infants and medically matched controls. However, the drug-exposed infants had significantly lower motor scale scores than did the controls, although average scores for both groups were within the normal range.

Representative samples of these infant groups were further evaluated at 5 years of age (Strauss, Lessen-Firestone, Chavez, & Stryker, 1979). Scores from the McCarthy Scales of Children's Abilities were examined, and no associations were found for prenatal drug exposure nor for sex of child. Recorded clinical impressions were also examined. The children of drug-dependent women were described as being more active and energetic, being less mature, and having poorer fine motor coordination than the control children. In general, the drug-exposed group was seen as exhibiting more task-irrelevant activity in the test situation but not in a free-play setting.

The difficulties inherent in studies of drug-dependent women and their offspring are such that no definitive behavioral effects from a particular drug are likely to be determined. Heroin-addicted women, whether or not they are maintained on methadone, are likely to use other drugs such as barbituates, amphetamines, and benzodiazepines (Ostrea & Chavez, 1979). In the studies reviewed here, the severity of withdrawal symptoms was frequently attributed to polydrug use. Although withdrawal symptoms are transient, some of the behaviors associated with withdrawal suggest more persistent problems. Sleep disturbances and especially disorganized sleep states in the neonate are reported for infants whose mothers used methadone and other drugs in pregnancy (Dinges, Davis, & Glass, 1980). The authors reported that this measure of the developing nervous system's basic capacity to organize adaptive states was closely associated with the severity of the behavioral problems associated with withdrawal. Several of the studies reviewed here have suggested some persistent disturbances in state organization and impulse control.

Infants of addicted mothers are likely to have a higher frequency of other neonatal complications that may predispose the infant to later problems. Olsen and Lees (1980) found that methadone-exposed infants had decreased ventilatory response to carbon dioxide for several days following birth (average 15 days) and suggest that, if such a condition persists, it could be a factor in the sudden-infant-death syndrome. Ostrea and Chavez (1979) compared 800 drug-exposed infants and 400 controls of similarly low socioeconomic status for neonatal problems other than withdrawal symptoms. They found that the drug-exposed group had a significantly greater incidence of problems such as jaundice, aspiration pneumonia, and hyaline membrane disease. In this study, the incidence of prematurity and of low birth weight was not greater for the drug-dependent sample. Such findings of indirect risks underscore the difficulties for the researcher who attempts to define adequate control groups for the assessment of fetal drug effects.

FETAL MALNUTRITION

Poor nutrition during fetal life is an obvious environmental variable to consider for its effect on the central nervous system and subsequent infant development. In many places throughout the world where maternal malnutrition is a serious problem, there are other factors to consider, such as poor medical care, disease, and inadequate infant nutrition. These problems contribute to the incidence of infant mortality and so, we can assume, to disabilities in survivors. However, in regions where resources are limited and priorities must be ordered for intervention efforts, the relative importance of fetal malnutrition *per se* is not an academic question.

In populations where maternal malnutrition is not endemic, fetal malnutrition is often identified by the low birth weight of the infant, relative to normative values. Normative data have been derived for pre-term as well as full-term infants, according to gestational age. Usually, infants are described as small for gestational age when their birth weight falls below the tenth percentile. Their subsequent development can be compared with that of other infants of comparable maturity at birth but with average birth weights for gestational age.

Maternal Malnutrition

The exact nutritional requirements for normal prenatal brain development are not precisely known. Although maternal diet is a critical factor, many regulatory events intervene between maternal food intake and brain cell nutrition. For example, maternal protein reserves may supply the fetus with some essential nutrients for a time when the maternal diet is inadequate. There is good evidence from studies of animals, however, that this kind of compensation has its limits. In a review of nutrition and prenatal brain development, Zamenhof and Van Marthens (1978) asserted, "It is now well recognized that the fetus is not a parasite which extracts from the maternal organism all it needs. When pregnant animals are faced with nutritional insufficiencies, it's the fetal development and/or survival that is affected, because the diet is inadequate to support two lives rather than one" (p. 152).

Another complicating factor in defining fetal malnutrition from maternal diet is the role of the placenta in fetal nutrition. Even with adequate diet, placental insufficiency may result in fetal malnutrition. However, the two conditions are not necessarily independent. Zamenhof and Van Marthens noted that in animal experiments, the placenta may be underdeveloped as a direct consequence of maternal protein insufficiency.

The foregoing examples indicate some of the difficulties in determining the precise correspondence between variables of fetal malnutrition and brain development for animal studies, even though maternal diet can be strictly controlled and specified throughout gestation, and fetal brains can be examined at precise gestational ages. In human studies, the variables are obviously much less well defined. Studies of infant malnutrition from infant diet have fewer methodological problems in both animal and human research. However, in many mammals (including man), cerebral neurons are undergoing their greatest proliferation and migration prenatally and so are most vulnerable to interference before birth. In addition, although most cell differentiation occurs postnatally, Zamenhof and Van Marthens suggest, "It is likely that the extent of ultimate [neuronal] differentiation depends on the integrity, nutritional status and size of cells at the end of the fetal period of development" (p. 163).

Leatherwood (1978) reviewed animal studies that examined the behavioral effects of fetal and neonatal undernutrition. Effects were often noted in early growth and development following fetal malnutrition, but little or no evidence was found for more permanent behavioral deficits in learning abilities. He concluded that evidence was lacking to link early malnutrition *per se* with long-term effects on brain function. Behavioral studies of rats and mice are, at best, only modest analogs of important human brain functions, and Leatherwood concurred that severe fetal nutritional deprivation was linked with structural brain changes in animals. However, he noted that the maternal behavior of previously malnourished animals differed from that of control animals and cautioned, "When differences [in offspring] are observed, they appear to be more closely linked to disruption of the early environment . . . than to undernutrition" (p. 204). Important cultural effects can be anticipated in natural studies of human maternal-fetal malnutrition, including the likelihood that malnutrition will continue for both mother and infant following birth.

Some evidence for the interaction between cultural effects and early malnutrition was found by Richardson (1976), although in this study malnutrition was defined in infancy. Jamaican boys who had been hospitalized for severe malnutrition during infancy were given intelligence tests in childhood (6–10 years) and compared with matched schoolmate or neighborhood controls. Social background histories were obtained for variables hypothesized to be associated with intellectual development. Height percentiles for age at testing were used to give some overall life history of nutrition. The boys who had been severely malnourished as infants had significantly lower IQs than did their matched controls. When both groups were combined, the social background score and measurement of height also yielded significant IQ differences, with social background accounting for the largest part of the IQ variance, followed by height, and last, by infant nutritional status. When interaction effects were examined, infant malnutrition was not a factor in the IQs of boys with subsequent good growth and favorable social backgrounds; for boys of short stature and unfavorable social backgrounds, those who had had infant malnutrition had lower IQs.

This study suggests that early, severe malnutrition made the boys more vulnerable to subsequent adverse biological and social events, but the author cautions, "The explanation that severe malnutrition in infancy causes central nervous system damage which then in turn causes mental retardation or impairment is too simple. A more complex conceptualization is needed which takes into account biologic and social variables" (Richardson, 1976, p. 61).

The effects of maternal malnutrition on neonatal behavior were examined directly in a Colombian study that randomly assigned pregnant women to nutritional treatment and control groups (Vuori, Christiansen, Clement, Mora, Wagner, & Herrera, 1979). The sample was selected from a densely populated slum district of Bogotá. Malnutrition was defined as 85% or less in weight-for-age according to Colombian standards. Free medical care was provided to all mothers; those receiving nutritional treatment were given a protein–calorie supplement plus vitamin A and iron beginning at 6 months of pregnancy.

One hundred unsupplemented and 144 supplemented infants were tested at 15 days of age for visual attention and habituation. No differences were found in the babies' initial state of arousal at testing, but supplemented babies were significantly better in their ability to respond to and process visual information, as measured by their orienting responses and the occurrence of habituation. No group differences were found for dishabituation. These results were considered consistent with a maturational effect, the supplemented babies functioning at a more mature level.

Sex differences were found in supplementation effects; habituation rate showed supplementation differences only among girls, while physical growth was associated with supplementation only for boys. The authors suggested that the difference for girls might be related to their relatively earlier neonatal timetable for central nervous system maturity (i.e., greater susceptibility of the organism that is maturing at a faster rate); it was suggested that the supplementation effect on growth for boys might be related to their faster rate of weight gain during the last trimester of pregnancy and the first few weeks of life. For boys but not for girls, the unsupplemented group had lower Apgar scores at birth and lower levels of initial response during testing. Such findings are in accord with other studies that note a greater vulnerability for males associated with their relative immaturity.

In a recent review, Gabr (1981) emphasized the difficulties in determining the influence of interacting factors associated with maternal malnutrition. He listed these as "racial and genetic background, age, health, education, nutritional habits and past nutritional status of the mother; parity, multiple pregnancies and fertility pattern; socioeconomic conditions, especially as they relate to sanitation, infection, and the availability of health services, climate, and possibly the health and nutritional status of the father" (p. 91). He reviewed studies that detailed deficiencies in both the quantity and quality of breast milk in malnourished lactating mothers and suggested that "marginal degrees of maternal malnutrition are more likely to be reflected on the breastfed infant than on the fetus" (p. 95).

In summary, the studies and discussions reviewed emphasize the complexity of the problems encountered in research designed to determine the specific effects of maternal malnutrition on brain and behavior. Even controlled animal experiments have not yet succeeded in resolving some of the major research issues, including the full significance of specific nutritional deficiencies, the verification of critical periods for brain damage from malnutrition, and the isolation of fetal malnutrition from other environmental variables that influence behavioral development. In a more general way, studies of human populations have presented evidence for some effects of malnutrition on early mental development, but these appear to be slight. However, even slight differences may be important in increasing the vulnerability of the malnourished infant to subsequent biological and social events that militate against normal intellectual development. It does not appear likely that prenatal nutritional enhancement can protect the infant from the effects of subsequent developmental adversity. The study by Richardson (1976) suggests a more optimistic interpretation, that the effects of severe infant malnutrition are not necessarily limiting when the subsequent environment is favorable.

Fetal Growth Retardation

The newborn infant who is unusually small for gestational age (SGA) would appear to have experienced fetal malnutrition whether or not the mother was malnourished during pregnancy. Presumably the infant either was not provided with adequate nutrition *in utero* or was not able to utilize nutrients normally. Birth weight is the usual criterion for establishing fetal growth retardation, but other body measurements such as length and head size are often considered. Two types of fetal growth retardation have been described by Miller and Merritt (1979), the short-for-date infant and the infant who is not short but is thin for length. They asserted that data from ultrasonic fetal measurements indicated that the long, thin infants had a milder form of growth retardation and one that had its onset later in pregnancy compared to the shorter infants. This conclusion has been challenged by Philip (1978), who presented evidence on epiphyseal ossification suggesting that chronic intrauterine growth retardation was associated with a low ponderal index (low weight for length) in full-term, growth-retarded infants.

Most behavioral studies of SGA infants have defined this population as infants whose birth weights fall below the 10th percentile for gestational age—that is, the most severe cases. This definition allows one to include infants of varied gestational ages and also to compare behavioral development for preterm infants whose birth weights are comparable but who differ in being either SGA or appropriate for gestational age (AGA). Behavioral studies often exclude or specifically examine conditions that in themselves are known to be associated with central nervous system damage, such as chromosomal abnormalities and fetal infections.

Some causes of growth retardation, such as placental abnormalities and twin pregnancies, are not modifiable environmental variables. However, these conditions allow us to study the behavioral effects of fetal malnutrition in populations that do not have widespread maternal malnutrition. In this way we avoid some of the more obvious confounding biological and social variables associated with poverty that may influence infant development. On the other hand, there may be different risks for the SGA infant, such as fetal and maternal diseases, maternal drug use, and perinatal risks that are not directly related to malnutrition but threaten the integrity of brain development. Although behavioral studies of SGA infants have not usually emphasized causality, it is unlikely that we can anticipate a high degree of homogeneity for causal factors. Koops (1978) reviewed studies of neurological and intellectual outcome for SGA infants, emphasized conflicting results across studies, and suggested some of the difficulties encountered in defining growth-retarded groups. Koops also presented data to show that many AGA infants who are born prematurely with birth weights less than 1,500 grams are below the 50th percentile for gestational age. She suggested that conditions may already have been present in such infants that ultimately might have led to more conspicuous fetal growth retardation in mature infants. This observation suggests that one major variable to be considered in behavioral studies is the relative maturity or prematurity of the SGA population.

Studies of Full-Term SGA Infants

At one end of the spectrum of neonatal status is the apparently normal, full-term SGA infant. Neonatal and infant behaviors were compared for 10 normal and 10 underweight full-term infants, all of whom appeared healthy at birth and were considered low risk on obstetrical and neonatal indexes (Als, Tronick, Adamson, & Brazelton, 1976). Underweight was determined by a ponderal index (weight-for-length ratio) below the 10% level, and the average for the underweight group was below the 3% level. All babies were caucasian, with intact families and no evidence of maternal malnutrition or maternal drug addiction during pregnancy. The groups did not differ on gestational age or length at birth. The Brazelton Neonatal Behavioral Assessment Scale was administered on days 1, 3, 5, and 10. Significant differences were found between groups on performance related to both motor and interactive processes, with thin infants tending to have poor scores on some reflexes such as rooting and sucking, and responding less well, with stress reactions, to stimulation and social interaction. The authors concluded that the assessment "differentiated the two groups on behaviors which are important for the caretaker of the baby: These are attractiveness, need for stimulation, interactive processes and motor processes" (Als et *al.*, 1976, p. 601). The thin infants were reassessed with the Denver Developmental Screening Test (Frankenburg, Dodds, & Fandal, 1970) once between 6 weeks and 9 months of age and the mothers were interviewed. All infants performed within normal limits on the test, but 8 of the 10 were described as difficult to live with, being, for example, easily overstimulated and unpredictable in eating and sleeping patterns. The authors speculated that caretaker response to the initial behaviors might have increased the level of caretaker anxiety and that this might further have influenced poor state organization in infancy.

The time of onset of intrauterine growth retardation has been of special interest to some investigators, particularly in relation to head growth. Parkinson, Wallis, and Harvey (1981) compared 45 elementary-school age children who had been SGA infants born at 37 weeks or more, whose fetal growth was measured during pregnancy by serial ultrasonic cephalometry; slow head growth was defined as a weekly increment in biparietal diameter below the fifth percentile over a period of two weeks or more, according to the standards of Campbell and Newman (1971). Four groups were identified: those with slow intra-uterine head growth beginning at or before 26 weeks gestation, those between 27 and 34 weeks, those after 35 weeks gestation, and those with no evidence of slow intrauterine head growth. Controls (AGA) were located for the first two groups, matched for age, sex, birth order, social class, and race. None of the children had a history of intrauterine infection or evidence for chromosomal or congenital abnormality at birth, all had standard neonatal care with attention to prevention of hypo-glycemia and hypothermia (common in SGA neonates), and there had been no serious neonatal problems.

A preliminary study of the same children at younger ages (range 28–84 months, average 4 years) had detected problems in the group with the earliest head-growth lags relative to all others (Fancourt, Campbell, Harvey, & Norman, 1976); these included a lower average score on the Griffiths extended scales (Griffiths, 1970) and significant differences in individual subquotients for practical reasoning, eye and hand coordination, motor development, and personal-social factors.

Parkinson *et al.* obtained school teachers' assessments in some detail when the children were between 5 and 9 years of age. The differences found related to social class, sex, and the prenatal age at which slow head growth began. Poor school achievement and behavioral problems were most prevalent in boys with head-growth slowing before 34 weeks' gestation; for both sexes (but especially for boys), those with head-growth lags before 26 weeks' gestation had problems with reading, writing, drawing, and concentrating. Children with slow head growth beginning before 34 weeks' gestation were smaller at 5 to 9 years and exhibited more extreme behavior than did the other SGA children and controls: Boys were described as clumsy, worried, fidgety, not very adaptable, and unable to concentrate; whereas girls were more likely to be rated as irritable, crying often, and bullying other children. Both the investiagtors and teachers noted overprotective parents in the SGA group. Three of the girls with earliest head-growth lags but with families from high socioeconomic categories were doing extremely well at school, suggesting that an advantaged environment could counteract the disadvantage of prolonged intrauterine growth retardation. The authors concluded that, by using ultrasonic fetal measurements, SGA children at highest risk for school problems could be identified at birth and that "we should now be thinking of how best to help them during their developmental years" (Parkinson et al., 1981, p. 48).

Head growth during infancy was not reported in the foregoing study, but there is some indication that this phase of growth may be related to subsequent intellectual status. Babson and Henderson (1974) measured intelligence and school performance for 10 full-term SGA girls between 7 and 11 years of age. Individual differences in head circumference at 1 year of age had been noted, with five girls having measurements at or below the 10th percentile and five having measurements between the 25th and 75th percentile for Denver standards. Although all the girls were functioning normally at childhood, the average IQ of those with smaller heads in infancy was 10 points below that of the girls with larger heads ($p < .1$), and the head-size difference remained significant in childhood.

A prospective study of biochemical placental function tests in late pregnancy was made for 29 pregnancies in which intrauterine growth retardation had been detected (Leijon, Finnstrom, Nilsson, & Ryden, 1980). The infants were classified at birth as severe (14 cases) or moderately severe SGA (15 cases) according to Swedish standards of birth weight for gestational age. Both groups were compared with a control group of 18 AGA infants on neonatal assessments of neurological and behavioral status. Some of the SGA infants were less than full-term because they were delivered early when declining growth curves were combined with significantly low placental tests. Allowances were made for varying gestational ages (some as low as 35 weeks) by giving the assessments at the equivalent of 40 weeks' gestation. No infants had serious asphyxia or major neurological abnormalities as neonates, suggesting that the time of delivery was optimal. Abnormal biochemical placenta tests were found to be of poor prognostic value for the newborn assessments. Both SGA groups (severe and moderate) showed lower muscle tonus compared with controls. The severely growth-retarded infants showed fewer optimal items in the neurological assessment and, in Brazelton Scale performance, they showed poorer orientation, motor function, and physiological stability than did the controls.

In summary, studies of full-term SGA infants and children suggest an increased incidence of some neurobehavioral abnormalities in populations for which other common risk factors were avoided or controlled. There is at least a suggestion of interaction effects between physical appearance and behavioral traits on the one hand and parental response on the other to influence the course of development.

Studies of Preterm, Low-Birth-Weight SGA Infants

When the SGA infant is also very tiny and premature, other associated risk factors can be anticipated. As noted by Koops (1978), behavioral studies of infants and children in this category have reported varied results, depending on the selection of the SGA population and the comparison populations used. In some studies, the SGA infants have constituted a subgroup in a population selected for low birth weight, with gestational age allowed to vary. This method of selection has been used because birth weight is often more reliable as a measurement than is gestational age, particularly for populations with little prenatal care. Also, some issues of survival and postnatal care for the smallest infants have been of interest and thus have dictated selection across a range of gestational ages.

The importance of considering gestational age as well as birth weight when evaluating the outcome of small preterm infants was emphasized by the publication of Lubchenco's tables of estimated risk (Lubchenco, Delivoria-Papadopoulos, & Searls, 1972). In this study, 91 infants with birth weights of 1,500 grams or less were evaluated for neurological and intellectual status at age 10 years. The incidence of moderate to severe handicaps was shown to be related both to gestational age and birth weight in a predictable way, with the highest incidence of handicap noted in the lowest birth weight/SGA groups and the lowest incidence noted for the larger infants who were AGA. Within this very low-birth-weight range, fewer handicaps were found for infants of greater gestational age when weight was held constant. These results indicated that the SGA infants had a developmental advantage over less mature infants of comparably low birth weight. It is important to consider the historical context of these results, since the infants were born before the advent of neonatal intensive care as it is currently practiced.

Vohr and colleagues (Vohr, Oh, Rosenfield, & Cowett, 1979) followed the development of 21 SGA infants with birth weights below 1,500 grams (average 1,220 grams) and of 20 AGA controls matched for birth weight, sex, and perinatal risk scores. All were born in 1975–1976 and both groups contained a mix of inborn and transferred infants. The average gestational age of the SGA infants was 33.4 weeks, 4 weeks more than that of the AGA group. There was a higher incidence of maternal hypertension of pregnancy and delivery by cesarean section in the SGA group. Parent occupation and education ratings were similar between the groups, with most being high-school graduates employed in skilled and unskilled labor. Nutrition, growth, neurological status, and behavioral development were evaluated at 3-month intervals during the first year, beginning at 3 months' chronological age; all measurements were adjusted for prematurity. No differences in weight, length, or head growth were found between SGA and AGA groups. Each group had similar numbers of normal, suspect, and abnormal neurological findings at each age. The proportion considered neurologically normal increased steadily across ages in both groups, from 7/20 at 3 months (both groups) to 14/16 (AGA) and 16/19 (SGA) at 12 months; most children had not yet been evaluated at 18 and 24 months. Average scores for Bayley mental and motor scales were significantly lower for the SGA group at all ages through 18 months, with more SGA infants testing in the abnormal range on both scales.

Commey and Fitzhardinge (1979) reported the outcome to 2 years of age for 71 SGA infants with birth weights more than 2 SD below the mean for gestational age (Usher & McLean, 1969) and gestational ages below 37 weeks. The average birth weight was 1,113 grams and all weighed less than 2,000 grams; the average gestational age was 31.8 weeks. The infants, born in 1974–1975, were delivered elsewhere and transferred to the neonatal intensive care facility. Average maternal weight and height were similar to national averages. Social class was predominantly middle class (65%), with a sizeable proportion of lower-class families (23%). Most infants were in stable, two-parent families. Both incremental and cumulative head growth were normal through 6 months post term but, in absolute terms, height and weight (especially for boys) and head circumference were below average through the first 2 years. Major neurological defects were diagnosed in 15 children (21%). The average Bayley mental and motor scores at 18 months were 86 and 78, respectively (adjusted for prematurity). A developmental handicap was considered present in 49% on the basis of a major neurological defect and/or Bayley scores at or below 80. Handicaps were not related to sex, socioeconomic status, degree of fetal growth retardation, degree of prematurity, method of delivery, or intrauterine or postnatal head growth, but they were strongly related to cerebral depression on admission to the intensive care unit. The authors concluded that the high incidence of handicaps in this SGA population was attributable to intrapartum hypoxia combined with the risks of premature delivery, and emphasized the importance of early diagnosis and surveillance of the SGA fetus and of immediate postnatal care.

The development of SGA infants from a more socially disadvantaged background was examined by Stave and Ruvalo (1980). Ninety-eight infants (61 AGA and 37 SGA) with birth weights between 1,000 and 1,500 grams were identified and evaluated, when possible, at the adjusted ages of 3, 6, and 12 months. Although the majority did not have parents of lowest socioeconomic status, a large proportion (more than 60%) of the mothers were unmarried. The racial distribution was 56% black and 44% white (including many Cubans). All infants had been discharged from the same neonatal intensive care center (no information was provided as to the proportion of inborn or transferred after being born elsewhere). Perinatal risk scores were determined, and AGA and SGA groups were further subdivided according to high and low perinatal risk; a control group of 29 full-term, low-risk AGA infants was also included in comparisons. Standardized neurological examinations were given at 3 months (117 infants) and 6 months (80 infants), and optimality scores were determined. All the low-birth-weight infant groups had significantly lower mean scores on this measurement at 3 months than did the full-term controls, but differences between low- and high-risk groups were not significant. Neurological optimality scores were improved at 6 months for all the low-birth-weight groups; again, trends for average score differences between low- and high-risk groups were not significant. The Bayley mental and motor scales were administered at 12 months to 67 infants. Average scores for all sub-groups were within the normal range for the mental scale, with a trend for high-risk groups to have lower scores. Average scores for the motor scale were normal and comparable for low-risk and control groups but fell below −1 SD for the two high-risk groups (average score = 83 for both). The psychomotor deficits found at 3 months were associated with Bayley motor scores at 12 months. Among the SGA infants, only those with an absence of catch-up head growth during infancy (7 infants) failed to achieve normal neurological optimality scores and Bayley scores.

The relative importance of birth weight, gestational age, and head circumference at birth and also the relative importance of being SGA or of having an unusually small head for age (SHA) in predicting develop-mental outcome in infancy were evaluated for 127 low-birth-weight in-fants at 7 months adjusted age (Lipper, Lee, Gartner, & Grellong, 1981). All infants were either below 37 weeks' gestational age at birth or below the 10th percentile of weight for gestational age. Of the infants, 41 were SGA and 35 were SHA (below the 10th percentile for head circum-ference); 27 infants were both SGA and SHA. Normal and abnormal outcome were categorized by neurological status and by Bayley mental and motor scores above and below 80. The incidence of abnormalities by both measurements increased with decreasing birth weight, gestational age, and head circumference and was significantly higher in SHA than in appropriate-head-size infants. SGA infants with appropriate head size were not different in outcome from AGA infants. The authors con-cluded that fetal head-growth retardation was a better predictor of ab-normal outcome than was SGA status, and that head size might be an unexamined factor in the varied findings reported in SGA studies.

The studies described above illustrate some of the difficulties en-countered when two high-risk groups are compared. A high incidence of severe to moderate developmental problems in childhood for low-birth-weight infants has been reported in our own (Hunt, 1981) and other longitudinal studies. Whether those who are also SGA have an added disadvantage would seem to depend on the problems of pre-maturity noted in both groups and on the management of the special problems of the SGA fetus and newborn. More recent studies do not appear to support the earlier evidence for better outcome in the low-birth-weight SGA infants because of greater maturity at birth. The improved care for preterm infants during the past decade has generally improved the outlook for normal development in the AGA group. Perhaps the better comparison is by gestational age rather than by birth weight. In our studies (unpublished) of low-birth-weight and extremely SGA infants (below the third percentile of weight for gestational age), those with gestational ages of 35–37 weeks (20 infants) had a high incidence of developmental problems in childhood (45%) that ranged from mental retardation to specific learning disabilities. Although the observed incidence of problems was not significantly higher than that determined for all low-birth-weight infants, it was certainly higher than would be anticipated for AGA infants who were likewise approximately 1 month premature. In this subsample, those who were the smaller of birth-weight-discordant twins and had no antenatal complications were normal in childhood; infants whose mothers had hypertension of pregnancy were at greater risk, and seven of nine such cases had developmental problems in childhood. These preliminary observations suggest, again, that fetal malnutrition *per se* may not be the most significant variable in determining the outcome of SGA infants even when, as with our data, only the most extreme cases of fetal growth retardation are considered.

Prematurity

Infants born prematurely are, as a class, disadvantaged in making the necessary adaptations to extrauterine life and are at risk for sustaining significant brain damage and developmental problems. The degree of risk, although strongly related to the degree of prematurity as measured by either gestational age or birth weight, is determined by other factors. These include the causes of prematurity, the quality of antenatal and neonatal care, and features of the postnatal environment. Although prematurity is defined as a gestational age at birth of less than 37 weeks, infants who are born as early as 3 months preterm (27 weeks gestational age) can now survive, and the lower limits are constantly being extended. Many research studies of behavioral outcome have defined preterm populations according to levels of prematurity so as to reflect the kinds and degrees of risks encountered. As noted, birth weight is frequently used for this purpose because it is more certain as a measurement than is gestational age, although SGA infants are often considered separately. Infants with birth weights below 1,500 grams constitute a relatively small portion of all preterm infants but are of special research interest because of their vulnerability. This is the group of infants whose survival rate has been most improved by the availability of neonatal intensive care.

The growing research literature on the mental development of preterm infants cannot be reviewed completely here; the interested reader is referred to some collected studies and reviews (Field, Sostek, Goldberg, & Shuman, 1979; Friedman & Sigman, 1981). This review focuses on research issues related to the broad determinants of risk within preterm populations.

The Causes of Prematurity

Some conditions that predispose the infant to preterm delivery also carry independent risks for abnormal outcome. They include certain congenital anomalies, maternal diseases, and fetal infections as well as some of the environmental variables reviewed previously. Such factors can be expected to influence the results of developmental studies of preterm infants and suggest one important source of variance across studies.

Drillien, whose seminal studies of preterm infants (Drillien, 1964) ushered in the current era of follow-up studies, also called attention to the problem of etiology (Drillien, 1972). She studied the relation between causal factors and developmental outcome to 1–3 years for 283 children with birth weights of 2,000 grams or less. Developmental quotients were divided into five levels, and outcome was compared for varied etiologies. She concluded that there were three main causes of premature birth and intrauterine growth retardation: developmental abnormalities of the fetus that were present during early gestation (associated with moderate or severe handicaps); adverse factors in late pregnancy, such as hypoxia and malnutrition, which carried an increased likelihood of mild neurological and intellectual handicaps; and infants with no apparent fetal complications, who were born "by accident" and were potentially normal at birth. Although preterm birth of the third type may be a medical accident, it is likely that there are environmental factors that contribute to the incidence of such an occurrence.

Miller and Merritt (1979) examined abnormal factors in prenatal life for their association with low birth weight. The total sample studied included approximately 6,000 infants born in 1973–1978 at the University of Kansas Medical Center, an urban special care center; there was a relatively high proportion of preterm infants in the population. They classified abnormal factors into four categories: fetal factors, medical complications of pregnancy, selected maternal behavioral conditions associated with pregnancy, and environmental factors (including high altitude and exposure to toxic substances). Of special interest is the third category, which included seven maternal conditions that the authors considered potentially modifiable: abnormally low prepregnancy weight for height, low maternal weight gain in pregnancy, lack of any prenatal care, delivery before the 17th birthday, delivery after the 35th birghday, cigarette smoking during pregnancy, and use of addicting drugs or large amounts of alcohol during pregnancy. Infants with abnormal factors were compared with control infants who had none (approximately 1,200 white and 473 black infants met all criteria for controls). Few preterm infants could be classified as controls and almost all who could had birth weights greater than 2,000 grams.

A subsample of 250 white infants with birth weights below 2,500 grams was identified. Gestational age was not controlled but the ratio of preterm to full-term infants in the group was reported (7 to 3). The incidence of low-birth-weight infants was low in the control group, higher among mothers with behavioral conditions, and highest in the group of mothers with medical complications; these differences were significant. Social-class effects were examined by grouping mothers into four socio-economic categories. More control mothers (with no abnormal factors) were found in the most advantaged group. There were no differences in the incidence of mothers with medical complications across social categories, but a difference was noted for the behavioral conditions, with highest incidence in the lowest socioeconomic group. The authors noted that birth-weight differences were not associated with socioeconomic status for the control group and that more than half of the low-birth-weight infants were born to women with the behavioral conditions in their pregnancies. They concluded that the maternal behavioral conditions studied were more important than socioeconomic status *per se* in influencing birth weight, but that these behaviors were significantly more prevalent in the lowest (poverty) class.

Both studies emphasized the diversity to be expected in low-birth-weight populations. Drillien determined that etiology and development

were associated; Miller and Merritt discriminated among certain categories of risk and identified a cluster of modifiable risks (some of which
have been reviewed elsewhere in this chapter as independent risk factors) that contributed to low birth weight and prematurity. Neither
study emphasized the distinction between SGA and AGA infants of low
birth weight within their populations, considering all to be potentially at
risk. However, it should be kept in mind that relative immaturity at birth
influences the incidence of some neonatal complications.

Neonatal Intensive Care

The introduction of neonatal intensive care in the 1960s resulted in
sharply decreased mortality rates for small preterm infants and also in
comparable decreases in the incidence of severe handicaps that can be
recognized in infancy (see review by Thompson & Reynolds, 1977).
Most follow-up studies have reported a similar decline in developmental
problems at older ages associated with the advent of intensive care
(Davies, 1976; Fitzhardinge, Kalman, Ashby, & Pape, 1978; Stewart &
Reynolds, 1974).

Continued research in fetal and neonatal medicine has resulted in a
rapidly changing technology and concomitant rapid changes in accepted
treatment. The long-term developmental effects of these innovations are
of intrinsic interest and also can be expected to influence the results of
behavioral studies of total preterm populations. In this sense, children
assessed at age 6 years already may be medically "out of date" and
belong to a different treatment epoch. This is not a trivial consideration
because data from various longitudinal studies (cf. Francis-Williams &
Davies, 1974; Hunt, 1981) have demonstrated that many intellectual
problems are not evident until childhood. Thus, in a comparison of
today's 1-year-olds and 6-year-olds, a difference in outcome in favor of
the younger group could be an artifact of age at evaluation or a true
indication of improving outcome. This problem results in a time lag
before all the developmental effects of modifiable treatment procedures
can be known. Comparisons across studies, already complicated by differences in social and medical risks, are further complicated by differences in treatment variables.

Another innovation has been the regionalization of neonatal intensive care. Three levels of treatment have been established, with special

tertiary care centers providing services for the most difficult medical problems. The transfer of infants to a special care center presents some risks, and the evening news often carries reports of infants who were met by a special medical team and flown to their treatment destination. There is another, less dramatic medical innovation that also benefits the preterm infant: It is the transfer of the high-risk pregnant woman to a special center before delivery. Fetal monitoring and control of the onset of preterm labor are some obvious benefits of this practice. The infants are subsequently delivered at the special care center, thus avoiding the risks of postnatal transfer. The regionalization of care can be expected to alter some etiological factors in preterm populations born in tertiary care centers, for example, by increasing the numbers of elective preterm deliveries associated with fetal distress, and delaying or preventing preterm birth when there are no fetal problems.

The increasing complexity and cost of intensive care has prompted some careful scrutiny of its effectiveness. Has the survival rate for small preterm infants continued to improve? If so, are the new survivors normal? These are questions of ethical as well as scientific interest and, increasingly, they are inserted into debates over the cost and value of neonatal care.

A 15-year experience with very low-birth-weight infants was reported for Hammersmith Hospital, London (Jones, Cummins, & Davies, 1979). All infants born alive at that hospital between 1961 and 1975 with birth wieghts of 501–1,500 grams were included (357 infants). Data were examined for three 5-year time periods: 1961–1965, 1966–1970, and 1971–1975. For the whole group, no differences were found across the three time periods for median birth weight, mean gestational age, proportion of SGA infants, sex distribution, or racial composition. The survival of SGA infants increased from the earliest to latest time period, but there was significant improvement neither in overall neonatal mortality nor in mortality for any birth-weight subgroup. There were no survivors below 750 grams, and the total neonatal mortality was 58.5%. The most frequent causes of death were intraventricular hemorrhage and/or respiratory distress syndrome, and the frequency of such illnesses was unchanged over time. Both qualitative and quantitative treatment differences were noted for the three periods; not surprisingly, those who died received significantly more items of special treatment.

The subsequent developmental status of 138 children was reported (93% of the survivors). There were no differences in major or minor handicaps across the three 5-year periods. The overall incidence of major handicaps was 13.1%; the incidence of minor handicaps was 14.5%. There were no differences in the socioeconomic status of survivors across the three time periods; two-thirds were from relatively low social-class categories. The authors mentioned that some of the most recently born children were not yet old enough for all minor handicaps to be identified and, conversely, some of the oldest children considered to have minor intellectual handicaps were now reported to be doing well in school, although they had not been retested. They concluded, "The nucleus of handicap (though not the causes of it and hence not necessarily the nature of it) probably remains unchanged. In addition, the long-term safety of some of the present effective forms of treatment has yet to be established" (Jones et al., 1979, p. 1335).

A different perspective is provided by Hack, Fanaroff, and Merkatz (1979). They categorized care for small preterm infants in this century by four epochs and summarized the developmental consequences associated with each. During the early decades of the century, when basic principles of newborn care were introduced and implemented, reports of long-term outcome were good but mortality rates were very high. Discouraging statistics were reported from the second stage (mid-1940s to the early 1960s), with a high incidence of handicaps reported, largely the result of misdirected therapeutic efforts. Pediatric and obstetric advances greatly reduced the incidence of major handicaps at selected centers between 1960 and 1970. Currently, the recombination of pediatric and obstetric skills and the regionalization of care were described as promoting a more positive clinical approach, including efforts to facilitate maternal–infant contacts and minimize the separation of mother, father, and baby.

Hack et al. reported data from their Cleveland sample (inborn and outborn) for two time periods, 1973–1975 and 1976–1978. Survival increased significantly in the more recent period for infants with birth weights of 501–1,000 grams and 1,001–1,500 grams. Greatest improvement was noted for the group with lower birth weights, for whom survival increased from 26% to 47%. Those weighing more than 1,000 grams had a survival rate greater than 80% during the second time period. The authors noted that increasing numbers of infants weighing less than 750 grams (gestational ages of 23 to 26 weeks) had been admitted, constituting 25% of those weighing less than 1,000 grams during the last 3 years of the study. Survival rates for these infants were low (16% to 30%).

Development status at the average age of 2 years was reported for 160 children born in 1975 and 1976 (85% of the survivors for those years). Severe handicaps, defined as gross neuromotor deficits or scores below 80 on the Bayley or Stanford–Binet scales, were noted in 18%. The incidence was 15% for those with birth weights greater than 1,000 grams. Within this low-birth-weight population, no difference in the incidence of handicaps was noted between AGA and SGA infants. The authors expressed a guarded optimism for the current status of regionalized intensive care.

The two studies reviewed have reported some similar data for comparable years. The mortality rate in the London study for 1971–1975 was 53%, and in Cleveland the rate was 43% during 1973–1975. The incidence of major handicaps was roughly comparable, being 13% (London) and 18% (Cleveland, 1975–1976). The difference in birth circumstances (inborn in the London study, inborn and transported in the Cleveland study) may be relevant in considering the comparative survival rates for infants with birth weights below 1,000 grams (13% in the London study, 57% in the Cleveland study). Many infants of such low birth weight would not have survived long enough to be transported. The different conclusions of the authors regarding the benefits of modern intensive care appear to be based on the relative improvement in outcome in each study rather than on absolute figures for the same time period.

Specific Environmental Risks of Prematurity

Newborn intensive care for preterm infants has two overlapping concerns: to prevent or modify specific acute complications of prematurity, and to meet the special needs of the preterm organism by providing a sustaining environment. Each concern includes a number of identifiable components at any given time. The specific components of care can be evaluated for their developmental consequences with more confidence than is possible when considering more global changes in the nursery across time. Appropriate comparison groups can be identified, usually even in the absence of a formal controlled study and therefore, subtle behavioral comparisons can begin early in infancy. Comparisons among centers become more meaningful for the specific situation.

Medical Complications

Hyaline membrane disease (HMD) is an example of a serious medical complication that afflicts a large proportion of small preterm infants in the first hours of postnatal life. This condition occurs because structural and physiological immaturity of the lungs prevents adequate ventilation. The infant becomes asphyxiated and, until recently, mortality was more than 50% for infants of very low birth weight. Without assisted ventilation, survivors experienced marked degrees of hypoxia and acidemia. With modern care, strenuous efforts are made to maintain lung function, stabilize blood levels of oxygen and carbon dioxide, and normalize the acid–base balance in the blood. Characteristically, HMD causes increasingly severe signs for 3 or 4 days, followed by slow improvement. Each aspect of intensive care has evolved over 2 decades of clinical and laboratory research. Currently, clinical research is focused on the prevention of HMD by enhancing lung maturity before delivery and by providing a synthetic substitute for the biological substances that cannot be produced in the premature lung.

The asphyxia associated with HMD is a potential cause of brain damage. There is an increased likelihood of subsequent damaging events, including cerebral hemorrhage and hydrocephalus in HMD infants. Also, the possibility of long-term iatrogenic problems related to aggressive treatment is a continuing concern. We have reported the incidence of intellectual problems in childhood for infants born in our hospital from 1965 to 1975 with birth weights at or below 1,500 grams (Hunt, 1981). A twofold increase in handicaps was observed for HMD infants compared with all others of the same birth weight. We have also found that most children who developed intraventricular hemorrhage and hydrocephalus had intellectual disabilities in childhood, ranging from severe to mild (Chaplin, Goldstein, Myerberg, Hunt, & Tooley, 1980). In this study, 20 of 22 infants also had HMD. Changes in the medical management of HMD in our nursery have been described for four time intervals between 1969 and 1978 (Tooley, 1979). Survival increased steadily during those years and the effect was particularly dramatic for infants with birth weights below 1,000 grams. The incidence of childhood handicaps has remained relatively constant in our population, but survivors include increasing numbers of the smallest and sickest infants.

A 2-year follow-up study of 43 infants with birth weights at or below 1,000 grams found no significant association between HMD and either neurological abnormalities or abnormal Bayley scores (Pape, Buncic, Ashby, & Fitzhardinge, 1978). The incidence of HMD was 39%. These infants were transferred to the intensive care nursery (born elsewhere), and more than half suffered severe birth asphyxia. Developmental problems were most closely associated with severe acidemia, intracranial hemorrhage, seizures, and very low birth weight.

In another study (Ruiz, LeFever, Hakanson, Clark, & Williams, 1981), 38 infants with birth weights below 1,000 grams were evaluated in infancy. All were born between 1976 and 1978; 17 were inborn and 21 were transferred. The Bayley scales were given at 8–10 months. Comparison on a number of variables was made between infants with mental scale scores above 84 and those scoring at or below 84 (scores adjusted for prematurity). The incidences of HMD, intracranial hemorrhage, and hydrocephalus were not significantly related to poorer outcome. Marked differences were found for use of mechanical ventilation, duration of oxygen therapy, and duration of ventilation. Of 10 severely handicaped children, 9 had been ventilated. The authors suggested that either the ventilated infants were too ill to respond to treatment or the use of ventilation was not optimal. They concluded, "Clarification of relationship between practice and results can only be achieved by the careful monitoring of developmental status . . . along with careful study of procedures and results used by others" (Ruiz et al., 1981, p. 334).

The outcome of larger infants (average birth weight, 2,473 grams; average gestational age, 34.7 weeks) was studied by Kamper (1978). Development in childhood was compared for 68 ventilated HMD survivors and matched controls born in 1966–1971 (inborn and transferred). No significant differences were found in the incidence of abnormalities or IQ level, except that HMD children younger than 5 years scored less well on one measure of verbal ability. For the HMD survivors, cerebral palsy was associated with greater prematurity; intellectual disabilities were associated with moderate intrauterine growth retardation and low milk intake during the first week. Of the HMD group, 17% had moderate to severe handicaps, compared with 7% for the control group. The author concluded that ventilator treatment was safe and could be recommended for the treatment of HMD in centers that were properly equipped and experienced.

Field, Dempsey, and Shuman (1979) examined developmental outcome in two cohorts of HMD infants born in 1972–1973 and 1974–1975. Each cohort was compared with a control group of full-term infants who had not experienced neonatal complications. All groups were assessed with the Bayley mental and motor scales at 1 and 2 years of age. The older children were also evaluated at age 4 years. Highly significant differences were found between HMD and control infants in each cohort for both mental and motor scores of the Bayley (scores adjusted for prematurity); in many comparisons, the score differences were approximately 20 points. At age 4, IQ differences were small and not significant but ratings of behavior problems and symptoms of dysfunction indicated significantly more problems in the HMD group.

The two HMD cohorts were compared. Although not matched initially, no differences were found for average gestational age (32 weeks), birth weight (slightly below 1,750 grams), or scores on obstetric and postnatal complication scales. Significant differences were found for Bayley mental and motor scores at 1 and 2 years, with the more recent survivors receiving higher scores. These infants also differed in having received a more advanced type of ventilatory support while in the nursery. The authors emphasized that it was speculative to attribute the improvement in development to "technology," citing other differences in nursery environment between these cohorts that would be expected to enhance social and sensory development in the later-born infants.

The studies of HMD reviewed here are illustrative of many published in recent years that have examined the developmental consequences of serious medical problems and intensive treatment methods. The growing concern with long-term effects of specific illness and treatment variables suggests that cross-study replication can result in important modifications in care.

Special Needs of Small Preterm Infants

When the very small infant is stabilized and not being treated for acute problems, the sustaining role of neonatal intensive care predominates. Attention now focuses on his basic needs for nutrition, rest, activity, sensation, and human contact. Many studies of preterm development in the nursery environment have examined the more mature, healthy infants, comparing their capacities and needs with those of full-term infants. However, recovering and fragile smaller infants have special needs that must be recognized and met. Also, because they spend proportionately more time in the nursery during the recovery phase than do their more mature counterparts, that environment plays a larger role in modulating early experience and organizing central nervous system functions. In this sense, all caretaking aspects of the nursery environment may be biological determinants of development for the very small infant, or at least may be inseparable from them.

A model of expanding infant organization within an infant–caregiver system has been suggested by Als, Lester, and Brazelton (1979) to characterize the early development of full-term infants. They describe the plight of the small preterm infant:

> Not only is the preterm infant an organism in an environment he has not yet evolved for, but he is an organism whose biological program is called upon prematurely so that the normal sequence of subsystem differentiation and integration generally found by term has not yet been executed. (p. 179)

The special needs of the small preterm infant were reviewed in detail by Gorski, Davison, and Brazelton (1979). They described the various stages of neurobehavioral organization, state control, and responsiveness; emphasized the negative effects of excessive and inappropriate stimulation; and suggested,

> Providing care geared to the state of neurobehavioral organization of these fragile, disorganized infants may be therapeutic. We have seen such infants begin to improve after prolonged periods of no weight gain, static oxygen requirements, and other indications of fixed respiratory disease after providing them programs of sensitively positive experience. (p. 61)

Experiences geared to the baby's ability to participate and reciprocate were described as increasing behavioral organization, and "such behavioral improvement might lead to better outcomes for high risk neonates by improving their neurologic and autonomic organization on which to build toward improved interaction with their environment" (p. 61).

Specific components of the nursery environment can be identified for studies of behavioral effects. An example is the research of Korner and her associates, who investigated the effects of vestibular-proprioceptive stimulation by means of oscillating waterbeds (Korner,

Kraemer, Haffner, & Cosper, 1975). The association of vestibular stimulation with increased soothing and increased visual alertness had been described for older infants (Korner & Thoman, 1972), then subsequently demonstrated for full-term newborns and newborn rats in experiments designed to isolate the specific effects of vestibular-proprioceptive stimulation (see review of this research by Korner, 1979). It was hypothesized that preterm infants, being deprived of the normal vestibular-proprioceptive stimulation associated with intrauterine life, might benefit from systematic stimulation. Waterbeds, which have other beneficial clinical applications for fragile infants, were designed to include gentle oscillations.

The result of a first study (Korner et al., 1975) was a significant reduction in the frequency of apneic spells (episodic cessation of breathing) during the first days of life for a stimulated group compared with controls. Similarly positive results were subsequently found for infants preselected for apnea who were used as their own controls during periods on and off the oscillating waterbeds (Korner, 1979). In a third study, infants with severe HMD were compared, with stimulated infants receiving gentle oscillations in the rhythm of maternal respirations (Korner, Forrest, & Schneider, 1981). Significant differences were found in neurobehavioral comparisons between small groups at 35 weeks of gestational age. Stimulated infants spent more time in a quiet and alert state, gave more responses to auditory and visual items of the Brazelton Scale, demonstrated more mature spontaneous motor behavior, and showed fewer signs of irritability and hypertonicity. First indications from sleep studies suggested that infants on oscillating waterbeds had longer periods of quiet sleep and less irritability than controls (A. F. Korner, personal communication).

In reviewing studies of the special needs of the small preterm infant, we have touched on two topics of very active current research. They are (1) the identification of normal and abnormal characteristics of prematurity across stages of neurobehavioral development, and (2) the modification of components of the nursery environment to enhance preterm development. The positive and negative behavioral and physiological responses of the infant to different physical, sensory, and social stimulation are apparent to the experienced observer. Research that incorporates planned intervention to modify the nursery environment requires a high level of sensitivity to the risks and needs associated with each stage of neurobehavioral organization.

Discussion

This review provides only a limited appraisal of the array of biological risks currently being investigated for their association with behavioral deficits in infancy and long-term intellectual disabilities. One self-imposed constraint on topic selection has been to consider only those variables that can be modified by environmental manipulation, but progress in medical and social-science research virtually guarantees that the list of variables will continue to expand. Another imposed constraint has been the exclusion of studies investigating the effects of the caregiver on neonatal and infant development, although the discussion of the environmental needs of the small preterm infant considers some associations between the caretaking environment and neurological organization. However, the fact remains that the behavioral organization of the infant takes place in a social context, as emphasized elsewhere in this volume. Topic selection has been illustrative, focusing on issues of contemporary research interest. The presentation of recent research publications has been deliberate, but the references to numerous review papers will perhaps partly compensate the reader for the omission of important earlier studies.

The substantive findings for each topic may seem so varied as to preclude any firm conclusions about the harmful effects of any specific variable. With some exceptions, such as acute maternal alcoholism and perhaps maternal drug addiction, the effects on behavioral development appear to be mild, variable within populations, and conflicting across studies. A conclusion that these risks are not substantial is unwarranted when we consider that most variables that have been considered are associated with increased fetal and neonatal death in both animal and human studies. In this sense, none can be considered harmless even when direct damage to the central nervous system cannot be demonstrated. The apparent relative immutability of the nervous system to some risks such as extreme malnutrition may be a reflection more of lethal effects on other organ systems; if the infant survives, the central nervous system is not grossly affected. There also may be undefined dose-related behavioral effects for some variables. Such effects (as were reported for maternal alcohol use) would contribute to differing results across studies and would also suggest that some observed differences among risks might be quantitative rather than qualitative. In sum, the unresolved issues make most conclusions about substantive findings tentative, but we become less tentative as evidence accrues through the replication of results.

One striking finding that has emerged from diverse studies is the subtlety of many of the significant behavioral correlates of risk states. This result is particularly obvious when standard test scores are reported, such as those from the Bayley Scales or IQ tests at older ages. We have frequently seen that, whereas there was a significant difference in scores between risk and control groups, both groups had average scores well within the normal range. Novel or nonstandard measurements could not be described in this way if no norms were available for comparisons, but in such cases, extreme group differences in measurements were rarely noted. The general finding of subtle effects argues persuasively for a detailed examination of behavioral data. Because problems are not always obvious, a dichotomous characterization of outcome as normal or abnormal may not be sufficient even though this categorical information is important from a pragmatic perspective. Our understanding of the full significance of many risk variables is enhanced by a more definitive analysis.

Methodological differences, apparent for studies within and across risk topics, certainly have contributed to the varied results and conclusions reported. These differences include the criteria used to identify a risk population, the variables deemed important in making controlled comparisons, the kinds of measurements used, and the ages at which outcome was defined. Such diversity is both inevitable and beneficial. No individual research effort can control for all relevant variables associated with behavioral effects in the study of human infants. In the studies that match risks and control infants the most rigorously, the select groups studied were often very small and usually represented only a fraction of the total risk population. Such studies were frequently directed toward specific issues of causality. Large demographic studies provide a different perspective and their findings have general applicability, but practical considerations have limited the range of variables that were investigated. Studies of individual differences, now reemerging as a valid research strategy, add another important dimension when their results are considered along with those of studies emphasizing central tendencies.

When we consider all studies reporting significant behavioral associations with risk, it is evident that none was able to demonstrate direct cause and effect between any specific risk and behavioral deficits. We must conclude that these risks, although significant for outcome, have explained only a small portion of the variability found in the different behavioral measurements at any age, from neonate to child. Such a result is commonplace in behavioral studies and we expect to find many factors, both independent and correlated, that influence behavioral status. We assume, for example, that other potentially damaging conditions may also be exerting effects on development when high-risk populations are studied. An hypothesis of cumulative vulnerability is often put forward, and in many studies, prenatal and perinatal risk scores have been used to provide some further differentiation within a specific risk population (see, for example, Stave & Ruvalo, 1980). Surprisingly, cumulative risk scores have not usually resulted in markedly improved prediction of behavioral status.

An alternative strategy is to focus on invulnerability rather than vulnerability. When a risk is found to exert a significant influence on development, we need a construct that can account for the exceptions—that is, for those infants and children who experienced the risk but escaped the effects. Shifts in relative vulnerability can be demonstrated by direct remedial intervention. For example, Zeskind and Ramey (1981) have shown that the added risk associated with SGA status for infants from low socioeconomic environments can be ameliorated by educational intervention. Aside from formal intervention, Sameroff and Chandler (1975) have suggested that transactional effects between the infant and his caretaking environment are critical in determining the "self-righting" tendencies of some risk infants. The origins of individual differences that influence such natural remedial effects can be investigated. Therapeutic efforts to organize and normalize the maturing neurobehavioral patterns of sick and premature infants have been described (Gorski et al., 1979; Korner, 1979), but the long-term effects of such efforts are not known. Studies that investigate the lasting effects of manipulations of the neonatal environment to meet the special needs of the most vulnerable infants can be anticipated.

References

Abel, E. L. Fetal alcohol syndrome: Behavioral teratology. *Pyschological Bulletin*, 1980, *87*, 29–50.

Als, H., Tronick, E., Adamson, L., & Brazelton, T. B. The behavior of the full-term but underweight newborn infant. *Developmental Medicine and Child Neurology*, 1976, *18*, 590–602.

ALS, H., LESTER, B. M., & BRAZELTON, T. B. Dynamics of the behavioral organization of the premature infant: A theoretical perspective. In T. M. Field, A. M. Sostek, S. Goldberg, & H. H. Shuman (Eds.), *Infants born at risk: Behavior and development*. New York: Spectrum, 1979.

BABSON, S. G., & HENDERSON, N. B. Fetal undergrowth: Relation of head growth to later intellectual performance. *Pediatrics*, 1974, *53*, 890–894.

BAYLEY, N. *Manual, Bayley scales of infant development*. New York: Psychological Corporation, 1969.

BRAZELTON, T. B. *Neonatal behavioral assessment scale*. London: Spastics International Medical Publications, 1973.

BROMAN, S. H., NICHOLS, P. L., & KENNEDY, W. A. *Preschool IQ: Prenatal and early development correlates*. New York: Halsted Press, 1975.

CAMPBELL, S., & NEWMAN, G. B. Growth of the fetal biparietal diameter during normal pregnancy. *Journal of Obstetrics and Gynaecology of the British Commonwealth*, 1971, *78*, 513–519.

CATTELL, P. *The measurement of intelligence in infants and young children*. New York: Science Press, 1940. (Reprinted by Psychological Corporation, 1960.)

CHAPLIN, E. R., GOLDSTEIN, G. W., MYERBERG, D. Z., HUNT, J. V., & TOOLEY, W. H. Posthemorrhagic hydrocephalus in the preterm infant. *Pediatrics*, 1980, *65*, 901–909.

CLARREN, S. K., & SMITH, D. W. The fetal alcohol syndrome: A review of the world literature. *New England Journal of Medicine*, 1978, *298*, 1063–1067.

COMMEY, J. O. O., & FITZHARDINGE, P. M. Handicap in the preterm small-for-gestational age infant. *Journal of Pediatrics*, 1979, *94*, 779–786.

DAVIES, P. A. Infants of very low birth weight: An appraisal of some aspects of their present neonatal care and of their later prognosis. In D. Hull (Ed.), *Recent advances in pediatrics*. London: Churchill-Livingstone, 1976.

DAVIS, M. M., & SHANKS, B. Neurological aspects of perinatal narcotic addiction and methadone treatment. *Addictive Diseases: An International Journal*, 1975, *2*, 213–226.

DINGES, D. F., DAVIS, M. M., & GLASS, P. Fetal exposure to narcotics: Neonatal sleep as a measure of nervous system disturbance. *Science*, 1980, *209*(4456), 612–621.

DOBBING, J., & SANDS, J. Quantitative growth and development of human brain. *Archives of Disease in Childhood*, 1973, *48*, 757–767.

DRILLIEN, C. M. *The growth and development of the prematurely born infant*. Baltimore, Md.: Williams & Wilkins, 1964.

DRILLIEN, C. M. Aetiology and outcome in low-birth-weight infants. *Developmental Medicine and Child Neurology*, 1972, *14*, 563–574.

DYKES, F. D., LAZZARA, A., AHMANN, P., BLUMENSTEIN, B., SCHWARTZ, J., & BRANN, A. W. Intraventricular hemorrhage: A prospective evaluation of etiopathogenesis. *Pediatrics*, 1980, *66*, 42–49.

FANCOURT, R., CAMPBELL, S., HARVEY, D., & NORMAN, A. P. Follow-up study of small-for-date babies. *British Medical Journal*, 1976, *1*, 1435–1437.

FIELD, T. M., DEMPSEY, J. R., & SHUMAN, H. H. Developmental assessments of infants surviving the respiratory distress syndrome. In T. M. Field, A. M. Sostek, S. Goldberg, & H. H. Shuman (Eds.), *Infants born at risk: Behavior and development*. New York: Spectrum, 1979.

FIELD, T. M., SOSTEK, A. M., GOLDBERG, S., & SHUMAN, H. H. (Eds.). *Infants born at risk: Behavior and development*. New York: Spectrum, 1979.

FITZHARDINGE, P. M., KALMAN, E., ASHBY, S., & PAPE, K. E. Present status of the infant of very low birth weight treated in a referral neonatal intensive care unit in 1974. In *Major mental handicap: Methods and costs of prevention* (Ciba Foundation Symposium 59). Amsterdam: Elsevier/Excerpta Medica/North-Holland, 1978.

FOGELMAN, K. Smoking in pregnancy and subsequent development of the child. *Child Care, Health and Development,* 1980, 6, 233–249.

FRANCIS-WILLIAMS, J., & DAVIES, P. A. Very low birthweight and later intelligence. *Developmental Medicine and Child Neurology,* 1974, 16, 709–728.

FRANKENBURG, W. K., DODDS, J. B., & FANDAL, A. W. *Denver developmental screening test.* Denver: University of Colorado Medical Center, 1970.

FRIEDMAN, S. L., & SIGMAN, M. (Eds.). *Preterm birth and psychological development.* New York: Academic Press, 1981.

GABR, M. Malnutrition during pregnancy and lactation. In G. H. Bourne (Ed.), *World review of nutrition and dietetics* (Vol. 36). Basel: Karger, 1981.

GESELL, A., & AMATRUDA, C. S. *Developmental diagnosis.* New York: Paul B. Hoeber, 1941.

GESELL, A., & AMATRUDA, C. S. *The embryology of behavior: The beginnings of the human mind.* New York: Harper & Brothers, 1945.

GORSKI, P. A., DAVISON, M. F., & BRAZELTON, T. B. Stages of behavioral organization in the high-risk neonate: Theoretical and clinical considerations. *Seminars in Perinatology,* 1979, 3, 61–72.

GRIFFITHS, R. *The abilities of young children: a comprehensive system of mental measurement for the first eight years of life.* London: Child Development Research Center, 1970.

HACK, M., FANAROFF, A. A., & MERKATZ, I. R. The low-birth-weight infant—evolution of a changing outlook. *New England Journal of Medicine,* 1979, 301, 1162–1166.

HARDY, J. B., & MELLITS, E. D. Does maternal smoking during pregnancy have a long-term effect on the child? *Lancet,* 1972, 2, 1332–1336.

HUNT, J. V. Longitudinal research: A method for studying the intellectual development of high risk preterm infants. In T. M. Field, A. M. Sostek, S. Goldberg, & H. H. Shuman (Eds.), *Infants born at risk: Behavior and development.* New York: Spectrum, 1979.

HUNT, J. V. Predicting intellectual disorders in childhood for preterm infants with birthweights below 1501 grams. In S. L. Friedman and M. Sigman (Eds.), *Preterm birth and psychological development.* New York: Academic Press, 1981.

HUTCHINGS, D. E. Behavioral teratology: Embryopathic and behavioral effects of drugs during pregnancy. In F. Gottlieb (Ed.), *Studies on the development of behavior and the nervous system (Vol. 4): Early influences.* New York: Academic Press, 1978.

JONES, K. L., & SMITH, D. W. Recognition of the fetal alcohol syndrome in early infancy. *Lancet,* 1973, 2, 999–1001.

JONES, K. L., SMITH, D. W., STREISSGUTH, A. P., & MYRIANTHOPOULOS, N. C. Outcome in offspring of chronic alcoholic women. *Lancet,* 1974, 1, 1076–1078.

JONES, R. A. K., CUMMINS, M., & DAVIES, P. A. Infants of very low birthweight. A 15-year analysis. *Lancet,* 1979, 1, 1332–1335.

KALTENBACH, K., GRAZIANI, L. J., & FINNEGAN, L. P. Methadone exposure in utero: Developmental status at one and two years of age. *Protracted Effects of Perinatal Drug Dependence, Pharmacology, Biochemistry and Behavior,* 1979, 11(Suppl.), 15–17.

KAMPER, J. Long term prognosis of infants with severe idiopathic respiratory distress syndrome. I. Neurological and mental outcome. *Acta Paediatrica Scandinavica,* 1978, 67, 61–69.

KOOPS, B. L. Neurologic sequelae in infants with intrauterine growth retardation. *Journal of Reproductive Medicine,* 1978, 21, 343–351.

KORNER, A. F. Maternal rhythms and waterbeds: A form of intervention with premature infants. In E. B. Thoman (Ed.), *Origins of the infant's social responsiveness*. Hillsdale, N.J.: Lawrence Erlbaum Associates, 1979.

KORNER, A. F., & THOMAN, E. B. The relative efficacy of contact and vestibular stimulation in soothing neonates. *Child Development*, 1972, 43, 443–453.

KORNER, A. F., KRAEMER, H. C., HAFFNER, M. E., & COSPER, L. M. Effects of waterbed flotation on premature infants: A pilot study. *Pediatrics*, 1975, 56, 361–367.

KORNER, A. F., FORREST, T., & SCHNEIDER, P. *Development of a longitudinal neurobehavioral assessment procedure for preterms: Preliminary results from an intervention study*. Paper presented at the meetings for the Society for Research in Child Development, Boston, 1981.

LEATHERWOOD, P. Influence of early undernutrition on behavioral development and learning in rodents. In G. Gottlieb (Ed.), *Studies on the development of behavior and the nervous system* (Vol. 4): *Early influences*. New York: Academic Press, 1978.

LEIJON, I., FINNSTROM, O., NILSSON, B., & RYDEN, G. Neurology and behaviour of growth-retarded neonates. Relation to biochemical placental function tests in late pregnancy. *Early Human Development*, 1980, 4, 257–270.

LIPPER, E., LEE, K. S., GARTNER, L. M., & GRELLONG, B. Determinants of neurobehavioral outcome in low-birth-weight infants. *Pediatrics*, 1981, 67, 502–505.

LODGE, A., MARCUS, M. M., & RAMER, C. M. Behavioral and electrophysiological characteristics of the addicted neonate. *Addictive Diseases: An International Journal*, 1975, 2, 235–255.

LUBCHENCO, L. O., DELIVORIA-PAPADOPOULOS, M., & SEARLS, D. Long-term follow-up of prematurely born infants. II. Influence of birth weight and gestational age on sequelae. *Journal of Pediatrics*, 1972, 80, 509–512.

McCARTHY, D. *Manual for the McCarthy scales of children's abilities*. New York: Psychological Corporation, 1972.

MILLER, H. C. & MERRITT, T. A. *Fetal growth in humans*. Chicago: Year Book Medical Publishers, 1979.

OLEGARD, R., SABEL, K. G., ARONSSON, M., SANDIN, B., JOHANSSON, P. R., CARLSSON, C., KYLLERMAN, M., IVERSEN, K., & HRBEK, A. Effects on the child of alcohol abuse during pregnancy. *Acta Paediatrica Scandinavica*, 1979, *Supplement 275*, 112–121.

OLSEN, G. D., & LEES, M. H. Ventilatory response to carbon dioxide of infants following chronic prenatal methadone exposure. *Journal of Pediatrics*, 1980, 96, 983–989.

OSTREA, E. M., & CHAVEZ, C. J. Perinatal problems (excluding neonatal withdrawal) in maternal drug addiction: A study of 830 cases. *Journal of Pediatrics*, 1979, 94, 292–295.

PAPE, K. E., BUNCIC, R. J., ASHBY, S., & FITZHARDINGE, P. M. The status at two years of low-birth-weight infants born in 1974 with birth weights of less than 1,001 gm. *Journal of Pediatrics*, 1978, 92, 253–260.

PARKINSON, C. E., WALLIS, S., & HARVEY, D. School achievement and behaviour of children who were small-for-dates at birth. *Developmental Medicine and Child Neurology*, 1981, 23, 41–50.

PHILIP, A. G. S. Fetal growth retardation: Femurs, fontanels, and follow-up. *Pediatrics*, 1978, 62, 446–453.

RAMER, C. M., & LODGE, A. Neonatal addiction: A two-year study. Part I. Clinical and developmental characteristics of infants of mothers on methadone maintenance. *Addictive Diseases: An International Journal*, 1975, 2, 227–234.

RANTAKALLIO, P. The effect of maternal smoking on birth weight and the subsequent health of the child. *Early Human Development*, 1978, 2(4), 371–382. (a)

RANTAKALLIO, P. Relationship of maternal smoking to morbidity and mortality of the child up to age of five. *Acta Paediatrica Scandinavica*, 1978, *67*, 621–631. (b)

RICHARDSON, S. A. The relation of severe malnutrition in infancy to the intelligence of school children with differing life histories. *Pediatrics Research*, 1976, *10*, 57–61.

ROSENZWEIG, M. R., & BENNETT, E. L. Experiential influences on brain anatomy and brain chemistry in rodents.In G. Gottlieb (Ed.), *Studies on the development of behavior and the nervous system* (Vol. 4): *Early influences*. New York: Academic Press, 1978.

ROSETT, H. L., SNYDER, P., SANDER, L. W., LEE, A., COOK, P., WEINER, L., & GOULD, J. Effects of maternal drinking on neonate state regulation. *Developmental Medicine and Child Neurology*, 1979, *21*, 464–473.

RUIZ, M. P. D., LEFEVER, J. A., HAKANSON, D. O., CLARK, D. A., & WILLIAMS, M. L. Early development of infants of birth weight less than 1,000 grams with reference to mechanical ventilation in the newborn period. *Pediatrics*, 1981, *68*, 330–335.

SAMEROFF, A. J., & CHANDLER, M. J. Reproductive risk and the continuum of caretaking casualty. In F. D. Horowitz, M. Hetherington, S. Scarr-Salapatek, & G. Siegel (Eds.), *Review of child development research* (Vol. 4). Chicago: University of Chicago Press, 1975.

SAXTON, D. W. The behaviour of infants whose mothers smoke in pregnancy. *Early Human Development*, 1978, *2*, 363–369.

SIMPSON, W. J. A preliminary report on cigarette smoking and the incidence of prematurity. *American Journal of Obstetrics and Gynecology*, 1957, *73*, 808–815.

SOKOL, R. J., MILLER, S. I., & REED, G. Alcohol abuse during pregnancy: An epidemiologic study. *Alcoholism: Clinical and Experimental Research*, 1980, *4*, 135–145.

STAVE, U., & RUVALO, C. Neurological development in very-low-birthweight infants. Application of a standardized examination and Prechtl's optimality concept in routine evaluations. *Early Human Development*, 1980, *4*, 229–241.

STEWART, A. L., & REYNOLDS, E. O. R. Improved prognosis for infants of very low birthweight. *Pediatrics*, 1974, *54*, 724–735.

STRAUSS, M. E., STARR, R. H., OSTREA, E. M., CHAVEZ, C. J., & STRYKER, J. C. Behavioral concomitants of prenatal addiction to narcotics. *Journal of Pediatrics*, 1976, *89*, 842–846.

STRAUSS, M. E., LESSEN-FIRESTONE, J. K., CHAVEZ, C. J., & STRYKER, J. C. Children of methadone-treated women at five years of age. *Protracted Effects of Perinatal Drug Dependence, Pharmacology, Biochemistry and Behavior*, 1979, *11*(Suppl.), 3–6.

STREISSGUTH, A. P. Maternal alcoholism and the outcome of pregnancy: A review of the fetal alcohol syndrome. In M. Greenblatt & M. A. Schuckit (Eds.), *Alcoholism problems in women and children*. New York: Grune & Stratton, 1976.

STREISSGUTH, A. P., BARR, H. M., MARTIN, D. C., & HERMAN, C. S. Effects of maternal alcohol, nicotine, and caffeine use during pregnancy on infant mental and motor development at eight months. *Alcoholism: Clinical and Experimental Research*, 1980, *4*, 152–164.

STREISSGUTH, A. P., LANDESMAN-DWYER, S., MARTIN, J. C., & SMITH, D. W. Teratogenic effects of alcohol in humans and laboratory animals. *Science*, 1980, *209*, 353–361.

STREISSGUTH, A. P., MARTIN, D. C., MARTIN, J. C., & BARR, H. M. The Seattle longitudinal prospective study on alcohol and pregnancy. *Neurobehavioral Toxicology and Teratology*, 1981, *3*, 223–233.

THOMPSON, T., & REYNOLDS, J. The results of intensive care therapy for neonates: I. Overall neonatal mortality rates: II. Neonatal mortality rates and long-term prognosis for low birth weight neonates. *Journal of Perinatal Medicine*, 1977, *5*, 59–92.

TOOLEY, W. H. Epidemiology of bronchopulmonary dysplasia. *Journal of Pediatrics*, 1979, *95*, 851–855.

USHER, R., & McLEAN, F. Interuterine growth of live-born Caucasian infants at sea level: Standards obtained from measurements in 7 dimensions of infants born between 24 and 44 weeks of gestation. *Journal of Pediatrics*, 1969, *74*, 901–910.

VOHR, B. R., OH, W., ROSENFIELD, A. G., & COWETT, R. M. The preterm small-for-gestational age infant: A two-year follow-up study. *American Journal of Obstetrics and Gynecology*, 1979, *133*, 425–431.

VUORI, L., CHRISTIANSEN, N., CLEMENT, J., MORA, J. O., WAGNER, M., & HERRERA, M. G. Nutritional supplementation and the outcome of pregnancy: II. Visual habituation at 15 days. *American Journal of Clinical Nutrition*, 1979, *32*, 463–469.

WILSON, G. S., DESMOND, M. M., & VERNIAUD, W. M. Early development of infants of heroin-addicted mothers. *American Journal of Diseases of Children*, 1973, *126*, 457–462.

WILSON, G. S., McCREARY, R., KEAN, J., & BAXTER, J. C. The development of preschool children of heroin-addicted mothers: A controlled study. *Pediatrics*, 1979, *63*, 135–141.

ZAMENHOF, S., & VAN MARTHENS, E. Nutritional influences on prenatal brain development. In G. Gottlieb (Ed.), *Studies on the development of behavior and the nervous system* (Vol. 4): *Early influences*. New York: Academic Press, 1978.

ZESKIND, P. S., & RAMEY, C. T. Preventing intellectual and interactional sequelae of fetal malnutrition: A longitudinal, transactional, and synergistic approach to development. *Child Development*, 1981, *52*, 213–218.

9 Social Intelligence and the Development of Communicative Competence

LOUISE CHERRY WILKINSON

INTRODUCTION

In this chapter, we consider social intelligence and its relationship to communicative competence. We share the position held by a substantial and growing number of psychologists that the construct of intelligence, as it has been defined and used, has been of limited utility and questionable validity in its ability to predict the ways in which individuals solve problems in everyday situations. It is necessary to develop and use an alternative construct of intelligence, which would guide research and practice, and we propose the term *social intelligence* to refer to the range of human competencies involved in functionally appropriate interpersonal behaviors. The ability to communicate effectively in social situations, *communicative competence*, is central to our conception of social intelligence.

The goals of the chapter include (1) defining and reviewing the history of the concept of social intelligence, with particular reference to assessment of infants; (2) explicating the communicative aspects of social intelligence, namely, those that involve learning to use language to communicate effectively; (3) considering the developmental course of the communicative aspects of social intelligence, in particular, those aspects that may be grounded in early parent–child interaction and those covering the continuities between the preverbal and verbal stages

LOUISE CHERRY WILKINSON • Department of Educational Psychology, University of Wisconsin—Madison, Madison, Wisconsin 53706.

of development; and finally, (4) suggesting possible relationships between early social intelligence and later communicative competence.

SOCIAL INTELLIGENCE

Critique of the Construct of Intelligence

The construct of intelligence is intended to reflect the individual's development of formal cognitive ability, and the IQ test was created to measure this ability. Intelligence is regarded as the most important component in predicting human problem-solving behavior. Within the last decade, both the construct of intelligence and the use of IQ tests have been under attack. This debate was stimulated by those in the scientific community as well as those who form and implement social policy in education (see Zigler & Trickett, 1978). McClelland (1973) challenged the validity of intelligence tests on two grounds: (1) performance on them bears little relationship with any other "real-life" performance, and (2) the predictive ability of IQ tests extends only to school performance, and even then considerable variability remains unaccounted for by IQ scores. McClelland notes a correlation of approximately .70 between test performance and school performance.

A consistent theme throughout these critiques of traditional concepts of intelligence testing is that they are two narrow. Most importantly, there is no acknowledgment of the role played by what can be broadly deemed "social" and "personal" factors inherent in human problem-solving behavior. As Zigler and Trickett summarized: "Stated most simply, we believe that one can obtain a very high IQ score and still not behave admirably in the real world that exists beyond the confines of the psychologists' testing room" (Zigler & Trickett, 1978, p. 791). Both Scarr (1981) and Zigler and Trickett (1978) have highlighted the limited predictability of IQ to school performance.

Social Intelligence: An Alternative Construct

During the 1970s some of the same psychologists who criticized the construct of intelligence suggested alternatives. We will discuss some of these alternatives, and consider each as a variant on the construct of social intelligence, defined as the human competence involved in functionally appropriate interpersonal behaviors.

It is interesting to note that while the experts in the intelligence testing movement have disregarded the social aspect of intelligence, nonprofessionals have not. Sternberg, Conway, Ketron, and Bernstein (1981) examined nonprofessionals' and experts' conceptions of intelligence and found both in agreement that a social competence underlies human intelligence. Nonprofessionals, however, were more likely to describe social intelligence as a separate and significant factor in addition to two other factors, verbal ability and practical ability. The factor of social intelligence included such test items as being sensitive to other people's needs and desires and displaying immediate interest in the environment. The factor of verbal intelligence included some clearly social-personal aspects of the use of language as evidenced by clear and articulate speech, verbal fluency, and effectively dealing with others.

Social intelligence can be regarded as human competencies that are involved in functionally appropriate interpersonal behaviors. One aspect of social intelligence is *communicative competence*. Both McClelland (1973) and Scarr (1981) have highlighted, in particular, the importance of communicative competencies and adaptation to real-life situations. Within the last decade, a new paradigm of research has developed on the study of communicative competence: the sociolinguistic approach (Wilkinson, 1982). This approach focuses upon descriptions of individuals' use of language and nonverbal behavior in interaction with one another, such as parents interacting with children in the home. The descriptions provide us with a rich view of the complexity of communication and the diversity of an individual's behavior in many situations. The use of vocal and verbal expression to communicate is the focus of language and communicative development from infancy through the school-age years.

Zigler and Trickett (1978) introduced the broader concept of social competence as the appropriate index of human functioning. They believe that two criteria should guide the use of assessments of social competence:

> The first is that social competence must reflect the success of the human being in meeting societal expectancies. Second, these measures of social competence should reflect something about the self-actualization or personal development of the human being. (Zigler & Trickett, 1978, p. 795)

They propose four broad areas for inclusion in assessment of social competence, including measures of health and physical well-being, measures of formal cognitive ability, measures of achievements, and the measurement of motivational and emotional variables.

Lee (1979) and Greenspan (1980), among others, commended Zigler and Trickett for drawing attention to the critical importance of social competence for the life adaptation of humans, but they also criticized the particular proposal of social competence. Lee (1979) believes that the Zigler and Trickett model does not take account of cultural variations in individuals, and she proposes an alternative definition of a more comprehensive model.

Lee refers to her concept as the *repertoire approach*, since social competence and development can be measured by the individual's repertoire and levels of complex strategies for dealing effectively with others in social situations. For her, repertoire refers to the range of different concepts and methods that an individual knows and can apply in particular situations. Her proposal incorporates some of the same aspects as does McClelland's (1973), including the belief that the individual's repertoire or social competence develops with learning and experience and that "the broader an individual's repertoire, the greater the potential for effective action with others, in diverse contexts, relative to diverse goals" (Lee, 1979, p. 795). Like McClelland, she believes that assessment of social competence necessitates observations of natural behavior during social exchanges in order to understand effective behavior, its goals, and the circumstances under which it occurs.

Greenspan (1980) does not believe that Zigler and Trickett's (1978) model of social competence is sufficiently "social" since it focuses on formal cognitive properties. Their proposal fails because it includes too broad a range of physical, intellectual, and interpersonal competencies. Greenspan believes that an adequate model of social competence could integrate a more focused range of elements identified by psychologists, including temperament, character, and social awareness. For Greenspan, *temperament* includes such elements as task orientation and emotional liability; *character* includes such elements as social autonomy and responsibility; and *social awareness* contains such elements as role-taking and social problem-solving. As the other theorists, Greenspan emphasizes the adaptation of the individual to a social context, but he emphasizes both the outcome and the process, what he calls the "whether" and the "why" individuals succeed.

Charlesworth (1976) has argued that human intelligence is best conceptualized as an adaptation of the individual to everyday life. His view is consistent with the others discussed in identifying life-success as the criterion against which assessment of intelligence, or social intelligence, ought to be conducted. Charlesworth believes that humans adapt through their intelligent behavior, which can be seen in their everyday interactions. Intelligent behavior refers to an individual's attempt to adapt to problematic situations and imbalances in interactions with social and physical environment. Cognitive processes control and organize intelligent behavior. Charlesworth believes that we must include observation of spontaneous behavior in natural circumstances in assessment of individuals' intelligence. Our goal should be to achieve accurate and comprehensive descriptions of individuals' intelligent behavior, as well as the cognitive demand characteristics of the environment, characteristics that not only daily challenge us to act intelligently but also play an important role in the development of intelligence.

In sum, it seems that the "experts" have been the ones deceived. Perhaps the critics have been right in their view of the construct of intelligence and the intelligence test itself as a "technological trap that has resulted in the calcification of our theoretical views and/or has misled us as to the essential nature of human development and the optimization of such development" (Zigler & Trickett, 1978, p. 790). It is clear that intelligence does include a social component, of which communicative competence is a part.

HISTORICAL PERSPECTIVE ON SOCIAL INTELLIGENCE: ADULTS AND INFANTS

The existence of a social component in intelligence has been debated since the beginning of the intelligence testing movement. Several substantive issues have recurred consistently in the 8 decades of research on adults and infants: (1) whether there is a separate factor for social intelligence, or whether the social dimension is merely the application of general intelligence to social situations: (2) whether it is best to measure social intelligence with standard tests or with observations of naturally occurring behavior; and (3) whether this aspect of intelligence is significantly affected by experience. There has been no resolution to these controversies. Social intelligence has played a minor role in most theories of intelligence and has been, to a large extent, ignored.

Adults

The traditional view of intelligence has focused on a general factor for intelligence, consisting of abstract, rational thinking assessed through standardized tests, and uneffected by experience (Lewis, this volume). There have been some exceptions to this general trend. Thorndike (1936) was among the first to propose that social intelligence existed as separate from abstract and concrete intelligence. He defined social intelligence as a facility in dealing with humans, in contrast with a facility in dealing with concepts (abstract, verbal intelligence), and manipulating objects (practical intelligence). Thorndike believed that the best way to assess social intelligence was to observe other people solving interpersonal problems in real situations. In one study, he analyzed tests of social and abstract intelligence. His results supported the existence of a separate factor for social intelligence.

Although most theories of intelligence have ignored any concept of social intelligence, there have been some exceptions. Woodrow (1939), examined several scales of social intelligence and scales of verbal, abstract intelligence. He found that the scales were related to verbal intelligence. Wedeck (1947) identified three factors: general intelligence, verbal intelligence, and behaviorial cognition. Guilford (1958) believed that there was a behavioral factor in intelligence, referring to the kind of knowledge involved in understanding the actions and thoughts of other people. In contrast to Thorndike (1936), Guilford believed that it was possible to design a test to assess social intelligence. He did not consider that social intelligence existed as a separate factor, but he believed that it represented the application of general intelligence to social situations.

The present status of research on intelligence and intelligence testing is consistent in some ways with the beginnings of the mental testing movement of the early decades of this century. For example, with one exception, the chapters in a recent edited volume on intelligence are concerned primarily with abstract and rational aspects of intelligence, and there is no treatment of social intelligence (Resnick, 1976). Keating (1978), in a recent review of factor analytic studies on intelligence, concluded that he had failed to demonstrate a separate factor for social intelligence. He offered several reasons for the failure to find confirmation, including the lack of a clear theoretical definition of social intelligence and the possible inadequacy of paper-and-pencil tests for assessing social intelligence. However, as noted previously, other recent work suggests new directions for the construct of social intelligence.

Infants

Since infant intelligence tests have their roots in the intelligence testing movement of the early part of this century, it is not surprising that the concept of a social aspect of intelligence was introduced in the early formulations but was largely ignored and remained undeveloped in subsequent test development. Brooks-Gunn and Weinraub (this volume) describe the history of infant intelligence in three "waves."

The most extensive assessment of infant intelligence was carried out by Gesell and his colleagues (Buhler & Hetzer, 1935). In his preliminary research on the development of an intelligence scale, he included data on several areas of human development, including language (comprehension and imitation) and social-personal behavior (reactions to people, initiative, independence, and play behavior). He also was the first to develop procedures for direct observation of children in their natural environments. Several years later, in his Gesell Developmental Schedules (see Gesell & Amatruda, 1941), he included more refined language items, including facial expressions, gestures, and vocalizations as well as personal-social items, including feeding, dressing, toilet-training, and play behavior. His schedules are considered to be highly significant in assessment of infant intelligence. Although they are considered to be less standardized and more subjective than other, more recent scales, they have been the primary source of material for subsequent assessments (Brooks-Gunn & Weinraub, this volume).

The second wave of infant intelligence tests developed during the 1940s. Cattell (1940, 1960, 1966) was very influential in this work. She indicated disappointment in the early Gesell work, particularly in the personal-social items, which were, in her opinion, heavily influenced by home experiences and lacking in objective procedures for administration and scoring. She, therefore, removed these items in her infant assessment test.

Griffiths (1954) provided a different criticism of the early Gesell work, arguing that not enough attention was given to that uniquely human competence, language. She included twice as many speech items in her test as had been in previous tests. For the first-year assessment, the items included measures of imitation and monologue, attention to sounds in conversation, repetition of single sounds, and babbling. Griffiths emphasized the social nature of many of the language items and the importance of social relationships to the infant's intellectual development. However, her test focuses less on language-interactional items in the second-year assessment. Measures of orientation toward, interaction with, and preference for other people were only included in the first-year assessment (Brooks-Gunn & Weinraub, this volume). During the second year, the personal-social items consisted only of self-help skills.

The third wave of tests differed from previous work in that they were designed not to test all aspects of infant intelligence, but to fulfill specific purposes and include particular types of measures. Brooks-Gunn and Weinraub (this volume) mention specific tests of infant language development and a renewed interest in tests of social behavior.

For example, Yarrow and Messer (this volume) examined clusters of items from the Bayley Scales applicable to infants. Two of the eight clusters are social responsiveness and vocalization-language. The cluster of social responsiveness is composed of items measuring the infant's early responses to people: smiling, vocalization, making anticipatory adjustments to being picked up, and engaging in simple play. Yarrow and Messer believe that this cluster of items reflects the infant's awareness of something other than its own feelings, and demonstrates the infant's reaching out to become involved with other people. The social responsiveness cluster is highly related to overall mental development, perhaps reflecting common psychological processes between both groups of measures, even though it did show an independence from such other clusters as cognitive and motivational. The vocalization-language cluster was composed of measures of rudimentary aspects, comprehension, responsiveness to own name, appropriateness of responses to words and expressive aspects of language, such as vocalizing syllables and attitudes. This cluster showed a low relationship with all other clusters except the social responsiveness cluster, an effect that probably reflects the common social basis for both clusters.

In sum, the conclusion of Brooks-Gunn and Weinraub (this volume) that the development of infant intelligence tests has mirrored the general progress of the mental testing movement in the United States is true, with two notable exceptions: (1) It appears that infancy researchers have been more attuned than mental testers to the importance of a factor for social intelligence and (2) infancy researchers have been more likely than mental testers to take on the difficult task of attempting to measure intelligence through observations in "real-life" situations as well as in restricted testing situations. Communicative competence has been virtually ignored by both adult and infant mental testers.

THE DEVELOPMENT OF COMMUNICATIVE COMPETENCE

Throughout this chapter, we have expressed our reservations about the traditional construct of intelligence, and we have proposed that a construct of social intelligence may be more useful in predicting human behavior, particularly in nontesting situations. We regard communicative competence as an important aspect of social intelligence, since it concerns the individual's use of language and nonverbal behavior to convey meaning in interaction with others.

In the remainder of this chapter, we consider some issues that are relevant to the development of communicative competence. Our goal is to summarize some aspects of it that are manifested in the preverbal infant and to suggest how these behaviors may change during infancy. Some of the infant behaviors that are described may be both manifestations of early social intelligence and may predict social intelligence in older children and adults. We also describe young children's use of requests to illustrate a communicative aspect of social intelligence in children who have acquired language. The skills involved in using requests may serve as the basis for assessments of social intelligence in young children.

Studies of Communicative Competence: The Preverbal Infant

A popular position in the infancy literature is that the infant is born with the ability to participate actively in social interactions with others (Schaffer, 1979), an ability that we consider social intelligence. The infant is born as a social being into a social context and is involved in communication with adults from birth.

Substantial research has been devoted to describing the patterns of responsiveness and communication between the infant and its caregiver. Descriptions indicate that these early interactions are well-timed and reciprocal, and they are thus presumed to be manifestations of social intelligence, from the first weeks of life (Bullowa, cited in Shatz, in press; Freedle & Lewis, 1977; Condon & Sander, 1974). Reported categories of behavior have included vocalization and such nonvocal behavior as smiling, gazing, and touching. Infants have been observed to be involved in well-timed, smooth turn-taking in vocal and nonvocal behavior with caregivers.

Turn-taking is defined as the exchange of turns between the caregiver and the infant, and it appears as evidence in support of the reciprocal and social nature of the caregiver–infant dyad. The earliest evidence of turn-taking is believed to occur in the first few weeks of life (Brazelton, Koslowski, & Main, 1974; Kaye, 1977). Brazelton *et al.* describe such patterns of alternation between caregiver and infant as attention and withdrawal of attention. Infants show the ability to differentiate their responses based on the caregiver's behavior, and the experimenters suggest that during the brief period when the infant has withdrawn attention, it is processing information just received during the attention phase.

Kaye's (1977) study of mother–infant interaction during feeding examined turn-taking in a burst–pause pattern. The burst occurs when the infant sucks, and the pause when it ceases to suck. During the pauses, mothers were observed to move the infant or bottle, and Kaye found that such movement reduced the probability of the infant's resuming sucking. He concluded that the burst–pause pattern exemplifies the earliest turn-taking pattern.

Ratner and Bruner (1978) examined the development of game-playing by infants and their mothers involving the appearance and disappearance of objects; mothers served as initiators of games and made them more complex as the infant got older. These adults believe that the games encourage turn-taking activity, which can form the basis for subsequent social exchanges.

Similar evidence is available from studies of infants' gazing and smiling. For example, Stern (1974) attempted to determine the effect of gaze of the mother and infant on their willingness to engage in and terminate interaction. They found that the infant's gazing elicited the same behaviors from the mother, which also included vocal behavior and facial expressions. Brazelton *et al.* (1974) reported that when an infant smiles, the mother follows with a smile. This contingent behavior results in the infant watching the mother and the mother smiling for longer periods of time. Stern also found evidence that the infant's smiling increased likelihood of a response from the mother.

Thus, the evidence from nonvocal behavior between mother and infant is clear. Either because of inherent patterns of behavior and/or rapid learning, the infant and mother are able to become involved in what appears to be social interaction, based on the reciprocal and well-timed nature of alternating behaviors.

The evidence for reciprocal vocal interaction presents greater complexity than for reciprocal nonvocal behaviors. The research literature consists of descriptions of infant vocalization and turn-taking, and parents' roles in initiating, maintaining, and facilitating vocal communication with infants.

Some research on young infants shows that they have a high rate of simultaneous vocalization with mothers (Anderson & Vietze, 1977; Kaye & Fogel, 1980). Anderson and Vietze found that 3-month-old infants were more vocal when their mothers vocalized, and their mothers were more vocal when the infants vocalized. However, other research has shown that older infants engage in reciprocal vocalization; 5-month-old infants vocalize more when their mothers are quiet, and the infants cease to vocalize when their mothers vocalize (Brazelton *et al.*, 1974). Davis (1978) found few instances of simultaneous vocalization with older infants. He found that these "turn-taking failures" occurred more often with mothers interrupting their infants than when the infants interrupted their mothers.

Condon and Sander (1974) reported the co-occurring action patterns that infants produced in connection with adults' speech sounds. They suggest that there is a direct relationship between these patterns during the preverbal stage and later conversational behavior during the verbal stage of development. Their own data and others show the predominance of overlapping turns and simultaneity between mothers and their young infants (Anderson & Vietze, 1977; Stern, Jaffey, Beebe, & Bennett, 1975).

In sum, there is substantial information in the field of infancy that primarily describes the early social behaviors of infants in interaction with caregivers. The communicative behaviors discussed here, such as reciprocal vocalizations of infants and mothers, are related to the personal, social, and communicative aspects of social intelligence. Any one or a combination of these behaviors may suggest bases for the construction of an assessment of social intelligence in infancy.

Continuity in the Development of Communicative Competence

A widely held belief among some researchers in infancy is that the reciprocal, turn-taking patterns observed between mothers and infants during the preverbal stage serve as precursors for later conversational behavior during the verbal stage, thus establishing a continuity between preverbal and verbal stages of development. In support of the continuity position, Bateson (1975) has argued that the alternating pattern of mothers' interaction with their 2- and 4-month-old infants are instances of "protoconversations" from which later verbal communication would emerge.

Shatz (in press) argues against a continuity position, citing primarily theoretical grounds. In her view, the issue of continuity in development from preverbal to verbal stages includes two aspects: (1) the notion of a precursor, of later behavior originating from early behavior, and (2) the establishment of operational criteria for continuity or mechanisms that specify the process whereby earlier forms/functions serve as the basis for subsequent ones. At present, no theory exists, and few data offer confirmation or contradiction of the view that verbal/vocal behavior may function as a precursor for subsequent communicative competence.

One problem concerns the choice of measures for behavior that could logically and theoretically serve as precursors to subsequent development. Shatz (in press) does not believe that there is any clear evidence that such infant behaviors as nonoverlapping turns or turn alternation index explicit preparation for later verbal communication, such as conversational skills. Even though much of the work cited here has been motivated by a belief in continuity from preverbal to verbal stages, the theory specifying the nature of the relationship between early preverbal development and subsequent verbal development is in preliminary stages (cf. Lewis & Cherry, 1977; Golinkoff, 1981; Golinkoff, in press), and adequate data confirming or contradicting the theories does not exist. Previous work consists of elaborate descriptions of verbal social behaviors that might be precursors (cf. Bates, 1976; Bower, 1974; Bruner, 1975).

There is no conclusive evidence however, that the infant's experience in protoconversations influences either the degree or onset of subsequent communicative competence as it emerges during the verbal stage of development. Consider the classic study by Freedle and Lewis (1977). They examined the opening and closing of conversation between 3-month-old infants and their mothers. Their data suggest a relationship between preverbal behavior and verbal behavior; 2-year-olds who had more advanced grammatical development than other children (as measured by mean length of utterance, MLU) were more likely to have mothers who initiated sequences of vocalizations with them when they were infants. One possible explanation for this finding is the concept of the "responsive mother," who is attentive to her infant, and whose toddler has greater grammatical complexity than other toddlers. However, Freedle and Lewis's (1977) data cannot be interpreted as conclusive support for the continuity position. Their findings showed positive correlations between the degree to which mothers were responsible for initiating vocalization sequences with their infants and the children's later language development as measured by MLU. However, MLU indicates morphophonemic (grammatical) rather than social development.

The configuration of communicative competence during the verbal stage, its relationship to social competence during that period, and the relationship of these two aspects to social intelligence have been ignored in the past and need more attention in the future. At present, we have no evidence to support the precursor role for the kinds of communication that exist in early infancy.

Parents' Role: Facilitating Communication or Competence

Some researchers believe that interaction between infants and parents facilitates the acquisition of communicative competence. Lewis believes that the development of communicative competence is facilitated by the social context in which communicative gestures, verbal and nonverbal, are used (cf. Lewis & Freedle, 1973). He formulated a conceptualization of the social context into which infants are born as the "social network" (Lewis & Feiring, 1977). This network consists of social objects, functions, and situations. Social objects refer to people and things, such as self, mother, father, siblings, peers, and stuffed animals. Social functions include all activities that occur within the social network, such as play, nurturance, protection, information-seeking. Social situations are the contexts of the network and are related to functions. For example, feeding behavior, a social function, occurs under specific conditions which are remarkably similar across cultures. However, play is a social function that is quite variable. Social objects and functions are also interrelated. Mothering (a function), and mother (an object), are usually considered to be the same. Lewis and Feiring (1977) argue that the relationships among functions, objects, and situations impart meaning, since they form discriminable units that the infant is capable of perceiving and responding to. They believe that early vocal production that refers to the here and now demonstrates the importance of the interrelationship of the infant's social and communicative intelligence; only after a period of development can the child understand and produce language that is independent of a social context.

Bruner (1975) has argued that the transition from the preverbal to verbal period is augmented by mothers' efforts to interpret their infants' vocalizations and treat them as communicatively significant, thus enhancing and building conversational skills. Bruner believes that early protoconversational experience of dialogue between parents and infants promotes language development and lays the foundation of the grammar.

Shatz (in press) does not believe that the rhythm and structure of the children's early interactive patterns necessarily indicate that infants are directly preparing themselves for later conversational turn-taking, or that parents are facilitating growth in skills by responding to them. Her alternative interpretation of the data is that parents work very hard to help their infants "look smart" in preverbal communication.

Several researchers have observed that mothers work their infants' repertoires into joint action patterns by accommodating their own behavior to that of the children. For example, Kaye and Charney (1980) have shown that it is the mother who takes the responsibility for maintaining conversation with the 2-year-old, beginning at the two-turn level. Mothers accomplish this support by "turnabouts," turns in which the speaker both responds to the listener's prior utterance and makes a request of the listener to respond. Wood, Bruner, and Ross (1976) have argued that the "scaffolding" parents create in interaction with infants is more elaborate and richer than structures that children create with themselves as the center and that such "scaffolding" may therefore facilitate linguistic and cognitive growth.

Snow (1977) examined mother–infant conversations of two female infants and their mothers from 3 to 20 months of age. When the infants were 5–7 months old their mothers' speech to them began to change by becoming more demanding of the infants as conversational partners, even though the grammatical complexity of the mothers' speech remained relatively stable over this time. The mothers increased the frequency of imperatives and declaratives as the infant got older while decreasing the frequency of utterances referring to the infant and increasing the frequency of utterances referring to the world. Snow's data suggest that the mothers were acting as though their children could take turns with them, and that they fine-tuned their behavior as the children matured linguistically.

In sum, there is a lack of theory and data regarding the possible facilitating role of parents in infants' development of communicative competence. There is no evidence that either the infants' experience in protoconversation or the parents' help influences either the onset or degree of skill in later conversational tasks. Even the Freedle and Lewis (1977) study, which comes closest, does not provide conclusive evidence. Their data did show positive correlations between the degree to which mothers initiated vocalization sequences with their 3-month-olds and the children's later language development as measured by grammatical, not communicative complexity.

Studies of Communicative Competence: The Verbal Child

We now turn to selected examples of research on the communicative competence of verbal children. Just as our prior discussion of research on early communication competence suggests measures of behavior and potential test items for assessments of intelligence in infancy, the following presentation of research has implications for alternative assessments of intelligence in early childhood.

In this section, we will discuss two studies of young children's production of requests. The child's development of both a structural (e.g., grammatical competence) and a functional system (e.g., communicative competence) differentiates the verbal from the preverbal child.

In addition to structural competence, there is knowledge of how to use language to communicate meaning and affect the behavior of others. Thus there appears to be psychological, social, and linguistic components underlying the social intelligence. The studies to be discussed concern two issues: understanding the relationship between language forms and language functions and the ability to apply relevant background knowledge to communication in a particular situation.

The idea that one can use a variety of different language forms to convey the same communicative function is an aspect of competence that develops early. For example, consider the request for action or directive. Any one of the following different forms can be used to convey the intention to a listener: "Give me a pencil" "I need a pencil" "Can I have a pencil." In addition, some recent work on children's comprehension of directives provides evidence that, very early, children use available cues from context to understand intention and adjust their messages to meet the demands of specific situations (Shatz, in press). Taken together these studies suggest that young children possess an extensive repertoire of forms to convey the directive function, and children are able to differentiate these forms according to a variety of contextual variables, such as the age or familiarity of the listener and the listener's probability of compliance. We conducted a study to examine the extent of the preschool child's communicative competence in producing directives.

In a study of preschool children, we expected that there would be individual differences in production of directives: that children of different ages (2, 3, and 4 years old) would not differ in the quantity of the directives produced but that they would differ in the quality of directives produced, that is, in the initial choice directive and subsequent revised directives (Read & Cherry, 1978). The task was introduced to each child as a game that he or she was going to play with the Cookie Monster, a popular character from the television show "Sesame Street," and the experimenter; a game in which the child was to request the puppet to provide a desired object, juice, a cookie, or a crayon. A series of interactions resulted in the children's producing a sequence of directives, when the puppet failed to comply to the directives. The directives were transcribed and coded according to the following categories: gestural directive (nonlexical nonverbal), imperative (command forms that are not embedded), embedded imperative (interrogative forms containing a modal verb), declarative statements (declarative statements with directives, intention, and utterances that name the desired object), "please" standing alone, and expression of want/need.

The data were analyzed using Kendall correlations. The results that older and younger children produce the same number of directive forms was not disconfirmed by a correlation with age of $+.13$. The absence of a developmental trend suggests that 2-year-olds already possess many directive strategies necessary to manipulate their environment just as older children do. A correlation of $-.49$ was obtained for a correlation of age and number of gestural directives, whereas the production of embedded imperatives correlated $+.56$ with age. Thus the data show individual differences in strategies of production of directives, but not in quantity of production.

The data provide evidence that children from 2 to 4 years of age possess extensive and flexible repertoires of directive forms. Older children were less likely than younger ones to produce gestural directives but were much more likely to use the polite embedded imperative form as well as the impolite imperative form. Children in the study did not have the full linguistic means to express the directive function; nevertheless, they conveyed sophisticated communicative behavior by varying their strategies and employing both their limited linguistic and nonlinguistic means.

The findings from this study are relevant to the notions of social intelligence discussed earlier in this chapter. We see children's developing sophistication in production of directives as part of their expanding repertoire of social strategies and skills. We see very young children acting in a socially competent way, while with their development we see an expanding repertoire of directives, as indicated by a more flexible use of the linguistic system. Even though linguistic knowledge was incompletely developed, young children used primitive nonverbal behavior as an adaptation to an environment that was not optimally responsive. All of the children in this study gave evidence of social intelligence in their attempting to solve a social problem in a real situation that was of relevance to them in which a particular outcome was desired, but individual differences were also apparent. These children all showed to varying degrees that they understood the use of vocal, verbal, and nonverbal communication in order to operate in a social world, to influence the behavior of another social being vis-à-vis the child's goal. Using language in this way to influence others is an aspect of communicative competence that shows continuity in development from the preverbal through the verbal years. The task and measures developed for this study could serve as the basis for a subtest of an inventory of social intelligence.

In recent work, we developed the concept of the *effective speaker*, who uses knowledge of language forms, functions, and contexts to achieve communicative goals (Wilkinson & Calculator, 1982; Wilkinson, Spinelli, Wilkinson, & Chiang, 1982). For example, in the case of directives—a request for action—effective speakers are successful in obtaining compliance for their request from listeners. We proposed the model of the effective speaker that characterizes the use of requests and responses by young children (Wilkinson & Calculator, 1982). Our model identifies the following characteristics of requests that predict compliance from listeners. First of all, speakers may express acts clearly and directly, in an attempt to minimize ambiguity and multiple interpretations of the same speech. For example, speakers may use direct forms and specifically designate them to one particular listener when making a request. Second, in the classroom, requests that are on-task, that is, those that refer to the shared activities in the teaching–learning situation, are most likely to be understood by listeners, and thus these types of requests are the most likely to be successful in obtaining compliance from listeners. Third, requests that are understood by listeners as sincere are most likely to result in compliance. Finally, effective speakers are flexible in producing their requests; for example, as we have seen with the preschool children, speakers should revise their initial requests when compliance from listeners is not obtained.

Our research on first and third grade children provides support for a model of the effective speaker (Wilkinson & Calculator, 1982). Data collected on 65 children interacting in their peer-directed reading groups (a combined data base of more than 3,600 requests and their responses) show that school-age children are, on the whole, effective speakers, since they obtain compliance to their requests from their listeners about two-thirds of the time. The typical child produced requests that were direct, sincere, on-task, and designated to a particular listener. When the listener did not comply with the speaker's request, the school-age children revised their requests two-fifths of the time. There were strong individual differences among the children. Not all children were effective in obtaining what they requested, and the requests of these children did not conform to our model.

Further analysis of our two data sets provided strong evidence for the predictive nature of our model of the effective speaker's use of requests. A hierarchy of log-linear models was used to fit the data. The model that best fit all the data assumed that there were associations among the five characteristics identified (direct, sincere, on-task, designated, and revised), and that these characteristics were associated with what was requested (action, information) and with whether the request obtained compliance. The major conclusions from our analysis are that these characteristics of requests are correlated, and whether a request obtains compliance depends on all the other characteristics identified by the model.

As with the preschool data, we see that the results from this study can be viewed as instances of children's social intelligence. In this case, we examined spontaneous behavior of children in their peer groups, in contrast with the "real-life" experimental situation we used with the preschool children. This research suggests that tests of social intelligence could be based on each of the components in our model of the effective speaker's use of requests in specific situations. Further research is needed to identify other components of social intelligence as manifested in other interactional situations. We believe it is valuable to use a variety of research methods in investigating social intelligence and communicative competence. We will now discuss some further directions future work may take.

CONCLUSION

This chapter concerns the relationship between social intelligence and communicative competence in infants and children. We have proposed that intelligence necessarily includes the unique human capacity to use language to communicate in social situations. Although ignored in the past, the constellation of behaviors and processes referred to as social intelligence is central to intellectual functioning.

A great deal of work is needed. We lack a firm definition of social intelligence. We have neither theory nor data explicating the issue of continuity in the development of social intelligence. Research should address the three issues that have remained as enduring controversies surrounding social intelligence since the early part of this decade: (1) whether there is a separate factor for social intelligence, (2) whether it is best to measure social intelligence with standardized tests or by observations of naturally occurring behavior, or both, and (3) whether this aspect of intelligence is significantly effected by experience.

The issues of the existence of a separate factor for social intelligence and its assessment can be addressed by two lines of work: (1) studies of the relationship of social intelligence in infancy, as assessed by standard tests such as the Bayley, to communicative competence, as measured by observations of infant behaviors such as vocalization and turn-taking and (2) studies of the relationship of verbal ability, as assessed by standardized tests, to communicative competence, as measured by behavioral observations and indices that were discussed in this chapter.

The issues of the development of social intelligence and the role of experience in development can be addressed by longitudinal and experimental studies. Future research should include descriptions of the configuration of communicative competence during the preverbal and verbal stages of development as suggested above, and the interrelationships between these stages. Studies of parent–child interaction during infancy as it relates to communicative competence during the verbal stage address the issue of the effect of experience on social intelligence.

However the controversies regarding the assessment of intelligence are addressed in the future, one thing remains clear: A socio-communicative component should be included in any effort. We need assessments of intelligence that predict performance in "real-life" situations from infancy to adulthood.

ACKNOWLEDGMENTS

The author thanks Michael Lewis, Alex Cherry Wilkinson, and Marilyn Shatz for the helpful comments in an earlier draft of this chapter.

REFERENCES

ANDERSON, B. J., & VIETZE, P. M. Early dialogues: The structure of reciprocal infant-mother vocalization. In S. Cohen & T. J. Comiskey (Eds.), *Child development: Contemporary perspectives*. Itasca, Il.: F. E. Peacock, 1977.

BATES, E. *Language and context: The acquisition of pragmatics*. New York: Academic Press, 1976.

BATESON, M. C. Mother-infant exchanges: The epigenesis of conversational interaction. In

D. Aaronson & R. W. Rieber (Eds.), *Developmental psycholinguistics and communication disorders—Annals of the New York Academy of Sciences* (263). New York: New York Academy of Sciences, 1975.

Bower, T. Repetition in human development. *Merrill-Palmer Quarterly*, 1974, 20, 303–318.

Brazelton, B., Koslowski, B., & Main, M. The origins of reciprocity: The early mother-infant interaction. In M. Lewis & L. Rosenblum (Eds.), *The effect of the infant on its caregiver*. New York: Wiley, 1974.

Bruner, J. The ontogenesis of speech acts. *Journal of Child Language*, 1975, 2, 1–20.

Buhler, C., & Hetzer, H. *Testing children's development from birth to school age*. New York: Farrar & Rinehart, 1935.

Cattell, P. *The measurement of intelligence of infants and young children*. New York: Psychological Corporation, 1940, 1960, 1966.

Charlesworth, W. Human intelligence as adaptation: An ethological approach. In L. Resnick (Ed.), *The nature of intelligence*. Hillsdale, N.J.: Lawrence Erlbaum Associates, 1976.

Condon, W., & Sander, L. Synchrony demonstrated between movements of the neonate and adult speech. *Child Development*, 1974, 45, 456–462.

Davis, H. A description of aspects of mother-infant vocal interaction. *Journal of Child Psychology and Psychiatry*, 1978, 19, 379–386.

Freedle, R., & Lewis, M. Prelinguistic conversations. In M. Lewis & L. Rosenblum (Eds.), *Interaction, conversation and the development of language*. New York: Wiley, 1977.

Gesell, A., & Amatruda, C. *Developmental diagnosis*. New York: Paul Hoeber, 1941.

Golinkoff, R. The influence of Piagetian theory on the study of development of communication. In E. Sigel, D. Brodzinsky, & R. Golinkoff (Eds.), *New directions in Piagetian theory and practice*. Hillsdale, N.J.: Lawrence Erlbaum Associates, 1981.

Golinkoff, R. The transition from preverbal to linguistic communication. Hillsdale, N.J.: Lawrence Erlbaum Associates, in press.

Greenspan, S. Is social competence synonymous with school performance? *American Psychologist*, 1980, 35, 938–939.

Griffiths, R. *The abilities of babies*. London: University of London Press, 1954.

Guilford, J. New frontiers of testing in the discovery and development of human talent. *The seventh annual western psychological conference on testing problems*. Calif.: Educational Testing Service, 1958.

Kaye, K. Toward the origin of dialogue. In H. R. Schaffer (Ed.), *Studies in mother-infant interaction*. New York: Academic Press, 1977.

Kaye, K., & Charney, R. How mothers maintain dialogue with two-year-olds. In D. Olson (Ed.), *The social foundations of language and thought*. New York, Norton, 1980.

Kaye, K., & Fogel, A. The temporal structure of face-to-face communication between mothers and infants. *Developmental Psychology*, 1980, 16(5), 454–464.

Keating, D. A search for social intelligence. *Journal of Educational Psychology*, 1978, 70, 2, 218–223.

Lee, L. Is social competence independent of cultural context? *American Psychologist*, 1979, 34, 795–796.

Lewis, M., & Cherry, L. Social behavior and language acquisition. In M. Lewis & L. Rosenblum (Eds.), *Interaction, conversation, and the development of language*. New York: Wiley, 1977.

Lewis, M., & Feiring, C. The child's social world. In R. Lerner & G. Spanier (Eds.), *Contributions of the child to marital quality and family interaction through the life-span*. New York: Academic Press, 1977.

LEWIS, M., & FREEDLE, R. Mother-infant dyad: The cradle of meaning. In P. Pliner, L. Krames, & T. Alloway (Eds.), *Communication and affect: Language and thought.* New York: Academic Press, 1973.

MCCLELLAND, D. Testing for competence rather than for intelligence. *American Psychologist,* 1973, *28,* 1–14.

RATNER, N., & BRUNER, J. Games, social exchange and the acquisition of language. *Journal of Child Language,* 1978, *5,* 391–402.

READ, B., & CHERRY, L. Preschool children's production of directive forms. *Discourse Processes,* 1978, *1,* 233–245.

RESNICK, L. *The nature of intelligence.* Hillsdale, N.J.: Lawrence Erlbaum Associates, 1976.

SCARR, S. Testing for children. *American Psychologist,* 1981, *36,* 1159–1166.

SCHAFFER, H. Acquiring the concept of the dialogue. In M. Bornstein & W. Kessen, (Eds.), *Psychological development from infancy: Image to intention.* Hillsdale, N.J.: Lawrence Erlbaum Associates, 1979.

SHATZ, M. Communication. In J. Flavell & E. Markman (Eds.), Cognitive development. In P. Mussen (Gen. Ed.), *Carmichael's manual of child psychology* (4th ed.), New York: Wiley, in press.

SNOW, C. The development of conversation between mothers and babies. *Journal of Child Language,* 1977, *4,* 1–22.

STERN, D. Mother and infant at play: The dyadic interaction involving facial, vocal, and gaze behaviors. In M. Lewis & L. Rosenblum (Eds.), *The effect of the infant on its caregiver.* New York: Wiley, 1974.

STERN, D., JAFFE, J., BEEBE, B., & BENNETT, S. Vocalizing in unison and in alternation: Two modes of communication within the mother-infant dyad. In D. Aronson & W. Rieber (Eds.), *Developmental psycholinguistics and communication disorders—Annals of the New York Academy of Sciences* (263). New York: New York Academy of Sciences, 1975.

STERNBERG, R., CONWAY, B., KETRON, J., & BERNSTEIN, M. People's conceptions of intelligence. *Journal of Personality and Social Psychology,* 1981, *41*(1), 37–55.

THORNDIKE, E. Factor analyses of social and abstract intelligence. *Journal of Educational Psychology,* 1936, *27,* 231–233.

WEDECK, J. The relationship between personality and psychological ability. *British Journal of Psychology,* 1947, *37,* 133–151.

WILKINSON, L. *Communicating in the classroom.* New York: Academic Press, 1982.

WILKINSON, L., & CALCULATOR, S. Requests and responses in peer-directed reading groups. *American Education Research Journal,* 1982, *19*(1), 107–122.

WILKINSON, L. C., SPINELLI, F., WILKINSON, A. C., & CHIANG, C. *Language in the classroom: Metapragmatic knowledge of school-age children* (Program Report 83-2). The Wisconsin Center for Education Research, University of Wisconsin-Madison, 1982.

WOOD, D., BRUNER, J., & ROSS, G. The role of tutoring in problem solving. *Journal of Child Psychology and Psychiatry,* 1976, *17,* 89–100.

WOODROW, H. The common factors in 52 mental tests. *Psychometrika,* 1939, *4,* 99–108.

ZIGLER, E., & TRICKETT, P. IQ, social competence, and evaluation of early childhood intervention programs. *American Psychologist,* 1978, *33,* 789–798.

10   *Culture and Intelligence in Infancy*
An Ethnopsychological View

CATHERINE LUTZ AND
ROBERT A. LEVINE

INTRODUCTION

Intelligence means many things to many people, but it is above all a Western cultural concept that has been incorporated into the language and theory of the social sciences. The continued usefulness of this term as currently conceptualized has been seriously called into question (Kamin, 1975; Lewis, 1976). Whether the debate on the existence, nature, and measurement of intelligence is conceived as a scientific, ethical, or political question (Cronbach, 1975; Samelson, 1979), it is also predominantly a debate embedded in American* cultural beliefs and social institutions. Gaining an understanding of the origins of our questions and preliminary theories is an important first step toward an understanding of intelligence. A comparative methodology becomes a necessary prerequisite to provide a perspective from which to view those questions.

Although questions of definition should balance questions of measurement, methodological issues surrounding measurement have been the focus of most of the comparative literature, including the cross-

---

*The term *American* will be used to refer to middle-class United States citizens of European descent. The wide variety of cultural patterns which exist in our society must, unfortunately, be glossed over in the interest of cross-cultural comparison.

CATHERINE LUTZ • Department of Anthropology, State University of New York at Binghamton, Binghamton, New York 13901.   ROBERT A. LEVINE • Laboratory of Human Development, Harvard Graduate School of Education, Cambridge, Massachusetts 02138.

cultural, developmental, and historical. This chapter will examine the cultural origins of intelligence by focusing on indigenous concepts of both the developing infant and intelligent abilities. This approach contrasts with that taken in most cross-cultural studies of cognition.

Cultural beliefs or theories (hereafter to be terms *ethnotheories*) about the nature of the developing child and of mental and social abilities are important for three reasons. First, ethnotheories serve an explanatory function. A universal need to classify, order, and correlate events is at the root of theories that are developed in every society to explain the nature of change with age. Second, cultural theories, insofar as they embody normative assumptions, are "models for" as well as "models of" social reality; they can serve as guides to and directors of behavior (Geertz, 1973, p. 93). Beliefs can be both maps of expectation and instigators of action (Quinn, 1979). Ethnotheories are developed in the context of culturally defined goals (Lutz, in press; Quinn, 1981) and thus aid socializers in realizing the end states that they desire for themselves and their children (LeVine, 1974). Finally, ethnotheories about infancy and intelligence reflect, to an important degree, the ecological, historical, and institutional conditions in which those theories have arisen (B. B. Whiting, 1974; J. W. M. Whiting, 1973). Thus, theories that develop in particular cultural contexts are not arbitrary, but rather reflect adaptations to specific environments. An important implication of the above-stated relationships among theory, behavior, and context is that we may expect ethnotheories of infancy and intelligence to bear an important resemblance to those things that they purport to describe and explain.

Anne Anastasi noted early on that "each culture . . . 'selects' certain activities as the most significant. These it encourages and stimulates; others it neglects or definitely suppresses" (1937, p. 511). Some of the abilities or behaviors seen as significant within a culture are recorded in the meanings of words, such as intelligence, that describe traits. Our first question then should be, "What kinds of characteristics are valued and when and how are they seen to emerge in the course of development?"

Because ethnotheories point to areas of significance or meaning in a culture, they reveal its value system. A parent's evaluation of infant behaviors occurs simultaneously with his or her ethnotheoretically aided understanding of that behavior. While two cultures may have roughly similar conceptualizations of and theories about intelligence, they may value its perceived appearance in infants differently. Conversely, certain infant behaviors that American parents find significant and emotionally rewarding as signs of intelligence may be seen as unimportant or may be valued for different reasons (and as part of a different and perhaps nonmental characteristic of the child) by parents in other societies. Thus, it is equally important to consider the connotative and affective meanings associated with infancy and intelligence as well as those that are more strictly referential.

ETHNOTHEORIES OF INTELLIGENCE

In any attempt to draw cross-cultural comparisons, the problems of definition and translation quickly arise. On what basis shall we compare concepts across cultures? What meanings does intelligence have in American English, and what is (are) the nearest equivalent concept(s) in each comparison culture? Which aspects of the meaning of intelligence shall we choose as the basis for comparison—its reference to particular traits or behaviors (e.g., intelligence as mental ability) or its evaluative force in use (e.g., intelligent as a very good way to be)? If we begin with the former, what kind of definition should be used? It is particularly important to set cultural concepts of intelligence in the context of other valued personal characteristics and in the context of the kinds of activities to which such attributes are typically applied in particular environments.

Intelligence needs to be examined, according to Neisser (1979), not for that concept's defining features or "true essence," but for prototypical instances of and resemblances to the "intelligent person." The intelligent person is conceptualized as someone who has one or more of a variety of traits. Two individuals may, in one person's view, both be identified as "quite intelligent and yet have very few traits in common [because] they resemble the prototype along different dimensions" (1979, p. 223). An examination of how such concepts as intelligence are applied to individuals in other cultures can help by providing concrete grounds for comparison.

The United States

Our own culture's concepts of intelligence, while an appropriate starting point for our comparison, present several surprising problems.

One is that, despite the American public preoccupation with intelligence, no one has studied our conceptions of it from an ethnographic point of view; thus, we lack a basic description of our own ethnotheories to compare with those of other peoples. Furthermore, our ideas about intelligence are so different from those of many other, particularly non-Western, cultures, that we have difficulty finding a terminology suitable for the description of both our own and other notions. A further complication arises from the fact that our own ethnotheories coexist with and have been influenced by a plethora of specialized writings in theology, philosophy, biology, and psychology that differ among themselves and that have been used for various political purposes (Gould, 1981). Fortunately, however, the task is not as difficult as it might seem, for the simple reason that behind the diverse concepts and controversies about intelligence generated by Western scholarly and folk traditions lie some fundamental and culturally distinctive assumptions.

The first of these is that intelligence as an attribute is located in an individual person rather than in a group, a situation, or a performance. Controversies abound in the West about the measurement and development of intelligence, but there is broad consensus that it is a personal attribute first and foremost. Indeed, were it not for our cross-cultural viewpoint, we would consider it unnecessary to state this, since it is usually taken as self-evident. This location is not universally subscribed to, however, as examination of other ethnotheories will show.

A second assumption is that intelligence is a faculty—a distinct aspect of the person, separate from morality, will, social skills, and other attributes. An older tradition derived from ancient Greece fused intellect with morality in such concepts as that of virtue, and in the 19th century, this early tradition was still evident in the lack of distinction drawn between ability and achievement (Calhoun, 1973). As late as the 1920s, many educators and parents explicitly stated their belief that "moral character [including effort or motivation], intelligence, and social worth were inextricably connected and biologically rooted" (Bowles & Gintis, 1972–1973, p. 80).* Currently, however, the concept of intelligence is seen as separable from that of effort and of morality.

---

*Bowles and Gintis's well-known critique of the IQ debate identifies this notion as the precursor of the current and widespread American belief in a relationship between success and intelligence.

Intelligence is also conceptualized as distinct from social and emotional skills. This view of intelligence originates, in the first place, with the differentiation of an inner self, in which mental activity takes place, from an outer self, which acts in the world. In addition, cognition is differentiated from emotion, with intelligent behavior originating in the former. The intelligence that Americans value over most other personal traits is a *technical* intelligence; "tact, sensitivity to social atmosphere, social and intersexual ease, unself-consciousness, [and the] guidance of behavior in accordance with accepted social values" (Mundy-Castle, cited in Putnam & Kilbride, 1980, p. 3) are not typically included as behaviors indicative of intelligence. The opposition of intelligent behavior to social behavior is common in American academic psychology and is evident in the following example. "In contrast to *social* behavior, there are little or no adequate data to test the many hypotheses generated by the evolutionary approach to *intelligent* behavior (Charlesworth, 1979, p. 213; emphasis added). This separation of moral, social, and intellectual skills is a point of great contrast with the ethnotheories of many non-Western peoples.

A third aspect of American ethnopsychology concerns the quality of thought presumed to constitute intelligence. Abstract formalism, or the separation of "form from content, and the structuring of experience in accord with that distinction" (Buck-Morss, 1975, p. 38), is defined as the intelligent mental activity par excellence. Other forms of thought, such as the spiritual mental disciplines, understanding of one's emotions and of the other, and animistic thought (Tulkin & Konner, 1973) are considered less advanced than abstract formalism. This is in contrast to the Greek conception of mental ability, which incorporated and valued

> intuition, imagination, depth of thought, speculative freedom, capacity to question assumptions, fluidity of thought, and its interpenetration with feeling, boldness of thought . . . and the capacity to be "open" to new values, ideas, and approaches. (Rosenthal, 1971, p. 49)

The quality of thought involved in intelligence in American ethnotheory can also be discerned in the implicit standards by which cross-cultural psychologists have judged performance on cognitive tests. Goodnow (1976) summarizes the intellectual performance criteria that cross-cultural researchers expect as part of intelligent completion of those tests. People are expected to give "an honest try" and to "attend to the task rather than to the people" (Goodnow, 1976, pp. 20–21; see also Harkness & Super, 1977). The intelligent person is expected to utilize particular methods in performing the set task. Speed is valued as are minimal moves, the use of universal principles, and definite, clear, yes-or-no responses. Abilities that are not valued include the exact replication of a taught item and the manipulation of the stimulus materials by hand rather than mentally. These expectations about how one exhibits intelligence reveal much about American ethnotheories. As Goodnow notes, the disappointments of cross-cultural researchers over the performance of their experimental subjects may say more about "us" than it does about "them."

Fourth, there is the assumption that intelligence develops (from innate or environmental sources) primarily in childhood. While it is recognized that adults acquire wisdom based on experience and that the wisest people may be quite old, their wisdom is not equated with intelligence. Indeed, intelligence developed in childhood is often seen as a necessary condition for the attainment of wisdom in adulthood; in other words, unintelligent persons are unlikely to become wise during their mature years. Intelligence is thus the most fundamental of intellectual attributes, the least bound to any specific social context or stage of life.

Finally, we assume that intelligence is an extremely salient attribute of the person, central to current and future evaluations and highly relevant to life chances in a world of invidious comparisons and intense competition. This assumption is most evident among the more highly educated groups in our society, but it is characteristic of most societies in which life chances are structured by Western-style schooling.

These five assumptions are taken for granted by most Americans—so much so that they seem as natural as the air we breathe—and they affect profoundly our views of infants. An examination of concepts of intelligence in three other cultures, however, will bring into relief the culture-specific quality of these assumptions and of their application to infant development.

Baganda

The Baganda of Uganda wish to see several behavioral dispositions develop in their children. These include, above all, obedience as well as respect; knowledge of kin, laws, beliefs, and customs; and achievement, with the latter being defined as "successful service to existing norms and values" (Wober, 1974, p. 263). An intelligent* person (*amagezi;* intelligence = *obugezi*) is wise and, simultaneously and consequently, has many of the other desirable characteristics listed above. Intelligence is a quality of the soul rather than of the body, and is not inherited. The Baganda hold a behavioral model of intelligence, with a common proverb stating that intelligence grows in people as they use it. Thus, *obugezi* is not seen so much as a passive trait, but rather as a "vector"; to the American English concept of intelligence is added a sense of direction (Wober, 1974, p. 263–264). The *amagezi* or intelligent person knows how to do and does the socially appropriate thing.

Wober collected semantic differential responses to the term *amagezi*. Unschooled Baganda see the concept as slow, careful, active, straightforward, happy, healthy, strong, steady, safe, friendly, hot, and public. The sharpest contrast with American ethnotheories lies in the latter's emphasis on speed. Several researchers have found that associations of intelligence with speed increase with exposure to the different ethnotheories of intelligence implicit in or suggested by Western-style schooling (Putnam & Kilbride, 1980; Wober, 1974). Wober found, in fact, that there are greater differences in the concept of intelligence held by groups of varying educational level within the same culture than there are between groups of the same educational level in two somewhat distinct cultures (the Baganda and the Batoro).

Another contrast to American ethnotheory is seen in the Baganda association of intelligence with *friendly* and *public*. Americans tend to associate intelligence with a form of introversion and an inward focus. Wober compares the Ugandan concept with the "Mediterranean-generated ideas of 'civility' and 'gravitas', of public-spirited orientation of the mind" (Wober, 1974, p. 267).

Putman and Kilbride (1980) note a similar social-altruistic component to the concept of intelligence (*amakesi*) used by the Samia of Kenya. In written essays, 144 Samia adolescents described the behaviors indicative of intelligence. These included especially innovative activities, such as the use of new farming techniques, and the surmounting of unusual difficulties, such as a physical handicap or poverty. In more cases than not, the goal or result of such behaviors was the benefit of the actor's

---

*The use of the English terms *intelligent* or *intelligence* in this and the following cultural descriptions is meant only as a preliminary gloss for an indigenous term. It would, we will demonstrate, be wrong to assume that the wider meaning and implications of the English gloss are identical with the indigenous term for which it stands.

social group. The adjectives most frequently found in the essays as descriptors of the intelligent person included (in descending order of frequency) clever, active, wise, brave, kind, and polite.

Ifaluk

The Ifaluk engage in subsistence horticulture and fishing in the western Pacific. The most central Ifaluk cultural values include food-sharing, hard work, nonaggression, obedience within a ranked social and political system, and cooperation. These values are seen in an exemplary and consistent way in those labeled *repiy*, the term most closely equivalent to intelligence. Moral and social knowledge and competence are at the heart of the concept of *repiy*. It is not simply the knowledge of good social behavior, but also its performance. *Repiy* persons are, above all, those who are constantly and carefully aware of their social wake, that is, they anticipate the impact of their own behavior (or their lack of appropriate action) on others. They monitor this social wake in accordance with the above-mentioned values, and in particular they try to anticipate and avoid conflict. The Ifaluk have a number of personality-trait terms, such as generous, even-tempered, and hard-working, that describe more specific aspects of valued behavior. To a large extent, these terms are subsumed by and may be implied by the term *repiy*. The intelligent person, by definition, has several of the valued personality traits and may, in fact, have a large number of them. The *repiy* person is a good person in one or more of the several ways in which good is defined by the Ifaluk.

Any valued behavior, from expert canoe design and building to particularly generous sharing of food with others, earns one the label of *repiy*. The more technological understandings and behaviors involved with such activities as navigation, weaving, horticulture, or canoe design are not, however, the primary referent of the term *repiy*. Although such skills are highly valued, a separate term (*sennap*) denotes the expert in such fields. A stingy and hot-tempered master navigator would not be considered intelligent.

Western-style schooling was introduced on Ifaluk about 25 years ago, and with it came other conceptions of intelligence. The term *repiy* has now been extended to include behavior that is rewarded in school. There are several important ways, however, in which indigenous conceptions of intelligence are at odds with those of the school, and some of those differences are explicitly perceived by adults. In particular, the school is seen as a place that sometimes encourages "show-off" behavior, whereas *repiy* is evidenced by a more self-effacing attitude. Calm, nonaggressive behavior is one of the quintessential intelligent behaviors; the school setting, it is said, produces boisterous children who are neither calm nor respectful. On the other hand, the child who is labeled *repiy* by the school system may go on for further training and eventually receive one of the few wage-earning positions in this society. A second standard for success has therefore developed and it is at least orthogonal to the traditional social and moral standards for achievement.

An additional important aspect of the traditional concept of intelligence is its reference to certain emotional behaviors. The emotionally mature person—that is, the person who expresses emotions in the Ifaluk manner and in the correct situations—is defined as *repiy*. This belief follows directly from one of the most basic ethnopsychological premises on the island, which is that thought and feeling are inextricably intertwined. The same basic process is involved in deciding whether to go fishing and in perceiving a rule infraction and coming to one's response ("justified anger") to that violation. It is said of someone who feels much compassion for others that they have "good thoughts/feelings" and, therefore, that they are *repiy*. Conversely, an individual who claims to be righteously indignant and creates a conflict over an event that is in fact culturally acceptable or nonexceptional is said to be "not *repiy*."

Cultural theories of intelligence are often revealed in daily life when *un*intelligent behavior is remarked upon. In the case of the Ifaluk, there are two main classes of person primarily defined by their lack of *repiy*. These include all children under the age of 5 or 6 and social deviants of all types. Young children are said to have no "thoughts/feelings" beyond those of eating and playing. (We will return to the nature of the infant in more detail below.) Like the young, social deviants do not behave in acceptable ways. Deviants include psychotic or mentally retarded individuals and, possibly, any adult who, momentarily at least, acts in a socially irresponsible or incapable manner. One of the primary characteristics that the child and the deviant share and that earns them both the description of being without intelligence is a lack of social awareness and a consequent lack of behavioral predictability. It is said of such people that others are afraid of them because it is not known what they will do next.

Northern Cheyenne

The Northern Cheyenne see the human person as composed of four parts—body, heart, breath, and spirit (Straus, 1977, 1982). "Breath" is "the human capacity for speech and understanding" (Straus, 1977, p. 328) and communication, and is related to aspects of our understanding of intelligence. It is through the breath that one acquires knowledge, culture, power, and individuality. A person's "spirit" consists of the potential for both good ("orderly, controlled, careful, thoughtful," Straus, 1977, p. 329) and "crazy" ("disordered, impulsive, animal-like," Straus, 1977, p. 329) behavior. It is also the ability to distinguish between the behaviors and to conform to the normative system. The spirit is where knowledge and experience accumulate.

Intelligence, if conceptualized as culturally defined success, is, for the Cheyenne, the acquisition of spiritual power. Power, or knowledge, is a sacred entity that is of the utmost importance for health, long life, and the individuals' achievement of cultural goals. It allows one to be good and follow the "straight" path although it can be used in crazy ways as well. Power or knowledge is acquired through communication, by means of which it is transferred from one person to another; it thus emerges only in relationships with others. All persons are intelligent in the sense that they all have the capacity for language, or breath.

If, however, the focus is on intelligence as a developed moral sense, then it is within the development of the spirit that intelligent behavior originates. The good part of the Cheyenne spirit is made of two subsystems that include wisdom and energy. The energetic male is active, bold, adventuresome, and strong. The wise male is reverent, prudent or careful, generous, quiet or peaceful, and a "medicine man." Which of these two behaviors people exhibit will vary with the company they keep since one's associates are seen as a primary source of motivation. Intelligence in this sense, then, is also a social product.

ETHNOTHEORIES OF THE NATURE AND DEVELOPMENT OF INFANTS

Despite the marked species-related similarities of human infants across cultures, there is some diversity in indigenous conceptions of the infant. Some of these reflect parallel diversity in the general conception of "person," irrespective of age (Geertz, 1976; Lutz, 1980; Straus, 1977). Thus, any treatment of ethnotheories of infant capacity must indicate the extent to which the young child is conceptualized as a person, and the nature of the person to which the infant is either analog, contrast, or prototype.

Most, if not all, cultures recognize and name stages or periods of development over the life course. The existence of culturally constituted stages results in greater attention being paid by socializers to developmental changes between, rather than within, stages (Harkness & Super, in press). Many cultures see the first 2 years as a special period during which the child has particular characteristics and needs. Two of the most common markers used as signs of the beginning of the end of infancy appear to be the onset of mobility and the development of language.

Interest in interpersonal variation and its nature, origin, and extent is also found in many societies (Lutz, 1980; Shweder, 1972; Valentine, 1963; White, 1978, 1980). Ethnotheories concerning the appearance and development of individual differences in infancy form the larger ethnopsychological context in which theories of infant intelligence are embedded. As Harkness and Super point out, judgments of individual differences are made "within the boundaries of culturally defined developmental stages" (in press, p. 3). Thus, stage-specific forms of traits such as intelligence can be expected to be postulated in many cultures.

In addition, the kinds of differences theoretically possible or attended to in infancy are variable. As will be seen, the phrase "infant intelligence" may be a contradiction in terms in some cultures. In others, the infant is seen as only potentially rather than actually intelligent. The conceived nature of intelligence will be a factor in the degree to which infant intelligence is seen as a crucial differentiator of individuals at that period of life.

The United States

In a model current in the middle classes, the American infant is considered a full person, with personhood defined biologically and legally. Theoretically, the infant has the same basic rights as the adult within the constraints of its ability to exercise them. The egalitarian ethic affects the adult's approach to infants; that approach is seen as an attempt to engage and interact with them (LeVine, 1980, p. 78). The infant must be named from birth, both to signal its humanness and to begin the process of identifying the child as a unique and independent constellation of traits.

In this ethnotheory, the experiences of the infant have crucial and long-lasting effects. It is the responsibility of the parents to provide the child with the environment that will "draw out" or develop those characteristics that are most desired—self-esteem and independence, for instance. Infants inherit certain characteristics literally and outright, while they are passed on only the potential for other behaviors. Intelligence is seen as a function of the brain, and is considered largely inherited. It is also a potential, however, and parents can help the infant and child actualize that potential.

The American infant is searched from birth for signs of intelligence. As noted above, this trait is seen as mental activity and, particularly, as problem-solving. Access to education and, through it, to economic security is believed to depend to some extent on the "amount" of intelligence that the child begins to exhibit. Social and emotional behaviors, which are considered outside of the domain of intelligence, are seen as less crucial indicators of future success. Less anxiety, attention, and effort are thus devoted by parents to the encouragement of many interpersonal and emotional abilities, except insofar as they are in the service of intellectual abilities. For example, the infant's interest and enthusiasm should be encouraged as they are necessary to an engagement with "problems to be solved." Similarly, social relationships are to be developed as sources of self-esteem and "individual resources for personal autonomy" (LeVine, 1980, p. 79), rather than as goals in and of themselves.

Baganda

The behaviors that Baganda adults particularly wish to see develop in children are obedience and respect (Kilbride & Kilbride, 1974). Children must be taught a code of social etiquette that includes compliance and respect as well as the proper ways of greeting others, noninterference in adult conversations, and congeniality. Although such behaviors should be exhibited by the good child, the primary characteristic that underlies them and is of the first necessity is that of being socially engaged. The code of social behavior "requires that one behave socially in the first instance, properly in the second" (Kilbride & Kilbride, 1974, p. 305).

This code is applied to infants from the earliest stage. Infants are smiled at and spoken to frequently in the first 6 months. Adults are concerned to elicit reciprocal smiles from the child. The effects of the high levels of social interaction on infants are evident at a very early stage. The Kilbrides (1974) report that sociability items on Bayley scales they administered show an early emergence in Baganda infants in comparison with United States norms. Thus, the kind of social intelligence that the Baganda value is in fact developed in infants as a result. Explicit training in proper behavior, however, does not begin until some time after the age of 1, when children are believed to be first able to understand what is said to them.

Ifaluk

On Ifaluk, the 7-month fetus may be termed a person, since it is at this point that miscarried fetuses first look physically human. It is taboo, however, to speak directly of children until they are between 4 and 10 days old. It is at this point, when the period of greatest life threat is passed, that the child is given a name and a social identity. The first major life stage extends from birth to about the age of 2, the end point of which is also the age of weaning from the breast and from the back. During this period, the infant is said to have only one thought, which is to eat. The infant's lack of *repiy*, or intelligence, is complete. Young children are not considered responsible for their actions. Because they do not know right from wrong, the aberrant behavior that emerges must be ignored or tolerated.

It is only after the age of 2 that the child becomes capable of learning, and it is only then that parents must seriously begin to "cause (them) to become intelligent" (*garepiy*). The weaned child is often instructed on proper behavior through stylized parental lectures. This is considered the central mechanism for learning because listening, to the Ifaluk, implies understanding, which in turn implies obedience. Because obedience is one of the central goals of child rearing, listening and language-understanding become crucial. It is the infant's inability to speak or understand that makes tuition and obedience impossible. Unlike the Cheyenne, as will be seen below, the Ifaluk see the infant as a person, albeit an "unintelligent" one.

Adults spend little time directly speaking to infants, since it is assumed that the child understands little. A person who has reached the age of intelligence is presumed to have little to say to those who have not, except insofar as it is necessary to correct their behavior or to lecture them on general principles of social behavior. The amount of time that parents spend talking and lecturing to their children goes up somewhat after the age of 2, and most dramatically at the age of 5 or 6, when the child is believed to first be able to approximate an adult-like intelligence.

Intelligence is not considered to be evenly distributed among the adults of Ifaluk. This is because parents' teaching plays a central role in the development of *repiy*, and adults are differentially diligent in their efforts to make children intelligent. All children are assumed to be basically the same until they reach the age of intelligence. Before that age, they are unsocialized and uniform in other respects that count. It is only when the child reaches the age of 2 that thoughts begin to emerge. It is consequently only then that individual differences arise, as the content of these thoughts begins to diverge in different individuals. Some children will acquire bad thoughts and feelings, and this constitutes a lack of intelligence, while others will come to know how to share their food and obey their elders. They will be termed intelligent and will be rewarded for their behavior.

Northern Cheyenne

The Northern Cheyenne see the infant as a liminal creature (Straus, 1977, 1982). For 1 to 2 years after birth, the child is not fully differentiated from the animals; it is not yet completely a person. The most crucial index of personhood that young children, like animals, lack is language. We have already seen that language, in Cheyenne ethnotheory, is integrally related to understanding. For this reason, the child is not named until she or he can speak. Infants are also seen as close to the spirit world from which they have just come, and they are considered liable to return there at any time.

Infants are born "empty of mind," and this ignorance is intimately related to their weakness. The infant's lack of full personhood is due to the absence of knowledge of the Cheyenne Way, a knowledge that can be acquired only through communication with those who know. Lacking speaking and listening skills, the infant is not truly incorporated within the human community. This ignorance, however, leaves children with no responsibility for their behavior.

Because speech is at the heart of intelligence or understanding, parents are very concerned to encourage language development in the infant. Much talk is directed to children by those around them. To the same purpose, infants were at one time fed the eggs of a bird whose call resembled the Cheyenne words for "Tongue River." The onset of speech in the child was cermonially marked by an ear-piercing ritual that celebrated the "opening up" of the ears and of verbal communication. This event in the child's life is still celebrated today, although in another form. At this point, the child is ready to begin to learn the Cheyenne Way.

The infant is also born with an undifferentiated spirit. Only later will it come to be divided into good, bad, wise, and active parts. Action is more typical of the young, and wisdom of the old. Individual differences in children may be inherited from parents. The Cheyenne do not see internal personality traits, however, as sources of motivation for behavior. "Relationships rather than internal dynamics are understood as the motivating force" (Straus, 1977, p. 333), and so the development of what would be considered intelligent behaviors depends on acquiring knowledge from others. Intelligence is a potential in the infant and can be realized only after the onset of communication. Speech allows the child to enter into the relationships in which intelligence has its only role.

CONCEPTS OF INTELLIGENCE AND INFANCY

This survey of the still scant cross-cultural evidence on conceptions of infancy and intelligence indicates that there is diverse understanding of the nature of the person and of individual differences and how they develop. There is also variation in cultural values about what behavior patterns are desirable and should be encouraged or sanctioned. One result of this variation is that the phrase *infant intelligence,* which has a specific, meaningful significance from our own view, is a contradiction in terms in two of the other societies examined here. Whether the traits valued are seen as social, behavioral, spiritual, or mental, the infant is often seen as incapable of exhibiting those traits. Frequently, little interest in infant individualism is shown by societies in which child-care patterns have evolved and continue to exist in response to high infant mortality rates (LeVine, 1974, 1977). With the infant's survival and health problematic, ethnotheoretical ideas will cast behavioral development as nonproblematic and/or as a task for the postinfancy period. Whiting has put forth the complementary suggestion that the wider opportunity that exists in many other cultures for potential parents to observe the variation across infants in such things as growth, size, and eating and sleeping habits results in a tendency for ethnotheories "to deemphasize psychological variables and dwell more on the child's physiological characteristics" (B. B. Whiting, 1974, p. 18).

Although non-Western cultures are as different among themselves as American culture is from each of them, there are certain aspects of ethnotheories of intelligence that appear to be shared by the three non-Western societies examined. First, higher value is placed on calm, deliberate behavior than on speed and quick cleverness. Thorough contemplation appears to be more highly valued than speed of solution. The connotations of wisdom might be more universally approbated than those of smartness.

Second, behavioral definitions of intelligence, as well as of other personal characteristics, are more common. Although the notion of *potential* that is prevalent in our own society is present to some degree in others, there is much greater interest in the behavioral enactment of valued qualities. Mental abilities that are not evident in everyday behavior may be of little concern or consequence. This may not be surprising if we consider that, in a non-surplus economy, there is most likely little leeway for awaiting the manifestation of "potential intelligence."

Third, and following from the behavioral emphases in many ethnotheories of intelligence, is an emphasis on the application of cognitive and emotional skills to social tasks. Intelligent behavior, like other behavior, is goal-directed. American cultural goals stress the attainment of individual autonomy from the demands of the group to an extent that is unusual cross-culturally. Many nonindustrial modes of human adaptation require close cooperation and personal contact among individuals. The ability to respond correctly in social situations and to be emotionally predictable is at a high premium in a context where the goals of individuals are to be achieved together with others, rather than alone.

A fourth common ethnotheoretical theme is the relation posited between language and intelligence. The prevalent connection made among speech, knowledge, and social communication and cooperation is consistent with the emphasis, noted above, placed on social cognitive skills. Language is often seen as the key to both knowledge and intelligence, and knowledge and intelligence are often seen, not as content and process, but as inseparable aspects of a person. The absence of language in infants is frequently cited as a crucial factor explaining their lack of intelligence or related characteristics. Where language is seen to differentiate the human from the animal, or the intelligent social adept from the unintelligent unsocialized being, the infant will then fall into a liminal category of the not-yet-human or not-yet-intelligent.

A valid definition of infant ability should arise out of the cross-cultural evidence of variation in ethnopsychological ideas. Intelligence need not be reified or internalized. The cross-cultural perspective may be successfully accommodated by some definitions that have already been suggested. Wober (1974) defines intelligence as that which is re-warded or esteemed within a culture. This is a definition that can be applied to any culturally valued behavior, but it is useful in that it points out the need to look at the entire range of traits that are positively evaluated in any particular culture and to examine the ranking of the relative importance of each trait and behavior. Rather than seeing intelligence as "some*thing* which one *has*," Fischer (1973, p. 16) calls for a view of it as an attribution about "the power of the way one does something" (p. 16). Neisser defines intelligence as "responding appropriately in terms of one's long-range and short-range goals given the actual facts of the situation as one discovers them" (Neisser, 1976, p. 137). An appropriate definition of intelligence might focus both on the skills necessary to attain culturally defined goals and on attributions of intelligence as interpersonal judgments made in a cultural context. Any definition that is more restrictive than these will necessarily reflect our cultural biases. Such restrictive definitions may "seem plausible to us as middle-class parents because we share with the psychologists who offer them a set of cultural premises about where development is going" (LeVine, 1980, p. 79). Other conceptions of the infant and of intelligence are possible, however, and reflect developmental environments, parental goals, and cultural meaning systems.

REFERENCES

ANASTASI, A. *Differential psychology: Individual and group differences in behavior.* New York: Macmillan, 1937.

BOWLES, S., & GINTIS, H. I.Q. in the U.S. class structure. *Social Policy,* 1972–1973, *3*(4&5), 65–96.

BUCK-MORSS, S. Socio-economic bias in Piaget's theory and its implications for cross-culture studies. *Human Development,* 1975, *18,* 35–49.

CALHOUN, D. *The intelligence of a people.* Princeton, N.J.: Princeton University Press, 1973.

CHARLESWORTH, W. An ethological approach to studying intelligence. *Human Development,* 1979, *22,* 212–216.

CRONBACH, L. J. Five decades of public controversy over mental testing. *American Psychologist,* 1975, *30,* 1–14.

FISCHER, C. Intelligence contra IQ. *Human Development,* 1973, *16,* 8–20.

GEERTZ, C. *The interpretation of cultures.* New York: Free Press, 1973.

GEERTZ, C. 'From the native's point of view': On the nature of anthropological understanding. In K. Basso & H. Selby (Eds.), *Meaning in anthropology.* Albuquerque: University of New Mexico Press, 1976.

GOODNOW, J. J. Some sources of cultural differences in performance. In G. Kearney & D. W. McElwain (Eds.), *Aboriginal cognition.* (Psychology Series #1, Australian Institute of Aboriginal Studies, Canberra, Australia.) Atlantic Highlands, N.J.: Humanities Press, 1976.

GOULD, S. J. *The mismeasure of man.* New York: Norton, 1981.

HARKNESS, S., & SUPER, C. Why African children are so hard to test. In L. L. Adler (Ed.), *Issues in cross-cultural research. Annals of the New York Academy of Science* (Vol. 285), New York, 1977.

HARKNESS, S., & SUPER, C. The cultural construction of child development: A framework for the socialization of affect. *Ethos,* in press.

KAMIN, L. F. *The science and politics of I.Q.* New York: Halsted Press, 1975.

KILBRIDE, P., & KILBRIDE, J. Sociocultural factors and the early manifestation of sociability behavior among Baganda infants. *Ethos,* 1974, *2,* 296–314.

LEVINE, R. Parental goals: A cross-cultural view. *Teacher's College Record,* 1974, *76,* 226–239.

LEVINE, R. Child rearing as cultural adaptation. In P. Leiderman, S. Tulkin, & A. Tulkin (Eds.), *Culture and infancy: Variations in the human experience.* New York: Academic Press, 1977.

LEVINE, R. Anthropology and child development. In C. Super & S. Harkness (Eds.), *Anthropological perspectives on child development.* San Francisco: Jossey-Bass, 1980.

LEWIS, M. What do we mean when we say "infant intelligence scores"? A sociopolitical question. In M. Lewis (Ed.), *Origins of intelligence: Infancy and early childhood.* New York: Plenum Press, 1976.

LUTZ, C. *Emotion words and emotional development on Ifaluk atoll.* Unpublished Ph.D. dissertation, Harvard University, 1980.

LUTZ, C. Parental goals, ethnopsychology and the development of emotional meaning. *Ethos,* in press.

NEISSER, U. General, academic and artificial intelligence. In L. Resnick (Ed.), *The nature of intelligence.* Hillsdale, N.J.: Lawrence Erlbaum Associates, 1976.

NEISSER, U. The concept of intelligence. *Intelligence,* 1979, *3,* 217–227.

PUTNAM, D., & KILBRIDE, P. *A behavioral understanding of intelligence: Social intelligence among the Songhay of Mali and the Samia of Kenya.* Paper presented at the annual meeting of the Society for Cross-Cultural Research, Philadelphia, February 1980.

QUINN, N. *A cognitive anthropologist looks at marriage.* Paper presented at the annual meeting of the American Anthropological Association, Cincinnati, Ohio, 1979.

QUINN, N. Marriage is a do-it-yourself project: The organization of marital goals. *Proceedings of the Third Annual Conference of the Cognitive Science Society,* Berkeley, Calif., 1981.

ROSENTHAL, B. *The images of man.* New York: Basic Books, 1971.

SAMELSON, S. Putting psychology on the map: Ideology and intelligence testing. In A. Buss (Ed.), *Psychology in social context.* New York: Irvington Publishers, 1979.

SHWEDER, R. *Semantic structure and personality assessment.* Unpublished Ph.D. dissertation, Harvard University, 1972.

STRAUS, A. Northern Cheyenne ethnopsychology. *Ethos,* 1977, *5,* 326–357.

STRAUS, A. The structure of the self in northern Cheyenne culture. In B. Lee (Ed.), *Psychosocial theories of the self.* New York: Plenum Press, 1982.

TULKIN, S., & KONNER, M. Alternative conceptions of intellectual functioning. *Human Development,* 1973, *16,* 33–52.

VALENTINE, C. Men of anger and men of shame: Lakalai ethnopsychology and its implications for sociopsychological theory. *Ethnology,* 1963, *1,* 441–477.

WHITE, G. Ambiguity and ambivalence in A'ara personality descriptors. *American Ethnologist,* 1978, *5,* 334–360.

WHITE, G. Conceptual universals in interpersonal language. *American Anthropologist,* 1980, *82,* 759–781.

WHITING, B. B. Folk wisdom and child rearing. *Merrill-Palmer Quarterly of Behavior and Development,* 1974, *20,* 9–19.

WHITING, J. W. M. *A model for psycho-cultural research.* Distinguished lecture address delivered at the annual meeting of the American Anthropological Association, New Orleans, 1973.

WOBER, M. Toward an understanding of the Kiganda concept of intelligence. In J. Berry & P. Dasen (Eds.), *Culture and cognition.* London: Methuen, 1974.

11   *Social Class and Infant Intelligence*

MARK GOLDEN AND
BEVERLY BIRNS

INTRODUCTION

There is generally widespread agreement that social-class differences in intellectual development are present in older preschool children, as reflected in their performance on standard intelligence tests beginning during the third year of life and later, in their academic achievement in school. However, there is disagreement about whether such differences are present in infancy. There has also been a great deal of controversy about how social-class differences in intellectual development can best be explained or interpreted, as well as strong disagreement about what can or should be done about such differences.

Before we explore these issues further, some discussion of what is meant by such terms as *social class* and *infant intelligence* may be useful. While there have been many definitions of social class, they seem to reduce to a relatively small number of factors. Studies of social class usually use the following indexes of socioeconomic status, either individually or combined: the education and occupation of the head of the household, income, source of income (wages or public assistance), residential area, and quality of housing. Hollingshead's (1957) Two-Factor Index of Social Position, for example, which is widely used, is based on the education and occupation of the head of the household. On the basis of such indexes, individuals or families are classified into two or more

MARK GOLDEN • Department of Psychiatry, Albert Einstein College of Medicine, 1300 Morris Park Avenue, Bronx, New York 10461.   BEVERLY BIRNS • Social Science Interdisciplinary Program, State University of New York at Stonybrook, Stonybrook, New York 11794.

social classes, ranging from upper class to lower class. Recent investigators, particularly those who have been involved in providing social services to poor families, have pointed out that such a category as *lower class*
is too broad and does not adequately reflect great differences in levels of
functioning between poor, stable, working families and socially disorganized, multiproblem families (Geismar & La Sorte, 1964; Miller, 1964;
Pavenstedt, 1965).

It is important to distinguish *socioeconomic status* (SES), in which
people are merely classified into various groups on the basis of the
foregoing indexes, from class-related *process variables,* such as nutrition,
patterns of mother–child interaction, language experience, etc., which
tend to be associated with social class and may have a more direct effect
on children's development. If we could identify such process variables
during infancy, perhaps we could discard the social-class concept in our
research. At the present time, however, social class indexes, such as
parents' education and occupation, predict the later intellectual development of children better than any other measure we now have available
during the infant period, including the child's own infant test scores
(McCall, Hogarty, & Hurlburt, 1972).

It is essential to distinguish between social class and race or eth-
nicity, since these factors are often confounded with social class. When
making comparisons among different racial or ethnic groups, even if
they are equated on certain SES indexes such as the number of years of
schooling, we cannot assume that their educational experience has been
equivalent, since we know that the quality of education has not been the
consistent for all ethnic and racial groups in this country. In reports of
results on social class, the particular racial or ethnic groups involved
should be identified because the presence or magnitude of SES differences found on a particular measure may depend on the population
studied.

It is also necessary to keep in mind that there may be an interaction
between social class and sex on some cognitive and personality measures (Birns & Golden, 1973). For example, there may be social-class
differences for one sex and not the other, or sex differences for one social
class and not another. Making social-class comparisons using only one
sex precludes studying such social class by sex interactions. In reports of
social-class differences, results on males and females should be reported
separately.

What about the term *infant intelligence?* Perhaps a definition of the term *infant* may be a good starting point, since the word has many meanings. According to Webster's unabridged dictionary, the term derives from the Latin word *infans*, which roughly translated means "one who cannot speak." By this definition, the infant period would be limited to the first year of life in humans, since most children usually say their first words by about 12 months of age. We would opt for a somewhat broader definition of infancy to include what Piaget refers to as the *sensorimotor period*, approximately the first 18–24 months of life. Intelligence has been defined in many ways, including the ability to learn, to remember, to solve problems, to reason, and to adapt to one's environment, but there is no generally agreed-on definition. When we speak of infant intelligence, we are referring to *sensorimotor intelligence*. Although children can understand and use language to a limited extent during the sensorimotor period, sensorimotor intelligence, according to Piaget (1952), operates predominantly on a presymbolic and preverbal level. While we have defined the infant period as roughly the first 2 years of life, we will review social class research on children up to 3 years of age since SES differences on some intellectual measures first manifest themselves during the third year.

Are there social-class differences in sensorimotor intelligence? This question is inseparably enmeshed with problems of measurement, with the particular intellectual measures used in SES comparisons, with the validity of these measures, and with the age of the infants being studied. Until recently, attempts to measure infant intelligence (and hence studies of SES differences) were largely based on standard infant tests or developmental scales, such as the Gesell, the Cattell, and the Bayley. These tests were initially conceived to be downward extensions of the standard intelligence tests for older children, such as the Stanford–Binet, and were expected to correlate with later measures of intelligence and with academic achievement. While intelligence tests for older preschool children are highly correlated with social class and with later school performance, the infant test scores of normal babies during the first 18 months of life do not seem to be related to later measures of intelligence, success in school, or social class. The validity of infant tests has been questioned on a number of grounds (Lewis & McGurk, 1973). Do infant tests fail to detect social-class differences because there are no SES differences in sensorimotor intelligence or because the tests are not valid?

During the 1960s, stimulated to a great extent by the "rediscovery" of Piaget, developmental psychologists began to question the value of standard infant tests and particularly, the usefulness of such global measures as DQ or IQ scores in understanding intellectual processes in babies. Using a variety of new cognitive measures that dealt with more discrete unidimensional aspects of behavior and that did not rely so much on children's motor abilities, psychologists began to take a "new look" at infant intelligence. The new measures included scales of cognitive development derived from Piaget's observations on the sensorimotor period; measures of attention to visual and auditory stimuli, which are assumed to reflect central information processes; instrumental learning and problem solving; and exploratory and play behavior. Do these new cognitive measures correlate more highly with social class and later intelligence than do standard infant tests? We will review studies of social-class comparisons of children's behavior in which these new measures were used.

It is also possible that there may be social-class differences in children's experiences during the sensorimotor period—for example, in their language experience—that may not be reflected in their intellectual development until later. Basil Bernstein (1960), a British sociolinguist, has hypothesized that there are social-class differences in the extent to which parents from different social classes use different linguistic codes or styles of communication with their children. He has described two different linguistic codes, *elaborated* and *restricted,* which are assumed to be characteristic of middle-class and lower-class speech patterns, respectively. Bernstein believes that the two linguistic codes arise out of fundamental differences in life style and outlook and that these different linguistic codes are reflected in social-class differences in cognitive style and intellectual development.

On the basis of Bernstein's ideas, Hess and Shipman (1965) have hypothesized that there may also be social-class differences in patterns of mother–child interaction for the teacher–pupil aspect of their relationship, which may subsequently be reflected in the child's performance on standard intelligence tests and in school. Middle-class children may be more successful in these situations because they have learned to assume the roles expected of them through earlier interactions with their mothers. Hess and Shipman found SES differences in patterns of interaction between black mothers and their 4-year-old boys on various learning tasks, and these patterns were related to the children's intellectual performance. How early can such social-class differences in patterns of mother–child interaction be observed? While there has been a great deal of speculation about the nature of SES differences in mother–child interaction during infancy, some investigators have made direct naturalistic observations of the interactions between mothers and infants from different social classes, with particular emphasis on those aspects of behavior assumed to be related to children's cognitive and language development. Are there SES differences in patterns of mother–child interaction during the sensorimotor period? And are the differences that may be present related to children's later cognitive and language development? We will review studies of social-class comparisons of mother–child interaction in an attempt to answer these questions.

Throughout the relatively brief history of the intelligence-testing movement, there has been a continuing controversy about what intelligence tests such as the Stanford–Binet actually measure. While there are reliable individual and group differences in performance on such tests, how do we explain these differences? Finally, what can or should be done about such differences? Historically, the major controversy has revolved around the heredity–environment issue. Recently, a new controversy has emerged among social scientists who have taken an environmentalist position. Can social-class differences (e.g., in intellectual performance, personality functioning, and maternal behavior) that seem to favor the middle class, judged by middle-class standards, be construed as deficits in the lower class or as merely cultural variations or differences?

First we will discuss the heredity–environment issue and then deal with the deficit–difference question. Binet, who developed the first successful prototype for most standard intelligence tests now in use, did not set out to measure intelligence. He developed his mental scale, which was not called an intelligence test, to identify children who were likely to do poorly in school, so that they could be given special educational treatment. Where Galton had previously failed to develop a useful measure of intellectual functioning (he had planned to use it for eugenic purposes), Binet succeeded admirably, since his scale correlated highly with academic achievement. While Binet (1909) did not assume that his scale measured innate intelligence, such an assumption was made by subsequent users and adaptors of his scale in the United States (Goddard, 1920; Terman, 1916; Yerkes & Foster, 1923) who were followers of Galton. They assumed that individual differences and social-class, ethnic, cultural, and racial differences in performance on such tests were largely due to heredity. As Kamin (1974) pointed out in a devastating critique of the intelligence-testing movement in this country, the leaders of this movement expressed such views long before there were any research data to support their ideas, that is, before the heritability studies were actually carried out. While the heritability studies later reported in the literature may provide evidence for a significant genetic contribution to individual differences in performance on such tests, they do not provide *direct* evidence that group differences can be explained on the same basis. The genetic point of view has certain educational implications. If one assumes that intelligence is fixed at birth and that intellectual development is predetermined, there is very little that can be done to increase a person's level of intellectual functioning through environmental enrichment or education.

In 1961 Hunt published *Intelligence and Experience,* a seminal work that has had a major impact on developmental psychology and early childhood education. Hunt presented evidence contrary to many of the assumptions of the hereditary view, particularly to the belief in fixed intelligence and predetermined intellectual development. While Hunt proposed an *interaction* point of view (i.e., intellectual development is a function of the interaction of heredity and environment), the thrust of his ideas seemed to shift the pendulum to an environmental position. He presented data from animal research and studies of institutionalized babies showing that a restricted environment and lack of intellectual stimulation during infancy may have permanent, irreversible, and detrimental effects on intellectual and problem-solving ability. Hunt's belief in the importance of early experience was also based on an epigenetic view of cognitive development, which assumes that later stages of intelligence are based on and are hierarchically related to early stages. He also assumed that as children grow older, their behavior patterns tend to become fixed and more difficult to modify. Therefore, early experience is likely to have a greater impact than later experience on children's intellectual development. In a paper that appeared at about the same time, Fowler (1962) presented evidence that infants and young children are capable of much more cognitive learning than has been expected of them. He argued that there should be greater emphasis on accelerating cognitive development as early as possible, so that children can achieve their full intellectual potential.

During the 1960s, as part of the War on Poverty, psychologists and educators became concerned with the educational problems of poor children in this country. The academic achievement of poor children has been considerably less than that of middle-class children. They seem to be behind middle-class children in intellectual development when they first enter school, as reflected in their lower performance on standard intelligence tests. They get further and further behind their middle-class peers in school as they grow older, a phenomenon that has been explained on the basis of a "cumulative deficit hypothesis" (Ausubel, 1964; Deutsch, 1964). Poor children drop out of school earlier and in greater numbers, and fewer of them go on to college. As a result, they are forced to take low-paying, menial, dead-end jobs or remain chronically unemployed. Helping poor children succeed academically was seen as a way of helping them break out of poverty.

Operation Head Start was established in 1965 on a national scale. Initially, the goals of Head Start were fairly modest. Four- and 5-year-old children from poverty areas were provided with a summer nursery school program. One of the major goals of such programs was to help poor children make a better transition from home to school by helping them to develop greater self-confidence, positive attitudes toward learning, and an expectation of success in school. The summer program had very little impact on their subsequent academic achievement in school. The programs were expanded and changed in a number of important respects. Children were enrolled in Head Start programs throughout the school year. The programs were extended downward to include younger children, with an increasing emphasis on stimulating cognitive and language development and pushing academic skills. The programs were also evaluated on a more stringent basis, in terms of measurable increases in children's intelligence test scores and later academic achievement in school.

In general, the evaluation studies of Head Start and similar programs have been disappointing. Even in programs that were judged to be excellent, a familiar pattern emerged: (1) In comparison to similar children who are not in the program, children in the program often show dramatic IQ increases after a relatively brief period of time, which Zigler and Butterfield (1968) attributed at least in part to changes in motivation. (2) These IQ increases tend to level off while the children are still in the program, generally to below the performance of middle-class children. (3) Children's IQ scores often drop to their original level after they leave the program, with some exceptions to be described later. (4)

While children who had been in a preschool program often showed initial superiority in their functioning in school over children without such experience, the differences seem to wash out by the end of the first or second grade.

These results have led to different conclusions and recommendations. Jensen (1969) concluded that compensatory education has failed because SES differences in intellectual development are genetically determined. He recommended that children from different social classes be given different kinds of education in line with their innate abilities. On the other hand, some proponents of compensatory education concluded that the programs did not begin early enough or continue long enough. On the basis of the similar disappointing results of his own intervention study with children between 15 and 36 months of age, Schaefer (1970) recommended that on the one hand the child's education should begin from birth at home, directly involving the mother in the educational process from the beginning, and that on the other hand the child should be provided with educational enrichment throughout the school years. Hunt (1969) and others (Denenberg, 1970) have also argued for preventive (rather than compensatory) intervention programs beginning in early infancy. Are such infant intervention programs necessary for lower-class children during the sensorimotor period? Are intervention programs that begin early in life, particularly during the first 2 years, more effective than those that begin later, in both their immediate and their long-term effects on children's intellectual development? We will review intervention studies with children under 3 years of age in order to attempt to answer these questions.

At about the time the heredity–environment issue flared up again with the publication of Jensen's (1969) explosive article in the *Harvard Educational Review*, the deficit–difference controversy emerged among environmentalists. While the environmental-deficit point of view has been fairly prevalent among proponents of early intervention for poor children, its most articulate spokesman has been Hunt, whose ideas appeared in a collection of articles entitled *The Challenge of Incompetence and Poverty* (1969). The basic assumptions of the environmental-deficit view seem to be as follows. The relatively low intellectual performance and school failure of poor children is no longer attributed to a genetic deficit but is explained on the basis of various environmentally produced psychological deficits—for example, in language and cognitive ability—as well as aspects of personality and motivation assumed to be related to learning and achievement. Hunt, for example, believes that this problem is particularly acute in a highly technological society, where the need for unskilled labor is diminishing. Many poor children may not develop the kinds of intellectual skills required to function in such a society. In this sense, they may be functionally incompetent and may be doomed to a life of poverty unless something is done to prevent it early in life. Since these intellectual deficits seem to be present long before these children enter school, as reflected in lower performance on standard intelligence tests, it is assumed that they are due to deficiencies in the children's home environments.

While less value-laden terms have been used, poor children have often been referred to as "culturally deprived." Oscar Lewis (1966) explained cultural deprivation on the basis of the "culture of poverty." Lewis believes that in most industrial capitalist societies exists a culture (or subculture) of poverty that has certain common characteristics. While most poor people do not have these characteristics, according to Lewis, the members of this subculture, who represent the most economically impoverished and socially disorganized segment of society, may perpetuate their own poverty as a result of self-defeating attitudes, values, and behavior patterns. These are socially transmitted from one generation to the next and prevent the children in these families from taking full advantage of educational and other opportunities to break out of poverty. For such children, "equal opportunity" may be an empty phrase.

Soon after Lewis published his ideas, they were strongly criticized by many social scientists. A volume edited by Leacock, *The Culture of Poverty: A Critique* (1971), contains a number of excellent articles critical of Lewis's views. From the viewpoint of a developmental psychologist, Tulkin (1972) questioned the validity of the assumption that social-class differences in intellectual performance on standard intelligence tests and differences in maternal behavior that seem to favor the middle class, judged by middle-class standards, can legitimately be considered deficits in the lower class. He pointed out that while anthropologists no longer view cultural differences from an ethnocentric point of view (one reflected, for example, in such terms as *primitive societies*), psychologists who interpret social-class differences in terms of deficits may still be operating on an ethnocentric basis. Tulkin also presented evidence that standard intelligence tests are culturally biased and tap a relatively narrow range of intellectual competence, specifically the range related to academic success in a middle-class-oriented school system. This does not mean that such tests may not be useful for some purposes. But if people who are not middle class in our society or who are from a different culture obtain lower scores on these tests, such scores do not mean that they are intellectually inferior, as a result of either a genetic or an environmental deficit. Tulkin argued for a relativistic approach similar to the one taken by cultural anthropologists, in the explanation and understanding of social-class differences in our society. He also believes that it may be more fruitful, in finding solutions to social problems such as poverty, to consider the possibility that the deficits may be in the educational system and in society rather than in poor children and their families. The position one takes on this issue is more than a semantic argument among academicians, for it may have profoundly different implications for dealing with educational and social problems in this country.

Social-Class Comparisons of Children's Behavior

First we will review studies of social-class comparisons of infants' behavior during the first year of life and then report similar research on children during the second and third years. Tulkin (1973) and Tulkin and Kagan (1972) compared the behavior of 56 10-month-old girls and mothers from white middle-class and working-class families. The children were studied in various situations, including at home under naturalistic conditions. Middle-class infants spent more time crawling and playing with objects than working-class babies, possibly because the middle-class children spent less time confined in a playpen and were given more toys to play with by their mothers. There were no SES differences in the amount of spontaneous vocalizing by the infants when they were not interacting with their mothers.

The children were also studied under laboratory conditions. In one situation, they were presented with two types of tape-recorded speech passages. The first type involved meaningful and nonmeaningful speech. There were no social-class differences in the children's responses to the passages themselves, in attention or vocal behavior to meaningful and nonmeaningful speech. However, the middle-class infants tended to look more than the working-class babies at the female coder, who was seated behind them, following the meaningful passages, as if looking for the speech source. This finding is similar to one by Kagan (1971) that upper-middle-class 8-month-old babies looked longer at the speaker baffle, when there was no coder in the room, following meaningful speech passages. The children in Tulkin's (1973) study were also presented with tape-recorded passages of a fairy tale being read by their own mothers and by a stranger. The middle-class infants showed greater differential responding to the two voices than the working-class babies. The middle-class children quieted more than the lower-class infants when they heard their mothers' voices ($p < .05$), and vocalized more when their mothers stopped speaking ($p < .10$). They also tended to look more at their mothers after hearing the mothers' voices ($p < .06$) and looked more at the coder after hearing the stranger's voice ($p < .03$).

The children were also observed in a standard laboratory play situation. The only significant SES difference in the infants' play behavior involved an experimentally elicited "conflict." After the babies had a chance to play with a few toys, a familiar and a new toy were presented to them. The middle-class infants seemed to be more "reflective" in making a choice. Before choosing a toy, they shifted their gaze more from one toy to another, as if trying to decide which one to play with. Once they had made a choice, the middle-class infants tended to play with the toy they had chosen longer and looked less at the rejected toy than the working-class babies.

In a similar standard laboratory play situation, Messer and Lewis (1972) found few social-class differences in the play behavior of 1-year-old children. The sample included 99 boys and girls from two different SES groups. There were no social-class differences in the children's play style as reflected in how frequently they shifted from one toy to another and how long they played with a single toy. There were significant SES differences in only two aspects of the children's behavior. Higher SES babies tended to use more floor space and vocalized more than lower SES infants.

Lewis (1969) has studied changes in children's attention as a function of age, social class, and intelligence. The procedure involved presenting children with the same visual stimulus for a number of trials and then introducing a new stimulus. An important measure of attention is fixation time—how long the child looks at the stimulus. In general, children show a response decrement (i.e., shorter fixation time) as the same stimulus is repeated, and an increase in fixation time when a novel stimulus is introduced. Older children show more rapid response decrement and recovery than younger children, suggesting that this response pattern reflects more mature behavior. Lewis believes that response decrement and recovery in attention reflect central information processes, with more rapid decrement and recovery indicating a higher level of cognitive functioning. How does this measure of attention relate to other intellectual measures and to social class? In a study of 32 3-month-old infants from five different social classes, lower SES babies showed greater response decrements than higher SES infants (Lewis & Wilson, 1972). Thus, on this measure, lower-class infants manifested a higher level of cognitive functioning than middle-class babies, at least when 3 months of age. In a study of 64 children of 1 year of age, Lewis (M. Lewis, personal communication) did not find social-class differences in a similar measure of attention.

In a follow-up study (Lewis, 1969), 40 of these children were tested on the Stanford–Binet at 44 months of age. There was a significant positive correlation between rate of response decrement at 1 year of age and IQ at 44 months ($r = .46$ and $.50$ for girls and boys, respectively; $p < .05$). These results support Lewis's contention that rate of response decrement in attention may be related to intelligence. At 12 months, it predicts intellectual performance at $3\frac{1}{2}$ years better than any other measure we now have available for normal babies of this age, including standard infant tests. In terms of social class, Lewis's research on attention indicated that middle-class infants do not show superior information-processing skills to lower-class babies on the sensorimotor level on a measure that has been shown to be related to later intelligence.

Wachs, Užgiris, and Hunt (1971) did a cross-sectional comparative study of 102 infants between 7 and 22 months of age from two different social classes. Most of the low SES babies were black. The high SES infants were from white middle-class families; they were matched with the poor children on the basis of age and sex. Children from the two SES groups were compared on Užgiris and Hunt's Infant Psychological Development Scales (1966), which were based on Piaget's observations of infants during the sensorimotor period. There were essentially no SES differences on "The Development of Object Permanence Scale" or on "The Development of Schemas for Relating to Objects Scale." There were social-class differences starting at 7 months of age on "The Development of Means Scale," which assesses an infant's ability to use tools for obtaining desirable objects that are out of reach. At 7 months of age, for example, middle-class infants were superior to lower-class infants in the use of a support to obtain a distant object. The children were also presented with "Learning and Foresight Tasks," which were designed to assess their ability to anticipate the consequences of their actions. No children were able to master any of the tasks before 15 months. Starting at 15 months of age there were consistent SES differences at all ages on one of these tasks, a stacking toy that involves placing rings on a pole. The children were presented with two sets of rings, one with a hole and one that was solid. The lower SES babies attempted to place the solid rings on the pole more often than the higher SES infants did. Starting at 15 months, there were also SES differences on a measure of "Vocal Imitation," which assesses both the number of appropriate words children spontaneously use for objects and the number of words that can be elicited through imitation. The middle-class children were superior to the lower-class babies on both language indexes.

White (1975) did a longitudinal study of 40 children between 1 and 3 years of age from five different SES groups (Hollingshead's Classes I through V). At 12 months of age, there was no relationship between social class and social competence, as measured by White's Index of Social Competence, or between social class and intellectual competence, assessed by various measures including the Bayley mental scale. At 3 years of age, however, social class correlated significantly with intellectual competence, as measured by the Stanford–Binet ($r = .65$, $p < .001$), but not with social competence ($r = .04$).

A research team at the Albert Einstein College of Medicine, including the present authors, has carried out a series of studies to discover how early social-class differences in cognitive development are present and which factors contribute to these differences. We knew that earlier studies had failed to find SES differences during the first few years of life (Bayley, 1965; Hindley, 1962; Pasamanick & Knobloch, 1961), but we questioned the results on two grounds. First, they were based on standard infant tests, whose validity has been challenged because they correlate so poorly with later measures of intelligence. Infant tests can detect gross developmental problems resulting from serious organic impairment or the kind of severe deprivation experienced by institutionalized babies, but such global measures may be relatively insensitive to social-class influences during the sensorimotor period. We assumed that the new scales of sensorimotor intelligence based on Piaget's observations (Corman & Escalona, 1969; Užgiris & Hunt, 1966) might be more related to later cognitive development and hence might detect social-class differences earlier than standard infant tests. Second, previous studies had compared children from middle- and lower-class families without differentiating between children from poor, stable, working families and those from socially disorganized, welfare families. Paven-stedt (1965) and Malone (1963) reported marked differences in patterns of child-rearing as well as in the intellectual and personality functioning of 3-year-old children from these two types of poor families. We assumed that if more infants from impoverished, socially disorganized families were studied, SES differences in cognitive development might be manifested earlier.

We did a cross-sectional study (Golden & Birns, 1968) that included a range of welfare to intact middle-class families in which 192 black children at 12, 18, and 24 months of age were compared on the Cattell Infant Intelligence Scale and the Piaget Object Scale. We hypothesized that there would be no social-class differences on the Cattell but that there would be SES differences on the Object Scale.

The Object Scale seemed an ideal alternative to standard infant tests such as the Cattell. Children's instrumental responses on the Object Scale remain the same from the easiest to the most difficult item; they are required to search for objects hidden under a cloth in increasingly complex conditions. Whereas tests such as the Cattell are largely concerned with increments in children's perceptual-motor skills, the Object Scale seems to assess their increasing knowledge about objects and laws of displacement. The development of object permanence also seemed to be an important precursor of later cognitive development.

There were no significant differences in either the Cattell or the Object Scale scores between black children from welfare families and those from middle-class families during the first 2 years of life. We did a longitudinal follow-up study (Golden, Birns, Bridger, & Moss, 1971) in which we retested on the Stanford–Binet at 3 years of age as many as we could of the children in the 18- and 24-month cross-sectional samples. Whereas at 2 years there was a nonsignificant 6-point mean IQ difference (96 versus 102) between children from the 24-month follow-up in our lowest and highest SES groups, at 3 years of age there was a significant 21-point mean IQ difference (93 versus 115) between the two extreme SES groups. The children in the 18-month follow-up sample showed the same pattern of IQ differences at 3 years (see Table I). The similar results obtained in two independent samples increased our con-

TABLE I

Mean IQ Scores of Black and White Children from Different Social Classes in Two Longitudinal Studies[a,b]

Sample (n)	Age in Months			
	18	24	30	36
Black welfare[c]				
18-month (10)	110	—	—	94
24-month (11)	—	96	—	93
Black middle-class[c]				
18-month (5)	106	—	—	115
24-month (11)	—	102	—	115
White low-education (30)	—	92	103	—
White high-education (30)	—	113	126	—

[a]Golden, Birns, Bridger, & Moss, 1971; Golden, Bridger, & Montare, 1974.
[b]At 18 and 24 months of age, the Cattell Infant Intelligence Scale was used with the black children. At 24 months, the Bayley Mental Scale was used with the white children. At 30 and 36 months, the Stanford–Binet was used with both groups.
[c]Represents only the two extreme SES groups in the follow-up sample in the study of black children. The total follow-up sample included 89 children. For purposes of data analysis, the IQ scores of the children from the 18- and 24-month cross-sectional samples (n = 89) were combined at 3 years of age.

fidence in the findings. Furthermore, the range in mean IQ scores for our sample of black children was almost identical to that reported by Terman and Merrill (93 to 116) in their 1937 Stanford–Binet standardization sample of 837 white children between $2\frac{1}{2}$ and 5 years of age. Finally, the Object Scale did not prove to be more related than the Cattell to later intellectual development. At 18 months, neither the Object Scale nor the Cattell correlated with the Stanford–Binet at 3 years of age. At 24 months, the correlations between the Binet and the Object Scale and Cattell were .24 ($p < .05$) and .60 ($p < .005$), respectively.

In the original cross-sectional study, children were also rated on seven behavior-rating scales while they were being tested on the Cattell and the Object Scale. These included such behaviors as delay capacity, attention, persistence, cooperation, and pleasure in task. While there were no SES differences on any of these measures at 18 or 24 months of age, most of them correlated with the Cattell and a few correlated significantly with the Stanford–Binet at 3 years of age (Birns & Golden, 1972). A particularly interested finding was that the "pleasure in task" rating at

18 months was the only measure at this age that correlated significantly with the Stanford–Binet ($r = .29$, $p < .05$). That is, at 18 months of age, the extent to which children enjoyed doing sensorimotor tasks was more predictive of later intelligence than was their success on the tasks themselves. The correlation between "pleasure in task" at 24 months and the Binet was even higher ($r = .43$, $p < .005$), with the correlation between the Cattell and the Binet ($r = .60$) partialed out. This finding suggests that while there may be discontinuity between sensorimotor intelligence and later verbal measures of intelligence, there may be continuity in certain aspects of personality or cognitive style that can be observed in infants and that may be related to later intelligence. However, these behaviors were not related to social class.

Since we did not find social-class differences in intellectual development or cognitive style during the first 2 years of life but did find significant SES differences in intellectual performance at 3 years of age, we focused our research on children between 2 and 3 years of age to see if we could discover factors that might explain why social-class differences emerge at this time.

We did a longitudinal study (Golden, Birns, Bridger, Montare, Rossman, & Moss, 1973) of 60 white males between 24 and 30 months of age from two different SES groups (Golden, Bridger, & Montare, 1974). Comparisons on a number of measures were made between 30 boys whose mothers had completed college and 30 boys whose mothers had not gone beyond high school (whom we will refer to as the high-education and the low-education groups). The measures included (1) *standard intelligence tests* (Bayley mental scale at 24 months and Stanford–Binet at 30 months); (2) *verbal inventory*, which assesses verbal comprehension and production in a standard test situation; (3) *verbal and nonverbal learning*, in which children were presented with identical perceptual-discriminatory learning tasks under verbal and nonverbal conditions; (4) *delay capacity, persistence, and embedded-figures tasks;* and (5) *mother–child interaction*, assessed by means of videotaped samples of mothers reading stories to their children. In addition, the Wechsler Vocabulary Scale was administered to the mothers.

On the basis of our previous research, we did not expect to find social-class differences in IQ at 2 years of age, but we expected SES differences on some of our other measures. We had selected measures that we believed might detect social-class differences earlier than standard intelligence tests and might correlate with later intelligence.

Contrary to our expectations, there was a large and significant mean IQ difference between our two SES groups at 2 years of age, as well as social-class differences on some of the other measures. Here, we will focus on the procedures and results at 2 years of age.

Whereas in our earlier study (Golden *et al.*, 1971) there was only a 6-point (96 versus 102) nonsignificant IQ difference between the two extreme SES groups at 2 years of age, in the subsequent study there was a significant 21-point IQ difference (92 versus 113) between the children in the high- and the low-education groups ($p < .01$). However, the two studies differed in several important respects. In the first study, the children were black, the study included both boys and girls, and the Cattell Infant Intelligence Scale was used. In the subsequent study, the children were white, only boys were included, and the Bayley mental scale was employed. While these factors may have contributed to the differences, we believe that the difference in results was largely due to a difference in the socioeconomic composition of the families in the two studies which we will briefly summarize.

1. *Black welfare families:* The families were drawn from Hollingshead's Class V and were all on public assistance. Of these families, 93% did not have a father living in the home.

2. *Black middle-class families:* These were selected from Hollingshead's Classes I, II, and III. All of the families included both parents. While in all of these families one or both parents had some schooling beyond high school, only 8% of the mothers had graduated from college.

3. *White low-education families:* The criteria for selecting children for this group were that the mother had not gone beyond high school and the father had not graduated from college. Most of the fathers were skilled blue- or white-collar workers or semiprofessionals, including sanitation workers, electricians, carpenters, policemen, and firemen.

4. *White high-education families:* The criterion for selecting children for this group was that the mother had graduated from college, whether the father had graduated from college or not. However, 90% of the fathers were college graduates, and most of them had professional, business, or managerial positions.

The IQ scores of the children in these four groups between 2 and 3 years of age are presented in Table I, where we also included the data on the 18-month follow-up sample of black children.

An examination of these data indicates that the difference in results of the two studies at 2 years of age is largely due to the fact that the children in the high-education group in the present study obtained substantially higher IQ scores than those from the black middle-class families in the previous study. We attribute this difference to differences in the parents' education. While 100% of the white high-education mothers completed college, only 8% of the black middle-class mothers in the earlier study finished college. However, even if the educational levels of the parents in the two studies were similar in terms of amount of formal schooling, we could not assume that their educational experience had been equivalent, since we know that the quality of education for blacks and whites in this country has not been equivalent.

Before we report some of the other data on the children in the present study, a social-class comparison of the mothers' Wechsler vocabulary scores may be of some interest. The mean score for the high-education mothers was 70 (with 80 as the maximum possible score), whereas the low-education mothers obtained a mean score of 44, with very little overlap in the vocabulary scores of these two groups of mothers. The difference was significant at the $p < .001$ level. While this is not surprising, it indicates that the mothers of the children in our two SES groups differed greatly, not only in terms of amount of formal schooling but also in their language competence.

Since we assumed that language plays a major role in explaining why social-class differences in intellectual performance first manifest themselves during the third year of life and not earlier, we wanted to assess the children's verbal ability. We felt that, for several reasons, standard intelligence tests were inadequate for this purpose. The tests include both verbal and nonverbal items. There is no clear-cut distinction between verbal production and comprehension. What is particularly lacking is an adequate measure of verbal comprehension of infants and young children. Since there were no language tests available for 2-year-old children, we developed a verbal inventory that we believed might detect SES differences earlier than standard intelligence tests, particularly in language comprehension.

The inventory consists of two subscales: (1) *comprehension,* in which the child is required to give a *nonverbal response* to a verbal statement by the examiner; and (2) *production,* in which the child is required to give a *verbal response* to a verbal statement by the examiner. Prior to the inventory's use in the present study, a reliability and validation study was carried out on 96 white children at 2½ and 3 years of age, equally divided by age, sex, and social class. There were highly significant SES and age differences, but no overall sex differences. The total verbal inventory was highly correlated with the Stanford–Binet at $2\frac{1}{2}$ ($r = .82$) and 3 ($r = .84$) years of age. While the inventory contains a few verbal reasoning items, most of the items are concerned with the child's knowledge of verbal concepts, so that for the most part, the inventory measures verbal knowledge rather than reasoning or problem-solving ability. The fact that the inventory correlates so highly with the Binet indicates that at least at this age, the Stanford–Binet may largely be a test of verbal knowledge rather than a measure of reasoning or problem-solving ability.

In the present study, the children in the high-education group obtained significantly higher scores than the low-education group on the total inventory (verbal comprehension and verbal production) at both 2 and $2\frac{1}{2}$ years of age. While the comprehension scores were higher than the production scores at 24 months, SES differences were present in both, and the correlations with the Bayley for comprehension ($r = .66$, $p < .01$) and production ($r = .61$, $p < .01$) were about the same. The correlation between the total inventory and the Bayley was somewhat higher ($r = .70$, $p < .01$), and its correlation with the Binet at 30 months was even higher ($r = .86$, $p < .01$). Verbal comprehension does not appear to be more sensitive in detecting social class differences at 2 years of age than verbal production or the Bayley mental scale. It is possible that with children under 2 years of age, measures of verbal comprehension may pick up SES differences earlier and correlate more with later measures of verbal intelligence than standard infant tests.

Several investigators have shown that language can facilitate perceptual discriminatory learning as early as the first year of life (Katz, 1963; Koltsova, 1960), but there have been no published studies on whether there are social class differences in this respect. In the present study (Golden et al., 1974), the children were presented with identical perceptual learning tasks under verbal and nonverbal conditions. The essential difference between the two conditions was that under the verbal condition the children were provided with verbal labels for the objects to be discriminated, whereas under the nonverbal condition they were not. Since there do not appear to be social-class differences in sensorimotor intelligence but there are SES differences in verbal intelligence, we hypothesized that there would be social-class differences under the verbal condition but not under the nonverbal condition. The results confirmed our hypothesis. While there were no social-class differences under the nonverbal condition, the children in the high-education group did significantly better than the low-education group under the verbal condition ($p < .01$). Furthermore, whereas nonverbal learning did not correlate with IQ, verbal learning at 24 months correlated with the Bayley at 2 years ($r = .37$, $p < .01$) and correlated even more highly with the Stanford–Binet at $2\frac{1}{2}$ ($r = .48$, $p < .01$).

Koltsova (1960) reported that providing a verbal label can facilitate perceptual discriminatory learning in children as young as 12 months of age, but to our knowledge this finding has not been replicated in this country. A replication study should, of course, be done. But in addition it would be of interest to see how early there are social-class differences in children's ability to use verbal information to facilitate learning, and whether such a measure in children under 2 years of age can predict later intelligence better than standard infant tests.

In the present study, *delay capacity* was assessed by means of the following "waiting games."

1. A *hiding game*, in which a reward (cookie) was hidden under one of three boxes and the child was asked to wait until a signal was given (the examiner blew a whistle) before he searched for the cookie. The children were not prevented from responding prematurely and were allowed to have the cookie whether they waited or not. Some children immediately searched for the cookie and eagerly gobbled it up before the signal was given. Other children seemed to enjoy the waiting game itself, anticipating the signal with mounting excitement, playfully pretending to reach for the cookie before the signal was given, and just as eagerly returning the uneaten cookie to the examiner to be hidden again.

2. A *train game*, in which the child was allowed to activate a miniature train set by means of a starting switch that he held in his hand. Again the child was told to wait until the signal was given before he started the train. On both tasks, the children were given a number of practice trials until they demonstrated that they understood the instructions. There were 10 delay trials given for each task, ranging from a 5- to a 50-second delay, randomly presented. The delay scores consisted of the number of trials on which children waited until the signal was given before responding.

There were significant social class ($p < 0.01$) and age ($p < 0.01$) differences in delay capacity. The children in the high-education group demonstrated greater delay capacity than those in the low-education group, and at 2½, the children showed greater delay capacity than they did at 2 years of age. At 24 months of age, the delay scores on the two tasks were highly correlated ($r = .75$, $p < .01$), which indicates that there was a relatively high degree of consistency in this trait, at least on these tasks and at this age. Delay capacity also correlated with the Bayley at 2 years of age ($r = .43$, $p < .01$). While the ability to delay was an important aspect in these tasks, other factors may have been involved. To some extent there may have been an element of compliance, but this did not appear to be essential.

The *persistence* tasks used in the present study also involved compliance, but the children's performance on these tasks was not at all correlated with social class, IQ, or age. Language also played a role, to the extent that through verbal instructions the examiner was able to get the children to inhibit a response to a very desirable goal object and respond when a signal was given later. Luria (1961) expressed the view that when children's behavior can be controlled or regulated by language they have reached a higher level of mental functioning. But more was involved than the verbal regulation of behavior or simply the ability to wait. Many of the children obviously derived a great deal of pleasure from the waiting game itself, from postponing gratification, which in Freudian terms also reflects a higher level of ego functioning.

Finally, the children's performance on the *embedded-figures tasks* in the present study were not related to social class or IQ. The *mother–child interaction* data will be presented later in this paper, with similar research by other investigators.

In reviewing studies of social-class comparisons of cognitive development during the first few years of life, we have focused on recent research using newer measures, since SES differences have not been demonstrated on standard infant tests. It is important to keep in mind, however, that there are very few such studies reported in the literature. The research evidence indicates that, in general, behaviors that involve language or children's responses to language are related to both social class and later intelligence, whereas with only a few exceptions measures of nonverbal behavior that do not involve language relate neither to SES nor to later intelligence.

The sensorimotor measures that relate either to social class or to later intelligence can be summarized briefly. In Tulkin's study (1973), middle-class 10-month-old girls seemed more "reflective" than lower-class girls in making a choice between a familiar and a new toy. In the study by Wachs *et al.* (1971), starting as early as 7 months of age, middle-class infants were more advanced than lower-class babies on the "development of means scale," a Piagetian measure that assesses children's ability to use tools to obtain distant objects, and middle-class babies were also superior to lower-class infants in one "learning and foresight task" starting at 15 months. We do not know how these measures relate to later intelligence. Lewis (1969) found that a measure of response decrement at 12 months of age, which he believes reflects information-processing ability, was significantly correlated with IQ at $3\frac{1}{2}$ years of age, but there were no SES differences on this measure at 1 year of age (Lewis, 1969). At 3 months of age there was an inverse relationship between rate of response decrement in attention and social class, with lower-class babies showing superior "information-processing" skills than middle-class infants.

The most consistent findings is that cognitive measures that involve language are related to social class as early as the first year of life, and that starting at 2 years of age such verbal measures are highly correlated with children's performance on standard intelligence tests, such as the Bayley mental scale or the Stanford–Binet. The relationship between infants' early vocal behavior or their response to language and later intelligence should be studied. In these early precursors to language we may find the roots of social-class differences, as well as individual differences, in later intelligence. Such behaviors in early infancy seem to manifest themselves most clearly in the infants' interactions with other people, and especially their mothers. For this reason, social-class comparisons of mother–child interaction, which we will review in the next section of this chapter, seem a particularly fruitful area of research.

SOCIAL-CLASS COMPARISONS OF MOTHER–CHILD INTERACTION

Research concerning the relationship between an infant's growth and development and the behavior of its primary caretaker is as elusive as it is compelling. How can we best select among all the events, actions, and nonactions that help shape the life of a growing human being—the smiles, words, and caresses, the responding and nonresponding to signals of distress or pleasure? We will begin this review with some history and theory. Then we will describe the very meager but important area of research that attempts specifically to relate patterns of mother–infant interaction to SES and to the child's cognitive development during this period and later.

In 1951, John Bowlby published an influential monograph, *Maternal Care and Mental Health,* which attempted to document the devastating and apparently permanent effects on children of being separated from their mothers in infancy, a condition referred to as *maternal deprivation.* The research he reviewed was primarily gleaned from institutionalized infants, who showed not only serious emotional problems but serious intellectual impairment as well. Babies reared in such institutions, deprived of both sensory and social stimulation, had difficulty relating to people, obtained low IQ scores, and did poorly in school during childhood and adolescence.

Bowlby's work had a strong impact. Wherever possible, orphanages have been replaced by early adoption, foster-home placement, or reforms in institutional care. A further positive effect was to stimulate research on the effects of the child-rearing environment, and in particular the mother's role, on the child's development. While earlier studies focused on the mother's behavior, more recent research has been concerned with the mother–child interaction itself and how it relates to the child's intellectual and personality development.

Unfortunately, Bowlby and others have drawn certain questionable inferences from research on institutionalized infants. Until recently, it was generally assumed in this country, without any direct evidence, that group day care for infants would have the same negative effects on children as residential institutional care. On the basis of this assumption, group day care for infants was prohibited by law in most places in the United States until a few years ago. It was also assumed by many people, without any direct evidence, that infants from impoverished families experience maternal deprivation, though perhaps less severe, similar to the deprivation of institutionalized children. A further assumption was that impoverished infants, who were believed to experience such early social deprivation, would demonstrate the same permanent and irreversible intellectual impairment often found in institutionalized children, although there is very little evidence that poor infants show intellectual deficits or emotional problems in comparison with middle-class babies during *infancy.*

On the basis of such assumptions, intervention programs, which may not be necessary, have been recommended for infants from impoverished families. However, Yarrow (1961) has pointed out that it is not possible to make generalizations about institutional care or its effects on children, since institutions vary greatly in terms of the quality of care provided and not all children are affected in the same way. Therefore, it is even less valid to generalize from studies of institutionalized infants to the experience of poor home-reared babies or those in group day-care centers.

While there has been a great deal of speculation about the nature of social-class differences in the ways mothers interact with their infants, only recently have a few investigators begun to make direct naturalistic observations of the patterns of interaction between mothers and babies from different social classes.

Lewis and Wilson (1972) described differences in mothers and infants from five social classes (Hollingshead's two-factor index). Thirty-two 12-week-old infants and their mothers were observed for a 2-hour period, and behaviors of both were coded. Differences in infant behavior demonstrated that lower SES infants vocalized twice as much as middle SES infants, smiled twice as much, and fretted and cried about half as much as their middle-class peers. Lower SES mothers touched and held their infants more than middle-class mothers. There were no SES differences in amount of maternal vocalization, but there were differences in the occasions that elicited maternal vocalization. The middle-class mothers were much more likely to vocalize responsively when their infants vocalized. They also touched and held their babies when they cried and more often watched them while they played. Lower SES mothers, on the other hand, were more likely to touch their infants when they vocalized and to vocalize when they fretted. It is possible that the middle-class mothers' vocal response to their infants may be a precursor to later responding when the infants speak and to answering the children's questions. Children whose questions are answered not only learn more about the world from the answers but also are encouraged to ask further questions.

Lewis and Freedle (1973) studied the patterns of vocal interactions between 3-month-old infants and their mothers. The authors pointed out that the phonetic aspect of babbling in infants, in the degree of advancement shown by the child in this respect, does not seem to be related to later verbal ability. They hypothesized that perhaps the semantic or communicative aspects of infants' vocal interactions with their mothers may be a precursor or anlage of later language development. Infants who are more advanced in these early prelinguistic patterns of interpresonal communication may show greater language competence later. The sample consisted of 80 mother–child dyads, including male and female, black and white infants distributed across Hollingshead's five social classes. Naturalistic observations were made of the mother–child interaction at home. While there were no SES differences in how much the mothers spoke to their children, there were social-class differences in how much the infants vocalized. Lower-class infants vocalized more than middle-class babies. The social-class differences in the patterns of the infants' vocal responses to their mothers' vocalizations were of particular interest. Whereas the lower-class infants were more likely to respond vocally when their mothers were speaking, the middle-class babies were more likely to stop vocalizing and listen. Early social-class differences in the extent to which babies listen when their mothers are speaking may be related to later SES differences in children's ability to use verbal information for learning (Golden *et al.*, 1974).

Lewis and Freedle (1973) also compared children on whether they responded differently when their mothers were speaking to them and to another person. Middle-class infants vocalized more when their mothers were speaking to them, whereas lower-class babies did not make this distinction. Lewis and Freedle hypothesized that such differentiated vocal responding reflects more advanced communication skill at this age. There is some support for this assumption in a follow-up study of these children at 2 years (Freedle & Lewis, 1977). The linguistic competence was then assessed on various measures, including their spontaneous use of language in a play situation, their performance on a verbal comprehension task, and the Peabody Picture Vocabulary Test. In general, the communication skills of the dyads at 12 weeks were related to their language ability at 2 years of age. Linguistic competence among the children at 2 years of age and the degree to which they responded differently when their mothers spoke to them and to another person was related to their language ability at 2 years. These results tend to support the authors' assumption that the communication skills of dyads may be precursors of later child language ability.

Tulkin and Kagan (1972) reported a study of 30 white middle-class and 26 white working-class mothers with their 10-month-old daughters. The results of this study showed that in the working-class homes, the babies experienced more noise, greater crowding, more interaction with many adults, and more time in front of the TV. Lower-class girls had less opportunity to explore and manipulate objects, they were more restricted, and they had fewer toys than their middle-class peers.

As in previous studies, SES differences were not found in the amount of affection or discipline demonstrated. However, significant SES differences were present in the mothers' attempts to keep the babies "busy" and in their verbal behavior. Every verbal behavior coded was more frequent among the middle-class mothers than among the lower SES mothers. Tulkin suggested that it was not that the two SES groups in his study differed so greatly in this respect, but rather that within the middle-class group there was a small subgroup of unusually verbal mothers. In addition, the middle-class mothers also responded more often and faster to their babies' frets.

A study of 106 2- and 3-year-old white children and their mothers was completed in our laboratory (Rossman, Golden, Birns, Moss, & Montare, 1973) investigating the ways in which mothers of two different social classes read to their children. In addition to videotaping the mother–child interaction in the laboratory, we administered a questionnaire to the mother about her reading practices at home. The questionnaire data supported previous findings that middle-class mothers have more books in their homes and read more frequently to their children than do working-class mothers.

In the laboratory, the mothers were instructed to "go through the book as you would at home." The results revealed social-class differences in several dimensions of behavior. The middle-class mothers and their children were more affectionate and seemed to enjoy the reading situation more than the lower SES pairs. However, the SES differences in verbal behavior were most striking. The middle-class mother was more verbally explicit, used more complex language, and gave more spontaneous explanations. She also expected more from her child, asked more questions, and was more likely to respond to her child's questions. Furthermore, most of the middle-class mothers finished reading the story at least once before discussing it, whereas very few of the working-class mothers finished reading the story. In discussing the pictures, the middle-class mothers tended to relate the pictures to the stories, while the working-class mothers did not. The data on the children also demonstrated clear SES differences, the most pronounced being that the middle-class children were more verbally explicit than the working-class children. These SES differences in language patterns in the mother–child interaction in general support Bernstein's ideas, as well as the research findings of Hess and Shipman (1965) described earlier, and those of Bee Van Egeren, Streissguth, Nyman, and Leckie (1969).

Arising out of a concern about the effects of the home environment of poor infants on their intellectual development, there have recently been a number of studies on mother–child interaction among poor families. These studies have shown that there are great differences in the child-rearing environment, and in particular in patterns of mother–child interaction, among poor families, which are related to the children's cognitive development. In an intervention study with black males between 15 and 36 months of age, Schaefer (1970) found that hostility in the mother correlated with hostility in the child and that these measures of hostility in turn correlated with low task-orientation and low mental test scores in the children. In a study of 36 poor mothers and their first-born children between the ages of 9 and 18 months, Clarke-Stewart (1973) demonstrated that there was a strong relationship between the infants' intellectual, language, and social competence at 18 months and earlier maternal behaviors. Such behaviors included verbal stimulation and provision of toys, as well as affection and the appropriateness of the mother's response to the baby. Such studies indicate that there are great differences in the ways poor mothers interact with their infants, and therefore it is not possible to make generalizations about the maternal behavior of poor mothers with infants.

The most consistent finding in the few studies of social-class comparisons of mother–child interaction during the infant period is that there appear to be SES differences as early as the first year of life in the area of language, but clear-cut and consistent social-class differences have not been found in other aspects of behavior. We are not saying that the mother's nonverbal behavior during infancy is not important in terms of the child's development. Most studies of mother–infant interaction have not been concerned with social class. These studies show that babies whose mothers are attentive, warm, stimulating, responsive, and encouraging develop better than infants whose mothers are inattentive, cold, hostile, rejecting, restrictive, and less sensitive to their babies' needs (Blank, 1964; White, 1975; Yarrow, Rubenstein, Pedersen, & Jankowski, 1973).

On the basis of such studies, perhaps we can learn from "successful" mothers of every social class the specific maternal behaviors that foster cognitive, language, social, and emotional development in infants.

EARLY INTERVENTION STUDIES

On the basis of research on animals and institutionalized babies suggesting that early environmental deprivation during infancy may have permanent, irreversible, and adverse effects on later intellectual and problem-solving ability, some researchers (e.g., Hunt, 1969) have recommended that in order to prevent later intellectual deficits in lower-class children, intervention should begin in infancy. The "failure of Head Start" has been explained on the basis that compensatory programs beginning at 3 years of age may be too late and that preventive programs beginning in infancy may be more effective (Schaefer, 1970). In reviewing early intervention studies, we will focus on programs that include children under 3 years of age in order to see whether early intervention is better than later intervention, in both immediate and long-term effects on children's intellectual development. In particular we are interested in seeing whether intervention programs that begin in infancy (i.e., during the first two years of life) are more effective than programs that start later.

In 1967, the Parent and Child Center (PCC) programs were established as a downward extension of Head Start. The programs were designed to serve low-income children from birth to 3 years of age, and their families. Thirty-six federally funded (Office of Equal Opportunity), community-controlled PCCs were established in various parts of the United States. The PCCs were designed to provide comprehensive child development, health care, and social services to poor families with children under 3 years of age. Each PCC developed its own child development program. Approximately half the centers provided both center-based and home-visiting programs; 11 provided only center programs, in which the emphasis was on working directly with the infants; and 6 PCCs provided only home-visiting programs, which emphasized facilitating the babies' development through education of the parents.

While the programs were primarily service-oriented, there was a plan to evaluate the effects of the various types of programs on children and their families. Thus far a report on only the first year of operation has been published (Costello, 1970). There is a great deal of descriptive information about the programs and the population served, which included 2,585 children during the first year of operation. However, quantitative data concerning the impact of the program on the children's development are limited. Children in six centers were given the Bayley Scales of Infant Development twice over a 10-month period, with 79 children pre- and posttested. There was no control group. The experimental children's development during this period was evaluated by a comparison of their pre- and posttest scores, as well as a comparison of the performance of the PCC children with the national norms provided by Bayley's standardization sample. On the pretest, when the children were about 1 year old, they obtained a mean Bayley mental score of 77.4, well below the national average. On the posttest, after 10 months in the program, they obtained a mean of 87.7, a 10.3-point gain. On the motor scale, the children showed a 7.3-point increment, from 91.1 to 98.4. There was no clear indication that one type of child development program was more effective than any other. The greatest effect seemed to occur in PCCs with well-organized programs in which the goals and methods were clear to the staff and the outside observers. Several of the PCCs have been carrying out more extensive evaluations of their own programs. When the results of these studies are published, they will provide more information about the impact of these programs on the children's development.

One of the most successful early intervention studies was carried out by Heber, Garber, Harrington, and Falender (1972). They pointed out that approximately four-fifths of the mentally retarded population in the United States fall within the 50–75 IQ range, that they show no evidence of organic impairment, and that a disproportionately large percentage are poor and are members of a minority group. Heber referred to this type of retardation as sociocultural mental retardation, which he believes may be environmentally determined. Heber's aim was to see whether it is possible to prevent this kind of retardation through environmental intervention beginning in infancy.

Prior to undertaking the intervention program, he did an epidemiological study of the residential area of Milwaukee having the lowest median income in the city. This area contained only 2.5% of the city's population but one-third of its educable mentally retarded schoolchildren. All families living in this area with a newborn infant and another child of at least 6 years of age were included in the epidemiological study. IQ tests (the Peabody Picture Vocabulary Test) were given to the mothers, the fathers, and the older siblings. While all of these were poverty families, they seemed to fall into two categories in risk or probability of producing a retarded child. Families in which the mothers had IQ scores of less than 80, which included less than half the study sample, accounted for four-fifths of the children with IQs of less than 80. Furthermore, only the children whose mothers had IQ scores of less than 80 showed a decline in intelligence as they grew older, while impoverished children whose mothers had IQs above 80 did not show a decline at 6 years of age. Heber assumed that a mentally retarded mother living in a slum creates a very different social environment for her children than does a poor mother of normal intelligence living in the same neighborhood.

Heber selected 40 black "high-risk" families for his intervention study. The selection critera were that the mother had a newborn infant and that her full Wechsler Adult Intelligence Scale IQ score was under 80. The 40 families were randomly assigned to the experimental and the control groups, with 20 families in each. While the experimental (E) and control (C) families were not matched beforehand, they did not differ in socioeconomic or family status or number of children, nor did the infants differ in terms of height, weight, or birth complications. The program began when the children were 3 months old and continued until they were 6 years of age. There were two major aspects of the program: (1) family intervention and (2) infant intervention. The family intervention program consisted of vocational training for the mothers, as well as helping them to improve their reading, homemaking, and child-rearing skills. The infant intervention program involved placing the babies in a day-care center, which they attended all day, 5 days a week, 12 months a year, from the age of 3 months to 6 years of age. The intervention program was divided into two periods: (1) the infant period (up to 24 months) and (2) the early childhood period (from 24 to 72 months). During the infant period, the program stressed facilitating perception–motor and cognition–language development, with a one-to-one teacher–child ratio maintained until 12–15 months of age. During the early childhood period, the children participated in more structured, organized group classroom instruction, with emphasis on three aspects of learning: language, reading, and mathematics/problem-solving.

The data show that at 66 months, the E group had a mean IQ of 124 and the C group a mean of 94, a most impressive difference of 30 points!

This difference is comparable to that between lower- and middle-class children at this age. At what age do the two groups diverge? Heber reported that the Gesell scores at 6, 10, and 14 months were comparable for the two groups. At 18 months, the E group exceeded the C group by a few months, but the C group was still close to the norm. By 22 months, the E group exceeded the C group, which was still close to the norm, by 4 to 6 months, and the gap between the two groups steadily increased from this point on. Heber also compared the IQ scores of the children in the experimental group with the scores of older siblings who were not in the program and younger siblings who entered the program later as part of the E group. The comparisons were made of the average IQ scores the children obtained between 36 and 66 months of age. Whereas the scores of the older and younger siblings in the E group were comparable, the IQ scores of the oldest siblings who were not in the program were 39 points lower! The IQ scores of the control children were also higher than those of their siblings who were not tested as regularly, which indicates that repeated testing alone may increase children's performance on standard intelligence tests.

What impact did the program have on other aspects of the children's development? The nutrition and medical care provided to the experimental children were greatly superior to what was available to the control children. Consequently, the health and the physical development of the E children were expected to be superior to those of the C children, a difference that in itself may have been responsible for the difference in their intellectual development. Independent medical evaluations of the children revealed no significant differences between the two groups in height, weight, or blood analyses. After 2 years of age, the children were evaluated on a more comprehensive battery of measures, which included (1) standard intelligence tests, (2) performance on learning and problem-solving tasks, (3) various language measures, and (4) several measures of personality and social development, one of which involved an assessment of the mother–child interaction in a series of learning situations similar to those used by Hess and Shipman (1967). The E children showed marked superiority to the C children on most of these measures. A particularly interesting finding in the mother–child interaction situation was that whereas the mothers in the two groups did not differ in teaching or verbal ability, the children in the experimental group substantially increased the level of verbal communication and exchange of information with their mothers (e.g., by asking more questions), which in turn resulted in faster and more successful learning. Heber's intervention program achieved a great deal more than raising IQ scores. He apparently succeeded in increasing the intellectual and language competence of high-risk children in a variety of situations, so that their performance appears comparable to that of middle-class children of this age. The real test of the program, of course, is how well the experimental children succeed in school.

Heber considered the intervention during infancy to be essential. During the first 18 months, the difference in intellectual performance between the E and the C children was slight. However, by 24 months of age there was a 25-point mean IQ difference between the two groups, which increased to 30 points at 66 months of age. Clearly, the most visible impact of the program on the children's intellectual performance, as measured by standard tests, occurred between 18 and 24 months of age, a period of rapid language growth. On the basis of similar intervention studies with older children, continued intervention during the early childhood period (from 24 to 72 months) was probably essential if the experimental children in Heber's study were to maintain their gains. However, the design of Heber's study does not allow us to determine how early it was necessary to begin the infant intervention program in order to achieve these results.

A well-designed intervention study by Gordon (1973) was concerned with the age that the child first enters the program and the length of time in the program. Gordon's approach stressed parent education as a way of stimulating cognitive development in poor infants during the first 2 years of the child's life. The sample of mothers and children consisted of 258 poor families. The program had two phases: (1) the home-based parent education program (for children between 3 and 24 months of age), and (2) the home-learning center program (for children between 2 and 3 years of age). Gordon trained a group of women who were of the same SES and ethnic group as the study families to function as parent educators. During the first phase of the program, the parent educators made weekly visits to the homes of the study families and spent 1 hour demonstrating a carefully designed Piagetian "curriculum" to the mothers. It involved teaching the babies various sensorimotor skills, such as searching for hidden objects. The mothers were also encouraged to label objects and actions for their babies. During the second phase of the program, when they were between 2 and 3 years of age, the children spent 2 hours twice a week in a home learning center. The center was simply the home of one of the study mothers, where five or six children were involved in small-group instruction and activities under the guidance of a home learning center director who had previously served as parent educator. The mother in whose home the center operated served as an assistant to the director. The center director also made weekly visits to the homes of the children in her program to demonstrate to the mothers the instructional materials used at the center. Eleven such home learning centers were operating at the same time.

To determine the effects of age of entry and length of time in the program, Gordon used the following eight treatment groups: intervention (1) from 3 to 36 months, (2) from 3 to 24 months, (3) from 12 to 36 months, (4) from 3 to 12 months and 24 to 36 months (with no intervention between 12 and 24 months), (5) from 3 to 12 months, (6) from 12 to 24 months, (7) from 24 to 36 months, and (8) a control group that was not provided with any intervention at all. Gordon hypothesized that the intervention program would have the greatest impact on the children who received the most intervention and that, given the same amount of intervention, the children who entered the program earlier would benefit more than those who began later.

Up to 24 months of age the children were evaluated on the Griffiths scales and the sensorimotor tasks used in the training program. At 3 years of age, the Stanford–Binet, the Peabody Picture Vocabulary Test, and the Leiter international scale were administered to all of the children. The Binet scores were factor-analyzed and yielded three factors: language, memory, and perceptual-motor ability. The children's scores on each factor were analyzed separately. The results were fairly consistent across all measures: (1) the more training the children had, the better their performance, with the greatest difference being between children with two or three years and those with only 1 year or no training; and (2) given equivalent periods of training, the age at which the children entered the program did not significantly affect their performance on any of these measures. In Gordon's study, the critical factor did not seem to be when the children entered the program but how long they were in it.

Painter (1969) was concerned with the question of what the most strategic age is for educational intervention with socially disadvantaged children. The subjects in her intervention study were the younger siblings of 4-year-old children attending a compensatory nursery school. She selected children between 8 and 24 months of age and provided them with a highly structured home tutoring program that stressed language and concept development. Painter chose to work with children of this age to see whether it is possible to prevent the intellectual "deficit" in low SES children that seems to emerge between 15 months and 3 years of age. On the basis of early results of compensatory studies with older preschool children, programs with a highly structured curriculum that emphasized cognitive and language development seemed to be the most effective in increasing the level of the children's intellectual functioning. Painter wanted to see whether such a structured curriculum could be devised for infants between 8 and 24 months and to determine how effective such a curriculum would be with children of this age.

There were 20 children in the study, including male and female, black and white infants. Painter randomly assigned 10 of the children to the experimental group and 10 to the control group. The E babies were tutored in their homes by female college graduates for 1 hour a day 5 days a week for a period of 1 year. The mothers were not directly involved in the tutoring sessions. The children in both groups were tested before and after the intervention program on a number of intellectual measures. The pretest scores of the E and the C groups on the Cattell Infant Intelligence Scale were similar: 98.8 and 98.4, respectively. On the posttest on the Stanford–Binet, the experimental group obtained a mean of 108.1 and the control group a mean of 98.8, a significant 9.3 IQ difference ($p < .05$). On the basis of these results, Painter concluded that a home tutoring program with a highly structured curriculum that emphasizes language and concept development can be effective in increasing the intellectual performance of socially disadvantaged infants of this age.

Early and late intervention studies could be compared to see which is more effective. If comparative studies during the later preschool and elementary school period indicate that these children do better than their older siblings, it would demonstrate that early intervention was more effective than later intervention. On the other hand, if the older siblings do as well or better, it would suggest that infant training may not be necessary.

Karnes, Teska, Hodgins, and Badger (1970) carried out an intervention program to facilitate intellectual development in low SES infants by working *only* with their mothers. There was no direct intervention with the children. In this respect the study is unique. Karnes believed that if feelings of dignity and self-respect were fostered in the mothers and if they were taught ways to facilitate their children's intellectual development, the cognitive and language skills of the children might be enhanced. The program for the mothers began when the infants were between 12 and 24 months of age. Karnes worked with a group of 15 mothers, who attended 2-hour weekly group sessions for a period of about 15 months. The training sessions included both mother-oriented and child-oriented topics. The child-centered aspect of the training program included demonstrations of how the mothers could use educational play materials with their children to stimulate their intellectual and language growth. The importance of establishing a positive relationship between mother and child was also emphasized. During the parent-centered discussions, the mothers were encouraged to become politically active, which many of them did, to reduce the feelings of powerlessness so often expressed by the poor.

At the end of the training period, the mean IQ scores of the children in the experimental and a matched control group at about 3 years of age were 106 and 91 respectively, a significant 15-point difference. An even more striking finding was based on a comparison of the IQ scores of six children in the experimental group with the scores obtained by their older siblings prior to the time their mothers were enrolled in the program. Whereas the mean IQ of the six children in the E group was 127 at age 3, the mean score for their siblings was 89, an impressive, significant 38-point difference! This is comparable to the 39-point mean IQ difference Heber et al. (1972) found at about the same age between children in his experimental group and their untreated older siblings. While the number of subjects was small, the results of this comparison between siblings before and after their mothers were enrolled in the program demonstrate that the program had a strong impact in changing the ways in which the mothers interacted with their younger children. What is even more impressive is that these results were accomplished through work with the mothers, without direct intervention with the children.

Schaefer (1970) carried out a home tutoring program for low SES black males, starting at 15 months and continuing until the children were 3 years of age. Schaefer's rationale for selecting this particular age period for his intervention program stemmed from research evidence indicating that early sensorimotor development does not predict later intelligence and that social class differences emerge during the second and third years of life, a period of rapid language growth. Since intellectual performance in older children and adults is highly correlated with language ability, Schaefer's intervention program stressed the facilitation of language skills. At the same time, personality characteristics (e.g., cooperation, curiosity, and perseverence) that might help the child succeed in school were reinforced. An attempt was also made to enhance the children's self-esteem and feelings of competence.

The tutors were college graduates who were trained to work directly with the children to facilitate their language skills. The tutors visited each child's home five times a week for approximately 1 hour. The parents were encouraged, but not required, to participate. There were approximately 60 children in the study, about half in the experimental group and about half in the control group. The children were not randomly assigned to the two treatment groups. The E and C children were selected from different poverty neighborhoods in Washington, D.C. The neighborhoods were comparable on reading readiness scores at school entrance. The E and C families were also comparable in family income (under $5,000), the mother's education (under 12 years), and the mother's occupation (unskilled or semiskilled if she was employed).

Intelligence tests were administered at a center by experienced psychologists to both experimental and control infants at 14, 21, 27, and 36 months of age. The Bayley mental scale was administered up to 27 months, and the Stanford–Binet was given at 3 years. Whereas the two groups did not differ in IQ at 14 months, before the intervention program began (105 versus 108 for the E and C groups, respectively), there was a significant 17-point difference favoring the experimental children (106 versus 89) at 3 years of age, when the program terminated. At 36 months, the children in the E and the C groups also differed significantly on other intellectual and verbal measures (the Peabody Picture Vocabulary Test, and the Johns Hopkins Perceptual Test, a nonverbal measure of intelligence). Children in the two groups also differed in measures of task orientation (derived from the Bayley Infant Behavior Profile). A finding of particular interest in this study was that the ratings of maternal hostility were significantly correlated with the ratings of the child's hostility at 36 months, and both were significantly correlated with the child's task orientation and IQ at 3 years of age. In subsequent follow-up testing after the program had terminated (Bronfenbrenner, 1974), the differences between the E and the C groups progressively diminished, and by the end of the first grade their IQ scores did not differ significantly (101 versus 97), nor did they differ in their academic achievement. On the basis of the "disappointing" follow-up data,

Schaefer (1970) concluded that intervention should begin at birth, involving the mother in her infant's education as early as possible, and that the enrichment program should continue throughout the school period.

Levenstein's, 1977 mother–child home program is considered by the American Institute for Research in the Behavioral Sciences (Wargo, Campeau, & Tallmadge, 1971) to be one of the 10 most successful early intervention programs in the United States. The program has been successfully replicated throughout the country. Levenstein believes that one of the most important factors leading to intellectual and educational deficits in disadvantaged children stems from the patterns of verbal interaction between poorly educated mothers and their children during the early preschool years. She attempted to change the mother–child verbal interaction pattern from what Bernstein (1960) called a "restricted code" to an "elaborated code," respectively, the dominant linguistic patterns of lower- and middle-class people. Levenstein also strongly emphasized the importance of fostering a positive emotional relationship between mothers and their children. The program was designed to facilitate children's intellectual, verbal, and psychosocial development. Levenstein works directly with the mother-child dyad in changing their patterns of verbal and social interaction.

While the program has evolved since its inception in 1967, the basic program consists of 46 semiweekly visits by toy demonstrators in each of 2 school calendar years, starting when the children are 2 years of age and continuing until they are 4. Levenstein considers this to be an optimal age for such a program because of the rapid growth of language during this time and because children are still quite emotionally involved with their mothers. The experience of interacting in a positive way with their mothers in a learning situation and on a one-to-one basis prepares children to learn in a group later. The toy demonstrators show the mothers how to stimulate their children's cognitive and language growth through the use of carefully selected, age-appropriate play materials and books (referred to as "verbal interaction stimulus materials" or VISM). The toy demonstrators, who follow detailed curriculum guidelines, encourage the mothers or other family members to interact with the children during the training sessions as early and as much as possible. Once the mother comfortably interacts with the child, the toy demonstrator fades into the background. The VISM are permanently left in the home, so that they can be used between visits. Initially the toy demonstrators were female social-work school graduates with master's degrees. Currently, many of them are paid former mother-participants with a high school education or less. The low SES toy demonstrators seem to be as effective as the more highly educated women, as reflected in the fact that the children whose mothers were trained by the demonstrators show comparable intellectual gains. The mother–child home program has served almost 400 low-income children and their families, most of whom are black.

Longitudinal data have been reported (Madden, Levenstein, & Levenstein, 1976) on 44 children in two treatment groups (T-68 and T-69). These children were in the program for 2 full years, between 2 and 4 years of age, and were in the first grade at the time of this follow-up. Of the children (T-68) who entered the program in 1968, 21 showed the following pattern of intellectual gains: (1) at 2 years of age, before they started the program, they obtained a mean Cattell infant intelligence score of 90.4; (2) by the end of the second year of the program, they obtained a mean Binet score of 108.9, a significant gain of 17.4 points; (3) in the first grade, they obtained a mean IQ of 105.4, essentially maintaining their intellectual gains 3 years after they had left the program. The other treatment group (T-69), consisting of 23 children who entered the program in 1969, showed a very similar pattern: (1) at 2 years of age they obtained a mean IQ of 88.8; (2) by the end of the second year of the program they obtained a mean IQ of 108.2, a significant gain of 19.4 points; and (3) in the first grade, they obtained an average IQ of 105.8. By contrast, 25 control children obtained a mean IQ of about 91 when they were first tested between 2 and 4 years of age and an average IQ of approximately 96 in the first grade, showing very little change in their intellectual performance. Whereas the children in the treatment and the control groups were functioning at about the same intellectual level during the early preschool period, by the first grade there was about a 15-point IQ difference between them. Of particular importance is the fact that the children in Levenstein's intervention program maintained their intellectual gains 3 years after they left the program, which is in sharp contrast with the usual decrement in intellectual performance of children in other preschool programs after they have left the program.

The mother-child home program has been replicated in more than 21 other agencies in the United States, with comparable though somewhat smaller intellectual gains. Levenstein has attributed the success of her program to the fact that she worked directly with the mother–child dyad, changing their patterns of verbal and social interaction at a critical time in the child's intellectual development. These changes in the patterns of mother–child interaction may be permanent and may continue to have a favorable effect on the children's intellectual development long after they have left the program.

One of the first model infant day-care centers in the United States was Caldwell's Children's Center at Syracuse University. Children are enrolled in the center as early as 6 months of age and remain until they enter school. They spend as much as 9 hours a day, 5 days a week at the center. The program strongly emphasizes cognitive and language development. Teacher training is intensive and continuous. The teachers are encouraged to verbalize frequently to the children, to label objects and actions, and to read to babies as early as the first year of life. Caldwell (1970) compared the intellectual gains of children who entered the program before and after 3 years of age. A relatively large group of children ($n = 86$) who entered the program prior to 3 years of age showed an average IQ gain of 14 points after approximately 2 years in the program. Most of these children started between 12 and 24 months of age. Another group of children ($n = 22$) who entered the program at about $3\frac{1}{2}$ years of age showed an average IQ gain of 18 points after approximately $1\frac{1}{2}$ years in the program. The two groups did not differ in their intellectual performance during the later preschool period, although the children with infant day-care experience had begun several years earlier and had been in the program more than twice as long as those who entered the program after 3 years of age. Caldwell's research demonstrates that children can benefit from a group day-care experience during the first 3 years of life, but in later intellectual development, children who begin this experience early do not have an advantage over children who start much later.

An intervention study by Palmer (1972) at the Harlem Research Center provides data on the question of whether early intervention is better than later intervention. On the basis of animal research demonstrating the importance of early experience to development, Palmer assumed that intervention designed to change intellectual, affective, or social development in children should begin as early in life as possible. In order to determine the effects of age on the success of intervention, Palmer provided children with 8 months of intervention beginning at 2 and 3 years of age.

The sample consisted of 310 black males: 240 of the children served as experimental subjects and 70 as controls; 120 of the experimental children were randomly assigned to the 2-year intervention sample and 120 to the 3-year sample. The experimental subjects at each age were randomly further subdivided into two treatment groups: (1) concept training and (2) discovery. Children in the concept-training group were taught age-appropriate concepts, such as big/little, open/closed, and same/different. The children were provided with individual instruction for 2 1-hour sessions a week over a period of 8 months. The discovery group was not taught these concepts but spent the same amount of time with instructors who played with the children with the same materials.

The children were assessed on a battery of tests, including the concept familiarity index (CFI), which tested the child's knowledge of the concepts taught in the training sessions, and the Stanford–Binet, as well as a number of other measures. All the experimental children were evaluated immediately at the end of training and 1 year after treatment. The children in the 2-year sample were also evaluated 2 years later. The children in the control group were assessed on the same measures at the same ages. Immediately after training, the two experimental groups performed significantly better on the CFI than the controls, and the concept-training group outperformed the discovery group. This was true for both the 2- and the 3-year samples. In the 1-year follow-up for the 2-year sample, the concept-training group did not perform significantly better on the CFI than the controls, whereas, surprisingly, the discovery group did. Two years after the training ended, neither of the experimental groups performed significantly better on the CFI than the controls. Thus, while concept training had a short-term effect on the 2-year sample, 1 or 2 years later the effects of such training at 2 years of age seemed to wash out. On the other hand, in the 1-year follow-up for the 3-year sample, the results obtained immediately after training were sustained a year later. The long-term effects of the intervention program were greater for the children who started later.

How did the intervention program generalize to other measures, such as the Stanford–Binet? Immediately after training, the experimental children in the 2-year sample obtained significantly higher Binet scores than the controls (96 versus 93), although the difference was quite small. A year after treatment, the IQ scores of the experimental and the control groups did not differ significantly (94 versus 92). Immediately after training, the experimental children in the 3-year sample performed significantly better on the Binet than did the controls (99 versus 93). A year later, the concept-training group maintained a significant superiority over the controls on the Binet (100 versus 92), whereas the discovery group did not (97 versus 92).

On the basis of the research reviewed here—with the exception of Heber *et al.* (1972), which differs from the others in many important respects—it would be difficult to conclude that, in general, early intervention is better than later intervention or that programs that begin in infancy are more effective than those that start later. With the exception of Heber's program, which we will discuss later, the short-term immediate effects of intervention on children's intellectual performance ranged in gain approximately 10–19 IQ points. There appears to be no relationship between the amount of intellectual gain children showed at the end of a program and the age of the children when they entered it. On the long-term effects of intervention on children's later intellectual development after they have left a program, we have data on only a few studies that involved children under 3 years of age. The only study in which children maintained their intellectual gains several years after they had left the program is Levenstein's mother–child home program (Madden *et al.*, 1976). However, these results may have been due more to the nature of her program than to the age of the children.

We also have follow-up data on Schaefer's (1970) home tutoring program, which was comparable to Levenstein's in many important respects. Schaefer's program began earlier than Levenstein's but continued for approximately the same length of time. In Schaefer's program, the children experienced more than twice the number of contacts as did those in Levenstein's program. Whereas the immediate gains in both programs were similar, the children in Levenstein's program maintained their gains in relation to the controls several years after they had left the program, while those in Schaefer's program did not. The major difference between the two programs is that Schaefer's tutors worked primarily with the children, whereas Levenstein's toy demonstrators worked with the mother–child dyad. Levenstein's results support Schaefer's conclusion that intervention programs involving mothers directly in the education of their children are likely to have more lasting effects than those that involve only the children. Levenstein assumes that the period from 2 to 4 years of age is the optimal time for intervention. She may be right, but her study does not permit us to draw this conclusion because she did not provide the same 2-year program to children who started at different ages. This brings us to a discussion of a serious problem in the comparison of intervention programs that are directed at children of different ages: It is very difficult to determine the effect of age when the programs differ in so many other respects.

The program of Heber *et al.* (1972), for example, differs from the others reviewed here in so many ways that it is difficult to make comparisons. Undoubtedly, this is one of the most effective early intervention programs reported in the literature for high-risk, very low SES children. It is also probably one of the most intensive, most extensive, and perhaps most costly (per capita) programs of its kind, which may make it impractical on a large scale. But Heber's intervention study may also differ from the others in its purpose. Heber wanted to determine whether sociocultural familial retardation (which constitutes four-fifths of the retarded population of the United States and is prevalent among impoverished minority groups in this country) was genetically or socially transmitted from one generation to the next. He wanted to see whether it was possible to prevent this kind of retardation through an educational program for children from high-risk families beginning in early infancy and continuing until 7 years of age. Heber succeeded admirably. He worked with children whose mothers and older siblings had very low IQ scores (below 80), and by 2 years of age these children were functioning on a middle-class level intellectually. At 24 months, they obtained a mean IQ of approximately 124, 25 points higher than the score of the control group and 39 points higher than the scores of older siblings who were not in the program.

What is particularly impressive about Heber's study is that the intellectual differences between the E and the C children were manifested not only in large IQ differences but in similar differences in functioning on a wide variety of measures and in a wide variety of situations. The ultimate test of the impact of the program, according to Heber, is how the children do in school after they have left the program. Of course, their performance may also depend on the school they attend.

While Heber's intervention study answers a number of important questions, it does not permit us to conclude that in order to achieve his results it was necessary to begin as early as he did, when the children were 3 months of age. During the first 18 months, the children in the experimental and the control groups differed very little in their intellectual performance. By 24 months of age there was a 25-point IQ difference between them, which increased relatively little after this age. These data suggest that the 18-24 months period is critical for social-class differentiation in intellectual development, perhaps because of the rapid growth of language during this period.

In contrast to most other intervention programs, Heber's succeeded in raising the intellectual level of high-risk, low-income children to that of middle-class children, almost 20 points higher than the scores obtained by the experimental children in Levenstein's highly successful program. Could Heber have achieved these results if he had started when the children were 12 or 18 months of age or even older? Engelmann (1970) reported the results of a 2-year intervention study on low SES black children, beginning when the children were 4 years old and using the Bereiter–Engelmann method. This approach stresses a highly structured, academically oriented language, reading, and mathematics program, in which small groups of children are drilled in these skills. Whereas at the start of the program the children's mean IQ was 95, at the end of 2 years in the program it increased to 121, a significant 26-point gain. The most dramatic effect of the program was shown by two children who started the program with IQs in the 80s and by the end of kindergarten were reading at the third-grade level! This result suggests that the program may have succeeded in doing more than increase the children's IQ scores: It seems also to have had a strong impact on their academic achievement. This intervention program, which began when the children were 4 years of age, produced intellectual gains as large as any reported in the literature, including those in Heber's study. This indicates that it may not be necessary to begin in infancy in order to achieve such results.

The intervention programs we have reviewed vary in a number of important dimensions, including the age of the child when intervention begins. They differ in such respects as (1) whether the program was carried out in a center or in the child's home; (2) whether it was directed only to the child, only to the mother, or to the mother–child dyad; (3) the frequency and intensity of contacts, ranging from 1 or 2 hours a week to all-day, 5-day-a-week programs; (4) the duration of the programs, which ranged from 1 to 6 years; (5) the content and methods employed; and (6) the population studied. All of these variables may interact with age and confound any interpretations that may be drawn of the specific effects on their subsequent development of the age of the children when intervention begins.

If a comparison is made between children in the same program who begin at different ages but who are in the program for the same period of time, the effect of other contaminating variables can be eliminated. In the few studies reviewed here that employed this experimental design, earlier intervention was not more effective than later intervention (Caldwell, 1970; Gordon, 1973; Palmer, 1972). In comparing programs, we have primarily looked at their effects on the children's performance on standard intelligence tests, which may not always be the most appropriate or most sensitive measure of the effects of a program. While there may be little difference in IQ scores between early and later intervention programs, it is possible that programs that begin earlier in life may have more pervasive and enduring effects on aspects of the children's personality or cognitive style that may be important for learning and academic success. Unfortunately, standard measures of personality and social and emotional functioning, which could be used to compare the effects of different intervention programs on these important aspects of development, are not available at present.

Standard intelligence tests are useful for comparative purposes. However, such tests can be construed only as intermediate criterion measures; they are useful only because they correlate with academic achievement, which must be viewed as the ultimate criterion of the success of an early intervention program. Intervention programs that succeed in increasing children's scores on standard tests but do not have a measurable impact on their school achievement may be doing little more than improving children's skills in performing on such tests. While early intervention programs must be judged by how well they prepare children to succeed in school, it must be remembered that how well children do in school depends not only on the quality of their preschool experience but also on the quality of the education provided to them once they enter school.

CONCLUSIONS

On the basis of the data available to us, we would have to conclude that, in general, social-class differences in infant or sensorimotor intelligence probably do not exist. This conclusion is based both on the results of earlier SES comparisons using standard infant tests and on more recent research in which newer cognitive measures were employed. While there may be SES differences on a few specific aspects of sensorimotor behavior, the relationship between these behaviors and later intelligence has not yet been demonstrated. Clear-cut, consistent, pervasive social-class differences in intellectual performance on a variety of measures emerge somewhere between 18 and 24 months of age. Since SES differences in cognitive development first manifest themselves during a period of rapid language growth, it is reasonable to assume that these differences may be due to language.

There may be a discontinuity between the sensorimotor and the verbal periods both in children's intellectual competence and in the environmental conditions that facilitate cognitive development on these two qualitatively different levels of intelligence. This discontinuity could explain the absence of a correlation between measures of sensorimotor and of verbal intelligence. While sensorimotor intelligence may be the foundation for later intelligence (as Piaget believes), there is no reason to assume that the rate of cognitive development or intellectual competence of normal children should be the same on the sensorimotor and the verbal levels.

Environmental conditions that facilitate intellectual development during the sensorimotor and the verbal periods may also differ. While children can understand and use language to a limited extent during the sensorimotor period, their knowledge about the world is acquired primarily through their own direct explorations of their immediate environment. Given an average expectable environment with an opportunity to explore and manipulate objects and a sufficient amount of attention and affection by care-giving adults, children reared under a variety of social conditions can acquire on their own the kinds of perception–motor skills and knowledge measured by infant tests and Piagetian scales. On the sensorimotor level, the child's construction of reality, in the basic dimensions described by Piaget (such as object permanence and spatial, causal, and temporal relations), does not appear to be socially transmitted but seems to be largely acquired through the child's own direct experience, and therefore, such knowledge may be universal. After 18 months of age, as children become increasingly capable of learning about the world through language, social class and culture begin to have a much greater impact in shaping children's cognitive development. That is, social class and cultural diversity in cognitive development may largely by mediated by language.

While social-class differences do not seem to be present in sensorimotor intelligence, there is some evidence that SES differences in the presymbolic vocal interactions between infants and their mothers may be present as early as the first year of life (Lewis & Freedle, 1973; Tulkin, 1973; tulkin & Kagan, 1972). These differences in vocal interaction may be the early precursors of later social-class differences in language and cognitive development. SES differences in language experience may begin very early in life without having much impact on children's intellectual development until later. Through such vocal interactions, middle-class children, more than lower-class children, may be primed to be responsive and to pay attention and listen when their parents speak to them. In this respect middle-class children may be better prepared to acquire knowledge through language, once they become capable of doing so, first from their parents and later in school from teachers.

In order to narrow the intellectual gap between children from different social classes, which emerges between 18 and 24 months of age, how early must we begin intervention? Heber *et al.* (1972) have demonstrated that through an intensive intervention program beginning at 3 months of age, it is possible to raise the intellectual performance of very low SES infants at high risk of retardation (those whose mothers and older siblings obtain IQ scores below 80) to the level of middle-class children by 2 years of age. In order to achieve such results, how early must we begin? The design of Heber's study does not permit us to answer this question. Engelmann (1970) reported similar results for low SES children in a 2-year intervention program starting at 4 years of age. One of the most effective and practical intervention programs reported in the literature, Levenstein's mother–child home program, is directed at children between 2 and 4 years of age. In the few intervention studies that had children in the same program beginning at different ages, there is no evidence that earlier intervention is more effective than later intervention (Caldwell, 1970; Gordon, 1973; Palmer, 1972). While infant experience may be extremely important, the impact of later experience on cognitive development may be underestimated. Studies reported by Skeels (1966), Dennis (1973), and Kagan and Klein (1973) indicate that even gross, environmentally produced intellectual retardation is not necessarily irreversible. In these studies, children showed significant intellectual retardation and apathy during infancy, while they were in a very restricted, unstimulating environment. However, when they were shifted to a more stimulating environment later, they seemed to catch up intellectually and appeared to function normally in other respects. While severe environmental deprivation in infancy may result in permanent and irreversible deficits in learning and problem-solving ability in animals, this may not be the case for humans. Compared with other animals, human beings may be much more plastic, may be more responsive to later environmental changes, and may retain greater flexibility and capability of changing their behavior throughout life.

We mention these studies only to show that even gross, environmentally produced intellectual retardation in infancy can be reversed later. Lower-class children in our society do not differ greatly from middle-class children in cognitive and personality development in infancy, nor is there evidence that the social environment of lower-class children is less favorable in fostering these aspects of development during the sensormitor period. While there may be SES differences in early language experience, it is possible that such differences can be made up at a later age. Although it may be possible to accelerate aspects of sensorimotor development through early environmental intervention, there would appear to be little point in doing so since the rate of sensorimotor development does not seem to be related to later intelligence for normal children. For these reasons, infant intervention for normal, home-reared children would seem to be unnecessary.

What must be done to improve the academic skills of lower-class children in our society? Whether one perceives social-class differences as deficits in lower-class children and their families, judged by middle-class standards, or as cultural variations or differences may make a great difference in educational goals and methods.

From a deficit point of view, one of the implicit goals of early childhood education seems to be to make lower-class children as much like middle-class children as possible, not only in intellectual and language competence but also in certain important aspects of their character makeup and personality. Such a goal may be questionable from several standpoints.

First, is it possible to achieve such a goal through education alone, even if the child's parents are included in the educational process? Bernstein (1960) believes that social-class differences in linguistic and cognitive style stem from and reflect fundamental differences in life-style and outlook. If this is true, is it possible to produce basic changes in the linguistic and cognitive style of poor children and their parents only through education, without materially changing their lives economically, socially, and politically?

And second, even if this were possible, would it be completely desirable? Early intervention programs for low SES children, to the extent that such programs are successful, may profoundly change the ways in which the children experience life, a change that in some ways may be undesirable. An essential aim in the rearing of middle-class children seems to be to develop distance from their immediate experience. This may be achieved in several ways. More of their experience, feelings, and impulses may be filtered through language and expressed verbally. Middle-class parents often place greater value on abstract than practical knowledge, although there may also be ethnic differences in this respect. Middle-class children are trained from a very early age to postpone gratification and to develop an orientation toward the future. All of these aspects of middle-class character development have adaptive and positive features, but there may be some negative side effects. For example, to the extent that experience and feelings are filtered through language, their intensity may be diminished and diluted. To the extent that a person places greater value on abstract knowledge, he may be highly competent in academic situations but incompetent in dealing with practical problems. To the extent that an individual becomes future-oriented, which may be society's way of keeping our noses to the grindstone, he may find it difficult to live in the present and to enjoy each moment.

Implicit in the deficit point of view is an assumption that lower-class children and their families are inferior in certain respects. While middle-class educators may attempt to cloak these beliefs in scientific terminology and to conceal them, perhaps even from themselves, poor children and their families may sense and resent these negative attitudes toward them. In their resentment, children may resist all efforts to "educate" them, and at the same time they may fulfill the teachers' underlying expectations of failure. This may be at least one of the reasons why our educational efforts with poor children in this country have on the whole been so spectacularly unsuccessful, despite the expenditure of large sums of money.

If educators could develop the belief—and this may require a great deal of self-examination—that poor children are not inferior to middle-class children but merely different as a result of very different life experiences, the educational process might take on a very different quality. It would require a recognition on the part of the middle-class educator of poor children that they have a common task—the child's education—and a mutual problem. There is a vast cultural and experiential gap between middle-class teachers and poor children, which becomes even greater if they also differ racially and ethnically. They hardly speak a common language. This gap must be narrowed if anything of educational value is to occur between them. From this point of view, the child would not have to do all of the accommodating, to use Piaget's term. Both middle-class educators and lower-class children have to accommodate to each other, to develop mutual respect, and to find a common meeting ground on which to carry out their joint educational task.

Under the best of circumstances, education may have a modest impact on the problem of poverty, but it is not a panacea, as some proponents of compensatory education originally had hoped. Poverty is basically an economic problem, which probably can be eliminated only through the political process.

REFERENCES

AUSUBEL, D. P. How reversible are the cognitive and motivational effects of cultural deprivation? *Urban Education,* Summer 1964, 16–42.

BAYLEY, N. Comparisons of mental and motor test scores for ages 1–15 months by sex, birth order, race, geographic location, and education of parents. *Child Development,* 1965, *36,* 379–384.

BEE, H. L., VAN EGEREN, L. F., STREISSGUTH, A. P., NYMAN, B. A., & LECKIE, M. S. Social class differences in maternal teaching strategies and speech patterns. *Developmental Psychology,* 1969, *1,* 726–734.

BERNSTEIN, B. Language and social class. *British Journal of Sociology,* 1960, *11,* 271–279.

BINET, A. *Les idées modernes sur les enfants.* Paris: Ernest Flammarion, 1909.

BIRNS, B., & GOLDEN, M. Prediction of intellectual performance at three years on the basis of infant tests and personality measures. *Merrill-Palmer Quarterly,* 1972, *18,* 53–60.

BIRNS, B., & GOLDEN, M. *The interaction of social class and sex on intelligence, language, personality, and the mother-child relationship.* Paper presented at the biennial meeting of the Society for Research in Child Development, Philadelphia, 1973.

BLANK, M. Some maternal influences on infants' rates of sensorimotor development. *Journal of Child Psychiatry,* 1964, *3,* 668–774.

BOWLBY, J. *Maternal care and mental health.* Geneva: World Health Organization, 1951.

BRONFENBRENNER, U. *A report on longitudinal evaluations of preschool programs* (Vol. 2): *Is early intervention effective?* Department of Health, Education, and Welfare publications, 1974, No. (OHD), 74–25.

CALDWELL, B. M. The rationale for early intervention. *Exceptional Children,* 1970, *36,* 717–726.

CLARKE-STEWART, K. A. Interactions between mothers and their young children: Characteristics and consequences. *Monographs of the Society for Research in Child Development,* 1973, *38*(Serial No. 153).

CORMAN, H., & ESCALONA, S. K. Stages of sensorimotor development: A replication study. *Merrill-Palmer Quarterly,* 1969, *15,* 351–363.

COSTELLO, J. Review and summary of *A national survey of the parent-child center program.* Prepared for the Office of Child Development, U.S. Department of Health, Education, and Welfare, August, 1970.

DENENBERG, V. H. (Ed.). *Education of the infant and young child.* New York: Academic Press, 1970.

DENNIS, W. *Children of the creche.* New York: Appleton-Century-Crofts, 1973.

DEUTSCH, M. Facilitating development in the preschool child: Social and psychological perspectives. *Merrill-Palmer Quarterly,* 1964, *10,* 249–260.

ENGELMANN, S. The effectiveness of direct instruction on IQ performance and achievement in reading and arithmetic. In J. Hellmuth (Ed.), *Disadvantaged child* (Vol. 3). New York: Brunner/Mazel, 1970.

FOWLER, W. Cognitive learning in infancy and early childhood. *Psychological Bulletin,* 1962, *59,* 116.

FREEDLE, R., & LEWIS, M. Prelinguistic conversations. In M. Lewis & L. Rosenblum (Eds.), *Interaction, conversation, and the development of language: The origins of behavior* (Vol. 5). New York: Wiley, 1977.

GEISMAR, L. L., & LA SORTE, M. A. *Understanding the multi-problem family.* New York: Association Press, 1964.

GODDARD, H. H. *Human efficiency and levels of intelligence.* Princeton, N.J.: Princeton University Press, 1920.

GOLDEN, M., & BIRNS, B. Social class and cognitive development in infancy. *Merrill-Palmer Quarterly,* 1968, *14,* 139.

GOLDEN, M., BIRNS, B., BRIDGER, W. H., & MOSS, A. Social class differentiation in cognitive development among black preschool children. *Child Development,* 1971, *42,* 37.

GOLDEN, M., BRIDGER, W. H., & MONTARE, A. Social class differences in the ability of young children to use verbal information to facilitate learning. *American Journal of Orthopsychiatry,* 1974, *44,* 86.

GOLDEN, M., BIRNS, B., MONTARE, A., ROSSMAN, E., & MOSS, A. *Symposium on social class and cognitive development.* Research presented at the Biennial Meeting of the Society for Research in Child Development, Philadelphia, March, 1973.

GORDON, I. J. A home learning center approach to early stimulation. In J. L. Frost (Ed.), *Revisiting early childhood education.* New York: Holt, Rinehart & Winston, 1973.

HEBER, R., GARBER, H., HARRINGTON, C. H., & FALENDER, C. *Rehabilitation of families at risk for mental retardation.* Progress report, 1972.

HESS, R. D., & SHIPMAN, V. Early experience and the socialization of cognitive modes in children. *Child Development,* 1965, *36,* 869.

HESS, R. D., & SHIPMAN, V. C. Cognitive elements in maternal behavior. In J. P. Hill (Ed.), *Minnesota symposia on child psychology* (Vol. 1). Minneapolis, Minn.: University of Minnesota Press, 1967.

HINDLEY, C. B. Social class influences on the development of ability in the first five years. In *Child and education: Proceedings of the XIV International Congress of Applied Psychology* (Vol. 3). Copenhagen: Munksgaard, 1962.

HOLLINGSHEAD, A. B. *Two-factor index of social position.* New Haven, Conn.: Author, 1957.

HUNT, J. McV. *Intelligence and experience.* New York: Ronald Press, 1961.

HUNT, J. McV. *The challenge of incompetence and poverty.* Urbana: University of Illinois Press, 1969.

JENSEN, A. R. How much can we boost IQ and scholastic achievement? *Harvard Educational Review,* 1969, *39,* 1.

KAGAN, J. *Change and continuity in infancy.* New York: Wiley, 1971.

KAGAN, J., & KLEIN, R. E. Cross-cultural perspectives on early development. *American Psychologist,* 1973, *28,* 947.

KAMIN, L. J. *The science and politics of IQ.* New York: Wiley, 1974.

KARNES, M. B., TESKA, J. A., HODGINS, A. S., & BADGER, E. D. Educational intervention at home by mothers of disadvantaged infants. *Child Development,* 1970, *41,* 925.

KATZ, P. A. Effects of labels on children's perception and discrimination learning. *Journal of Experimental Psychology*, 1963, 66, 423.

KOLTSOVA, M. M. *The formation of higher nervous activity of the child.* Berlin: Veb Verland, Volk and Gesundheit, 1960.

LEACOCK, E. B. (Ed.). *The culture of poverty: A critique.* New York: Simon & Schuster, 1971.

LEVENSTEIN, P. The mother-child program. In M. C. Day & R. K. Parker (Ed.), *The preschool in action* (2nd ed.). Boston: Allyn & Bacon, 1977.

LEWIS, M. A developmental study of information processing within the first three years of life: Response decrement to a redundant signal. *Monographs of the Society for Research in Child Development*, 1969, 34(Serial No. 133).

LEWIS, M., & FREEDLE, R. The mother-infant dyad. In P. Pliner, L. Kranes, & T. Alloway (Eds.), *Communication and affect: Language and thought.* New York: Academic Press, 1973.

LEWIS, M., & McGURK, H. Infant intelligence scores . . . True or false? In S. Chess & A. Thomas (Eds.), *Annual progress in child psychiatry and child development.* New York: Brunner/Mazel, 1973.

LEWIS, M., & WILSON, C. D. Infant development in lower-class American families. *Human Development*, 1972, 15, 112.

LEWIS, O. The culture of poverty. *Scientific American*, 1966, 215, 19.

LURIA, A. R. *The role of speech in the regulation of normal and abnormal behavior.* New York: Liveright, 1961.

MADDEN, J., LEVENSTEIN, P., & LEVENSTEIN, S. Longitudinal IQ outcomes of the mother-child home program, 1967–1973. *Child Development*, 1926, 47, 1015–1025.

MALONE, C. A. Some observations on children of disorganized families and problems of acting out. *Journal of the American Academy of Child Psychiatry*, 1963, 2, 22.

McCALL, R. B., HOGARTY, P. S., & HURLBURT, N. Transistors in infant sensorimotor development and the prediction of childhood IQ. *American Psychologist*, 1972, 27, 728.

MESSER, S. G., & LEWIS, M. Social class and sex differences in the attachment and play behavior of the year old infant. *Merrill-Palmer Quarterly*, 1972, 18, 295.

MILLER, S. M. The American lower classes: A typological approach. In F. Riessman, J. Cohen, & A. Pearl (Eds.), *Mental health of the poor.* New York: Free Press, 1964.

PAINTER, G. The effect of a structured tutorial program on the cognitive and language development of culturally disadvantaged infants. *Merrill-Palmer Quarterly*, 1969, 15, 279.

PALMER, F. H. Minimal intervention at age two and three and subsequent intellective changes. In R. K. Parker (Ed.), *The preschool in action.* Boston: Allyn & Bacon, 1972.

PASAMANICK, B., & KNOBLOCH, H. Epidemiological studies on the complications of pregnancy and the birth process. In G. Caplan (Ed.), *Prevention of mental disorders in children.* New York: Basic Books, 1961.

PAVENSTEDT, E. A comparison of the child-rearing environment of the upper-lower and very low-lower class families. *American Journal of Orthopsychiatry*, 1965, 35, 89.

PIAGET, J. [*The origins of intelligence in children.*] (Margaret Cook, trans.) New York: International Universities Press, 1952.

ROSSMAN, E., GOLDEN, M., BIRNS, B., MOSS, A., & MONTARE, A. *Mother-child interaction, IQ, and social class.* Paper presented at the biennial meeting of the Society for Research in Child Development, Philadelphia, March 1973.

SCHAEFER, E. S. Need for early and continuing education. In V. H. Denenberg (Ed.), *Education of the infant and young child.* New York: Academic Press, 1970.

SKEELS, H. M. Adult status of children with contrasting early life experiences. *Monographs of the Society for Research in Child Development*, 1966, 31(Serial No. 105).

TERMAN, L. M. *The measurement of intelligence.* Boston: Houghton-Mifflin, 1916.

Terman, L. M., & Merrill, M. A. *Measuring intelligence.* Boston: Houghton-Mifflin, 1937.

Tulkin, S. R. An analysis of the concept of cultural deprivation. *Developmental Psychology,* 1972, *6,* 326.

Tulkin, S. R. Social class differences in infants' reactions to mothers' and stranger's voices. *Developmental Psychology,* 1973, *8,* 137.

Tulkin, S. R., & Kagan, J. Mother-child interaction in the first year of life. *Child Development,* 1972, *43,* 31.

Užgiris, I. C., & Hunt, J. McV. *An instrument for assessing infant psychological development.* Unpublished manuscript, University of Illinois, 1966.

Wachs, T. D., Užgiris, I. C., & Hunt, J. McV. Cognitive development in infants of different age levels and from different environmental backgrounds: An exploratory investigation. *Merrill-Palmer Quarterly,* 1971, *17,* 283.

Wargo, M. J., Campeau, P. L., & Tallmadge, G. K. *Further examination of exemplary programs for educationally disadvantaged children.* Palo Alto, Calif.: American Institute for Research in Behavioral Sciences, 1971.

White, B. Critical influences on the development of competence. *Merrill-Palmer Quarterly,* 1975.

Yarrow, L. Maternal deprivation: Toward an empirical and conceptual re-evaluation. *Psychological Bulletin,* 1961, *58,* 459.

Yarrow, L. J., Rubenstein, J. L., Pedersen, F. A., & Jankowski, J. J. Dimensions of early stimulation and their differential effects on infant development. In S. Chess & A. Thomas (Eds.), *Annual progress in child psychiatry and child development.* New York: Brunner/Mazel, 1973.

Yerkes, R. M., & Foster, J. C. *A point scale for measuring mental ability.* Baltimore, Md.: Warwick & York, 1923.

Zigler, E., & Butterfield, E. C. Motivational aspects of changes in IQ test performance of culturally deprived nursery school children. *Child Development,* 1968, *39,* 1.

12 Temperament–Intelligence Reciprocities in Early Childhood
A Contextual Model

RICHARD M. LERNER AND
JACQUELINE V. LERNER

INTRODUCTION

In the more than quarter century since Alexander Thomas and Stella Chess initiated a longitudinal study of the implications of individual differences in temperament for psychosocial adaptation—the New York Longitudinal Study—there has been considerable interest in the study of temperament among psychiatrists, developmental psychologists, and pediatricians. In part, this interest derives from the fact that the Thomas and Chess approach to the study of temperament was both a product and a producer of the contextual zeitgeist that has characterized American social science for more than a decade (e.g., Jenkins, 1974; Lerner, Hultsch, & Dixon, in press; Mischel, 1977; Sarbin, 1977).

Preceding Bell's (1968) influential paper on the direction of effects in socialization, Thomas and Chess (e.g., Thomas, Chess, Birch, Hertzig, & Korn, 1963) emphasized that different children may have differential

RICHARD M. LERNER AND JACQUELINE V. LERNER • College of Human Development, The Pennsylvania State University, University Park, Pennsylvania 19802. Much of the first author's work on this chapter was done while he was a Fellow at the Center for Advanced Study in the Behavioral Sciences. He is grateful for financial support provided by National Institute of Mental Health Grant #5-T32-MH14581-05 and by a grant from the John D. and Catherine T. MacArthur Foundation, and for the assistance of the Center's staff. Both authors' work on this chapter was supported in part by another grant from the John D. and Catherine T. MacArthur Foundation.

effects on others in their social world as a consequence of their specific characteristics of behavioral individuality; by affecting those who affect them, children may provide a source of their own development—they may contribute to their further individual trajectory of development. This view about the potential role of the child in parent–child relations was involved in the recognition among social scientists that both parent and child were developing organisms and were active in their social exchanges with each other (e.g., see Lerner & Spanier, 1978; Lewis & Rosenblum, 1974). One example of this view is that the nature of the parent's own development (as a parent or spouse, for instance) places limits on the effects of the child's individuality on him or her but at the same time, the child influences the parent's further individual and social development (Bell, 1974; Lerner, 1978, 1979; Spanier, Lerner, & Aquilino, 1978). Thus, temperamental characteristics have been seen as useful dimensions of individuality with which to study child effects on others, to study transactional relations between children and their social context, and to study the means by which children can promote their own development.

Given the theoretical importance attached to the study of temperament, it is not surprising to learn that more and more published (and proposed) studies include measures of various temperamental attributes in their repertoire of psychosocial assessments. Quite typically, the study of temperament in these investigations involves an attempt to appraise the relation between some set of temperament attributes and other indexes of psychosocial functioning, for example, measures of adjustment or of intellectual/cognitive functioning. Such studies, as well as others pertinent to the temperament literature (for instance, temperament scale development research, e.g., Carey & McDevitt, 1978; McDevitt & Carey, 1978; or studies of the heritability of temperament, e.g., Matheny, 1980; Plomin & Foch, 1980), have been discussed in recent reviews by Bates (1980), Dunn (1980), and Goldsmith and Campos (1981). There is no need to replicate these three reviews here. However, and quite apropos of the focus of this paper (i.e., a discussion of the links between temperament and cognition or intelligence in early life), we may briefly summarize the status of the literature devoted to studies interrelating temperament attributes, measured alone, with other indexes of psychosocial functioning, assessed contemporaneously with temperament and/or at later points in development.

Simply, some data sets indicate the existence of links between temperament and other psychosocial variables, and in other data sets no support is found for such relations. Studies involving the assessment of temperament–cognition links provide an excellent example of this disagreement within the literature. Field, Hallock, Ting, Dempsey, Dabiri, and Shuman (1978) report that the more caregivers perceive their 4-month-olds as having temperamental attributes often labeled as difficult (low rhythmicity of biological functions, negative mood, low adaptability to new stimuli or people, withdrawal, and high intensity reactions), the lower the infants' mental test scores at 12 months. Sameroff (1974) reports a similar relation between temperament ratings at 4 months and mental test scores at 30 months. Similarly, Carey, Fox, and McDevitt (1977) report a relation between temperamental difficulty at 6 months and early-school-age level scores on a cognitive style measure (i.e., a measure of reflectivity–impulsivity). Moreover, Gordon and Thomas (1967) report that temperament and teacher ratings of intellectual ability are related in a sample of kindergarten children, and Lerner and Miller (1971) report that temperament among seventh grade students is related to both rated ability and an independent measure of actual intellectual ability.

In turn, however, Vaughn, Crichton, Taraldson, and Egeland (cited in Bates, 1980) report that caregiver ratings of 6-month-old infants' temperamental difficulty are not related to later mental development scores. In addition, both Chess and Korn (1970) and Baron (1972) report that temperament does not often discriminate between mentally retarded and normal children.

Because of such findings, some reviewers (e.g., Bates, 1980) have implied that the study of temperament alone is not likely to provide a means to account reliably for variation in other psychosocial variables. Certainly, as demonstrated by the discrepancy of conclusion in the above-noted studies attempting to link temperament attributes with cognitive-intellectual measures, Bates's conclusion may appear warranted. However, as we have argued elsewhere (J. Lerner & Lerner, in press), the point is that an appropriate understanding of the significance of temperament for cognitive functioning, and vice versa, derives from an appreciation of the contextual interpretation of the role of temperament advanced by Thomas and Chess (e.g., 1981) as well as by some of their collaborators (J. Lerner, in press; Korn, 1978).

Thomas and Chess (1977) initiated their study of temperament not because they believed that the possession of particular characteristics of organismic (temperamental) individuality *per se* was the main contributor to a person's psychosocial adaptation. In their view, the import of a particular repertoire of temperamental characteristics for adaptation lies instead in whether the particular attributes are congruent, or provide a "goodness of fit," with the adaptational presses present in a person's context at a given point. It is this "goodness of fit" model that, in the view of Thomas and Chess (1977) and the present authors, gives the study of temperamental individuality its theoretical significance and its relevance to cognitive/intellectual functioning. This model allows one to specify when temperament should relate to cognitive functioning *and when it should not be so related*. In other words, this model facilitates the explanation and prediction of temperament–context relations. Before we present the model and describe how it allows us to conceptualize the relations between temperament and cognition, it will be useful for us to discuss our definition of temperament.

THE DEFINITION OF TEMPERAMENT

There are several definitions of temperament found in the literature (e.g., Buss & Plomin, 1975; Rothbart & Derryberry, 1981; Strelau, 1972). However, the Thomas and Chess (1977) definition, as the stylistic components of behavior—not *what* the person does, but *how* he or she does whatever is done—is predominant in the field (see J. Lerner & Lerner, in press, for a review). Because of its use in providing an empirical base for the theoretical view of temperament that we will argue is most compelling—the contextual interpretation of temperament—we, too, are attracted to the Thomas and Chess (1977) conception.

Thomas and Chess (1977) conceptualize nine categories of behavioral style: rhythmicity, activity level, attention span–persistence, distractability, adaptability, approach–withdrawal, threshold, intensity of reaction, and mood. To illustrate the meaning and use of their definition of temperament as the stylistic component of behavior, consider that one could describe a large portion of the content of a young infant's behavioral repertoire as eating, sleeping, and excreting. Inasmuch as all infants engage in such behaviors, one could not easily discriminate among infants by use of these descriptors. But if one infant ate at regular intervals while another ate at irregular times, discrimination between the two could be achieved. Because the terms *regularity* and *irregularity* qualify eating and specify the distinction between otherwise identical eating behaviors, the terms are used as designators of the stylistic aspect of behavior.

In our view, it is this stylistic component of behavior that gives a normative behavioral repertoire its distinctiveness, especially at early ages (cf. Dunn, 1980). To presage our later argument, it is this distinctiveness that in turn serves to mark the young organism as a unique individual—to parents (especially experienced ones), to teachers, and to others in the social network of the young person who must discriminate him or her from among many in order to interact most appropriately. It is this distinctiveness that may serve to channel the child along interindividually different trajectories, trajectories that may be associated with interindividual differences in cognitive development. We are implying, then, that the significance of the definition of temperament as behavioral style lies in its impact on the social context of the person.

This impression is precisely what we want to convey. Indeed, the impression is quite consistent with the position advanced by the authors of recent reviews of the temperament literature. For example, Goldsmith and Campos (1981) note that, "rather than referring merely to differences in the infant's *susceptibility* to experiences, temperament is now generally considered to include processes which help to regulate the child's social relationships," (p. 3) and that "the temperament phenomena of prime importance during infancy are those which have socially communicative functions" (p. 21). Similarly, Bates (1980) contends that "temperament has its main impact on socially-relevant outcomes through a process of transaction between the child and the social environment" (p. 316). Just such a role for temperament is found in the present authors' contextually based goodness of fit model (see J. Lerner & Lerner, in press).

A GOODNESS OF FIT MODEL OF TEMPERAMENT–CONTEXT RELATIONS

The conception that the relationships between an organism and its context must involve congruence, must match, or simply must fit in order for adaptive transactions to exist is an idea traceable at least to Darwin (1859). As explained by White (1968), this idea has permeated the thinking of American and to some extent European social science, albeit in formulations as seemingly diverse as those of G. S. Hall (1904), Clark Hull (1952), and George Herbert Mead (1967). A version of this idea that has been attracting increasing attention in the human development literature was introduced by the psychiatrists Thomas and Chess (1977) and has been elaborated by the present authors (J. Lerner, in press; J. Lerner & Lerner, in press).

The Thomas and Chess goodness of fit model indicates that adaptive developmental outcomes occur when the physical and behavioral characteristics of the person are consonant with the demands of the physical and social context within which the person is developing. In other words, as a consequence of characteristics of physical distinctiveness (e.g., in regard to sex, body type, or facial attractiveness; Berscheid & Walster, 1974) and/or psychological individuality (e.g., regarding conceptual tempo or temperament; Kagan, 1966; Thomas & Chess, 1977), children promote differential reactions in their socializing others; these reactions may feed back to children, increase the individuality of their developmental milieu, and provide a basis for their further development.

Through the establishment of such "circular functions" in ontogeny (Schneirla, 1957), children may be conceived of as producers of their own development (Lerner & Busch-Rossnagel, 1981). However, in order to understand the specific characteristics of the feedback (e.g., its positive or negative valence) children will receive as a consequence of their individuality, one must recognize that there are also demands placed on children by the social and physical components of the setting. These demands may take the following form:

1. attitudes, values, or stereotypes held by others in the context regarding children's attributes (either their physical or behavioral characteristics);
2. the attributes (usually behavioral) of others in the context with whom children must coordinate, or fit, their attributes (also, in this case, usually behavioral) for adaptive interactions to exist; or
3. the physical characteristics of a setting (e.g., the presence or absence of access ramps for the motorically handicapped) or the "affordances" (Wohlwill, in press) of the stimuli in the child's context, which require the child to possess certain attributes (again, usually behavioral abilities) for the most efficient interaction within the setting to occur.

Children's individuality, in differentially meeting these demands, provides a basis for the feedback they get from the socializing environment. Considering the demand "domain" of attitudes, values, or stereotypes, for example, nonfamilial caregivers (e.g., those in day-care settings or teachers of preschool or early grade levels) and parents may have relatively individual and distinct expectations about behaviors desired of their students and children, respectively. Teachers may want students who show little distractibility, since they would not want attention diverted from the lesson by the activity of other children in the classroom. Parents, however, might desire their children to be moderately distractible, for example, when they require their child to move from playing with toys or from watching television to eating dinner or to bed. A child whose behavioral characteristics are such that he or she is either generally distractible or generally not distractible would thus differentially meet the demands of these two contexts. Problems of adaptation to school or to home might thus develop as a consequence of a child's lack of match (or goodness of fit) in either or both settings. As will be elaborated below, the development of cognitive variables may be influenced in either case, for instance, differences in attention to a teacher or other caregiver may lead to differences in knowledge attained in a setting (J. Lerner, Lerner, & Zabski, in press).

Similarly, considering the second type of contextual demands— those that arise as a consequence of the behavioral characteristics of others in the setting—problems of fit might occur when children who are highly irregular in their biological functions (e.g., eating, sleep–wake cycles, and toileting behaviors) interact in a family setting composed of highly regular and behaviorally scheduled parents and siblings, or in a school setting wherein arrhythmicity (e.g., in toileting behaviors) is distracting to a teacher. In addition, children who are "fussy"—for example, withdrawing from and slow to adapt to new stimuli and/or people, having negative mood and high intensity reactions—might not fit well in a behavioral exchange with a person whose behaviors must be routinized and/or precise in order for an appropriate interaction to occur. Two examples of people involved in such behaviors would be a pediatrician conducting an infant physical examination and a psychological examiner conducting an infant psychomotor assessment.

In turn, considering the third type of contextual demands—those that arise as a consequence of the physical characteristics of a setting—a child who has a low threshold for response and who also is highly distractible might find it difficult to perform efficiently in a setting with high noise levels (e.g., a crowded home, a preschool situated near the street in a busy urban area) when tasks necessitating concentration and/or attention are required. That is, the physicosocial characteristics of a situation have what Wohlwill (in press) terms "affordances"—distinctive features that permit but also require (for successful use of, or appropriate interaction with, them) specific behaviors on the part of the child.

For example, a problem-solving task, if it is solvable, affords (i.e., permits and requires) approach and attention; if one could not come near to and remain focused on a stimulus array because of that array's characteristics, one could not acquire information about it. However, while affording such behaviors as approach and attention, a solvable information-containing task or situation also requires such behaviors of the interacting child *if* the child is to deal with it adaptively. This being the case, a withdrawing, highly distractible child would not fit with the psychosocial demands imposed by such task affordances.

We will return again to considering the implications of the types of demands present in a child's context. Here, however, we should note that Thomas and Chess (1977, 1980, 1981) and J. Lerner (in press) believe that adaptive psychological and social functioning do not derive directly from either the nature of the person's characteristics of individuality *per se* or the nature of the demands of the contexts within which the person functions. Rather, if a person's characteristics of individuality match (or fit) the demands of a particular setting, adaptive outcomes in that setting will ensue. Those people whose characteristics match most of the settings within which they exist should receive supportive or positive feedback from the contexts and should show evidence of the most adaptive behavioral and cognitive development. In turn, of course, mismatched children, whose characteristics are incongruent with one or most settings, should show alternative developmental outcomes.

The major support for these ideas about goodness of fit comes from research in the temperament literature. Thus, before discussing the relevance of the goodness of fit model for understanding the link between temperament and cognition in the early years of life, it will be useful to summarize the details of this research.

Studies of the Goodness of Fit Model

In a series of cross-sectional studies in our laboratory, the usefulness of the goodness of fit model has been confirmed among samples in the late childhood to early adolescent age range. J. Lerner (in press) used a version of the Dimensions of Temperament Survey (DOTS) to measure eighth-graders' temperaments. This instrument, develped by Lerner, Palermo, Spiro, and Nesselroade (1982), assesses multiple dimensions of temperament: activity level, rhythmicity, adaptability/approach–withdrawal, attention span–persistence/distractability, and reactivity—an attribute composed of items relating to threshold, activity level, and intensity. J. Lerner also assessed the demands for behavioral style in the classroom maintained by each subject's classroom teacher and peer group. Those subjects whose temperaments best matched each set of demands had more favorable teacher ratings of adjustment and ability, better grades, more positive peer relations, fewer negative peer relations, and more positive self-esteem than did subjects whose temperaments were less well matched with teacher and/or peer demands.

There is evidence that temperament–context fit also covaries with actual abilities—that is, the J. Lerner (in press) study demonstrated a relation among peer and teacher ratings, teacher-assigned grades, and goodness of fit. However, no relation between actual academic abilities and fit was seen. Such a relation was found by J. Lerner *et al.* (in press), however. That is, for several dimensions measured by the DOTS, and most notably for reactivity, fourth-grade students whose self-rated temperament best fit the teacher demands scored better on two standardized achievement tests—the Stanford Achievement Test for Reading and the Comprehensive Test of Basic Skills—than did less well-fit children.

Moreover, in a study by Palermo (1982), fifth-graders' self-ratings of temperament were found to be interchangeable with their mothers' ratings of their temperament in the prediction of teacher evaluations, peer relations, and parental identification of problem behaviors in the home. Again, better-fit children had more favorable scores on these measures than did less well-fit children.

In essence, these data indicate that at a given point in development, neither the organism's attributes *per se* nor the demands of the setting *per se* are the best predictors of children's adaptive functioning. Instead, the *relation* between organism and context seems most important. Moreover, there are longitudinal data, pertinent to infancy, that indicate the relevance of the goodness of fit model for the long-term predictability of adaptive functioning.

In their New York Longitudinal Study of the psychosocial significance of temperamental individuality, Thomas and Chess (1977; Thomas, Chess, Sillan, & Mendez, 1974) have prospectively studied for over 25 years a core sample of 133 white, middle-class, largely Jewish children of professional parents. In addition, a sample of 97 New York City Puerto Rican children of working-class parents have been followed for about 10 years. Each sample was studied from at least the first month of life onward. Although the distribution of temperamental attributes in the two samples was not different, the import of the attributes for psychosocial adjustment was quite disparate. One example may suffice to illustrate this distinction.

Let us consider the impact of low regularity or rhythmicity of behavior, particularly in regard to sleep–wake cycles. The Puerto Rican parents studied by Thomas and Chess (1977; Thomas et al., 1974; and see Korn, 1978) were quite permissive. No demands in regard to rhythmicity of sleep were placed on the infant or child. Indeed, the parents allowed the child to go to sleep at any time the child desired and permitted the child to awaken at any time as well. The parents molded their schedule around the children. Thus, because parents were so accommodating, there were no problems of fit associated with an arrhythmic infant or child. Indeed, neither within the infancy period nor throughout the first 5 years of life did arrhythmicity predict adjustment problems. In this sample, arrhythmicity remained continuous and independent of adaptive implications for the child (Thomas et al., 1974).

In the white, middle-class families, however, strong demands for rhythmic sleep patterns were maintained. Thus, an arrhythmic child was not fit with parental demands, and, consistent with the goodness of fit model, arrhythmicity was a major predictor of problem behaviors both within the infancy years and across time through the first 5 years of life (Thomas et al., 1974). However, the parents in the white, middle-class sample took steps to change their arrhythmic children's sleep patterns, and since most of these arrhythmic children were also adaptable, low rhythmicity tended to be discontinuous for most children.

Thus, in the white, middle-class sample, early infant arrhythmicity tended to be a problem during this time but proved to be neither continuous nor predictive of later problems of adjustment. In turn, in the Puerto Rican sample, infant arrhythmicity was not a problem during this time of life, but it was continuous and—because in the Puerto Rican context it was not involved in poor fit—it was not associated with adjustment problems in the first 5 years of life. However, to underscore the importance of considering the context of development, we should note that arrhythmicity did begin to predict adjustment problems for the Puerto Rican children when they entered the school system. Their lack of a regular sleep pattern interfered with their getting sufficient sleep to perform well in school and, in addition, often caused them to be late getting to school (Korn, 1978).

In summary, then, data from the New York Longitudinal Study and from our laboratory suggest that it is not a child's temperamental attributes *per se* that play the major role in contributing to the child's adaptive function. Rather, it is the degree of congruence, match, or simply, fit, between attribute and contextual demand that carries the major import for psychosocial adaptation. Put more generally, temperamental attributes represent a feature of an organism's functioning that, if and when they meet the contextual presses impinging on the organism, afford adaptive interchange with the context. Seen in this way, temperament has its meaning for adaptation when it is involved in an organism–environment relation that meets contextual demands. Of course, temperament is not the only feature of the organism's functioning that achieves meaning for adaptation in this way. Indeed, another prime example is cognitive functioning. This recognition allows us now to specify some of the major dimensions of relation between temperament and cognition, especially as they may exist in the early years of life.

DIMENSIONS OF THE TEMPERAMENT–COGNITION/INTELLIGENCE RELATION

A contextually based model can be argued to suggest at least three ways in which temperament relates to cognitive or intellectual functioning.

Temperament and Cognition as Adaptive Processes

Both intelligence and temperament are instances of organism adaptation, when adaptation is understood as involving organism–environment relations that meet the demands of the organism's environment. Our view of intelligence is thus a biosocial one, that is, intelligence pertains to those mental and behavioral functions of the organism that allow it to meet the demands of its context (i.e., adapt); such adaptation occurs by an organism's altering the context to fit its internal needs, individual characteristics, and capacities and, reciprocally, by the alteration of these needs, characteristics, and capacities to fit the demands of the organism's contexts. The best-known example in developmental psychology of this general view regarding the nature of intelligence is the structural cognitive development theory of Piaget (1950, 1970, and especially 1971, 1978), but this view is also consistent with ideas about the evolutionary significance of organism–environment interaction that are presented by Lewontin and Levins (1978). They note:

> As a preliminary analysis, the separation of organism and environment or of physical and biological factors of the environment . . . have proved useful. But they eventually become obstacles to further understanding: the division of the world into mutually exclusive categories may be logically satisfying but in scientific activity there seem to be no non-trivial classifications that are really mutually exclusive. Eventually their interpenetration becomes a primary concern of further research. (p. 71)

and thus that

> the activity of the organism sets the stage for its own evolution . . . the labor process by which the human ancestors modified natural objects to make them suitable for human use was itself the unique feature of the way of life that directed selection on the hand, larynx, and brain in a positive feedback that transformed the species, its environment, and its mode of interaction with nature. (p. 78)

The Impact of Temperament on Cognitive Development

By virtue of children's possession of temperaments that are differentially congruent with the demands of their contexts, children may experience interindividual differences in their histories of organism–environment relations. By being differentially exposed to quantitative and qualitative aspects of their world, and thus by having different developmental presses imposed on them, different children will have different experiential histories from which to derive the interactive, stimulative bases of their intelligence. Such different histories may provide a basis for eventual intraindividual differences among mental and behavioral abilities. This effect of temperament on the history of the organism–environment relations involved in cognitive development may occur in at least two ways.

The Appraisal of Intellectual Ability

First, children whose temperaments are differentially fit with the demands of the contexts within which they exist may (1) elicit different reactions from socializing others, (2) receive differential feedback as a consequence of these reactions, and (3) be differentially channeled within and between contexts.

In other words, differentially fit children may be judged or rated differently, and such ratings may be of cognitive functioning. If different temperaments are believed to be differentially associated with intellectual ability (and there are data indicating that this is indeed the case, Gordon & Thomas, 1967; Lerner & Miller, 1971), and if the believed association may be greater than the actual one (and this may be especially the case in the early years of life, Gordon & Thomas, 1967), then the basis for a self-fulfilling prophecy process may exist.

Several illustrations of this type of process may be offered. As an example in early infancy, consider the differential situations that may be created when a temperamentally "easy" child is assessed on an infant "intelligence" test, such as the Bayley Scales. As summarized by Thomas and Chess (1977), temperamentally easy children are characterized by such attributes of behavior style as approach and relatively rapid adaptability to new people and stimuli, positive mood, and moderate-intensity reactions. In turn, the difficult child tends to be characterized by such attributes as withdrawal and relatively slow adaptability to new people and stimuli, negative mood, and high-intensity reactions. Two such infants would differentially meet the demands of the testing situation, by virtue of the fact that the examiner must standardly and efficiently introduce the infant to various stimuli and tasks in order to conduct the assessment appropriately.

In other words, one may see in this example that there are demands placed on the infant (the testee) by virtue of the behaviors required of the tester. The tester needs to perform his or her test administration behaviors in a way consonant with the proper, standard conduct of the assessment. The behavior of the difficult child thus creates a behavior–behavior mismatch, that is, an instance of the second type of temperamental attributes–contextual demands situation discussed earlier.

In short, the difficult child's withdrawal and slow adaptability might lead to scores lower than those of the approaching, rapidly adapting, easy child, when in fact, no actual "ability" differences exist. Evidence that such a possibility is plausible is found in data pertaining to kindergarten children's estimated and actual intelligence. Gordon and Thomas (1967) found that experienced kindergarten teachers overestimated the intelligence of temperamentally easy children and underestimated the intelligence of difficult children.

If such an over- and underestimation process is operative in kindergarten, it is conceivable that it may also be present in preschool and daycare settings. Increasing proportions of today's children spend many of the daytime hours of their early years in such programs. This being the case, the intellectual "training" that might have been provided by familial caregivers during the day is now more likely to be provided by caregivers in such settings. To the extent that such caregivers base and convey intellectual appraisals on what may be, at any early age, a relatively indirectly intellective variable (such as behavior style) the labeling/categorization and differential treatment that may ensue might initiate a self-fulfulling prophecy process.

In sum, difficult infants may not score well on assessments of their abilities because their behavioral style attributes make them troublesome to test. Somewhat older children may also encounter stereotypic beliefs that their temperament is highly linked to their abilities. Such effects of temperament on the appraisal of intelligence may lead to differential treatment and channeling of behavior, experiences that would constitute a history of organism–environment relations eventually affecting actual cognitive development—thus a self-fulfilling prophecy process would be created. Temperament may affect cognition in another way, however; this second type of impact is more direct than the first one.

The Acquisition of Knowledge

Particular temperamental attributes may diminish or enhance a child's acquisition of information in specific settings. Developmental psychologists have not paid as much attention to the differentiations or dimensions of the child's ecology as they have to those of the child's psychological status (Wohlwill, in press). It may, however, be both possible and useful to partition the character of any situation into the presses that situation imposes on the person, for example, for adaptive cognitive, social, and emotional functioning. For instance, at least two contextual demands are imposed on a child if he or she is to acquire information about the skills (e.g., fine motor behaviors) that an adult present in a room with the child is demonstrating: to show both (1) social approach (a close personal space) and (2) high attention and low distractability.

The more general point we are making, then, is that in order for children to acquire information about their physical and social world, particular temperamental constellations may be required by situations. The affordances of the stimuli must be matched by the child's temperamental attributes.

As a general expectation, we would advance the ideas that approach, high attention, low distractability, and moderate threshold and intensity would best fit most contexts requiring information acquisition. Children who withdraw, do not attend, are easily distracted from engaging any focal stimulus, and are easily highly aroused by stimulation would seem to have behavioral style characteristics not well matched with a situation affording relatively long engagement with an object, task, or person in order for information about it to be acquired.

Of course, situations are not static over time. With change, the temperamental attributes that do not facilitate information acquisition will similarly interfere with the child's ability to utilize information. Stimulus changes that herald the occasion for an alteration in ongoing behavior cannot be detected appropriately if not attended to, if incompletely dealt with due to distraction, if withdrawn from, etc. Thus, as was the case in regard to the acquisition of knowledge in a particular situation, the child may possess temperamental attributes that preclude fit with the demands imposed by a situation affording information utilization.

The Impact of Cognition on the Development of Temperament

An intelligent child may be construed to be one who is able to adapt to his or her environmental niche. In other words, adapting to one's context and the degree of fit thereby attained may be indexes of a child's intelligence. This, of course, is an instance of the first way that we have described temperament and intelligence as being related—as involving fits between organism and context. In turn, however, differences in intellectual abilities might result in a child's being differentially able to attain a temperament–context fit and, that being the case, engage in adaptive interchanges with his or her setting. Again, in other words, intellectual functioning that is adaptive may lead to changes in temperament–context fits. Here, then, arises a third way in which temperament and intelligence may be related.

Children may use their developing mental and behavioral abilities to (1) select contexts within which to interact that possess demands that best fit their repertoire of temperamental characteristics (Snyder, 1981) and (2) alter the demands that exist in the contexts within which they interact (Snyder, 1981). In addition, particular instances of intellectual functioning may be important in determining how one engages a context within which one may be potentially fit. For example, particular combinations of behavioral style characteristics (e.g., high attention span, persistence, approach, and low distractibility) may be adaptive to "couple" (Lewontin & Levins, 1978), or combine, in particular contexts (e.g., as discussed above, situations requiring the acquisition of new information on the appropriate use of already acquired information). But it may be more adaptive at another time (e.g., when the requirement is to change from one situation to a distinctly different one) to uncouple these attributes; for instance, across time in the home, it may be most adaptive for the child to show moderate distractibility (as when a parent tries to divert the child's activity from playing with toys to a meal or to bed), withdrawal from such alternative activities, and a continuation of high attention (in this case to the parent). However, children may not have knowledge about how the coupling and/or uncoupling of attributes they possess may be used to meet demands of their context; or, while having such knowledge, children may not have the skills to couple or uncouple appropriately; or they may not perceive themselves as efficacious in either performing the appropriate coupling–uncoupling behaviors or in otherwise doing what they know is necesarry to meet the demands of their context.

Thus, because of limitations of their knowledge (e.g., about coupling–uncoupling), because of limitations in their behavioral skills (e.g., to couple or uncouple appropriately), or because their perceived self-efficacy (Bandura, 1978, 1980a,b) is not sufficient to lead them to engage their context adequately, children may not adaptively interact in (fit the demands of) contexts to which they are actually potentially temperamentally matched. Here, therefore, cognitive deficits affect temperament–context fits. Cognitive interventions designed to alter coupling–uncoupling knowledge or skills or to change perceived self-efficacy may thus lead to a change in the nature of a child's fit in his or her context. Such an alteration would thus change the nature of the child's experiences in the context, and, as a consequence, feed back to change the *history* of experiences and thereby affect further intellectual development. Because interventions aimed at enhancing goodness of fit through enhancing cognitive functions are so importantly implicated in this third instance of the relation between temperament and cognition, it will be useful to discuss these procedures in greater detail.

Enhancing Goodness of Fit

Interventions aimed at altering cognitions in order to enhance temperament–context relations may be targeted at the level of the child and/or at the level of the context. For example, one may attempt to modify any or all of the three demand domains (attitudes, values, and stereotypes of others; behavioral characteristics of others; and physical characteristics of the setting) that we have discussed. Among infants and very young children, for instance, such intervention targets may be the only ones available. Caregivers may have to be informed about the contextual nature of temperament and the role of fit with contextual demands in the infant's psychosocial development. Here, the cognitive developmental level of the caregiver is itself an issue (Sameroff, 1975), since such educational interventions with the caregiver may be expected to vary in their success as a consequence of the caregiver's cognitive abilities to deal with such concepts as "bidirectional influences," "goodness of fit," and "behavioral individuality." This being the case, a complete intervention repertoire aimed at altering the context of individual children through affecting their caregivers, might need to include behavior modification and parent-education techniques as well as cognitive or cognitive-behavioral ones. Through such procedures, both the attitudes and the behaviors—the first two of the the three demand domains we discussed—of these significant others may be altered.

The physical context of the infant and young child may also be altered in order to increase the match between temperament and contextual demands. Here, Wohlwill's (in press) concept of affordances may again prove to be a useful conceptual guide. By analyzing a young child's stimulus context in terms of the behaviors afforded by it, one will be able to (1) gain greater precision in determining what specific behaviors are required of the child for adaptation and (2) make data-based decisions about whether it is easier and/or less costly to institute interventions targeted at the stimulus context or the child.

That is, it is of course the case that one may focus one's efforts on the child and, by altering his or her actions *in or on* the context, change the goodness of fit. This latter alternative is quite important, for two reasons. First, children have different levels of goodness of fit because of the impact of their characteristics of individuality on the context. Since it is the individuality of the child that is the initiator of the "circular functions," and a key basis of the level of fit that results, work focused on the child is therefore directed to a major basis of the developmental process with which we are concerned. Second, with the individuals so central in their own developmental processes, it is important to enhance their ability to regulate their own further development. If we changed a particular context for the individuals, but did not give them those behavioral and/or cognitive abilities to continue to alter *either self or context*, then it is unlikely that they would be able to have appropriate self-regulation when new contexts and/or demands were encountered. Simply, one cannot anticipate all the contexts and all the demands someone may encounter in life; thus, it may be most efficient to focus one's intervention efforts on providing bases for the child to change self *or* context. Moreover, a child's perceptual "filtering" of contextual demands is an important component of the ways in which demands create presses for fit (J. Lerner, in press), and this role of the child, as a key part of his or her own context, underscores the importance of targeting interventions at the individual level of analysis.

Thus, by enhancing those self-regulatory functions involved in allowing children to become active producers of their own development, we believe we would be appropriately and efficiently providing means to enhance goodness of fit. Note, however, that we are *not* saying that we wish to eliminate individual differences in order to allow children to meet contextual demands. Rather, our goal in such individual interventions is to provide the means by which children could alter themselves *or* the contextual demands imposed on them. For example, our goal would be to give children the cognitive skills necessary to (1) detect the affordances of the stimuli impinging on them, (2) evaluate whether there was a fit between their temperaments and the behaviors required for adaptation, and (3) if not, alter self or the affordances to promote a better fit.

Are there procedures useful for enhancing fit in infancy and early childhood? Some available intervention strategies, specifically many of the cognitive-behavioral ones, have been applied for the most part only to children from 8–12 years of age (see Kendall, 1981, for a review). However, while such age boundaries may be quite appropriate for some target behaviors, they are not necessarily applicable to other target behaviors. For instance, cognitive-behavioral procedures involving training children to label internal states verbally and out loud may be useful in training 2- or 3-year-olds to gain the self-control necessary for successful toilet training. Similarly, interventions using either live or televised modeling procedures may be used for some simple motor behaviors (e.g., of the face or hands) in the first year of life, and during the second and third years of life, these modeling procedures may be used in regard to quite complex behavioral sequences (e.g., McCall, Parke, & Kavanaugh, 1977). In addition, of course, behavior modification procedures, less dependent on advanced cognitive development, can usually be instituted at early age levels. Accordingly, we may conclude that depending on the specific behaviors involved in a particular instance of a temperament–cognition relation, there are procedures that may be realistically applied to infants and to children in their early childhood years.

For example, Kendall (1981) notes that there is an array of behavioral and cognitive-behavioral interventions useful for teaching behavioral self-regulation to young children. Children with problems in self-control are often referred to an interventionist because their characteristic style of behavior—a low threshold for response initiation (often described as "impulsive" behavior), high approach, and, often, high-intensity responses and high activity levels—does not meet with the approval of parents and/or teachers. This lack of fit between their behavior and the values of others in their contexts often causes both intellectual problems (most notably in school-based information acquisition and utilization) and problems in interpersonal relations. Kendall (1981) describes various modeling, behavioral contingency management, role-playing, verbal self-instruction, and self-evaluation procedures that are useful to institute successively in children ranging in age level from infancy through late childhood in order to eliminate self-control problems.

Conclusions

There are at least three ways in which temperament and cognition are related in early life. First, both are instances of organism adaptation, at least when conceptualized from the basis of a paradigm leading to ideas stressing that adaptation arises as a consequence of organism–environment relations that enhance the individual's fit with his or her context. The second and third instances of the temperament–intelligence relation view these two aspects of organism–environment relations as potentially reciprocally related—as potentially dynamically interactive (Lerner, 1978, 1979, 1980) with each other over the course of development. The second instance regards temperament–context goodness of fit as an antecedent of the interactive, stimulative experiences that affect the development of those mental and behavioral abilities we label intellectual ones. The third instance views intellectual abilities as potentially affecting the nature of the fit that exists between temperament and context and hence, the implications of such fits for adaptation. Temperament–context fit can affect intellectual functioning, and intellectual functioning can in turn affect temperament–context fit.

Moreover, the third instance of the temperament–cognition relation suggests an array of strategies for giving a person those cognitive and behavioral skills necessary to change self, context, or both. That is, one may capitalize on the nature and/or potential modifiability of a child's abilities and cognitions to effect change in temperament and hence goodness of fit; such a change in fit will feed back to the person, affecting his or her further cognitive (and other psychosocial) developments. Thus, the links between temperament and cognition are then potentially simultaneous, reciprocal, dynamically interactive ones.

Thus, rather than demanding that children be "passive recipients" of the fit immediately afforded them as a consequence of their characteristics of temperamental individuality, interventions associated with the goodness of fit idea can provide children with those abilities necessary to create a good fit actively for themselves, and thus enhance their own further cognitive and other psychosocial developments. A contextual view of temperament thus not only allows us to understand adequately how this facet of human individuality contributes to development, but in so doing, it also provides an excellent illustration of the ways in which individuals themselves contribute to their own development. This being the case, the contextual view also offers an important example of the potential plasticity of human development and, as a consequence, of the potential for successfully intervening to enhance human life.

ACKNOWLEDGMENTS

The authors thank Anneliese F. Korner, Rolf Oerter, Marion E. Palermo, and Alexander Thomas for their helpful reviews of an earlier draft of this chapter.

REFERENCES

BANDURA, A. The self system in reciprocal determinism. *American Psychologist*, 1978, *33*, 344–358.

BANDURA, A. Self-referent thought: A developmental analysis of self-efficacy. In J. H. Flavell & L. D. Ross (Eds.), *Cognitive social development: Frontiers and possible futures.* New York: Cambridge University Press, 1980. (a)

BANDURA, A. The self and mechanisms of agency. In J. Suls (Ed.), *Social psychological perspectives on the self.* Hillsdale, N.J.: Lawrence Erlbaum Associates, 1980. (b)

BARON, J. Temperament profile of children with Down's syndrome. *Developmental Medicine and Child Neurology*, 1972, *14*, 640–643.

BATES, J. E. The concept of difficult temperament. *Merrill-Palmer Quarterly*, 1980, *26*, 299–319.

BELL, R. Q. A reinterpretation of the direction of effects in studies of socialization. *Psychological Review*, 1968, *75*, 81–95.

BELL, R. Q. Contributions of human infants to caregiving and social interaction. In M. Lewis & L. A. Rosenblum (Eds.), *The effect of the infant on its caregiver.* New York: Wiley, 1974.

BERSCHEID, E., & WALSTER, E. Physical attractiveness. In L. Berkowitz (Ed.), *Advances in experimental social psychology.* New York: Academic Press, 1974.

BUSS, A., & PLOMIN, R. *A temperamental theory of personality development.* New York: Wiley, 1975.

CAREY, W. B., & McDEVITT, S. C. Revision of the infant temperament questionnaire. *Pediatrics*, 1978, *61*, 735–739.

CAREY, W. B., FOX, M., & McDEVITT, S. C. Temperament as a factor in early school adjustment. *Pediatrics*, 1977, *60*, 621–624.

CHESS, S., & KORN, S. Temperament and behavior disorders in mentally retarded children. *Archives of General Psychiatry*, 1970, *23*, 122–130.

DARWIN, C. *The origin of species.* London: John Murray, 1959.

DUNN, J. F. Individual differences in temperament. In M. Rutter (Ed.), *The scientific foundations of developmental psychiatry.* London: Heinemann Medical Books, 1980.

FIELD, T., HALLOCK, N., TING, G., DEMPSEY, J., DABIRI, C., & SHUMAN, H. H. A first-year follow-up of high-risk infants: Formulating a cumulative risk index. *Child Development*, 1978, *49*, 119–131.

GOLDSMITH, H. H., & CAMPOS, J. Toward a theory of infant temperament. In R. N. Emde & R. Harmon (Eds.), *Attachment and affiliative systems: Neurobiological and psychobiological aspects.* New York: Plenum Press, 1981.

GORDON, E. M., & THOMAS, A. Children's behavioral style and the teacher's appraisal of their intelligence. *Journal of School Psychology*, 1967, *5*, 292–300.

HALL, G. S. *Adolescence.* New York: Appleton, 1904.

HULL, C. L. *A behavior system.* New Haven: Yale University Press, 1952.

JENKINS, J. J. Remember that old theory of memory? Well forget it. *American Psychologist*, 1974, *29*, 785–795.

KAGAN, J. Reflection-impulsivity: The generality and dynamics of conceptual tempo. *Journal of Abnormal Psychology*, 1966, *71*, 17–24.

KENDALL, P. Cognitive-behavioral interventions with children. In B. Lahey & A. E. Kazdin (Eds.), *Advances in child clinical psychology* (Vol. 4). New York: Plenum Press, 1981.

KORN, S. *Temperament, vulnerability, and behavior.* Paper presented at the Louisville Temperament Conference, Louisville, Kentucky, September 1978.

LERNER, J. V. The role of temperament in psychosocial adaptation in early adolescents: A test of a "goodness of fit" model. *Journal of Genetic Psychology*, in press.

LERNER, J. V., & LERNER, R. M. Temperament and adaptation across life: Theoretical and empirical issues. In P. B. Baltes & O. G. Brim, Jr. (Eds.), *Life-span development and behavior* (Vol. 5). New York: Academic Press, in press.

LERNER, J. V., LERNER, R. M., & ZABSKI, S. Temperament and elementary school children's actual and rated academic performance: A test of a "goodness of fit" model. *Journal of Child Psychology and Psychiatry*, in press.

LERNER, R. M. Nature, nurture, and dynamic interactionism. *Human Development*, 1978, *21*, 1–20.

LERNER, R. M. A dynamic interactional concept of individual and social relationship development. In R. L. Burgess & T. L. Huston (Eds.), *Social exchange in developing relationships.* New York: Academic Press, 1979.

LERNER, R. M. Concepts of epigenesis: Descriptive and explanatory issues: A critique of Kitchner's comments. *Human Development*, 1980, *23*, 63–72.

LERNER, R. M., & BUSCH-ROSSNAGEL, N. A. Individuals as producers of their development: Conceptual and empirical bases. In R. M. Lerner & N. A. Busch-Rossnagel (Eds.), *Individuals as producers of their development: A life-span perspective.* New York: Academic Press, 1981.

LERNER, R. M., & MILLER, R. D. Relation of students' behavioral style to estimated and measured intelligence. *Perceptual and Motor Skills*, 1971, *33*, 11–14.

LERNER, R. M., & SPANIER, G. B. (Eds.). *Child influences on marital and family interaction: A life-span perspective.* New York: Academic Press, 1978.

LERNER, R. M., HULTSCH, D. F., & DIXON, R. A. Contextualism and the character of developmental psychology in the 1970s. *Annals of the New York Academy of Sciences*, in press.

LERNER, R. M., PALERMO, M., SPIRO, A., III, & NESSELROADE, J. R. Assessing the dimensions of temperamental individuality across the life-span: The Dimensions of Temperament Survey (DOTS). *Child Development*, 1982, *53*, 149–159.

LEWIS, M., & ROSENBLUM, L. A. (Eds.). *The effect of the infant on its caregiver.* New York: Wiley, 1974.

LEWONTIN, R. C., & LEVINS, R. Evolution. *Enciclopedia, V: Divino-Fame.* Torino, Italy: Einaudi, 1978.

MATHENY, A. P., JR. Bayley's infant behavior record: Behavioral components and twin nalyses. *Child Development.* 1980, *51*, 1157–1167.

McCALL, R. B., PARKE, R. D., & KAVANAUGH, R. D. Imitation of live and televised models by children one to three years of age. *Monographs of the Society for Research in Child Development*, 1977, *42*(5, Serial No. 173).

McDEVITT, S. C., & CAREY, W. B. The measurement of temperament in 3-7 year old children. *Journal of Child Psychology and Psychiatry*, 1978, *19*, 245–253.

MEAD, G. H. *Mind, self, and society from the standpoint of a social behaviorist.* Chicago: University of Chicago Press, 1967.

MISCHEL, W. On the future of personality measurement. *American Psychologist*, 1977, *32*, 246–254.

PALERMO, M. E. *Child temperament and contextual demands: A test of the goodness-of-fit model.* Unpublished doctoral dissertation, The Pennsylvania State University, 1982.

PIAGET, J. *The psychology of intelligence.* London: Routledge & Kegan Paul, 1950.

PIAGET, J. Piaget's theory. In P. H. Mussen (Ed.), *Carmichael's manual of child psychology* (Vol. 1). New York: Wiley, 1970.

PIAGET, J. *Biology and knowledge.* Chicago: University of Chicago Press, 1971.

PIAGET, J. *Behavior and evolution.* New York: Pantheon Books, 1978.

PLOMIN, R., & FOCH, T. T. A twin study of objectively assessed personality in childhood. *Journal of Personality and Social Psychology*, 1980, *39*, 680–688.

ROTHBART, M. K., & DERRYBERRY, D. Development of individual differences in temperament. In M. E. Lamb & A. L. Brown (Eds.), *Advances in developmental psychology* (Vol. 1). New York: Lawrence Erlbaum Associates, 1981.

SAMEROFF, A. J. *Infant risk factors in developmental deviancy.* Paper presented at the meeting of the International Association for Child Psychiatry and Allied Professions, Philadelphia, July 1974.

SAMEROFF, A. J. Transactional models in early social relations. *Human Development*, 1975, *18*, 65–79.

SARBIN, T. R. Contextualism: A world view for modern psychology. In J. K. Cole (Ed.), *Nebraska Symposium on Motivation, 1976.* Lincoln: University of Nebraska Press, 1977.

SCHNEIRLA, T. C. The concept of development in comparative psychology. In D. B. Harris (Ed.), *The concept of development.* Minneapolis: University of Minnesota Press, 1957.

SNYDER, M. On the influence of individuals on situations. In N. Cantor & J. F. Kihlstrom (Eds.), *Cognition, social interaction, and personality.* Hillsdale, N.J.: Lawrence Erlbaum Associates, 1981.

SPANIER, G. B., LERNER, R. M., & AQUILINO, W. Future perspectives on child-family interactions. In R. M. Lerner & G. B. Spanier (Eds.), *Child influences on marital and family interaction: A life-span perspective.* New York: Academic Press, 1978.

STRELAU, J. A diagnosis of temperament by nonexperimental techniques. *Polish Psychological Bulletin*, 1972, *4*, 97–105.

THOMAS, A., & CHESS, S. *Temperament and development.* New York: Brunner/Mazel, 1977.

THOMAS, A., & CHESS, S. *The dynamics of psychological development.* New York: Brunner/Mazel, 1980.

THOMAS, A., & CHESS, S. The role of temperament in the contributions of individuals to their development. In R. M. Lerner & N. A. Busch-Rossnagel (Eds.), *Individuals as producers of their development: A life-span perspective.* New York: Academic Press, 1981.

THOMAS, A., CHESS, S., BIRCH, H., HERTZIG, M., & KORN, S. *Behavioral individuality in early childhood.* New York: New York University Press, 1963.

THOMAS, A., CHESS, S., SILLAN, J., & MENDEZ, O. Cross-cultural study of behavior in children with special vulnerabilities to stress. In D. F. Ricks, A. Thomas, & M. Roff (Eds.), *Life history research in psychopathology.* Minneapolis: University of Minnesota Press, 1974.

WHITE, S. H. The learning-maturation controversy: Hall to Hull. *Merrill-Palmer Quarterly*, 1968, *14*, 187–196.

WOHLWILL, J. F. The physical and the social environment as contrasting environmental modes relevant to the development of the child. In D. Magnusson & V. Allen (Eds.), *Human development: An interactional perspective.* New York: Academic Press, in press.

13 Looking Smart
The Relationship between Affect and Intelligence in Infancy

JEANNETTE HAVILAND

INTRODUCTION

Without being wholly conscious of it, parents, pediatricians, and particularly psychologists have used facial expression to infer the existence and development of intelligence. We use facial expression or "affect" to denote consciousness, interest, surprise, intention, fear, or frustration; then we use these states of emotions to determine motivation, knowledge, and ability. This may surprise some readers, since we claim to be looking at *behavior* to determine motivation, knowledge, and ability. By behavior we usually mean actions accomplished—"grasped the block," "looked under the pillow," "turned away." Awareness of these actions accomplished seems to block awareness of the affect accompanying, preceding, substituting for, or following the action, even though we use affect to interpret the action.

This blockage of psychological awareness has had serious consequences for the direction of research in this century. Those who test infant intelligence and observe infant behavior use affect continuously to infer intelligence and knowledge, but they do not acknowledge it. Hence in research, we use as primary variables the behaviors that the testers and observers believe to be crucial; these behaviors actually form only a portion of their unacknowledged system for assessing knowledge.

The present research view of intelligence is much like that of a television picture with strong interference, so that random points are

JEANNETTE HAVILAND • Department of Psychology, Rutgers—The State University, New Brunswick, New Jersey 08903.

missing, making the picture lose its form. The impact of coherent theories of cognitive development that emphasize sensorimotor development has been so great that less and less attention has been given to other portions of infants' behavior that influence and motivate their interactions with the social world, including the influential aspect of looking smart.

It is time for us to examine ourselves and our methods of assessing intelligence and intelligent behavior. Just to be objective and honest, we must admit that almost never do we assess infants' intelligence solely by observing their actions or by observing their sensorimotor behaviors. We infer infants' awareness, understanding, and knowledge by an intuitive, unstructured, unsystematic method that takes into account affect, environment, interpersonal relationships, and sensorimotor behaviors. Isolating a single variable invariably results in questions about the validity of the assessment and the predictive quality of the assessment. Furthermore, isolating sensorimotor behaviors results in assessments that give minimal information about infants' social interactions, which largely determine their cognitive environment.

In this chapter, we will show that in practice no one really assesses infant intelligence by assessing motor or sensorimotor behaviors, neither on tests such as the Bayley Scales of Infant Development nor on assessments of intellectual growth and change such as Piaget's. However, the systematic study of affect in the assessment of intelligence has only begun and then primarily in the assessment of atypical infants.

We will show from a brief review of observations of twins that it is possible to study affect and that it has a significant place in our interpretation of infant behaviors. Further, we will show that infants can be strikingly different in facial expression—even the twins show strong, measurable individual differences from the second week of life—and that these differences may be psychologically meaningful.

Last, we will show how the study and assessment of affect may lead to new or renewed understanding of infant cognitive-affective disorders. In the case of autism and in the case of retardation, cognitive development is so clearly involved with social and emotional development that study of these children may lead to a new interpretation of infant cognition.

If we would determine and understand infant intelligence, we must study how infants reveal their intelligence and how the adult community responds to their looking smart.

USING AFFECT ON MEASURES OF INFANT INTELLIGENCE

At one time or another we all use a shorthand method of assessing intelligence. We say, in effect, "That child looks smart." What do we ordinarily mean by this? Usually we use this shorthand method when we are unable to observe a person's problem-solving abilities systematically. We may use it in spite of these observations when a person acts stupid, but we keep giving him or her another chance because we cannot quite believe that a person who looks that smart could act so dumb. Why do we persist? In part, it is because we have a fairly reliable system of inferring intelligent behavior from intelligent affect. If a student sits in class and gazes attentively at the lecturer, the blackboard, and other relevant objects, laughs appropriately, looks puzzled during confused explanations, and is somewhat excited during animated discussions, we expect that student to do well in the course. When a student lacks affect, gazes into space, looks sleepy, or reacts inappropriately, we do not expect him or her to do well. We have been wrong, particularly in the latter case, because lectures are the kinds of events that inspire stupid affect. However, the affect system is commonly useful for shortcut assessments. None of these observations is particularly startling, but they all gain in importance when we systematically review psychological tests of intellectual behavior and problem solving and find that a significant variable in the determination of infant cognitive ability is infant affect, not alone "manual-tactile" ability, as Piaget, among others, claimed and as most developers of tests, such as Bayley and Gesell, emphasize.

A clear example of this is contained in the frequently used Bayley Scales of Infant Development: "The Bayley Scales of Infant Development . . . are designed to provide a tripartite basis for the evaluation of a child's developmental status in the first two and one-half years of life. The three parts are considered complementary, each making a distinctive contribution to clinical evaluation" (Bayley, 1969, pp. 3–4). The Mental Scale assesses sensory-perceptual abilities and other bases of abstract thinking. The Motor Scale measures control of the body, coordination, and locomotion. The Infant Behavior Record assesses the "child's social and objective orientations toward his environment" (p. 1). Even the briefest of reviews reveals that affect plays a major role in enabling the examiner to determine the infant's abilities. The scale titled the "Infant Behavior Record" is most conspicuous in this regard. Let us take each category of the Behavior Record and observe how affect is used to assess intelligent behavior.

In the first category of the Infant Behavior Record, "Social Orientation," the examiner must discover whether the infant behaves differentially toward objects, mother, and examiner. The first indication is "interest," as with our student in the lecture above, but then more specifically the examiner notes "freezing, frowns, wariness, brightening, smiling, laughing, vocalization, fussing" (p. 1). These are commonly called affects. From the occurrence of these affects and presumably their appearance in what seems to the examiner to be an appropriate sequence (although that is not specifically mentioned), the examiner infers that the infant can discriminate objects from persons and his or her mother from a stranger. Presumably, one could set up a conditioned learning situation in which it is proved that the infant can reliably make such a discrimination, but in all probability the inference is correct and takes considerably less time than devising a situation in which affect is not used. It is our unacknowledged "shorthand." Note also that the examiner assumes that these frowns and fussings are meaningful statements about the infant's cognitive development, a point to which we will return.

The next items—"Cooperativeness" and "Fearfulness"—involve a stronger dichotomy between positive and negative affect than the first category. The first primarily involves an assessment of "enjoyment" in the tasks at hand or "resistance" to them. One should point out that this is not a trivial measure of intellectual competence. If infants "enjoy" the testing situation, they probably enjoy many activities that will relate to academic achievement and possibly language in particular—that is, they enjoy working with puzzles, making discoveries, and interacting with new people in goal-oriented tasks.

Fearfulness is used as an indication of many things during infancy. If it is very strong, it is cause for concern about development because it interferes with normal exposure to the environment. If fear is not apparent at all, the examiner looks for its interaction with age and situational variables. Such children could be reckless or unaware of danger, or they might not show affect in general and fear in particular, or any one of several possibilities might apply. Of course, fear of strangers is "expected" and is part of the developmental norm for infants between 8 and 12 months. In any case, fearfulness can serve as an indicator of recognition of difference, although it is not the sole indicator and probably not even the predominant one.

The next area examined on the Bayley scale is "Tension." This covers tension in the whole body and may be part of affect. An inert infant lacks affect and would inspire the examiner to rely heavily on manual–tactile indications of cognition. The extreme lack of tension greatly inhibits ability to predict intelligence, except as the lack of it predicts lack of intelligence. On the other hand, "tautness," "startle," "quiver," and "trembling" (Bayley, 1969, p. 2) are indications of extreme affect, although they do not determine the particular affect. Only when combined with other affects and behavior sequences do they gain meaning. For example, infants who startle easily and are tense may also be fearful or may be excited. Facial expression and determination of approach or avoidance would tell the examiner which was more likely. These affects in combination with manual-tactile abilities are determined differently. Two-month-old infants who startle at noises in the testing room but do not cry, who tense when they reach for the red ring, and who stiffen when being picked up by the examiner may be designated as tense infants, but their tenseness is in each instance a response to something and an indication of awareness. On the other hand, infants who are taut and quivering and do not react easily and intensely with testing stimuli are thought of in a different category. In each instance, the categorization produces a part of the assessment of intelligence.

The next category is wholly affective and is called "General Emotional Tone." This assesses distress or contentedness during the testing procedure. There is no doubt that this category is highly related to "cooperativeness," although one describes a general state and the other relates more directly to interaction with the examiner. It is clear, though, that a distressed infant will not be a cooperative one. Infants may be happy and noncooperative, or cooperative and only moderately content, and various changes may occur during which they are in one or the other of these conditions. Again the relevant behavior includes affects— crying, fussing, and whining, or smiling, laughing, crowing, and animation (Bayley, 1969, p. 3).

We could go on indefinitely, but it is apparent that affect is the primary determinant in the assessment of infant development from the Infant Behavior Record. Every single item on this test relies on an assessment of infant affect and an inference about which affect it is and whether or not it is appropriate. It is not possible to rate an infant without affect on this scale. However, the reader must be reminded that this assessment, a conspicuous phenomenon, occurs in the absence of systematic knowledge of what is being assessed.

Let us turn to the other scales, which rely more heavily on motor behaviors or language and problem-solving. Here, the examination of the use of affect becomes more complex. Part of the problem in describing the use of affect to determine development and intelligence from these two scales is that the use of affect by the examiner changes rapidly as the range of the child's abilities widens. During the first 3 months, 40%* of the items cannot be assessed in the absence of appropriate affect, 58% might be so assessed but probably are not, and 2% (one item "head turns") could be assessed without affect. From 3 to 8 months, 25% of the items need affect to be assessed, 17% probably need affect, and 57% could be assessed even if the examiner could not recognize affect. From 8 to 18 months, only one item ("jabbers expressively") requires affect, another 33% probably need affect, and 75% can be assessed without affect. Obviously, the best cue to early infant development is affect. When infants can manipulate their environment by turning their heads, picking up blocks, and removing pellets from bottles, then we can begin to look at behavior without assessing affect. Even then it is an unusual and uneconomical thing to do. Again, one must note that even in infants under 3 months of age, the examiner relies on the functional meaningfulness of the affect signal.

Our assessment of affect usage in these scales is dependent on our definition of *affect*. Some items are difficult to bisect into affect or sensorimotor and would not be bissected in a real testing situation. Categories that begin with "regards," as in "regards person," "regards cube," and so on, are especially difficult. In some sense, "regards" may include "shows interest"; "prolonged regard" has this quality. As soon as the child's interest is inferred, we have an affect state. Therefore, in practice, the infant could usually be described as having interest, and a glance without interest or without affect would be very strange (a cold glance or a glance that appears to go "through" an object). "Responds" also has this quality. Occasionally, one may mean "responds" in the sense of a reflex response, such as the orienting reflex. The relationship among the orienting reflex, the orienting response, and interest is not known. It is probably not a direct or one-to-one relationship, but involves considerable cognitive awareness and discrimination during the process of responding, attending to, and showing interest. On the other hand, in-

---

*Two testers who regularly use the Bayley in research evaluation of infants categorized the items with 100% agreement.

terest may be only the conscious prolonging of the orienting reflex, which ontologically becomes separate but is not separate in the infant. Can an infant orient without having an adult infer interest or other affect? It seems improbable, and if it does occur, it would require unusual circumstances that do not pertain to the present testing situation. Therefore, all items that begin "regards" or "responds" have been classified as indicating affect. The justification for this becomes easier to understand later when we examine Piaget's use of "regards" in denoting the infant's awareness. In any case, the percentages are clearly boosted by the inclusion of "responds" and "regards" in the category of affective behaviors.

Most other categories are clearly affectual or not. These include "smile" (item 18), "anticipatory excitement" (item 22), "play" (item 36), and expression of "attitudes" (e.g., item 42). Items that do not require affect to score include "picks up cube," "turns head," "eye-hand coordination," and so forth. However, a child who performed these behaviors correctly but had no affect or even affect unrelated to the behavior would give cause for grave concern (Bayley, 1969).

In every instance, affect is either necessary to or is the common method for determination of awareness, interest, recognition, cooperation, anticipation, and change of state. All of these words reflect cognitive states in the infant that cannot be determined or inferred from any source other than affect.

The motor scale has no affect items on it, although again the infant who performs tasks without affect would be very strange indeed. Also, one should note that the motor scale is the least good predictor of later development and in particular of later cognitive development. Although it is correlated to some degree with the mental development scale, much of that correlation is of a noncausative nature. There is some relationship between the ability to move oneself and the ability to bring objects to the self that is reflected in correlation of motor and mental scales in infant scores, but this is incidental to the present discussion.

Which scale is the best predictor of later mental or cognitive development? Birns and Golden (1972) have suggested, after many examinations of infant intelligence-test items, that one of the best predictors of later intelligence test scores is positive affect during testing. To generalize this suggestion, one would predict that an infant who is interested in problems (the interest would be determined from affect, primarily) at the first testing is more likely to have mastered solutions to these problems at a later date than an infant who shows no interest at the first testing. Further, Cicchetti and Sroufe (1976, 1978) find that the age of the onset of smiling is one of the best predictors of later cognitive ability among Down's syndrome infants.

Thus, to summarize the importance of affect in infant intelligence testing, it seems that a very good predictor of later intelligence may be affect, but even if it is not a good predictor, it is the main index of intelligence in infancy. It probably is at any time, but we systematically ignore it in testing situations.

PIAGETIAN USES OF AFFECT TO INFER COGNITION

No one maintains that infant intelligence tests tell us much about how infants learn to behave intelligently in their world. For this, we turn to the cognitive epistemologist Jean Piaget, who has described the development of logical thinking and the structure of knowledge in young children (e.g., Piaget, 1971). In his assessment of cognitive structure one finds unacknowledged but extensive use of affect to infer knowledge, even though Piaget (1954) saw affect as the motivation of cognition.

Although Piaget certainly never intended that one should go through his observations of infant behavior and pull out descriptions of affect, it is possible to give a sketch of how he used affect to determine mental states and sometimes even uses affect sequences from his diary observations to resolve theoretical positions.

To illustrate the impact of affect on observation of cognition, we have reexamined Piaget's *The Construction of Reality* (1971), noting each affect and its use. The reader must keep in mind that these observations cannot be used directly to develop a developmental scale of affect or even to "prove" that the child has an awareness of the affect described. They indicate only that we, the observers, are using affect to describe cognition, and they show to some extent how we do it even when we are not aware of so doing.

In his observations of the very young infants (2–4 months), Piaget relied primarily on interest, recognition pleasure, and crying to determine the infant's beliefs about the world. For example, Lucienne sees Piaget at the extreme left of his visual field and smiles vaguely; he then looks away but constantly returns to the place where he sees him and dwells on it. If the infant had merely turned his head with a vacant gaze and no facial mobility, Piaget probably would have found it difficult to say that Lucienne recognized him or was "bringing to himself the image of his desires" (p. 12). There is something in the smooth turn of the head accompanied by a searching gaze, the stopping of the head, the fastening of the gaze on the "desired" object, and the looking with *interest* and even *pleasure* that aroused in the heart of the trained genetic epistemologist a belief that the child had desires and recognized a particular object, and also, though this is more complex, that the child did not dissociate these objects from himself.

The notion that Lucienne did not dissociate objects from his own behavior is illustrated by "crying at random or by looking at the place where it [the desired object] disappeared or where it was last seen" (p. 12). Piaget illustrated the idea that the object has no motion of its own but is associated with the child; he inferred this from the child's *astonishment* when the object disappeared without any activity on the part of the child. When Jacqueline is watching Piaget's watch, he drops it; it falls too fast for Jacqueline to follow the trajectory, or so Piaget suggested, so that she does not follow the movement and then she "looks at my [his] empty hand with *surprise*" (p. 16, emphasis mine). Here, it is the surprise that indicated to Piaget that the child thought she had control over the watch and was surprised when it disappeared without any behavior change on her part. Without the affect of surprise, Piaget would not have been able to come to the conclusion that the child had any such expectation—there was no reported change in reaching, head direction, or any other nonaffective behavior.

Disappointment was also a cue to Piaget that the child expected an occurrence and had no ready explanation for its not occurring. More precisely though, the disappointment itself is not as important as the sequence of affects. At 7 months, 5 days, Laurent loses a cigarette box that he has just grasped and swung to and fro. Unintentionally, he drops it outside the visual field (it is not clear how Piaget determined that the dropping was unintentional, but he could have done so by assessing affect). He then immediately brings his hand before his eyes and looks at it for a long time with an expression of "*surprise, disappointment, something like an impression of its disappearance*" (p. 25, emphasis mine). After a short time, Laurent again swings his hand and "looks intently at it." Piaget interpreted this to mean that Laurent either wishes that the box would reappear or that he believes that it might if he repeats box-swinging behaviors. The box does not appear, of course, and Laurent stops searching altogether. How Piaget knew that Laurent had stopped searching is a good question. Again, it may have been affect; he may have shown interest in another activity, or there may have been practically no affect. (We will return to the meaning of "no affect" shortly.)

Here we have an excellent example of the interaction of affect and motor behavior that enables the observer to reach reasonable explanations of the infant's cognitive process. To emphasize the importance of affect, consider the sequence without affect. Laurent swings a box to and fro but shows no interest or excitement in this activity. The box falls from his hand, but he continues to gaze in the direction of his hand with no change in affect, resumes swinging the empty hand, and after a while, stops. In this case, one would wonder whether he had realized that the box had dropped; the continued arm swinging might be self-stimulatory as in the case of infants who rock themselves. It would be very difficult to distinguish between the possibilities (1) that the object did not matter to the swinging behavior, (2) that the child was not aware that the object was missing, simply another version of the first possibility, or (3) that the child was wishing that the object would return. As soon as one adds that he was interested in swinging the box, was surprised at its disappearance and then disappointed, and began swinging the arm with an air of attentive expectation, then Piaget's explanation seems reasonable, if not obvious. And so it is with much intelligent, problem-solving behavior: The affect is necessary to interpretation.

The interaction between affect and behavior is used many times. The affect "curiosity" is frequently used in this way. Laurent shows "curiosity when he is first shown a whole pencil which is then lowered partly behind a screen" (Piaget, 1971, p. 30). Piaget writes that "he looks at this extremity with curiosity, without seeming to understand" (p. 30). The curiosity expression occurs only when Laurent can see less than three centimeters of the pencil. If he can see more than that, he grasps the pencil. Here the interaction is between *curiosity* to indicate misunderstanding and *grasping* to indicate understanding of the wholeness and continuity of the object. There must have been some affect other than curiosity expressed when the child grasped and reached for the pencil, but it probably seemed too obvious to report at the time.

Crying and anger seem to be related in several instances of cry, no cry, and cry again—followed by "rage." For example, at 6 months, 19 days, Laurent begins to cry from hunger and impatience upon seeing his bottle. He often frets before the appearance of the bottle but cries only when it appears. But when the bottle disappears, he stops crying. Piaget wrote, "I repeat the experiment four more times; the result is constant until poor Laurent, beginning to think the joke bad, becomes violently *angry*" (p. 33, emphasis mine). Once again Piaget's interpretation of this behavior sequence is determined in part, if not primarily, by affect. First there is a whimper of hunger, denoting a motivational state prior to the appearance of the bottle; then crying and fretting when the bottle appears, denoting recognition of the bottle and desire for the bottle; and a return to the hunger state without crying when the bottle disappears, denoting belief in the nonexistence of the bottle. And finally there is rage after several presentations and disappearances of the bottle, denoting belief that the bottle can be commanded to "remake" itself. It is not clear in this instance that the rage signifies belief in the ability to command its appearance. That possibility exists from other examples Piaget gives, including the one concerned with shaking a box. But the rage could be a reaction to the sequence of affects, to internal events, or to inability to control the bottle, rather than an affect-laden demand for the bottle. Note that "manual-tactile" events play no part in this observation. Piaget claims that the child did not even reach for, much less grasp, the bottle since he was not accustomed to holding it himself.

The particular affect that accompanies the disappearance of an object seems to vary. Laurent cried at the appearance of his bottle and quieted at its disappearance; he looked surprised and then disappointed when he dropped the cigarette box; Jacqueline kicked in anger and impatience when Piaget hid her bottle; Lucienne laughed at the disappearance of her stork; Jacqueline "whimpered" once when her doll was hidden but laughed when her stork was hidden. The affect seems to occur as the object is in the process of disappearing. This example seems to verify the idea: Laurent is on a diet and screams when the bottle disappears, but calms when it is gone. During stage III, the child may temporarily "forget" that the object exists when it is not apparent. During later stages, Piaget presented evidence that memory is not the primary factor in the discovery of displaced objects. It seems to be the act of disappearing and not the disappearance itself that is of interest, but it is difficult to determine from Piaget's reports since the mode of disappearance varies. Sometimes it disappears without the child's being able to determine the procedure (it drops too rapidly for the infant to perceive it), or it may disappear gradually behind a screen or hand or be placed under a cloth while the child is watching. The sudden disappearance seems to be the one most likely to produce surprise, and the deliberate and observed disappearance more likely to produce pleasure, especially if the object continues to reappear as it often does in the games Piaget was playing with his children. The rage reaction to the disappearance of the bottle was a reaction both to the object that disappeared repeatedly and to unfulfilled expectation.

Affect is also used to determine when an infant expects an activity to occur, a cause-and-effect relationship. Between the eighth and the ninth months Jacqueline uses arching herself to attempt to induce a reoccurrence of some activity. Piaget knows that she intends that "arching" to cause an event because she shows pleasure when she "produces" the activity and she shows "surprise" or "constant surprise" when she is unable to do so.

"No affect" is one of the more interesting events that Piaget describes. It is difficult to decide when he meant that the child did not reach or grasp and when he meant that the child actually registered no reaction at all, including no affect. The exceptions to this are few, and I may have misinterpreted some. But it seems as though Piaget meant that the child is not aware or does not remember an event or object when "nothing" happens or there is "no reaction." For example, Piaget hides the watch from his 8-month-old daughter. He hides it behind his hand, behind the quilt, etc. She does not react and forgets everything imediately. Several times Piaget stated that there was "nothing" and the child did not understand or was forgetful. This use of no affect may be the strongest indication that we are using affect to determine an aware and understanding state in the child (or the older person for that matter). This signifies that at all other times, we use affect to determine whether the child is attending and dealing cognitively with events, as the case below illustrates.

THE CONSEQUENCES OF "NO AFFECT"

The observations of Shirley illustrate the interaction of affect and intelligence in yet another way. Shirley exhibited very little affect. The reaction of caretakers and observers was that she was not very intelligent. She performed almost the same sensorimotor behaviors as did Piaget's children with no affect. She was not testable on the Bayley, nor could we use the Gesell developmental scales to describe her behavior. Her lack of affect had a profound effect on her intellectual environment.

Shirley was brought by her mother, who was a high-school student, to our day-care center in a local high school when she was 8 months old. She stayed during school hours 5 days a week from January until June. When she first arrived at the center, she appeared to be a plump, large, not unattractive baby. The well-baby clinicians had stated that she was in good health and had no abnormalities. Her mother described her as an "easy" baby. She never cried or fretted, even if it was considerably past her mealtime or if she were left alone and awake in her crib for a long period of time. On the other hand, she never smiled, cooed, or looked intently at anything but was always extremely relaxed and passive.

Shirley was almost untestable on the developmental scales commonly used for demonstration in the center. Her responsiveness to persons was almost indiscernible—she would look longer at her mother and move her torso slightly. She was neither cooperative nor uncooperative, fearful or reckless—she was nonresponsive. Her body was either inert or relaxed. Her constant emotional tone was "nonexpressive." She was not blind. She would follow her bottle from the caretaker's hands to her own. She was not deaf. Her body would give a startle response if there were a loud noise, although her facial expression would not change, nor would she ever cry. She could be persuaded to grasp a cube in one hand if it were directly over her, but not if it were on the side; she could bring objects to her mouth, and she held her own bottle routinely both in the center and at home. She could sit with support sometimes, but more frequently she would "melt" back into a prone position.

This behavior, or rather lack of behavior, had a profound effect on her caretakers. The supervisors soon noticed that Shirley was being ignored, if not avoided. It was extremely difficult to play with her or to care for her because she never signaled her desires, nor, in a different frame, did she reinforce her caretakers for their care. Systematic observations revealed that Shirley, although called a "good" baby, was becoming more and more isolated. She was fed, bathed, changed, and placed. She was not talked to, walked, rocked, sung to, placed on the exercise mats with other infants, or even approached unless someone had a task with her or was instructed to talk or play with her.

We could find nothing "wrong" with her, nor could her pediatrician, except lack of affect. She had an extremely low developmental profile, but this was primarily owing to our inability to assess motor or cognitive behavior because we could not assess affect and could not "motivate" her to participate.

Through a systematic but humanistically oriented behavior-modification program, we intervened and changed Shirley's affect–communication patterns. When she left at the end of the school year, 4 months later, she was a somewhat "depressed" but normal 1-year-old. Still placid except for rare intervals when she would tear around the nursery in a walker with a small smile on her face and her legs churning, she seemed more interesting to strangers in the nursery, who no longer avoided her. Her mother said she had learned to be "smart." Returning student caretakers said she "looked a whole lot smarter and more lovable," and on our developmental tests she was now "testable."

We cannot know, but the literature on autistic children suggests that we intervened in the development of an autistic child. Reports from mothers on the infancy of autistic children include descriptions that are similar to our description of Shirley. "'I never could reach my baby.' 'He never smiled at me.' 'He never greeted me when I entered, he never cried or even noticed when I left the room.' 'She never made any personal appeal for help at any time'" (Mahler, 1968, p. 67).

Shirley was very definitely separating herself from the human world. Although she was not so "difficult" as many autistic children are remembered by their parents to have been, still she did not communicate with her mother or caretakers in a lovable way; she seemed "strange." Her mother has married and moved away beyond our knowledge of her, so we do not know the long-term effects of our intervention, but we suspect that she is still relatively normal because her mother felt that she was lovable and smart and not just "easy" and simple to ignore.

The consequences of Shirley's lack of affect and general inactivity, combined with mothering that reinforced her pattern of inactivity, led to the development of a child whose abilities were difficult to assess and whose demands, if she made any, were difficult if not impossible to interpret at 8 months. From our point of view, the main problem was one of affect. If her inactivity had been combined with smiling and intense gazing, it would not have led to the same concern or the same lack of interaction as did inactivity combined with lack of affectivity. Indeed, our first demand from her was affect—interest as revealed in eye-to-eye contact. Intuitively we knew that affect was more important than rolling over or grasping a block without aid. Indeed, we suspected that she could do both those things if she wanted to. Here, as always, motivation and awareness had to be measured in affect.

Without intervention, it is likely that Shirley would have become more and more isolated from social events. Unlike some autistic children, she showed no motoric precocity, and most likely she would have been categorized as severely retarded in another year or so because her abilities would not have been assessable at all if she were completely withdrawn.

This case illustrates familiar issues in the history of intelligence measures. What is measured by the intelligence tests? How do examiners know whether they observed the right behaviors or organization of behaviors? How do children determine their interactive environment, and what are the consequences of their particular environment for their development? This case and this chapter argue that these are all part of the same question: What is the nature and meaning of the infant's interaction with his social environment?

THE NATURE AND MEANING OF INFANT AFFECT

One of the essential problems—perhaps the essential problem—hindering our further understanding of why affect and cognition are signals for each other is that there have not been acceptable categories of events to call "affect" until very recently (e.g., Izard & Dougherty, 1982). We have not known how to say, "This baby has a trait or habit of being 'interested' or 'cheerful' or 'frightened'" in a way that would prove satisfactory for research or behaviorally based theory.

Several years ago, while home-bound and musing on the vast difference between the knowledge of parents and that of psychologists that exists in this realm, I began to videotape my own twins. At that time I was interested in discovering whether I could capture identifiable emotion expression independent of context. I was also interested in whether my own dizygotic twins would have consistent and similar modes of emotional expression.

With all the wisdom of hindsight, I regret not having filmed the babies in a series of different situations. However, at that time, I was still reeling from studies of emotion demonstrating that emotional response was not directly keyed to the stimulus. In revolt, I created a situation that was quite boring—unstimulating. Each baby was placed in an empty, white room, about 6' × 6' × 8'. The baby was in an infant seat facing a camera. No other object or person was in the room during the filming. In adult terms, it was a good opportunity for boredom or fantasy. The twins, a boy and a girl, were videotaped from 2 weeks of age to $2\frac{1}{2}$ years.

The videotapes have aided immeasurably in the development of an ethogram (behaviorally based coding system) of infant affect. Affect, like all communication systems, requires time and movement for analysis. Still photographs yield unreliable and often meaningless data, whereas videotapes or films are easily, although tediously, coded. The differences in affect between the twins can be easily assessed from an examination of the ethograms.

Even the briefest examinations of the tapes show that the twins are quite different in the portrayal of affect. These differences are reliable and stable from the second week after birth to the second year. An examination of each area of the face indicates the nature of these differences most clearly.

As shown in Table I, brow positions are nearly identical. The most common position is the relaxed or normal position. The next most common position is the slightly raised brow. Slightly raised brows give the face a quizzical or curious look and when combined with widely opened eyes, indicate interest to the observer. Although the twins do not differ on brow position, other young children are quite different. In particular, a series of films from our laboratory taken in Chiapas, Mexico, of Zina-

TABLE I

Eyebrow Positions: Percentages of Frequently Occurring Eyebrow Positions for Two Infants Aged 2 Weeks to 6 Months[a]

Alex		Lizbeth	
1. Relaxed	43%	1. Relaxed	45%
2. Raised	29%	2. Raised	30%
3. Weak frown	12%	3. Weak frown	16%
4. Contraction	12%	4. Contraction	8%
5. Strong frown	2%	5. Strong frown	2%

[a]The descriptions of facial positions are adapted from Blurton-Jones (1971).

TABLE II

Eye Openness: Percentages of Frequently Occurring Degrees of Eye Openness for Two Infants Aged 2 Weeks to 6 Months

Alex		Lizbeth	
1. Normal	33%	1. Bit wide	36%
2. Bit narrow	18%	2. Normal	21%
3. Very narrow	14%	3. Bit narrow	15%
4. Bit wide	14%	4. Wide	8%
5. Upper lid down	3%	5. Upper lid down	2%
6. Wide	2%	6. Very narrow	2%
7. Other (closed)	17%	7. Other (closed)	16%

canteco Indian children shows that a weak frown was extremely common and that the raised or slightly raised brow was never seen.

Eye openness is a feature that distinguishes the twins (see Table II). Lizbeth's eyes are most frequently a bit wide, whereas Alex's eyes are either normal or a bit narrow. In the case of eye openness, Alex is more variable, with more even distribution across all categories than Lizbeth.

The direction of glancing is also quite different (see Table III). Of course both babies most frequently look straight ahead, but the second most frequent directions for Lizbeth are "side" and "up," and the second most frequent directions for Alex are "down" or "down and to the side."

Table IV shows that mouth positions are more variable for Lizbeth

TABLE III

Eye Direction: Percentages of Frequently Occurring Eye Directions for Two Infants Aged 2 Weeks to 6 Months

Alex		Lizbeth	
1. Ahead	34%	1. Ahead	34%
2. Down-side	28%	2. Side	19%
3. Down	19%	3. Up	14%
4. Side	14%	4. Down	7%
5. Up	5%	5. Down-side	23%
6. Other	3%	6. Other	23%

TABLE IV

Mouth Expressions: Percentages of Frequently Occurring Expressions for Two Infants Aged 2 Weeks to 6 Months

Alex		Lizbeth	
Mouth positions			
1. Relaxed	37%	1. Relaxed	28%
2. Squared upper lip	27%	2. Squared upper lip	17%
3. Squared lower lip	10%	3. Corners raised	15%
4. Corners lowered	10%	4. Corners lowered	10%
5. Corners raised	8%	5. Lips retracted	8%
6. Lips retracted	1%	6. Squared lower lip	7%
7. Other	7%	7. Other	15%
Lip positions			
1. Relaxed, slight separation	65%	1. Relaxed, slight separation	36%
2. Contraction	8%	2. Lips pressed together	25%
3. Lips pressed together	5%	3. Contraction	16%
4. Lengthening upper lip	3%	4. Lengthening upper lip	8%
5. Two-lip pout	3%	5. Lips rolled in	4%
6. Lower-lip pout	1%	6. Lower lip bit	3%
7. Lower lip bit	0%	7. Lower-lip pout	1%
8. Lips rolled in	0%	8. Two-lip pout	0%

than for Alex. Either Alex has a relaxed mouth or his upper lip is slightly square. The slightly squared upper lip may also be a relaxed position for infants. Lizbeth's mouth movements are fairly evenly distributed over all possible positions. This is also true of lip positions. Alex's lips are almost always in the relaxed position (75% of the time), whereas Lizbeth's movements are more variable. Tongue positions are nearly identical for both infants, with the positions appearing in this order: tongue invisible, visible, pushed forward, out of mouth.

In more recent years, we have changed the coding system. We now use the MAX coding system developed by Izard and Doughterty (1982). Since this system is theoretically based and anatomically linked to inferred emotion states, it allows for better predictions of temperamental traits than does the more atheoretical ethological scale. Even so, the two scales complement each other. We have selected portions of the original tape to recode. For every month of the taping we selected the most positive and the most negative expressive segment lasting 1 minute and recoded using the MAX system. The recoding confirms that Lizbeth is more labile in that she changes her expression more often and uses a wider variety of codes. Alex is more labile in the eye region than Lizbeth, but her versatility of expressiveness in the mouth region is much greater. When one looks at the affect combinations, both babies were "interested" and "joyful" more than half the taping time, even though we selected negative affect sequences in a biased way. However, Alex used a gaze aversion—eyes-down expression—that Lizbeth never used during the first year of taping. On the other hand, Lizbeth commonly prefaced a fret-cry with the anger expression in both the mouth and brow regions, whereas anger codes did not appear in Alex's section.

The microanalyses using MAX confirm the previous ethological analyses. Both describe stable but subtly different expressive patterns for each twin during the first year. The research is being continued by continued taping and by using more sophisticated sequential analyses so that more complex statements can be made about the socialization of expression. One hypothesis generated from the clear difference in gaze aversion and anger expressions would be that Alex's gaze aversion may represent an ability, often linked to "temperament," to cut off aversive or too stimulating an experience before it produces anger, whereas Lizbeth's anger may occur because she is not yet able to control her level of response and so gets frustrated or overwhelmed somewhat more frequently. Another hypothesis, not yet confirmed by analysis, suggests that the affect expressions generally become muted during the third year either by becoming more restricted to one part of the face, such as mouth or brow, or by becoming briefer. The sequential analyses will allow us to examine developmental hypotheses in the coming years. Similar developmental results at a later age have been reported by Zivin (1982), who finds muting of threat-faces through restriction of the signal to a part of the face (i.e., only a partial threat-face is made).

The infants' expressions are not only responses but also stimuli for caretakers and casual observers. It is in the interaction that the subtle differences become significant. To a mouth-oriented observer it would appear that Lizbeth is the more active infant; her movements are more variable and the mouth area of the face is seldom seen at rest. Her interest and responses are easily determined. On the other hand, Alex's mouth is commonly still and relaxed. Little information about his interest, motives, and response comes from the mouth area of the face. This relaxation gives rise to several reactions from observers. One's first impression is that Alex is a relatively calm and passive infant. However, this impression is contrasted with the fact that mouth movements appear to be very dramatic on Alex's face and command an unusual amount of attention. Consequently his reactions are sometimes seen as more extreme than Lizbeth's, even though they are also seen as occurring less frequently—a contrast effect. In relation to the observation that "anger" expressions occur fairly frequently only for Lizbeth, this seems an accurate conclusion. Lizbeth may, in fact, be more easily stimulated to move from an interested, moderately pleasurable state into extremes. Thus, the observation that Alex seems dramatic, but that Lizbeth actually exhibits many more extreme movements highlights both coding problems and conception of interpretive issues. Do infrequent behaviors contain more information or are they more salient than frequent behaviors? In coding terms, is one affective gesture from Alex equivalent to several from Lizbeth, or not? In conceptual terms, what is the motivational component for isolated gestures versus repeated gestures? With our current methods of observation, we may be able to answer questions such as these very soon. The answers will be important both for understanding the origins of personality and for the development of affective communication.

To the eye-oriented observer Alex appears to be more active and variable; he may also appear to avoid eye contact because he looks down and to the side so frequently. On the other hand, Lizbeth, with eyes directed ahead or up and opened a bit wide, is very appealing and often seems to be asking for interaction in the traditional supplicant manner with eyes raised.

Differences in areas of the face used to express interest and awareness, curiosity and understanding do not seem to be significant indicators of intellectual differences in infants. At this age, they reflect a style of responding more than a quantitative difference in responsiveness. The attentive observer can "read" affect from partial cues of either mouth or eye. But the "still" examples of the most frequent expressions only begin to describe the differences. Another aspect of facial communication is the relationship of "figure to ground." This involves both descriptions of "figure" and "ground" and the relationship between figure and ground. The phenomenon being described is easy to illustrate but difficult to analyze. In the instance of the babies Alex and Lizbeth, we note that Alex is most frequently "at rest" or "normal"— that is, his facial expression is most passive, his "ground" is very bland. In Izard's system, he is mostly attentive or interested. Lizabeth seems also frequently resting, but the frequency is significantly lower than Alex's. Lizbeth as she grows older is less frequently seen as merely interested. This accounts for the description by observers of Alex as more extreme in affect. His change in facial expression occurs less frequently and occurs on a predominantly bland background, giving an observer the impression that his smiles are happier, his frowns unhappier. Lizbeth seems to be more even-tempered because the "ground" expression is more active than Alex's.

It is interesting and invites speculation when the spontaneous descriptions of relatives and passers-by are noted in conjunction with the schema presented above. Lizbeth was called by fanciful, endearing names such as "little pumpkin," "woozel," "sweetie," and so on; Alex was called "the judge," a "cool customer," and other rather unusual names. It would be rash to suggest that facial expression completely determines people's view of their intelligence and personality in any sense, but it would be foolish to ignore the possibility that the infant's facial expression has some control over the social responses of his observers and caretakers.

It seems to be the informal consensus that Alex had an interest in moving things and that even as an infant he was critical and thoughtful about the cause and effect of events. The "evidence" for this comes from his prolonged and calm contemplation of things coupled with his gaze aversion and relative lack of affective responsiveness around people. Relatives predicted that he would do well in academic fields in which people are to be either avoided or used, such as physical science or law. As a matter of record, however, 8 years later, this prediction lacks precision. Alex demonstrates considerable ability in the mechanical arts, including model building and artistic types of drawing. However, he is well known at school for his imaginative adventure stories and is much appreciated for a strong humorous streak.

On the other hand, Lizbeth was seen as creative and affectionate. Her father and grandfather reported that she had musical affinities and abilities. These expectations are formed from observations of her quick and mobile affective responses, both her attentiveness to people and her quick response to them. The prediction about musical affinity possibly originates from her clear affective responses to music and to her own humming and singing. Now, 8 years later, her strongest talents are academic rather than affective–social. In all the school tasks, she excells—reading, writing, spelling, and arithmetic. Temperamentally, she is somewhat more easily distressed and angered than Alex, but she recovers faster. The specific predictions of parents and other caretakers seem to be quite imprecise in retrospect. However, the temperamental-affective qualities derived from the analysis of expressiveness in the first year appear robust.

Although the differences are not significant when scored on the Bayley, Lizbeth at first maintained a higher developmental score on the motor scale and, at first, also on the mental scale. After 1 year, Alex began to score consistently higher on the mental scale because of his verbal abilities. No one seemed surprised to see Lizbeth sit up, walk, feed herself, and be toilet-trained first. Neither was anyone surprised at Alex's relatively large vocabulary and precise grammar. In fact, the differences in affect were nicely paralleled by differences in speaking style. Alex spoke carefully, searching for his words and arranging them correctly. He never used a word invented by Lizbeth, nor did he invent one himself. Lizbeth played with sounds, liked to rhyme, and thought mispronounced words were hilarious. Her own speech was difficult to understand because she used many letters interchangeably (e.g., "teddy bear" becomes "tebby dare"). She recognized the interchange but did not correct it. She also used grammatical shortcuts and did not use prepositions or conjunctions as soon as Alex. It seems that relatives' predictions, self-fulfilling prophecies or not, were quite accurate—more accurate than the Bayley development scale. However, these cognitive measures reversed themselves during preschool, at which time Lizbeth's verbal abilities advanced geometrically. She maintains this lead through elementary school on verbal tests. In this case, pleasure in verbal play was a better predictor of later verbal development than early accomplishment, a finding akin to general predictability of the Bayley scales.

DEVELOPMENTAL CHANGES

Developmentally, a few changes can already be noted from a preliminary analysis of the twin data. More different kinds of movement are noted as the babies grow older, although the changes are very small. Out of the 41 possible featural changes that we coded, 30 were seen during the first 72 days, 1 more during the next 50, and 2 more during the next 100 days. The 8 not seen, but used by Blurton-Jones (1971), may

occur only under unusually stressful circumstances. These small differences suggest that the very young infant has a nearly complete repertoire of facial movement, and perhaps a complete repertoire. On the
Izard scale, no additions in type of movement were seen after 72 days.
Certain ones are entirely missing, including those linked to fearful
expressions.

The study of the twins in combination with the earlier analysis of
cognitive assessment techniques takes us back to a point raised earlier in
this chapter. Initially, it seemed appropriately cautious to point out that
observations of an examiner's use of affect to infer cognition proved
nothing about the development of the affect system. However, at this
point, we have accumulated enough pieces of evidence to construct a
hypothesis: Not only do we, the parents, pediatricians, and psychologists, use affect to infer cognition, but we do so appropriately. Affective expression indicates, from the early moments of life, an affective
and motivational state for the human being. When a mother sees her
baby gazing intently at her, she is behaving appropriately when she
responds to the message of interest and regard quite naturally imparted
by this behavior. In this sense, the affective expression is "meaningful"—it is part of a meaningful state and actually conveys meaningful
information. Only a few studies have courageously tackled the problems
of meaning and coherence in neonatal and infant affective expression
(e.g., Oster, 1981, and in our laboratory, Lelwica, Lelwica, & Haviland,
1983). Quite appropriately, several researchers have noted that there are
serious limits to what can be inferred about the meaning. Most of these
concerns are with the socialization of affect rather than with the origin of
affective expression. For example, Lewis and Michaelson (1982) propose
classification of affect development that distinguishes neonatal or infant
state from later aware "experience." These concerns go well beyond the
limits of this discussion into later cognitive-social developments that
have parallels in the affect system (for reviews, see Lamb & Sherrod,
1981).

Affect Signals and the Observer

From the analysis of the Bayley test and from the examination of
Piaget in this chapter, it is clear that observers use affect as a signal of
cognitive response from the first moments of life. However, studies of
how naive observers interpret early infant affect indicate that it is a fuzzy

area. For example, the sex of the infant biases an observer's interpretation of what is being expressed (Haviland, 1977); observers' own temperaments bias their responsiveness to their infant's affect (Malatesta &
Haviland, 1982). Adults in particular use situational cues as well as facial
cues to infer affect, and their actual labeling is influenced by many
variables including social class and family relationships (e.g., Lewis &
Michalson, 1982; Zivin, 1982). In other words, it is not enough to know
that parents, pediatricians, and psychologists use affect, because they
do not use it completely reliably. The observer may have a fuzzy decod-
ing system or the baby may have a fuzzy encoding system, or both.
Thus, it is easy to demonstrate that there is a phenomenon—people use
affect signals to understand cognition—but it is difficult to know exactly
how this process occurs or how "fuzzy" it is.

Affect Signals and the Cognition

Why should affect be either a signal for cognition or a representation of an infant's cognitive process? One obvious and simple answer is
that affect is merely an indication of the psychophysiological integrity of
the infant organism. But is it really that simple? Piaget (1954), Tomkins
(1962), and Zajonc (1980) hypothesized that affect could precede or motivate cognitive processes, if affect is an independent system as they
hypothesize. Zajonc reviews evidence indicating that affect responses
occur without conscious perceptual and cognitive encoding. He concludes that affect and cognition are independent sources of information
and partially independent processing systems in adulthood. If Zajonc is
correct and if his hypothesis is correct for infants as well as adults, then
the affect and cognition may interact simultaneously or in any order. For
an example, consider an infant with affect deficits such as Shirley, the
baby described earlier in this chapter. She would be likely to develop
cognitive deficiencies as a consequence of the affect deficit for several
reasons. If the affect system "amplifies" messages of all types (Tomkins,
1962)—drives, affects or cognition—then deficient affect would lead to
less urgency or salience in cognitive processes, and that may mean less
problem solving in general. Additionally, certain types of cognitive process would be less likely to occur. Memories and perceptions strongly
associated with emotions would be less likely; this might easily result in
unusual or deficient self-conceptions and preferences. Inferences about

infants' knowledge, perception, interpretation, or understanding of their own affective responses (see Lewis & Brooks, 1978; Lewis & Michalson, 1982) are not necessary to the psychologist who uses the affective response as a cue to cognitive processing. These are only two possibilities from the many suggesting that social experiences are also much affected. Generally, however, we would predict that less affective response would result in less cognition and hence a deficient development of the affective system as well. So, a measure of affective responses that indicates little affect response would accurately reflect a deficient learning system because of the interaction; it might even precede and predict a deficient learning system.

On the other hand, suppose the affect system is somehow intact but the cognitive system is not. Would the affect system then reflect the original deficiencies in the cognitive processes? If the cognitive system does not process well, there is less to be excited about, less to enjoy, and on occasion less to be distressed about. Some affects will become unlikely to occur. Also some affects may not be "socialized" by learning. The affect expressions may be slow, less separated from vocal channels, and less likely to be attenuated by cognition. In fact, this is at least part of the description of Down's syndrome infants (Cicchetti & Sroufe, 1978). Down's infants are difficult to arouse to any affect peak, are unlikely to have certain affective responses such as fear or anger, and are difficult to interrupt when a strong negative affect expression has occurred. However Down's children may actually have cognitive, affective, and/or motoric difficulties to varying degrees.

This discussion of how affect and cognition may separately influence the development and expression of both leads naturally to the question of affect socialization and self-knowledge. Of course, cognitive development will modify affect expression. This means that changes in affect expression can be a reflection of cognitive development. For example, if babies smile at a familiar person, it indicates that they "know" who is familiar, know the person's relationship to them, and know a bit about the self (Lewis, 1981). Smiling indicates "knowing" and may be a useful index of knowing in early socialization, but it is not the same as learning to smile as if smiling were a response that required strict socialization conditions—it is an indicator and then serves other functions as well. It may be a relatively sensitive indicator of learning or socialization, but we must be careful not to confuse the indicator of what is learned with the learning itself. The development of any affective system is actually much more complicated than this, of course. It is the complications with which most attempts to explain the socialization of affect deal (Haviland, 1981). The primary concerns are with the knowledge of one's own feelings as well as others' feelings. A secondary but quite important concern is with the integration of that experience into issues of identity (Tomkins, 1979).

THE USES OF AFFECT

The production of affect and its meaning or its relationship to intelligence, personality, and the network of the infant's interactions are just beginning to be studied. This is true even though we have used affective cues habitually in an unaware manner to indicate cognition.

To draw a parallel with verbal communication, there are three areas of concern in the study of affect: *syntactics* (signs and relations between signs), *semantics* (relations between signs and their designata) and *pragmatics* (aspects which involve sign users). In this chapter we have dealt primarily with syntactics and pragmatics and only partially with semantics. We have documented the facial movements of two infants and noticed differences in production of movements of different sorts. We have noted that particular configurations of these facial positions have been given semantically meaningful names (Izard, 1971; Tomkins, 1962), but we have hesitated to impart meaning in the instances studied. Taking an entirely different set of observations and tools, we argued that regardless of semantics, sign users—parents, caretakers, and psychologists—interpret facial expressions in infants as if they were meaningful indicators of motivation and intellectual involvement, and there is a wealth of common sense to suggest that this is the case. Finally, we argued that the semantics of affect are interlocked with cognition and are meaningful, but fuzzy, indicators of cognitive development.

REFERENCES

BAYLEY, N. *Manual for the Bayley scales of infant development.* New York: The Psychological Corporation, 1969.
BIRNS, B., & GOLDEN, M. Prediction of intellectual performance at 3 years from infant test and personality measures. *Merrill-Palmer Quarterly,* 1972, *18*(1), 53.

BLURTON-JONES, N. G. Criteria for use in describing facial expressions in children. *Human Biology*, 1971, *43*(3), 365.

CICCHETTI, D., & SROUFE, L. A. The relationship between affective and cognitive development in Down's syndrome infants. *Child Development*, 1976, *47*, 920–929.

CICCHETTI, D., & SROUFE, L. A. An organizational view of affect: Illustration from the study of Down's syndrome infants. In M. Lewis & L. A. Rosenblum (Eds.), *The development of affect*. New York: Plenum Press, 1978.

HAVILAND, J. M. Sex-related pragmatics in infants' nob-verbal communication. *Journal of Communication*, Spring 1977, *27*, 80–84.

HAVILAND, J. M. *The origins and early socialization of affect*. Unpublished manuscript. Paper presented at University of Delaware Departmental Colloquium, 1981.

IZARD, C. W. *The face of emotion*. New York: Appleton-Century-Crofts, 1971.

IZARD, C. E., & DOUGHERTY, C. M. Two complementary systems for measuring facial expressions in infants and children. In C. E. Izard (Ed.), *Measuring emotions in infants and children*. New York: Cambridge University Press, 1982.

LAMB, M. E., & SHERROD, L. R. (Eds.). *Infant social cognition*. Hillsdale, N.J.: Lawrence Erlbaum Associates, 1981.

LELWICA, M., & HAVILAND, J. M. Ten-week-old infants' reactions to mothers' emotional expressions. Paper presented at a meeting of the Society for Research in Child Development, Detroit, April 1983.

LEWIS, M. Self-knowledge: A social cognitive perspective on gender identity and sex-role development. In M. Lamb & L. Sherrod (Eds.), *Infant social cognition*. Hillsdale, N.J.: Lawrence Erlbaum Associates, 1981.

LEWIS, M., & BROOKS, J. Self-knowledge and emotional development. In M. Lewis & L. Rosenblum (Eds.), *The development of affect*. New York: Plenum Press, 1978.

LEWIS, M., & MICHALSON, L. The measurement of emotional state. In C. E. Izard (Ed.), *Measuring emotions in infants and children*. New York: Cambridge University Press, 1982.

MAHLER, M. S. *On human symbiosis and the vicissitudes of individuation* (Vol. 1): *Infantile psychosis*. New York: International Universities Press, 1968.

MALATESTA, C. Z., & HAVILAND, J. M. Emotion socialization in the infant: Age and sex differences and the influence of maternal emotional traits. *Child Development*, 1982, *53*, 991–1003.

OSTER, H. "Recognition" of emotional expression in infancy? In M. E. Lamb & L. R. Sherrod (Eds.), *Infant social cognition*. Hillsdale, N.J.: Lawrence Erlbaum Associates, 1981.

PIAGET, J. *Les relations entre l'affectivité et l'intelligence dans le développement mental de l'enfant*. Paris: C.D.U., 1954.

PIAGET, J. *The construction of reality*. New York: Ballantine Books, 1971.

TOMKINS, S. S. *Affect, imagery, consciousness* (Vol. 1): *The positive affects*. New York: Springer, 1962.

TOMKINS, S. S. Script theory: Differential magnification of affects. In H. E. Howe, Jr. & R. A. Dienstkier (Eds.), *Nebraska Symposium on Motivation* (Vol. 26). Lincoln: University of Nebraska Press, 1979.

ZAJONC, R. B. Feeling and thinking; preferences need no inferences. *American Psychologist*, 1980, *35*, 151–175.

ZIVIN, G. Watching the sands shift: Conceptualizing development of nonverbal mastery. In R. S. Feldman (Ed.), *The development of nonverbal communication in children*. New York: Springer-Verlag, 1982.

14   *Motivation and Cognition in Infancy*

LEON J. YARROW AND DAVID J. MESSER

Throughout the history of psychology there has been lively controversy about the concept of intelligence. A recurrent issue has been whether it is meaningful to conceptualize intelligence as a global attribute or whether it is more meaningful conceptually to think of it in terms of its component functions (Burt, 1972; Guilford, 1956; Spearman, 1927; Wechsler, 1950). For some time, however, there has also been unease about the isolation of cognitive characteristics from other aspects of functioning (Dember, 1974; Rapaport, 1951; Wechsler, 1950). A basic question is whether cognitive behaviors are a separate domain, an isolated segment of functioning, or whether these abilities are integrally related to other personality and motivational characteristics. Controversy on these issues has been sharpened in recent years as our concepts of motivation have changed, and our view of cognitive functioning has become more differentiated.

The single index of intellectual development, the IQ, proved to be a useful device for cataloging children and for adapting educational experiences to children's abilities. The simplicity of the measure was, however, misleading; in concentrating on a single index of ability, psychologists lost sight of the varied aspects of competent functioning. Over 50 years ago one of the pioneers of infant testing, Gesell (1925), pointed out

Leon Yarrow died in July 1982. His work on motivation and cognition demonstrated a continuing concern with answering some of the fundamental questions of developmental psychology. It is hoped that this is reflected in my contribution to our chapter (D.J.M.).

LEON J. YARROW • Late of the Child and Family Research Branch, National Institute of Child Health and Human Development, Bethesda, Maryland 20205. DAVID J. MESSER • Psychology Division, The Hatfield Polytechnic, Hatfield, Herts AL10 9AB, England.

forcefully the limitations of global measures: "A single summative numerical value cannot do justice to the complexity and variability of infant
development. Any adaptation of our tests and methods which, for psychometric convenience, would affix IQ's to infants is undesirable, and is
inadequate for the scientific study of growth processes" (p. 29). Despite
these admonitions, the global measure of intelligence still dominates
thinking about development. In infancy, it seems more meaningful to
consider specific developmental competencies rather than global intelligence. Recently, traditional views of intelligence have begun to be
replaced by more process-oriented conceptions. In infancy, intelligence
is no longer seen simply as a global aggregate, but as being made up of
many component processes such as prehension, object permanence,
motor control, and symbolic abilities. The common measures of intelligence in infancy have not been successful in predicting later functioning. Essentially these measures assess limited functions that show
marked transformations. The underlying motivation for competence
that may have significance for later functioning has been largely ignored.

Since 1960, there have been important shifts in our thinking about
motivation. The new model of motivation views infants, children, and
adults as active seekers of stimulation who are motivated to explore the
environment, to process stimulation actively, to have an impact on their
surroundings, and to become competent. From this perspective, the
desire to have effects on and control the environment is as compelling a
motive as the drive to reduce basic physiological tensions. This new
view on motivation came from a variety of sources, such as Piaget's
observations on young children and the research on animals that pointed to the existence of a motive to explore, to be active, and to solve
manipulative problems (Berlyne, 1955, 1957, 1958; Butler & Harlow,
1957).

A variety of views about the motivation to be competent and to
master the environment have been articulated; there is a common core
in these theories, but there are also important differences. Because the
concept lacks clarity, the theoretical base for the development of assessment procedures in this area has been weak. Moreover, despite recognition of the importance of tracing the beginnings of mastery motivation,
there are almost no techniques for assessing mastery motivation during
infancy. In this chapter we shall attempt to clarify different theoretical
positions about the concept of effectance or mastery motivation. We

shall also describe the procedures that we have developed to assess mastery motivation during infancy and shall consider the relations of mastery motivation to cognitive development.

THEORETICAL POSITIONS ON COMPETENCE MOTIVATION

Piaget's (1936/1952) great contribution to psychological theory has been called a theory of cognitive development, but Piaget, perhaps more than any other cognitive theorist, has emphasized the goal-oriented character of behavior. Although he does not talk about motivation *per se*, it is clear that Piaget did not draw the neat boundaries of the academic psychologist between thinking and motivation. The concept of adaptation that is central in Piagetian theory emphasizes the dynamic interaction between the infant and the environment. His finely detailed descriptions of infant behavior, in sharp contrast to global measures of development, highlight the varied psychological processes in cognition and their interplay with motivational factors.

In Piaget's view, the motivation to find out, explore, and solve problems is inseparable from the cognitive process of achieving a better adaptation to reality. The disparity between the child's schemata and perceptions of reality is regarded as the source of motivation. The implication is that cognitive development is facilitated by an environment that provides an optimal level of disparity between the children's own schemata and the problems and events that the children encounter.

In a seminal article on effectance motivation, R. W. White (1959) pointed out that many behaviors cannot be explained simply in terms of deficiency motives. The processes governing hunger, thirst, or sex are inadequate to explain play, exploration, and efforts to interact effectively with the environment. White hypothesized the existence of a *competence motive* manifested in exploration, curiosity, mastery, and the seeking of an optimum level of stimulation. He postulated that behavior in the service of this drive is directed, selective, and persistent. Later, White (1963) introduced the term *sense of competence* to represent the subjective feelings that individuals develop as a consequence of past successes or failures. White also pointed to the reaction, in the writings of Hendrick (1943) and Erikson (1952), against Freudian notions of instincts based on sex or aggression. Hendrick (1943) postulated an instinct to master the environment, indicating that the child shows a motive to use and improve each new ability, with pleasure arising from competent performance. In a similar vein, Erikson (1952) suggested that the child has a sense of industry, a disposition to refine and develop new skills in the gross motor, fine motor, and intellectual spheres.

White's (1959, 1963) views about a motive for competence brought some conceptual order to the diverse and apparently unrelated activities of children. He did not, however, attempt to make the concept operational. Moreover, because his focus was on developing a brief for the existence of a competence motive, no consideration was given to measuring differences in the strength of this motive.

Hunt's (1963, 1965) views of intrinsic motivation are basically in accord with White's formulations. He questions drive-reduction as the sole source of motivation. Hunt proposes that not all behavior is motivated by deficit drives and suggests the existence of an intrinsic motivation to attend to people and objects, to learn about them through exploration, and to have an impact on them. With regard to the development of purposive behavior, he notes that at first the neonate orients to objects and visually inspects them. As infants' behavioral repertoires increase, manipulative exploration is added to their way of learning about the environment. This manipulation in turn leads to active attempts to have an impact on the environment and to secure feedback from objects and people.

Regarding the process by which mastery behavior is elicited and maintained, Hunt suggests that the earliest form of intentional behavior occurs when spontaneous acts are responded to, as when the adult imitates the infant's vocalizations. Very soon the infant anticipates that vocalization will lead to a response by the mother or other caregiver. Early in life, infants explore objects and repeat these activities until the object becomes familiar; in the course of exploration, they develop a generalized expectation that things should be recognizable. After this expectation has developed, the appearance of novel stimuli elicits the infant's visual attention to novel objects, which in turn leads to inspecting and exploring them. An optimal degree of discrepancy from the familiar is thought to be most conducive to attention and active exploration. Too little discrepancy is associated with apathy, and too great discrepancy is associated with withdrawal and sometimes with fear. Hunt believes that when an infant has fully explored and mastered the stimuli available to him, interest wanes and he seeks situations of greater complexity.

Intrinsic motivation, according to Hunt, becomes a significant motive force only when hunger, thirst, and pain are absent. As long as homeostatic needs are the only source of motivation, the infant remains largely a responsive organism, reacting to drive stimuli or to changes in the character of receptor inputs. Hunt suggests that as the memories of an object or event are established, children become capable of instituting actions to regain perceptual contact (Piaget's "reversal transformation" or "secondary circular reaction"). Other data (Užgiris & Hunt, 1975) indicate that infants engage in activities in anticipation of specific outcomes before many objects and events achieve recognitive familiarity. This shift from responsiveness to stimuli to anticipation of effects marks the beginnings of intentionality and is an early manifestation of the distinction of means from ends (Piaget, 1936/1952). The processes that Hunt outlines provide a bridge between White's views of competence in the human organism and the beginnings of mastery motivation in infancy. Moreover, Hunt relates to theory and research findings on learning and motivation the mechanisms by which mastery develops and becomes consolidated.

Harter (1978a, 1980) uses the term *effectance motivation* to describe a motive toward producing effects on and dealing effectively with the environment. Effectance motivation is made up of several components—cognitive, interpersonal, and motor competence—and associated with these competencies are feelings of efficacy. Initially, the association of praise with nurturance leads to praise becoming a secondary reinforcer. Independent mastery attempts resulting in success and social reinforcement lead to increased strength of effectance motivation. Social reinforcement of the striving toward mastery serves a number of functions: an incentive function that leads a child to actions for the sake of the reward, an affective function that elicits feelings of satisfaction, a general information function that helps the child identify which mastery behaviors are important, and a specific information function that provides feedback about success or failure. The reinforcement of success leads to a decreasing need for external approval because the child develops an internal self-reward system and an internal set of mastery goals. At some time during middle childhood, these functions become internalized, and children are able to reward, maintain, and control their own behavior. Harter also notes that success with too easy or too difficult tasks is less pleasurable than success on tasks that present an optimal level of challenge.

Effectance motivation, according to Harter, is weakened by the failure of mastery attempts, the reinforcement of dependency, and disapproval or lack of reinforcement for independent mastery. All of these circumstances lead to dependence on external approval and on external goals, and because the child has not experienced approval, it does not become internalized. Disapproval and lack of success are likely to result in children seeing themselves as lacking in competence, becoming anxious and adhering to a belief in an external locus of control in mastery situations.

Harter concedes that this model may not adequately capture the complexity of behavior. She points out that the model may become much more complicated if one considers all possible combinations of success/failure with reinforcement/lack of reinforcement. Harter recognizes that the positive and negative aspects do not operate in isolation, but interact. She raises questions about the optimal conditions for the development of intrinsic motivation, that is, whether it is more likely to arise in an environment where success is maximized than in one where there is a reasonable balance between success and failure. Although Harter differentiates two forms of intrinsic motivation, she does not develop the very different implications of these two views. One form, similar to White's, is motivation that exists from birth and is relatively stable with respect to environmental influences. The other is motivation whose origins lie in early experience; it is intrinsic only in the sense that it eventually becomes internalized. Harter has given us a more differentiated concept of effectance motivation and has also emphasized the importance of social factors in the development of motivation.

Using a slightly different perspective, a number of studies have focused on the importance of responsiveness to the infant and to the infant's effect on others. Watson (1966, 1972, 1979) proposes that detection by the infant of a contingency between stimulus and response is a releasing stimulus for vigorous smiling and cooing. This expression of positive affect can be seen as one aspect of White's notion that satisfaction results from mastery of a task. Watson (1966, 1972) also proposes that having an effect on the environment is associated with a generalized contingency awareness that facilitates learning in other circumstances. This awareness, he indicates, is similar to White's effectance motivation. Similarly, Lewis and Goldberg (1969) have suggested that maternal responsiveness to the infant leads to a generalized expectation of having an effect on the environment. Contingent social stimulation by the mother motivates the infant to engage in new behaviors and try new skills. Goldberg (1977) also emphasizes the relevance of predictable and contingent responses in developing feelings of competence in the child. An unresponsive and unpredictable parent can decrease an infant's feelings of effectiveness (and an unpredictable infant can have a similar effect on a parent's feelings of effectiveness).

In studying the growth of competence, Bronson (1971, 1974) sees young children's behavior as being organized toward four motivational goals, one of which is the attainment of skills that are contingent on the child's actions. Bronson regards this goal as being particularly important for the later development of an "orientation for competence." Contingent relations between actions and consequences in the inanimate environment, as well as patterns and contingencies in mother–infant interaction, are precursors of an orientation for competence. Children's awareness of having effects on the environment through their actions leads not only to a sense of competence but to the acquisition of new skills.

The use of learning paradigms has suggested that early exposure to contingent stimulation has an influence beyond that of forming an association between two events. Contingent simulation, by facilitating learning and the development of cognitive skills, reinforces the infant's sense of competence and thus may influence mastery motivation. Ramey, Starr, Pallas, Whitten, and Reed (1975) found that tutoring families to foster children's feeling of mastery influenced later performance on a learning task. Use of a more controlled procedure, (Finkelstein & Ramey, 1977; Ramey & Finkelstein, 1978) demonstrated the beneficial influence of exposure to contingent stimulation on subsequent learning of different responses in new situations.

Observations suggest that the contingent responsiveness of adults to children has effects on the development of competence. Clarke-Stewart (1973) has found a relationship between responsiveness of the mother and the child's later social and intellectual development. Lewis and Goldberg (1969) have obtained a positive correlation between the contingency of responding to young infants' vocalizations and their habituation to stimuli. This finding can be linked to a later study (Lewis & Brooks-Gunn, 1981) that suggests there is a relationship between young infants' functioning on a habituation test and later scores on a developmental test at 2 years. Lewis and Coates (1980) report a positive relationship between parents' responsiveness to a variety of child behaviors and the child's developmental competence. Morgan, Busch, Culp, Vance, and Fritz (1982) found that the mother's responsiveness to the infant's signals was associated with mature behavior in free play, for example, appropriately combining objects and producing effects with toys. Beckwith, Cohen, Kopp, Parmalee, and Marey (1976), on the other hand, found no strong relationships between contingent response to distress and later scores on a developmental test.

The presence of noncontingent stimulation appears to have a disruptive influence on development. Watson (1971) exposed one group of infants to a mobile that turned in response to their head movements; another group received noncontingent stimulation. Both groups were exposed 6 weeks later to a mobile that was contingent on their response. Only the infants who had previously received contingent responses learned this relationship. Experience with noncontingent stimulations appears to have *interfered* with later learning of the same response. The literature on learned helplessness contains a similar theme. Dogs that could not escape from electric shocks were found to lack this ability when escape was later possible (Seligman & Maier, 1967). Clearly, these results have implications for the early development of mastery motivation. They also raise questions about the extent to which the lack of contingent responsiveness that is often associated with child-care institutions is a factor in the retardation of infants (Dennis, 1973).

The common thread running through these studies is that mastery behavior or striving for competence is engaged in for its own sake. The motive to master has been related to many of the cognitive and motoric developments that occur during infancy and childhood. The origins of such motivation are still not completely clear. There may be congenital differences in the strength of the drive to master. It also seems clear that the responsiveness of the environment contributes to the development of mastery motivation; both the responsiveness of inanimate objects and responsiveness of people seem to play a role. Since White (1959) first called attention to the importance of motivation for competence, there has been gradual refinement in the models and in the attempts to understand the underlying processes.

ASSESSMENT OF MASTERY MOTIVATION

For the most part, procedures for assessing mastery motivation have not kept pace with the theorizing. Little systematic attention has been given to assessment procedures; there have been few guidelines for selecting techniques to elicit mastery behavior or to define the behaviors that indicate mastery. In the discussion of the research that follows, we shall consider in some detail a few of the measures that have been used to examine mastery motivation and shall discuss some procedures that we have developed for assessing mastery early in infancy.

Research by Harter and her colleagues has used several methods for studying effectance or mastery motivation. In one investigation, Harter, Shultz, and Blum (1971) examined children's signs of pleasure when they completed a task. Children 4 and 8 years old smiled more frequently when they were able to give correct responses to the pictures on the Peabody Picture Vocabulary Test than when they produced incorrect answers. In a subsequent study using anagrams varying in levels of difficulty, Harter (1974) found that smiling in 11-year-old children was directly related to the difficulty of the anagrams. More recently, a curvilinear relationship was found between pleasure and task difficulty; the lowest and highest levels of difficulty produced less pleasure than an optimal level of difficulty (Harter, 1978b). Although pleasure derived from a sense of mastery may be associated with smiling after successfully completing a task, the expression of pleasure is probably not an especially sensitive measure of a motive for mastery. Many factors influence smiling, thus obscuring the relationship between mastery and the expression of pleasure. Moreover, this method can assess only the presence or absence of smiling, not the degree of pleasure experienced.

In another study, Harter and Zigler (1974) defined effectance motivation in terms of four characteristics: (1) the use of different responses on subsequent encounters with a problem that can be solved in different ways, (2) preference for novel stimuli, (3) preference for having competence acknowledged rather than being given tangible rewards, and (4) preference for challenging tasks. Four tasks were designed for use with first- and second-grade children to measure these dimensions of mastery. The measure of response variation was a paper and pencil maze with alternative paths. The score was the number of different maze segments traversed on different trials. The measure of curiosity for novel stimuli consisted of a number of cardboard houses. One had the same picture behind the door as on the front; the other was blank in front and had an unknown picture behind it. The score was the percentage of trials on which the child chose the novel picture behind the blank door. As a measure of mastery for the sake of competence, a pegboard was used that had a graduated series of holes differing in depth in which pegs of different lengths could be placed so that they all would be at the same height. The fourth task, preference for challenging tasks, had puzzles varying in difficulty. The difficulty level of the tasks was varied by changing the number of puzzle pieces that had to be replaced. On each of three trials, the children were given an opportunity to choose the puzzle on which they wished to work. Findings supported the expectation that retarded children with deprived life histories would be less strongly motivated to master tasks than would normal children.

Harter (1981) has recently presented methods for assessing the various elements of effectance motivation based on the use of questionnaires. A self-report scale compared children's responses to questions on the dimension of intrinsic-extrinsic motivation in the classroom. Five characteristics are identified: (1) preference for challenging rather than easy work, (2) incentive to work for one's own satisfaction or interest rather than for external approval, (3) desire to work independently rather than to seek help, (4) independence of opinions rather than reliance on external judgments, and (5) internal criteria for success or failure rather than external criteria. Reasonable intercorrelations were found between these dimensions. As the children progressed from third to sixth grade, a shift from intrinsic to extrinsic motivation occurred for dimensions 1, 2, and 3, which appears to reflect an increasing motivation to work for the teacher's approval rather than the child's own interest; and from extrinsic to intrinsic for dimensions 4 and 5, which appears to reflect a development of independence in evaluating the child's own efforts.

The studies considered so far have all dealt with children who have had a reasonable command of language. With preverbal children, the difficulty in setting up tasks and explaining the tasks is increased considerably. Perhaps because of these problems, striving for competence during infancy has not been studied extensively.

Wenar (1972, 1976) studied "executive competence" in 25 mother–infant pairs between 12 and 20 months, observing behavior in the home. *Executive competence* was defined as the ability to "initiate and sustain locomotor, manipulative and visually regarding activities at a given level of complexity and intensity, and with a given degree of self-sufficiency" (Wenar, 1976, p. 191). Three 15-minute periods selected from the observation sessions were coded in terms of the child's locomotor, manipulative, and visual encounters with the environment. The behaviors were rated on duration, intensity, complexity (except for vision), and affect accompanying the activity. The observers also gave an overall rating of the child's behaviors. Wenar found acceptable consistency over time for highly competent children, whereas low competence either improved or improved and then declined. Relationships were found between the environment and general competence. Stimulating mothers raised the level of competence, whereas restrictive mothers depressed it.

Our interest in mastery motivation was stimulated by the findings of a study in which we examined selective relations between parameters of the environment and clusters derived from the items of the Bayley Scales (Yarrow, Rubenstein, Pedersen, & Jankowski, 1972). Some clusters were classified according to the psychological processes they tapped, for example, *goal-directedness, object permanence,* and *secondary circular reactions.* Still others dealt with the more conventional classifications of early developmental functions: *gross motor, fine motor, visually directed reaching and grasping, social responsiveness,* and *vocalization and language.* Our review of this study summarizes the interrelations among these clusters. We found differing degrees of relationship among separate aspects of development; at the same time, we were struck by the interdependence of cognitive, motor, and motivational functions.

Three clusters were labeled "cognitive-motivational" to emphasize their dual character: *goal-directedness, reaching and grasping,* and *secondary circular reactions.* The setting of goals and persistence in activities required to attain goals are core manifestations of motivated behavior. One of the earliest evidences of goal-directedness is the infant's visual orientation to novel objects, followed soon by attempts to secure interesting objects and to explore them. Six items in the *goal-directed* cluster measured the infant's focused and persistent attempts to make contact with objects: obtaining a cube out of reach, pulling on a string to secure a ring that was beyond the infant's immediate grasp, attempting to secure a toy by unwrapping the paper around it, reaching for and trying to pick up three cubes, and attempting to grasp a small pellet. The high correlation of this cluster with the Bayley Mental Developmental Index supported our view of the centrality of motivation in early development. Giving additional support to our central thesis of a link between cognition and motivation was the high relationship between gaol-directedness and object permanence. The high correlation between goal-directedness and the fine motor cluster emphasizes the difficulty in sorting out the relative importance of skill and motivation.

The nine items in the *reaching and grasping* cluster measured primarily the coordination of vision and prehension. Reaching for and grasping an object involves a desire to obtain it, persistent attention to the object, and adapting fine motor skills to secure and manipulate it. This cluster was also highly related to all the other variables except vocalization and language.

Giving further support to the central importance of the motivational aspects of cognitive development was the high relationship between *secondary circular reactions* and all the other clusters except vocalization and language. The behaviors subsumed under the category of secondary circular reactions were activities directed toward producing interesting results.

The three cognitive-motivational clusters were highly interrelated. The boundaries between the skill and the purposeful use of the skill are not sharp; the distinction is a subtle one. These cognitive-motivational activities appear to indicate the earliest manifestations of attempts to master and to obtain feedback from the environment. They may well be precursors in infancy of effectance motivation in later childhood. It would seem that these cognitive-motivational functions are closely dependent on the development of perceptual, motor, and cognitive skills; in turn, these functions may influence significantly the development and elaboration of these skills.

Measurement of Mastery Motivation

The findings of the previous study, which was not primarily concerned with the relations between goal-oriented behavior and cognitive development, encouraged us to attempt to develop techniques for assessing mastery motivation (Morgan, Harmon, Gaiter, Jennings, Gist, & Yarrow, 1977; Yarrow, Morgan, Jennings, Harmon, & Gaiter, 1982). Mastery motivation in infancy was conceptualized in terms of three components: producing effects with objects, practicing emerging sensorimotor skills, and problem-solving. Eleven tasks were chosen to elicit these behaviors. The *effect-production* tasks involved manipulating objects to secure feedback or to produce visual or auditory effects—for example, pushing a button to make an animal come out of a door. The *practicing emerging sensorimotor skills* items included tasks that require the use of skills just emerging at the developmental period studied, for instance, at 1 year, placing objects in a container. The *problem-solving* items elicited behavior directed toward solving detour problems and using means–end relationships—reaching around a plexiglass screen to obtain an object for example. The major measures of behavior obtained from each of these different tasks were *persistence, competence,* and *positive affect. Persistence* was the length of time the child engaged in task-directed behavior. Ratings were also made of persistence during the structured mastery sessions and during administration of the Bayley Scales; in addition, it was noted when a child spontaneously attempted to repeat a problem on the structured task or the Bayley Scales. The measure of *competence* was simply the number of trials in which the infant correctly produced the effect, combined the objects, or secured the goal. *Positive affect* was rated on a 5-point scale.

On the whole, persistence in task-directed behavior was fairly high in these 1-year-olds; 60% of the session was spent trying to accomplish the task. More than half of the infants (63%) solved the tasks (Jennings, Harmon, Morgan, Gaiter, & Yarrow, 1979). There were, however, great individual differences in persistent behavior. Examination of the correlations among persistence in the three tasks revealed different patterns of relationships for boys and girls. For boys, persistence was highly correlated across the three components (i.e., effect-production, practicing emerging skills, and problem-solving), with only one relationship failing to reach significance. However, the same correlations for girls were lower, with only two reaching significance (Jennings, Yarrow, & Martin, 1981). *Positive affect* shown while working on the tasks was also measured on the assumption that it might be an index of the child's feelings about being confronted with a challenging situation. Although Bronson (1974) and Watson (1972) have reported that infants smile when they have handled tasks successfully, expression of positive affect was rare.

In this study we had a subsample, 25 cases, on whom we had data at 6 months of age, thus enabling us to investigate precursors of mastery motivation. A significant relationship was found between the Goal directedness cluster from the Bayley scale at 6 months of age and the overall measure of mastery motivation at 1 year ($r = .36, p < .05$). The finding can be interpreted as indicating some continuity in infants' motivated behavior during the second half of the first year.

In a more recent study, we have attempted to investigate the first manifestations of mastery motivation in infancy, and have looked at its origins and the course of its development in a sample of 68 children from 6 months to 2½ years (Yarrow, McQuiston, MacTurk, McCarthy, Klein, & Vietze, 1983). The three components of mastery were similar to those in the previous study: effect-production, practicing sensorimotor skills, and problem-solving. The measures of the child's behavior with the three types of mastery tasks included the time spent in *task-unrelated behavior,* a measure of inattention; the duration of *visual attention* to the toys without manipulation; the duration of general *exploratory behavior* not specific to the task (e.g., shaking the toy); and *persistence,* the total time spent in manipulative behavior directed toward solving the task. To assess the child's eagerness to become involved with the materials, we also measured *latency to task involvement.* (These measures are described in Vietze, Pasnak, Tremblay, McCarthy, Klein, & Yarrow, 1981.)

Analyses of the interrelations among the measures of mastery at 6 and 12 months show some interesting patterns of relationships. As early as 6 months, there seem to be stylistic differences in ways of handling challenging tasks. Some infants are cautious in approaching test materials and tend to become involved with them in more superficial ways— they spend more time off task and look at the materials without manipulating them—whereas those who are eager to make contact with the materials quickly persist in their efforts to solve the tasks.

Analyses of the relationships between measures of mastery at 6 and 12 months show changes. These changes do not reflect discontinuity in these functions but represent theoretically meaningful transformations of mastery behavior. Interpretation of these changes as transformations is based on the assumption of a hierarchical organization of the components by the level of skill required. Producing effects requires only simple motor and low-order cognitive skills. The component practicing sensorimotor skills requires the coordination of fine and gross motor skills with simple cognitive abilities. It involves a more sophisticated level of response than simply making contact with objects and securing feedback. Problem-solving, in addition to demanding the coordination of motor skills, requires rudimentary symbolic capacities, such as an appreciation of means–end relationships.

The changes in mastery behavior between 6 months and 12 months can be divided into two patterns. Either task involvement predicts the same behavior on a higher component; for example, at 6 months, persistence on effect production predicts later persistence on tasks involving practicing sensorimotor skills. Alternatively, a higher level of task involvement is predicted on the same component. For example, at 6 months, exploratory behavior on the component practicing sensorimotor skills predicts later persistence on this component. These findings imply developmental changes in capacities. Such developmental transformations are consistent with the views of McCall, Eichorn, and Hogarty (1977), who suggest that during periods of rapid development there is no simple continuity for most behaviors, but there is some predictability for functionally equivalent behavior patterns.

INTERDEPENDENCE OF COGNITION AND MOTIVATION

A number of studies have examined the relationship between cognition and motivation; the findings from these studies will be reviewed in this section. Before we consider these data, a brief presentation will be made of the theoretical connections between these two dimensions.

The interdependence of motivation and cognition has been noted by theorists from many different perspectives (Hunt, 1965; Piaget, 1936/1952; Rapaport, 1951). The unifying theme in Piaget's (1936/1952) discussion of infant activities is the intentionality of behavior. Pervading his vivid descriptions of young infants are accounts of their active attempts to process stimuli and to have an impact on the environment. Young children's responses and sensitivity to stimuli are influenced by their level of competence. Thus, cognitive development occurs when there is disequilibrium between external information as the stimulus and the subject's schema or internal structure of activities.

Peter Wolff (1960), in attempting to clarify the central role of motivation in Piagetian theory, points out, "The concept of motivation in sensorimotor theory is inextricably tied to its structural conception. The schema, as the basic structural unit, and the need to function, as the motivational concept, are indissolubly linked to sensorimotor theory; all need to function results from disequilibria in structure and there is no motivation which does not refer specifically to a schematic imbalance" (p. 62). Hunt (1965), in elaborating on Piaget's ideas, discusses the notion of motivation inherent in information-processing. He believes that the infant's first orienting response to stimuli and to changes in stimulation is the earliest indication of the young child's motivation to learn about the environment. The infant first tries to maintain perceptual contact with familiar objects; later, there is a shift in interest to visual and tactile exploration of novel objects. The essence of his view is summarized in the conclusion: "A basic source of motivation is inherent within the organism's information interaction with its circumstances" (p. 270).

Ulvund (1980, 1981), reviewing the work of White (1959), Piaget (1936/1952), Hunt (1965), and others, raises the question of the extent to which cognition and motivation can be identified as separate processes. He concludes that it is difficult to separate them during the first 2 years. He reiterates Piaget's conclusion that cognitive development will progress most smoothly when there is an optimal level of discrepancy between children's perception of the world and their cognitive structures. In examining the implications of this view for early environmental stimulation, Ulvund suggests that variety of environmental stimulation is particularly important for cognitive development because "variation increases the probability that individuals at different levels of cognition encounter situations which represent a moderate discrepancy" (1980, p. 27). He also suggests that cognitive development may be related to specific qualities of stimulation in the infant's environment, a hypothesis supported by some research findings (Gaiter, Morgan, Jennings, Harmon, & Yarrow, 1982; Wachs, 1979; Wachs, Francis, & McQuiston, 1979; Yarrow, Rubenstein, & Pedersen, 1975).

Empirical support for a cognitive-motivational thesis comes from several studies. Two factor analytic studies of the Gesell Developmental Examination were carried out, one by Richards and Nelson (1938) and the other by McCall, Hogarty, & Hurlburt (1972). Both studies were based on data obtained in the Fels longitudinal study. In the first investigation (Richards & Nelson, 1938), there were only 17 items (probably because the intercorrelations had to be arduously computed on a manual calculator). Richards and Nelson found that such items as "reaches for dangling ring," "pats table," and "secures cube" had high loadings on the first factor. (These items are very similar to those used in the goal-directedness, secondary circular, and reaching and grasping clusters by Yarrow, Rubenstein, & Pedersen, 1975.) Although Richards and Nelson recognized that the first factor accounted for a substantial proportion of the variance in test performance at this age, in accord with the conceptual biases of their time, they labeled the factor "testability" or "halo effect."

Similar findings, reported 34 years later in another factor analysis of the Gesell test based on a larger sample from the Fels study, were interpreted from a different perspective (McCall *et al.*, 1972). The first factor replicated Richards and Nelson's finding in its essential features, but it included other items of a manipulative and exploratory nature. Over half the items with higher loadings on the first factor identified in this study are comparable in content to the items in the cognitive-motivational clusters identified in our studies (Yarrow *et al.*, 1972; Yarrow, Rubenstein, & Pedersen, 1975). McCall *et al.* (1972) noted that many of these items involve perceptual contingencies and suggested that these behaviors may be related to the development of internal control. This view emphasizes that an important dimension of early competence is exploration of the environment.

Further evidence of the interdependence of cognitive and motivational functions in early infancy comes from the findings of a study by Matheny, Dolan, and Wilson, (1974). Analyzing the relationships between the Mental Development Index (MDI) from the Bayley scales and ratings of behavior on the Bayley Infant Behavior Record, they found significant relationships between the behavior ratings and the MDI at 6 months; the range of correlations was from .50 to .79. Although these relationships were somewhat lower at later ages and were not significant at 12 months, both goal-directedness and object-orientation were significantly related to the Bayley MDI at 18 and 24 months. Several of the ratings assess characteristics similar to those previously identified as cognitive-motivational clusters, goal-directedness, object-orientation, attention span, and manipulation (Yarrow *et al.*, 1972; Yarrow, Rubenstein, & Pedersen, 1975). Furthermore, the relationships at 6 months were quite similar in magnitude to those found by Yarrow, Rubenstein, and Pedersen (1975) between the cognitive-motivational variables and the Mental Developmental Index.

The findings of a recent study (Yarrow *et al.*, 1982) show that some aspects of mastery motivation are significantly related to scores on the Bayley scales. (The measures and tasks are described on page 462.) The overall measure of mastery, persistence on the 11 mastery tasks, is significantly related to concurrent Bayley MDI ($r = .60$, $p < .01$) at 1 year. An especially high relationship is found between the Bayley MDI and one component of mastery, persistence in practicing emerging sensorimotor skills ($r = .78$, $p < .001$). Another measure of mastery, latency to task involvement (which is simply the speed with which the child becomes involved in the tasks), is also related to the Bayley MDI ($r = -.56$, $p < .01$), signifying that children who show task-orientation more rapidly have higher scores on the Bayley. The especially high relationship between the Bayley and the practicing emerging sensorimotor skills component indicates that infants who work assiduously at perfecting emerging skills are likely to become more competent. However, an alternative explanation might be that more competent infants derive greater satisfaction from working on skills and, therefore, are more likely to practice them.

In a follow-up study of these children at $3\frac{1}{2}$ years of age (Jennings *et al.*, 1981), relationships from infancy to early childhood were found between developmental competence and mastery. For the subjects on whom data were available at 6 months ($n = 15$), the Bayley MDI was significantly related to the overall measure of mastery at $3\frac{1}{2}$ years that is persistence on challenging tasks ($r = .44, p < .05$). Different patterns of relationships between 1 and $3\frac{1}{2}$ years were found for boys and girls. For girls, early mastery and early competence predicted later *competence*: Overall persistence and the Bayley MDI were related to one aspect of the $3\frac{1}{2}$ year McCarthy General Cognitive Index, perceptual performance ($r = .60, p < .05$ and $r = .61, p < .05$, respectively). On the other hand, early mastery and early competence predicted later *mastery* for boys. Persistence in producing effects at 1 year was significantly related to two measures of mastery at $3\frac{1}{2}$ years: the overall measure of mastery ($r = .44, p < .05$), and task orientation ($r = .57, p < .05$). In the case of the relationship between competence and mastery, for boys, the Bayley MDI was significantly related to later persistence ($r = .54, p < .05$). These results support previous findings of greater predictability for measures of cognition for girls than for boys (Bayley & Schaefer, 1964; Honzik, 1976; McCall *et al.*, 1972). Similarly, other studies show greater stability in task-oriented behaviors for boys during the early years (Hunt & Eichorn, 1972; Matheny *et al.*, 1974).

In a more recent study (Yarrow *et al.*, 1983) a detailed examination of these relationships was made (the tasks and measures are described on pages 463 and 464). At 6 months, the total persistence across the three mastery components was not significally related to concurrent Bayley MDI scores. However, analysis showed that mastery on two of the components was meaningfully related to the MDI. At 6 months, persistence on effect-production tasks was related to the MDI ($r = .43, p < .01$). Furthermore, exploratory behavior on practicing emerging sensorimotor skills was related to the MDI ($r = .34, p < .05$). These findings suggest that during early infancy, producing effects with objects and exploring the properties of objects are especially important ways of interacting with the environment. In a previous study (Yarrow, Rubenstein, & Pedersen, 1975), a significant relationship was found between the number of responsive toys available to the infant and cognitive development, thus giving support to the importance of experiences with responsive objects and opportunities to learn about the environment through exploration.

At 12 months, the children who showed a higher level of task-involvement were more competent. Significantly related to concurrent Bayley MDI were overall measures of time off task ($r = -.35$, $p < .01$) and total persistence ($r = .59$, $p < .01$). Exploratory behavior on the two lowest components was negatively correlated with the Bayley MDI (effect-production, $r = -.33$, $p < .05$; practicing sensorimotor skills, $r = -.63$, $p < .01$), but persistence on the highest level component, problem-solving, was positively related to the competence ($r = .28$, $p < .05$).

The findings of a positive relationship between exploratory behavior and competence at 6 months and of a negative relationship at 12 months deserve some comment. Exploratory behavior signifies interest in learning about the characteristics of the environment, but it is not necessarily oriented to successful completion of the task; it does not represent striving to complete a task successfully. The finding that at 6 months the more competent children engage in exploratory behavior on tasks involving practicing sensorimotor skills suggests that exploratory behavior may be an especially salient behavior at this age because the active exploration of objects may contribute to the infant's level of competence; this behavior may be more important to developmental progress than achievement. By 12 months, exploration may be a less appropriate way of interacting with the environment. This change might be related to the growing awareness of means–ends relationships and the development of secondary circular reactions. These data suggest that the investigation of mastery requires taking into consideration both the type of task used and the form of the behavior that is observed, two dimensions that appear to change with the children's development. The findings that concern persistence further strengthen this interpretation.

Persistence is goal-oriented behavior; it reflects striving to achieve a goal. At both ages, persistence has a stronger relationship with the MDI than any of the other levels of task involvement. However, the relationship between persistence and competence differs at 6 and at 12 months; this change appears to be a meaningful transformation. At 6 months, persistence on the lowest level component was correlated with the MDI; at 12 months, persistence on the middle level component was correlated with the MDI. As the infants become older, the strongest relationship between persistence and competence occurred with more demanding tasks.

Our findings indicate that motivated behavior is not a static entity; as the children develop, new skills are introduced and new levels of involvement are manifested. Considering the different strategies employed in our studies and the studies of Richards and Nelson (1938), McCall et al. (1972, 1977), and Matheny et al. (1974), there is a notable similarity in findings: All investigations show that there are common motivational components in measures of early infant development. Motivational characteristics such as persistence, attentiveness to objects, and the desire to interact with and elicit feedback from objects were highly correlated with the Mental Development Index of the Bayley scales.

In examining the relation between motivation and cognition, the question of which comes first has no clear answer. However, there are some indications that with handicapped children, mastery motivation and the associated feelings of efficacy may have an important impact on the development of competence. Provence and Lipton (1962), comparing institutionalized and home-reared infants, observed major differences between the motive to use a skill and the emergence of the skills; the institutionalized children were lacking in motivation to try new skills. The importance to developmental progress of motivation to master the environment is further emphasized by recent findings. Meyers and Howard (1982) have reported that contingent auditory stimulation has a remarkable influence on the linguistic progress of handicapped children. Brinker and Lewis (1981, 1982) have evidence that exposure of young infants to contingent stimulation leads to generalized developmental progress. A study of children with Down's Syndrome (Vietze, McCarthy, MacTurk, McQuiston, & Yarrow, 1980) identified patterns of mastery in these children that are similar to those in nondelayed children. Although Down's infants were slow in becoming involved with objects and spent a long time looking at materials before manipulating them, they showed persistence in goal-directed activity when they did become involved. The implications of these findings are that the parents and teachers of a child with Down's syndrome may interpret the child's slow involvement with objects as disinterest and consequently, may not provide him or her with sufficient opportunities to explore and master new stimuli. This lack in turn may influence the infant's cognitive development.

SUMMARY AND CONCLUSIONS

Research findings buttress the conception of cognition as an active process. Stimuli do not simply impinge on the young infant; young children do not perceive stimuli and passively register discrepancies between a familiar and a novel stimulus. When an interesting object is placed within the visual field, they explore it with their eyes and attempt to learn about its properties through touch and manipulation. When infants discover that an object gives some feedback, that sound or shape changes are made when they hit or squeeze it, they repeat the actions that elicited the responses. When they become aware of an obstacle to obtaining a desired object, infants actively attempt to circumvent it. All these active efforts to reach out and to have an effect on the environment are indications of motivated, directed behaviors. These activities require some level of cognitive awareness, a capacity to process stimuli, and the beginnings of the ability to handle symbols—for example, the capacity to compare the present stimulus with a stimulus to which they have been exposed a few seconds earlier, or the capacity to retain an image of an object at least momentarily. The reconciliation of discrepancies, mastery of the environment, and repetition of activities that produce interesting results involve more than cognition; they require the coordination of cognitive skills with motivation.

There are a number of theoretical perspectives on the origins of mastery motivation and on its relationship to cognition. The term *intrinsic motivation* suggests that the origins are within the infant. There are several ways in which an intrinsic motive is thought to develop. Piaget's model of cognitive development suggests that a motive to master is inherent to the cognitive structure. Attempts to find out and master new aspects of the environment arise from the disequilibrium between accommodation and assimilation. Another view is that the development of contingency awareness can be seen as the product both of the infants' actions and of responses to their behavior. This is intrinsic motivation in the sense that the infants have to produce actions to elicit responses from others. A further viewpoint (Harter, 1980) holds that the parent's response to mastery attempts leads to an internalization of external values. Infants who are praised for attempting difficult problems will come to value such activities. These perspectives about the origins of mastery are not necessarily incompatible. Each describes important features of the infant's early development. However, they also raise questions about which forms of influence are most important to the development of a motive to master. Although we have made progress in our conceptual thinking, our understanding of the origins of mastery motivation remains limited.

When considering the origins of mastery motivation, it is important to remember that the relationship between infants and their environment is not a simple unidirectional one. Early differences in motivation do not merely reside in the infant; rather, they represent the infant's complex interrelationship with the environment. As we have noted elsewhere in a discussion of the relations between the early environment and later development (Yarrow, Klein, Lomonaco, & Morgan, 1975), if the infant interacts actively with people and explores objects, a sequence of interactions may be set in motion that is in some measure self-reinforcing and self-perpetuating. "The infant affects his environment, not simply by selectively filtering stimulation through his individualized sensitivities, but also by reaching out and acting on his environment. He learns about the world through his active manipulation and exploration of inanimate objects, and he elicits stimulation from caregivers and others in his environment by his signals and the quality of his responsiveness to their responsive behavior to him" (Yarrow, Rubenstein, & Pedersen, 1975, p. 501). By active exploration of objects, by showing initiative, and by responsiveness to people, infants exert a powerful effect on the environment while at the same time influencing their own cognitive development; in this sense they determine the continuity of their environment.

The recognition of the intertwining of cognition and motivation in early infancy has implications for improving the predictive efficacy of infant tests. For a long time, psychologists operated on the assumption that measures of an infant's developmental status were predictive of later intellectual functioning. Most studies, however, found low- or zero-order correlations between test scores in the first year of life and IQ at 5 years and beyond. It is reasonable to suppose that there might be little continuity between a single score based on a variety of early sensorimotor functions in infancy and the verbal and symbolic skills tapped by intelligence tests. However, if one examined the relationship between specific early skills and some aspects of later intelligence, there might be more meaningful bases for prediction. We might expect that certain aspects of early development would not be related to later intellectual abilities, while others might be essential building blocks for later symbolic capacities. If measures of motivation for competence were added to the measures of specific skills, the bases for predicting later development would be broadened and prediction might be enhanced.

The research on cognition and motivation emphasizes the complex interrelationships between motivation and cognitive development. They are so closely interdependent that the question of which comes first is meaningless. The interactions between mastery motivation and developmental competence are so subtle that it is not possible to identify one as a precursor of the other; rather, the interactions are reciprocal. The few studies we have reported constitute only a small beginning in unraveling the relations between cognition and motivation. Further research is needed to identify the relevant dimensions of both areas of functioning. We believe that infant development needs to be conceptualized in more complex ways than as a taxonomy of skills; we must be also be aware of the motivational components of infant behaviors. Sensitivity to motivational components of infant's activities should help in conceptualizing infant behavior in dynamic terms; ultimately such conceptualization might lead to more adequate measures of infant functioning.

ACKNOWLEDGMENTS

The contribution of ideas, advice, and critiques from past and present members of the Child and Family Research Branch is gratefully acknowledged. This paper was written while the second author was on an NIH visiting fellowship to the Child and Family Research Branch.

REFERENCES

BAYLEY, N., & SCHAEFER, E. S. Correlations of maternal and child behaviors with the development of mental ability: Data from the Berkely Growth Study. *Monographs of the Society for Research in Child Development*, 1964, 29(97).

BECKWITH, L., COHEN, S. E., KOPP, C. B., PARMALEE, A. H., & MAREY, T. G. Caregiver-infant interactions and early cognitive development in preterm infants. *Child Development*, 1976, 47, 579–587.

BERLYNE, D. E. The arousal and satiation of perceptual curiosity in the rat. *Journal of Comparative Physiology and Psychology*, 1955, *48*, 238–246.

BERLYNE, D. E. Attention to change, conditioned inhibition (S.R.) and stimulus satiation. *British Journal of Psychology*, 1957, *48*, 138–140.

BERLYNE, D. E. The present status of research on exploratory and related behavior. *Journal of Individual Psychology*, 1958, *14*, 121–126.

BRINKER, R. P., & LEWIS, M. Cognitive intervention in infancy. In J. Anderson (Ed.), *Curriculum materials for high risk and handicapped infants*. Chapel Hill, N.C.: Technical Assistance Development System, 1981.

BRINKER, R. P., & LEWIS, M. Discovering the competent handicapped infant: A process approach to assessment and intervention. *Topics in Early Childhood Special Education*, 1982, *2*(2).

BRONSON, W. C. The growth of competence: Issues of conceptualization and measurement. In H. R. Schaffer (Ed.), *The origins of human social relations*. New York: Academic Press, 1971.

BRONSON, W. C. Mother-toddler interaction: A perspective on studying the development of competence. *Merrill-Palmer Quarterly*, 1974, *20*, 275–301.

BURT, C. Inheritance of general intelligence. *American Psychologist*, 1972, *27*, 175–190.

BUTLER, R. A., & HARLOW, H. F. Discrimination learning and learning sets to visual exploration incentives. *Journal of General Psychology*, 1957, *57*, 257–264.

CLARKE-STEWART, K. A. Interactions between mothers and their young children: Characteristics and consequences. *Monographs of the Society for Research in Child Development*, 1973, *38*, (6–7, Serial Number 153).

DEMBER, W. N. Motivation and cognitive revolution. *American Psychologist*, 1974, *29*, 161–168.

DENNIS, W. *Children of the creche*. New York: Appleton-Century-Crofts, 1973.

ERIKSON, E. H. *Childhood and society*. New York: Norton, 1952.

FINKELSTEIN, N. W., & RAMEY, C. I. Learning to control the environment in infancy. *Child Development*, 1977, *48*, 806–819.

GAITER, J. L., MORGAN, G. A., JENNINGS, K. D., HARMON, R. J., & YARROW, L. J. Variety of cognitively-oriented caregiver activities: Relationships to cognitive and motivational functioning at 1 and 3½ years of age. *Journal of Genetic Psychology*, 1982, *141*, 49–56.

GESELL, A. *The mental growth of the pre-school child*. New York: MacMillan, 1925.

GOLDBERG, D. Social competence in infancy: A model of parent-infant interaction. *Merrill-Palmer Quarterly*, 1977, *23*, 161–177.

GUILFORD, J. P. The structure of intellect. *Psychological Bulletin*, 1956, *53*, 267–293.

HARTER, S. Pleasure derived from cognitive challenge and mastery. *Child Development*, 1974, *45*, 661–669.

HARTER, S. Effectance motivation reconsidered: Toward a developmental model. *Human Development*, 1978, *21*, 34–64. (a)

HARTER, S. Pleasure derived from optimal challenge and the effects of extrinsic rewards on children's difficulty level choices. *Child Development*, 1978, *49*, 788–799. (b)

HARTER, S. A model of intrinsic motivation in children: Individual differences and developmental change. In A. Collins (Ed.), *Minnesota symposium on child psychology*. Hillsdale, N.J.: Lawrence Erlbaum Associates, 1980.

HARTER, S. A new self report scale of intrinsic versus extrinsic orientation in the classroom: Motivational and information components. *Developmental Psychology*, 1981, *17*, 300–312.

HARTER, S., & ZIGLER, E. The assessment of effectance motivation in normal and retarded children. *Developmental Psychology*, 1974, *10*, 169–180.

HARTER, S., SHULTZ, T., & BLUM, B. Smiling in children as a function of their sense of mastery. *Journal of Experimental Child Psychology*, 1971, *12*, 396–404.

HENDRICK, I. The discussion of the 'instinct to master.' *Psychoanalytic Quarterly*, 1943, *12*, 561–565.

HONZIK, M. P. Value and limitations of infant tests: An overview. In M. Lewis (Ed.), *Origins of intelligence* (1st ed.). New York: Plenum, 1976.

HUNT, J. McV. Motivation inherent in information processing and action. In O. J. Harvey (Ed.), *Motivation and social interaction*. New York: Ronald Press, 1963.

HUNT, J. McV. Intrinsic motivation and its role in psychological development. In D. Levine (Ed.), *Nebraska symposium on motivation* (Vol. 13). Lincoln: University of Nebraska Press, 1965.

HUNT, J. V., & EICHORN, D. M. Maternal and child behaviors: A review of data from the Berkeley growth study. *Seminars in Psychiatry*, 1972 4(4).

JENNINGS, K. D., HARMON, R. J., MORGAN, G. A., GAITER, J. L., & YARROW, L. J. Exploratory play as an index of mastery motivation: Relationships to persistence, cognitive functioning, and environmental measures. *Developmental Psychology*, 1979, *15*, 386–394.

JENNINGS, K. D., YARROW, L. J., & MARTIN, P. P. *Mastery motivation and cognitive development: A longitudinal study from infant to three and one half years.* Unpublished manuscript, University of Pittsburgh, Penn., 1981.

LEWIS, M., & BROOKS-GUNN, J. Visual attention at three months as a predictor of cognitive functioning at two years of age. *Intelligence*, 1981, *5*, 131–140.

LEWIS, M., & COATES, D. L. Mother-infant interactions and cognitive development in 12-week-old infants. *Infant Behavior and Development*, 1980, *3*, 95–105.

LEWIS, M., & GOLDBERG, S. Perceptual-cognitive development in infancy: A generalized expectancy model as a function of the mother-infant interaction. *Merrill-Palmer Quarterly*, 1969, *15*, 81–100.

MATHENY, A. P., DOLAN, A. B., & WILSON, R. S. Bayley's infant behavior record: Relations between behaviors and mental test scores. *Developmental Psychology*, 1974, *10*, 696–702.

McCALL, R. B., HOGARTY, P. S., & HURLBURT, N. Transitions in infant sensorimotor development and the prediction of childhood IQ. *American Psychologist*, 1972, *27*, 728–748.

McCALL, R. B., EICHORN, D. H., & HOGARTY, P. S. Transitions in early mental development. *Monographs for the Society for Research in Child Development*, 1977, *42*(3, Serial Number 171).

MEYERS, L. F., & HOWARD, J. *Mastery motivation in handicapped toddlers: The effect of control over speech output on language acquisition.* Paper presented at the International Conference on Infant Studies, Austin, 1982.

MORGAN, G. A., HARMON, R. J., GAITER, J. L., JENNINGS, K. D., GIST, N. F., & YARROW, L. J. A method for assessing mastery motivation in one year old infants *JSAS Catalog of Selected Documents in Psychology*, 1977, *1*, 68 (Ms. No. 1517).

MORGAN, G. A., BUSCH, N. A., CULP, R. E., VANCE, A. K., & FRITZ, J. J. Infants' differential response to mother and experimenter: Relationships to maternal characteristics and infant functioning. In R. Emde & R. J. Harmon (Eds.), *Attachment and affiliative systems: Neurobiological and psychobiological aspects.* New York: Plenum Press, 1982.

PIAGET, J. *The origins of intelligence in the child.* New York: International University Press, 1952. (Originally published, 1936.)

PROVENCE, S., & LIPTON, R. C. *Infants in institutions.* New York: International University Press, 1962.

RAMEY, C. T., & FINKELSTEIN, N. W. Contingent stimulation and infant competence. *Journal of Pediatric Psychology,* 1978, *3,* 89–96.

RAMEY, C. T., STARR, R. H., PALLAS, J., WHITTEN, C. F., & REED, V. Nutrition, response-contingent stimulation and mineral deprivation syndrome: Results of an early intervention program. *Merrill-Palmer Quarterly,* 1975, *21,* 45–53.

RAPAPORT, D. Toward a theory of thinking. In D. Rapaport (Ed.), *Organization and pathology of thought.* New York: Columbia University Press, 1951.

RICHARDS, T. W., & NELSON, V. L. Studies of mental development, II: Analysis of abilities tested at the age of six months by the Gesell Schedule. *Journal of Genetic Psychology,* 1938, *52,* 327–331.

SELIGMAN, M. E. P., & MAIER, S. F. Failure to escape traumatic shock. *Journal of Experimental Psychology,* 1967, *74,* 1–9.

SPEARMAN, C. *The abilities of man.* New York: Macmillan, 1927.

ULVUND, S. E. Cognition and motivation in early infancy: An interactionist approach. *Human Development,* 1980, *23,* 17–32.

ULVUND, S. E. The psychological basis for the identification of physical environment parameters in the development of early cognitive competence. *Scandinavian Journal of Educational Research,* 1981, *25,* 125–140.

UŽGIRIS, I. C., & HUNT, J. McV. *Assessment in infancy: Toward ordinal scales of psychological development in infancy.* Urbana: University of Illinois Press, 1975.

VIETZE, P. M., McCARTHY, M. E., MacTURK, R. H., McQUISTON, S., & YARROW, L. J. *Exploratory behavior among infants with Down's syndrome.* Paper presented at the American Association of Mental Deficiency Annual Meeting, San Francisco, 1980.

VIETZE, P. M., PASNAK, C. F., TREMBLAY, D., McCARTHY, M., KLEIN, R. P., & YARROW, L. J. *A manual for assessing mastery motivation in 6 and 12 month old infants.* Unpublished manuscript, Child or Family Research Branch, NICHD, NIH, Bethesda, Md., 1981.

WACHS, T. D. Proximal experience and early cognitive-intellectual development: The physical environment. *Merrill-Palmer Quarterly,* 1979, *25,* 3–41.

WACHS, T. D., FRANCIS, J., & McQUISTON, S. Psychological dimensions of the infant's physical environment. *Infant Behavior and Development,* 1979, *2,* 155–161.

WATSON, J. S. The development and generalization of contingency awareness in early infancy: Some hypotheses. *Merrill-Palmer Quarterly,* 1966, *12,* 123–135.

WATSON, J. S. Cognitive-perceptual development in infancy: Settings for the seventies. *Merrill-Palmer Quarterly,* 1971, *17,* 139–152.

WATSON, J. S. Smiling, cooing and "The Game". *Merrill-Palmer Quarterly,* 1972, *18,* 323–329.

WATSON, J. S. Perception of contingency as a determinant of social responsiveness. In E. B. Thomas (Ed.), *Origins of the infant's social responsiveness.* Hillsdale, N.J.: Lawrence Erlbaum Associates, 1979.

WECHSLER, D. Cognitive, conative and non-intellective intelligence. *American Psychologist,* 1950, *5,* 78–83.

WENAR, C. Executive competence and spontaneous social behavior in one-year olds. *Child Development,* 1972, *43,* 256–260.

WENAR, C. Executive competence in toddlers: A prospective, observational study. *Genetic Psychology Monographs,* 1976, *93,* 189–285.

WHITE, R. W. Motivation reconsidered: The concept of competence. *Psychological Review*, 1959, *66*, 297–333.

WHITE, R. W. Ego and reality in psychoanalytic theory. *Psychological Issues*, 1963, *3*(11), 1–40.

WOLFE, P. H. The developmental psychologies of Jean Piaget and psychoanalysis. *Psychological Issues*, 1960, *5*, 1–105.

YARROW, L. J., RUBENSTEIN, J. L., PEDERSEN, F. A., & JANKOWSKI, J. J. Dimensions of early stimulation and their differential effects on infant development. *Merrill-Palmer Quarterly*, 1972, *18*, 205–218.

YARROW, L. J., KLEIN, R. P., LOMONACO, S., & MORGAN, G. A. Cognitive and motivational development in early childhood. In B. Z. Friedlander, G. M. Sterrit, & G. E. Kirk (Eds.), *Exceptional infant* (Vol. 3). New York: Brunner/Mazel, 1975.

YARROW, L. J., RUBENSTEIN, J. L., & PEDERSEN, F. A. *Infant and environment: Early cognitive and motivational development*. Washington, D.C.: Hemisphere/Wiley, 1975.

YARROW, L. J., MORGAN, G. A., JENNINGS, K., HARMON, R., & GAITER, J. Infants' persistence at tasks: Relationships to cognitive functioning and early experience. *Infant Behavior and Development*, 1982, *5*(2), 131–141.

YARROW, L. J., McQUISTON, S., MacTURK, R. H., McCARTHY, M. E., KLEIN, R. P., & VIETZE, P. M. The assessment of mastery motivation during the first year of life. *Developmental Psychology*, 1983, *19*, 159–171.

15   *Mental Retardation*
Developmental Issues in Cognitive and Social Adaptation

SHARON LANDESMAN-DWYER
AND EARL C. BUTTERFIELD

To study intelligence, investigators have compared people of different ages, different biological conditions, and different environments and have examined the results of different tests of intelligence. Such comparisons have been guided by five assumptions: (1) intelligence is distributed along a continuum, (2) intelligence can be measured, even though present tests are imperfect, (3) everyone has some intelligence, (4) a person's intelligence is fairly stable over time and settings, and (5) intelligence matters for individuals, groups, and societies. The assumption of a continuum means that some people have more intelligence and some people have less. This chapter is about less intelligence. It is about the children who constitute the lower range of the continuum. *Mental retardation, mental deficiency,* and *mental subnormality* are the terms most frequently used to denote the condition of much lower-than-average intelligence.

The definition of mental retardation has varied with historical context, intended use, performance standards within a given society, and basic beliefs about the role of intelligence in determining successful adaptation to everyday demands. In the past century, the concept of intel-

SHARON LANDESMAN-DWYER • Department of Psychiatry and Behavioral Sciences, University of Washington, Seattle, Washington 98195. EARL C. BUTTERFIELD • Department of Education, University of Washington, Seattle, Washington 98195. Work on this chapter was supported in part by grants from the National Institute of Child Health and Human Development (HD-11551, HD-00346, and HD-16241) and by the Child Development and Mental Retardation Center at the University of Washington and the Mental Retardation Research Center at the University of Kansas.

ligence has changed from a unidimensional reference to performance on intelligence tests or scholastic achievement to a multidimensional reference to functioning in many domains. The assumptions listed above and the multidimensional conception of intelligence are apparent in the current definition of mental retardation given by the American Association on Mental Deficiency:

> Mental retardation refers to significantly subaverage general intellectual functioning existing concurrently with deficits in adaptive behavior, and manifested during the developmental period. (Grossman, 1977, p. 11)

"Significantly subaverage general intellectual functioning" is defined as performance that is more than two standard deviations below the mean for an individually administered test of general intelligence. Thus, all children are assumed to have some intelligence that can be estimated by giving a test which yields scores along a quantitative continuum. In addition to earning a low IQ score, a child must demonstrate significant deficits in adaptive behavior before he or she is judged mentally retarded. This requirement reflects the multidimensional nature of intelligence and the assumption that intelligence matters: Truly low intelligence will be associated with poor functioning in many areas, especially adaptive behavior. Adaptive behavior is described as the "effectiveness or degree to which an individual meets the standards for personal independence and social responsibility for age and cultural group." (Grossman, 1977, p. 11)

Adaptive behavior encompasses a wide range of functional abilities, from sensorimotor, self-help, communication, and social skills in infancy and early childhood, to academic skills, reasoning, community survival skills, social judgment and responsibility, and vocational aptitude in later years. Although adaptive behavior and measured intelligence are closely related theoretically, their observed correlation is not perfect. Accordingly, assessing intelligence and adaptive behavior independently can minimize the incorrect labeling of a child who scores poorly on an intelligence test for such noncognitive reasons as test anxiety, language differences, poor social compliance, and motivational problems or who lags in adaptive behavior skills due to physical, emotional, or environmental constraints rather than lack of basic intelligence. It remains a real challenge to devise accurate and useful measures of adaptive behavior that are sensitive to the effects of both age and cultural norms (Meyers, Nihira, & Zetlin, 1979; Sundberg, Snowden, & Reynolds, 1978).

The purpose of this chapter is threefold. First, we will answer some general questions about mental retardation. How many children are mentally retarded? Who are they? What caused their mental retardation? Is retardation a permanent condition? How homogeneous is the mentally retarded population? Second, we will highlight findings from behavioral research, giving particular attention to learning and cognition and to the social behavior of retarded infants and young children. To what extent do retarded and nonretarded children differ in their perceptual awareness, social responsiveness, cognitive processing, learning strategies, and overall adaptation to changes in their environments? Third, we will consider the reasons for studying mentally retarded children from a theoretical and developmental perspective.

THE EPIDEMIOLOGICAL PICTURE

The epidemiological picture of mental retardation is neither static nor sharply focused. Because mental retardation is fundamentally a social and relativistic concept about incompetence and low intelligence, obtaining an accurate picture of its incidence (true occurrence in a population) and its prevalence (detected cases in a population) is extremely difficult. Two well-established facts underscore the problem:

1. The prevalence of mental retardation varies with age.
2. Milder degrees of mental retardation, unlike the more severe forms, have markedly different prevalence rates across different cultures, countries, and time periods.

As Figure 1 shows, age-specific prevalence rates from community-wide surveys are extremely low during infancy, reach a dramatic peak among 10- to 15-year-olds, and then decline in the adult years (Gruenberg, 1964; Mercer, 1973). The shaded area, which represents an idealized average of the various curves, highlights this trend. These age-related changes are consistent with the view that *incompetence is relative to the demands of the environment*. In school settings, where heavy demands are placed on children for task-oriented, academic, and socially compliant behavior, the likelihood of identifying poor performers is far greater than when the same children are younger or older and are therefore in less structured and less supervised contexts. In the United States, most mildly retarded children are identified after they enter school (Birch, Richardson, Baird, Horobin, & Illsley, 1970; Mercer, 1973). The rates of mild retardation drop by almost 50% when children leave school

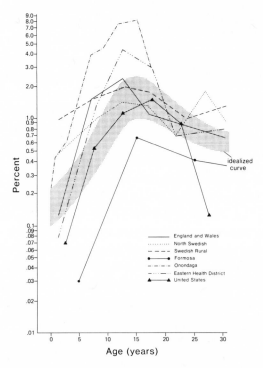

FIGURE 1.   Age-specific prevalence rates for mental retardation, based on data provided by Gruenberg (1964) and Mercer (1973).

and enter young adulthood (Susser, 1968). The phenomenon of higher prevalence during the school years has been labeled "the six-hour re-tarded child" (President's Committee on Mental Retardation, 1970). In both inner-city and rural areas, such children are usually from poor families. It is unknown whether "six-hour retarded" children could have been identified as retarded during preschool years if they had been tested and observed systematically. Whether they continue to function in the retarded range during adulthood is also unknown, because inves-tigators seldom have assessed the adult adjustment of individuals who lose their mental retardation label when they leave school (Henshel, 1972).

One of the first portrayals of the community adjustment of mildly retarded adults was provided by Edgerton (1967), an anthropologist who gathered data by participant observation (getting to know subjects personally and participating in their daily lives). He discovered that mentally handicapped adults use diverse "passing behaviors" to conceal their prior stigmatized identity. When they find themselves in difficult situations, they often solicit help from others, most often from a benefactor friend. Edgerton's observations suggest that many of these adults simply donned "cloaks of competence," without acquiring new cognitive skills. In a 12-year follow-up study, Edgerton and Bercovici (1976) found that some of the least competent of these adults had made remarkable gains in their adaptive behavior. In other words, reliable predictions could not be made from early to late adulthood. Apparently, predictive failure is not limited to infancy and early childhood! What Edgerton and his colleagues recognized is that, in certain environments, individuals may not show the full range of their abilities or may have no reason to acquire new skills. At present, there are no assessment or observational methods for accurately predicting an individual's future resourcefulness or potential for adaptive change.

Estimates of the prevalence of mild mental retardation differ by as much as 30-fold from country to country (Grunewald, 1979; Stein & Susser, 1975). In contrast, the prevalence rates for severe and profound retardation are relatively constant (Abramowicz & Richardson, 1975). Stein and Susser conclude that accurate identification of mental retardation, especially the milder forms, is complicated in part because mental retardation has *organic, functional,* and *social* components:

> The organic component we term impairment; the functional component we term disability; the social component we term handicap. There is by no means a one-to-one relationship among these three criteria. Hence, epidemiological understanding of the condition will be quite different depending on which among these three components is counted. For each component, age distribution, treatment, and service needs will differ. It follows that for many epidemiological purposes an undifferentiated concept of mental retardation is futile. (1975, pp. 63–64)

Moreover, many variables influence how prevalence is judged in different societies. Among the influential variables are a society's degree of industrialization, urbanization, reliance on individual decision-making, use of extended family networks, and dependence on symbols and abstract concepts in everyday communication. Thus, in the People's Republic of China, where the central government tries to match individuals to school and work settings and to provide such necessities as shelter, food, clothing, medical care, transportation, and social support, the opportunities for individual failure are minimized. Consequently, mental retardation is an unused and foreign concept (Robinson, 1978). Presumably, children in such societies show the full range of cognitive abilities and individual differences, but they are not classified as having "more" or "less" of what intelligence tests measure. That is, the adaptive behavior skills associated with success in the People's Republic of China seem to be less correlated with intelligence than they are in Western, industrialized, and individually competitive societies.

How many children in the U.S. are considered mentally retarded? Data from the more recent surveys suggest that the rate of mental retardation lies between 1% and 2% (Birch et al., 1970; Mercer, 1973; Robinson & Robinson, 1976; Tarjan, Wright, Eyman, & Keeran, 1973). In 1981, 4.2 million handicapped children received special education programs in the U. S.; for 738,650 of these, mental retardation was considered a primary handicap, while another 59,512 had retardation as one of several handicaps (Fourth Annual Report to Congress on the Implementation of Public Law 94-142: The Education for All Handicapped Children Act, 1982). The lack of uniform screening procedures prevents any precise determination of school-age children's distribution of cognitive and adaptive behavior skills.

Several interesting facts emerge from the population-wide surveys. First, the rates for mild and moderate retardation approximate those predicted by the assumption that intelligence has a normal bell-shaped (Gaussian) distribution. In contrast, there is a huge excess of severely and profoundly retarded children relative to the number expected under a Gaussian curve (Dingman & Tarjan, 1960). Second, the overwhelming majority (75% or more) of the retarded population is mildly retarded (Stein & Susser, 1975) in the sense that they fall two to three standard deviations below the mean on intelligence and adaptive behavior tests. Third, mildly retarded children come disproportionately from lower socioeconomic families, while severely and profoundly retarded children do not (Abramowicz & Richardson, 1975; Ramey, MacPhee, & Yeates, in press; Susser, 1968). And fourth, level of mental retardation is highly correlated with degree of sensory and motor impairment. The majority of severely and profoundly retarded children suffer multiple physical impairments (Berkson & Landesman-Dwyer, 1977), while only 20% to 40% of the mildly retarded have any recognized clinical pathology (Hagberg, Hagberg, Lewerth, & Lindberg, 1981; Stein & Susser, 1975; Tarjan, Dingman, & Miller, 1960). Although literally hundreds of specific malformation syndromes and biological factors have been identified in the etiology of mental retardation (cf. Warkany, 1971), most retarded children are diagnosed as having "cultural-familial" or "sociocultural" retardation, which is a way of saying that their retardation cannot be traced to any specific biological cause.

These and other differences between the mildly and severely retarded groups are summarized in Table I. The existence of these differences has led investigators to question the value of considering mental retardation a single condition. Zigler (1967, 1969) and others (e.g.,

TABLE I

Characteristics of Milder versus More Severe Degrees of Mental Retardation

	Level of mental retardation	
	Mild to moderate	Severe to profound
Incidence	1.0–3.0%	.7%
Prevalence	1.0% or less	.25%
Proportion of retarded population	75%–90%	10%–25%
IQ range (Modal IQ)[a]	50–70 (57)	below 50 (17)
Mortality during childhood	Normal	Three or more times normal rate
Age at diagnosis	School-age	Preschool
Duration of condition	Variable (function of environment and treatment)	Lifelong
Out-of-home placement rates	Low	High
Physical size	Normal to slightly below average	Much below average
Medical/neurological signs and symptoms	Usually absent	Often present
Usual genetic causes[b]	Polygenetic interaction, sex chromosome aberrations	Rare autosomal genes, autosomal chromosome aberrations (e.g., Down's syndrome)
Primary environmental causes	Inadequate social and/or environmental stimulation, childhood cerebral disease	Cerebral disease or injury in very early life (pre- and postnatal)
Parents' intelligence	Often low IQ	Normal IQ
Sibs' intelligence	Often low IQ	Usually normal; sometimes severely affected (e.g., recessive disorders)
Social class	Lower social classes over-represented	All classes

[a]Based on studies prior to 1960.
[b]Account for only minority of cases.

Bijou, 1963; Ramey, McPhee, & Yeates, 1982) have proposed a two-way classification to distinguish familial retardation or developmental retardation from clinical or organic forms of mental retardation. These investigators assume that familially retarded children represent the lower range of a natural distribution of intelligence. They are developmentally delayed, but unlike the more severely retarded, they are not qualitatively different from children with higher IQ scores. While this seems theoretically reasonable, such a classification scheme presents practical problems (Ellis, 1969; Milgram, 1969). For instance, mildly retarded children may appear to have no organic condition because medical science has not yet devised ways to detect subtle organic defects. The validity of this possibility is suggested by the relatively recent discoveries of X-linked mental retardation (Larbrisseau, Jean, Messier, & Richer, 1982) and of the fetal alcohol syndrome (Streissguth, Landesman-Dwyer, Martin, & Smith, 1980), both of which indicate that presumedly undifferentiated forms of milder retardation may have biological causes. For both syndromes, there is a strong family pattern, often among lower socioeconomic classes or minorities. These children typically function in the mildly retarded range and they seldom used to be diagnosed as having major central nervous system damage. Later in this chapter, we will argue that selection and classification of a person for scientific study must be closely related to a theoretical formulation of the person's conditions and/or to the practical purposes of the research. No single classification system will suffice because different objectives require different classification schemes.

In sum, mentally retarded children are an extremely heterogeneous group. Most retarded children are identified after early childhood. Those who are identified earlier tend either to have specific patterns of malformation, the most common of which are Down's syndrome and neural tube defects (Center for Disease Control, 1982), or to be multiply handicapped and severely delayed in their development. Mildly retarded children who have no clinically detected pathology are likely to live in poverty and to have parents of low intelligence. Many of these children will be considered mentally retarded only when they are in school settings. The current definition of mental retardation emphasizes the interplay between environmental expectations or standards and a given child's personal competencies. Thus, mental retardation is not necessarily a permanent condition, even if a child's IQ score remains stable over the years. The incidence and prevalence of mental retardation are difficult to estimate accurately. The best estimates are that between 1% and 2% of the U.S. population is mentally retarded.

CHARACTERISTICS OF RETARDED CHILDREN

Mentally retarded children share two common features: a failure to accomplish as much as their age-mates and a label that conveys their lack of a highly valued trait—good intelligence. Beyond this, what is known about the behavioral development of mentally retarded children?

At the turn of the century, the prognosis for retarded children was so grim that few efforts were made to enhance their learning opportunities, and scientific inquiry into the nature of mental retardation was minimal (Crissey, 1975; Windle, 1962). Vigorous efforts were made to improve intelligence testing, but these efforts were fuelled by pragmatic rather than scientific concerns:

> The ideal would be that in one-half hour we could get any case so settled that we should know accurately what to expect from him in the years to come and that we should know just what treatment to give him so that we should waste neither his time nor our energies in training. (Goddard, 1909, p. 49)

By the 1960s, psychology had developed an impressive technology for training skills (Estes, 1970; Kazdin, 1973) and had recognized the importance of early experiences for later cognitive and social development (e.g., Hunt & Kirk, 1971; Yarrow, 1961). Then investigators of retarded children began to ask more diversified questions: What factors contribute to retarded children's poor performance on intelligence tests? Are retarded children merely delayed in their development, or do they show genuine deficits and differences in their behavior? How is social behavior affected by mental retardation? What effects does the environment have on the behavior of retarded children? How much can retarded children learn? As investigators gathered data to answer these and related questions, they established a valuable descriptive base for understanding human potential and adaptation.

The answers to such questions about retarded children's development have come primarily from carefully controlled studies of learning and cognition, from observations of children in natural settings, and from systematic evaluation of the effects of intervention or treatment programs. Before highlighting the major conclusions from these diverse fields of inquiry, we will describe a conceptual framework for discussing the research findings.

A Concept of Child–Environment Interaction

We start with the assumption that in order to understand the behavioral patterns of retarded children, we must consider the *contexts* in which they are studied. This can be seen in the details of scientific studies, but it can be seen more readily in changes in treatment philosophy, social support services, and educational programs that undoubtedly have affected retarded children and their families in recent decades. Twenty-five years ago, many infants born with central nervous system defects were relegated to institutional care, largely because professionals believed that retarded children had no significant learning potential. That belief has changed dramatically with innovative applications of scientific principles of learning and social behavior (Berkson & Landesman-Dwyer, 1977; Bijou, 1963, 1966; Butterfield, 1983). As a consequence, retarded children now are more often reared at home and educated in public schools instead of in institutions. Accordingly, the "typical" behavioral profile of Down's syndrome children or severely retarded children in the 1950s and 1960s (e.g., Berkson, 1973; Farber, 1959, 1960) is likely to be markedly different from that of such children observed in the 1980s.

We continue with the assumption that the fundamental heterogeneity of the retarded population presents major obstacles to the search for common patterns and principles of development. We must judge "when exceptions obscure the rule" (Berkson, 1966) and when the "rules" simply do not exist. We must be prepared to judge, as Estes (1970) did about profoundly retarded, multiply handicapped children, that "these pathological cases must necessarily be so heterogeneous in character that one can scarcely expect to accomplish much in attempting to deal with them on general theoretical principles" (p. 186). Retarded children's behavior may be even more variable than the behavior of nonretarded children in comparable settings (Baumeister, 1967). Answers to the question "What are retarded children like?" must be related to the characteristics of the children, as well as to the contexts and time periods in which they are observed.

Figure 2 illustrates our conception of child–environment interac-

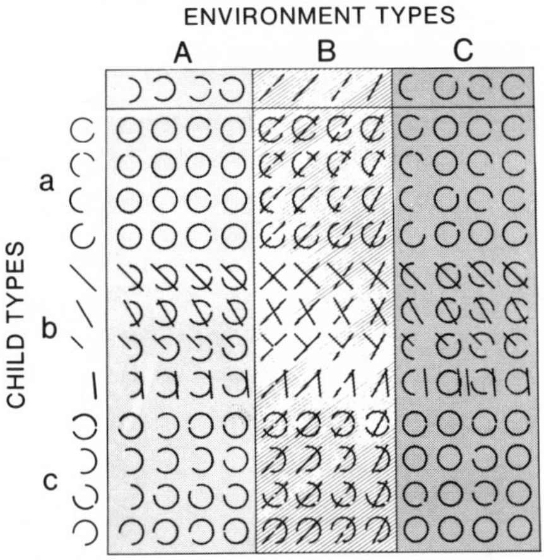

ENVIRONMENT TYPES

FIGURE 2.   An illustration of the principles of child–environment interaction.

tion. It reflects the assumption that retarded children are heterogeneous by including three child types—a, b, and c. The nature of the types of children will change with the phenomena being studied and the theoretical assumptions guiding the research. No single classification scheme serves all purposes (Ellis, 1969). For most research studies, retarded children are grouped on the basis of their general intelligence or their etiology. So, in research studies, the child types in Figure 2 might be a = average intelligence, b = mildly retarded, and c = severely retarded, or a = cultural-familial, b = Down's syndrome, and c = brain-damaged due to postnatal trauma. But in many situations, intelligence and etiology are far less important in determining adaptation to environment than the child's sociability, temperament, ability to anticipate consequences before responding, physical appearance, and so forth.

Even within types of children, there is individual variation. Children of the same type differ, and individual children change over time. This is indicated in Figure 2 by the presence of four rows of symbols for each of the three child types. Note that while within types the symbols are similar, they are not identical. The examples (rows) of each child type can represent either four different children or the same child at four different stages. Since children can never be studied independently of environments, investigators face serious difficulties in valid classification of children. Some environments serve to mask or obscure individual differences, while other environments permit certain aspects of the child's nature to be detected more readily. This is indicated in Figure 2 by environment B, in which the three types of children can be distinguished from one another more readily than they can in environments A or C.

Just as subjects vary, so do environments. We define environments in a broad sense, including conventional physical environments (such as homes, schools, and playgrounds), treatments or interventions (such as behavior modification programs, mainstreamed classrooms, drug therapies, and structured cognitive training), and experimental situations (such as different test protocols, systematic presentation of stimuli, and standardized social situations). In all environments, there will be natural variation, associated with particular settings, times, or changes in the constellation of social and nonsocial variables within a given context. Such differences, even seemingly minor ones (such as administration of the same test by two different examiners, the same classroom in the morning versus the afternoon, or a mealtime with father absent or with visitors), can contribute to significant differences in child behavior. This is illustrated in Figure 2 by the inclusion of four environments (columns) of each type, A, B, and C. Notice that the first child (row) of type a appears similar (i.e., a closed circle) in three of the four settings or occasions (columns) in environment A. The same is true for the second example of child type a, except that the particular setting (or time) that fails to produce the "closed circle effect" differs for these two people (or the same person at two different ages).

Describing environments in an absolute sense is as difficult as trying to evaluate pure traits in children. If environments A and C were studied only when child types a and c were present, it would be impossible to portray these environments accurately. In contrast, when only type b children are present, it is much easier to perceive the distinctive features of these environments (i.e., the circles are open on the left side in environment A and on the right side in B). The search for a multipurpose typology of environments is likely to be futile: Classification of settings, like that of people, should vary with its intended theoretical or practical uses (Mischel, 1977).

Since different combinations of children and environments result in different developmental patterns, investigators would like to define optimal environments as those that foster positive outcomes for particular types of children. Hunt (1961) labeled this "the problem of the match."

The concept of matching children to environments is based on the belief that environments are not inherently good or bad. Environments can be assessed only in terms of the needs of the particular children in them. The variables that promote cognitive growth in some types of children may not be suitable for others; and what constitutes an ideal environment at one time is not likely to remain ideal as the child changes. Because children and environments are multidimensional, it is likely that the environmental variables that contribute to positive developmental outcomes in one domain (such as cognitive abilities or independence) will not do so in other domains (such as social maturity or creative expression). This principle is illustrated in Figure 2. If the definition of a good outcome were "looking like a letter of the alphabet," then the best overall fit of children to environments would be child type a with environment A, b with B, and c with C. Although this formula works for the majority of examples, there are some exceptions—such as the fourth example (row) in child type b, who only looks like a letter when in the fourth setting (column) of environment C, or the third example of child type a who also looks like a letter in the fourth setting of environment C. Alternatively, if another criterion were established for judging good outcomes, such as "having intersected lines" or "having no openings," then different conclusions would be reached about what constitutes an optimal child-to-environment match.

Psychology is largely the study of person × environment effects (Cronbach, 1975; Pervin & Lewis, 1978). Our guiding conception is simple and old. We present it because mental retardation research often has focused disproportionately on one of the two sides of the person × environment interaction. Many investigators have studied children's characteristics, many have evaluated the effects of environmental variables, but few have focused on child × environment interactions. We find our guiding conception helpful in explaining findings that at first appear contradictory.

Learning and Cognition

Almost everyone assumes that intelligence influences a child's learning and thinking. More intelligent children appear to learn more from their everyday experiences than do less intelligent children. Do bright children really learn more about their environments in an absolute sense, or do they focus selectively on learning what is more likely to be useful and important? Are there features that clearly differentiate retarded and nonretarded children's learning? For more than 30 years, investigators have probed systematically for a description of what constitutes efficient learning. The research is extensive, ranging from microanalysis of components of classical and operant conditioning to more global assessment of problem-solving and memory strategies. This diverse literature provides impressive support for five general conclusions about retarded children's learning and thinking.

Conclusion #1: The principles governing the acquisition, maintenance, and extinction of simple responses and basic memory functions are the same for retarded and nonretarded children, with the exception of those who have significant neurological impairment. Estes (1970) performed a comprehensive analysis of the principles of learning and the experimental studies of how these principles apply to normal and retarded people. His conclusions are very much like those of Denny (1964), who also carefully reviewed the experimental literature. Their conclusions are still valid. Concerning classical conditioning, Estes noted that the most striking finding "is the absence of any conspicuous differences between acquisition rates of normal and retarded groups" (Estes, 1970, p. 60), regardless of whether the groups are matched on chronological age or mental age. Small differences in favor of normal people matched on chronological age are found in speed of extinction of classically conditioned responses, but the interpretation and importance of this difference are unclear (Estes, 1970). Concerning operant conditioning, Denny (1964), Estes (1970), and Spradlin and Girardeau (1966) all concluded that normal and retarded children behave remarkably alike: "Even severely retarded individuals respond in qualitatively much the same way to various reward schedules and contingencies as do animals and normal human beings" (Estes, 1970, p. 61).

It should not be surprising that the same principles of learning apply to normal and retarded people, because the principles were derived from studies of animals, and many experiments show that the principles apply equally to species (e.g., worms, rats, pigeons, and apes) known to differ greatly in the nature of their intelligence. Only when central nervous system damage is extensive do behavioral irregularities appear to defy the laws of learning (cf. Landesman-Dwyer & Sackett, 1978). For intact organisms, learning by both classical and operant conditioning is simple and requires little cognition. The fact that normal and retarded people differ in abstract and complex cognition does not require that they learn differently when environmental demands are simple. Remember our guiding scheme: People will seem the same or different depending on the environment in which they are viewed. Thus, retarded and normal children differ hardly at all in simple human memory functions (Ellis, McCartney, Ferretti, & Cavalier, 1977; Pennington & Luszcz, 1975), even though they differ markedly in more complex memory processes (Belmont & Butterfield, 1969; Ellis, 1963).

Conclusion #2: Mental age exerts a powerful influence on many aspects of learning and cognition. When retarded children are compared with normal children of the same chronological age, they do learn and think differently in some situations. In fact, retarded children often resemble younger normal children. Intelligence tests yield two scores; the more commonly mentioned is the IQ, but intelligence tests also yield mental age scores, making possible the comparison of normal and retarded children who are of the same mental age rather than the same chronological age. Such comparisons preserve IQ differences between the compared groups, and they show that there are many fewer differences between retarded and normal children of the same mental age than between retarded and normal children of the same chronological age. In other words, mental age matters more for many kinds of behavior than IQ does. Remember our argument that different classification schemes are useful for different purposes. For the purpose of seeing the importance of IQ, classifying children by their mental ages is more helpful than classifying them by their chronological ages.

Zeaman and House (1963) showed that retarded children acquire certain kinds of discriminations particularly slowly because they take longer to attend to the relevant dimensions of the stimulus material, not because they learn more slowly once they have attended. This sort of slowness is also typical of younger normal children. When retarded and normal children are equated for mental age, they are equally slow to attend to the appropriate dimensions. Once they attend, all children learn discriminations at the same rate.

While matching normal and retarded children on mental age eliminates many performance differences between them, some still remain. Some of these differences result from noncognitive factors.

Conclusion #3: Noncognitive factors affect learning and thinking. Noncognitive factors include temperament or personality, motivation, and sensory and motor capacities. Retarded and normal children frequently differ in such noncognitive factors, and these in turn produce differences in learning and cognition. Some noncognitive characteristics appear to be innate, while others are determined in part by environmental experiences. For example, investigators of learning used to compare institutionalized retarded children with noninstitutionalized normal children (Zigler, 1969). When they found differences, as they frequently did, they attributed such differences to the different intelligence levels of the children, ignoring differences in the children's living environments. When comparisons are made between normal and retarded children with comparable experiences, and therefore fewer noncognitive differences, the children perform more similarly. Some environments affect retarded and normal children in comparable ways. Moreover, when experience is manipulated so as to change noncognitive variables, children frequently behave more intelligently.

Various studies have shown that when children from economically and educationally deprived families are provided with enriched nursery school experiences, their IQs increase (Bereiter, 1966; Deutsch, 1963; Gray & Klaus, 1965; Ramey & Haskins, 1981). Zigler and Butterfield (1968) sought to determine whether such IQ increases were due in part to changes in noncognitive factors. They administered two intelligence tests to children from deprived home environments at the beginning of their first year of nursery school. One test was administered under standard testing conditions, and it showed that the children had IQs on the borderline between the normal and retarded range. The other test was administered so as to reduce the influence of noncognitive factors such as wariness of strangers and low expectations for one's performance, and it showed that the children had IQs very near average intelligence. The conclusion is that changing the testing environment to minimize the debilitating effects of noncognitive factors results in higher IQs for young deprived children. Both types of IQ test administration were given again at the end of the first year of nursery school. The children earned higher IQs on the standard administration than they had previously and were well within the normal range of measured intelligence. Under the administration designed to minimize the contribution of noncognitive factors, the children's IQs were unchanged. The conclusion is that an enriching nursery school experience can raise deprived children's IQs by changing their noncognitive characteristics without changing their general level of intellectual abilities.

Conclusion #4: Retarded children's intellectual deficiencies are due more to higher-order cognitive processes than to subordinate processes. While IQ increases due to enriching experiences such as nursery school may reflect changes in noncognitive factors, experiences designed to teach intellectual skills also raise the performance of mentally retarded children (Belmont & Butterfield, 1971). Psychologists have shown that, having analyzed the cognitive processes underlying a particular intellectual performance, they can teach retarded children the skills needed to perform as do nonretarded children (Butterfield, Wambold, & Belmont, 1973). Unfortunately, such teaching seldom helps retarded children solve problems that are closely related to the ones on which they were taught, and this kind of instruction has never changed their everyday behavior significantly.

The failure of teaching to improve a child's performance on a related problem or in everyday situations is called a failure of transfer. Failures of transfer are common for both retarded and young normal children (Borkowski & Cavanaugh, 1979; Stokes & Baer, 1977) and are best understood as failures of the children during learning to abstract the rules from the lesson, to learn how to learn, to acquire helpful strategies, or to alter their superordinate skills and metacognitive processes (Brown, 1975, 1978). What children learn readily are the basic, lower, or more subordinate skills required for good performance on the problems used during teaching. Thus, while retarded children can easily be taught the subordinate mechanisms of attention and rehearsal needed to memorize well, they fail at the superordinate level to see the utility of attention and rehearsal outside of the environment in which they learned these skills. Immaturities or deficits in superordinate processes seem to prevent both retarded and young normal children from transferring newly learned subordinate skills to related situations (Routh, 1982) and from solving many novel problems. Superordinate processes seem particularly important to good performance in school during late childhood and adolescence, the ages at which retardation is most likely to be diagnosed.

Scientific study of superordinate processes is new and has not been extended to early childhood. Nevertheless, some cognitive problems have been analyzed well enough to allow us to infer the general outlines of superordinate processes that probably contribute to transfer and to problem-solving in general. One such outline of superordinate pro-

TABLE II
Steps in Efficient Problem Solving[a]

Step 1: Identify problem	
1a	Perceive problem (situational demands)
1b	Decide to respond
1c	Define best possible outcome (set goals)
Step 2: Plan action	
2a	Review what is known about similar situations
2b	Assess resources available to solve problem (e.g., personal abilities, environmental support)
2c	Develop strategies: Design one strategy, consider alternative strategies
2d	Estimate outcomes of each strategy
2e	Compare estimates with goals and resources
2f	Select the best strategy (i.e., one that yields the smallest goal–estimate discrepancy and is possible to implement)
Step 3: Take action	
3a	While implementing strategy, monitor differences between plan and action
3b	Estimate response accuracy or outcome if implementation were stopped
3c	Compare current estimate of outcome with above probability (Step 3b)
3d	Decide whether to continue or to respond
Step 4: Review outcome	
4a	Assess response accuracy
4b	Compare outcome to original estimates (Step 2d)
4c	Determine how closely outcome matched estimate
4d	Consider reasons for outcome–estimate differences: e.g., strategy selected was not the best, strategy was not implemented as planned, parameters of problem changed or were identified incorrectly, estimates were inaccurate, outcome was not measured correctly, resources available were not assessed adequately
4e	Store information derived from experience, so that subsequent problem-solving will be more successful

[a]Adapted from Belmont, Butterfield, & Ferretti (1982) and Butterfield (1981).

cesses is presented in Table II as a series of steps involved in efficient problem-solving. The steps begin with the recognition of a problem and goal setting (Step 1), move through planning for action (Step 2) and executing and monitoring the best available strategy (Step 3), and end with evaluating the outcome of strategic action and probable reasons for success or failure (Step 4).

The model outlined in Table II acknowledges that ideal goals and perfect outcomes are seldom achieved. The best strategy available to most people is the one that minimizes the discrepancy between their goals and their estimate of its outcome. Their goals and outcome estimates are necessarily influenced by their abilities, preferences, and environmental resources. Even a person whose superordinate processes are as complete as those outlined in Table II must use many subordinate processes in complex and timely sequences in order to perform well in cognitively demanding situations. Whether a child succeeds in solving a problem is determined by, among other things, memory abilities (initial encoding from prior experiences and recall strategies), self-awareness while thinking and acting, ability to generate hypotheses and to estimate outcomes abstractly (conducting "mind experiments"), discrimination of elements of the problem, readiness to inhibit responses so that monitoring and reevaluation can occur, and adaptability (willingness to discard ideas and behavior that fail to accomplish the desired outcomes). At each step in this problem-solving chain, a child's success is co-determined by personal traits and by characteristics of the environment. Theoretically, children's overall success in solving problems could be altered by training them to use superordinate or executive processes such as those in Table II. Such cognitive training should yield behavioral changes that are more durable than the teaching of specific skills in highly controlled stimulus situations.

While no one has tried to teach all of the steps in Table II to mentally retarded children, some of the steps have been taught experimentally. To evaluate the effects of teaching superordinate skills, investigators first instructed children to use particular subordinate processes, such as rehearsal or categorization. They also taught some of the children to use certain superordinate processes. Then, they tested all of the children to determine how well they solved a new problem. For example, Brown and Barclay (1976) taught mentally retarded children to use a rehearsal strategy to memorize lists of words. Some of their retarded subjects were also taught to assess their own recall readiness. The readiness instruction provided a means by which children could determine if they needed to study more before trying to recall. In the terms of Table II, this additional instruction should have taught the processes of Step 3. Brown, Campione, and Barclay (1979) tested the same children 1 year later with a reading comprehension task. Those who had learned the superordinate readiness-to-recall showed significantly greater comprehension than those who had been instructed only in subordinate skills. This is an unusually long time interval over which to find positive transfer between such different tasks. Such findings suggest that important general benefits can be given to mentally retarded people by teaching them to use superordinate cognitive processes. Belmont, Butterfield, and Ferretti (1981) review other studies, including some with children of quite low mental ages, in which instruction to use superordinate processes improves subsequent performance.

Conclusion #5: Even with cognitive training and supportive environments, retarded children continue to differ from nonretarded children. Although the magnitude and nature of differences between normal and retarded children depend on the children's ages, the environments in which they are assessed, the environments they have experienced previously, and their noncognitive characteristics, no one yet has found a combination of these factors that eliminates all differences between retarded and nonretarded children. Even though sophisticated prosthetic aids, instructional approaches, and supportive social environments foster marked behavioral change in retarded children, no combination of such special accommodations has made retarded children the same as normal children who never needed such accommodations. Whether science will ever enable us to "cure" retarded children is an open question. Most professionals doubt it. A few are guardedly optimistic. It is too soon to tell whether our society will try seriously to apply the techniques that we now know can reduce differences between normal and retarded children. Complex societal, rather than scientific, issues will decide the matter (Kauffman, 1981).

In sum, retarded children's behavior is similar in many ways to that of normal children, especially younger children of like mental ages. The disparity between retarded children's mental and chronological ages often creates an atypical pattern of life experiences relative to those of nonretarded children. These life experiences sometimes contribute to added noncognitive differences between normal and retarded children. Noncognitive factors such as motivation and self-concept sometimes further reduce the cognitive performance of retarded people. This result argues for the type of "relativistic appraisal" of intelligence advocated by Wechsler (1975) and is consistent with the view that mental retardation represents nonoptimal functioning in many domains.

It is noteworthy that many children fail cognitive tasks for reasons other than low intelligence and thus are never labeled mentally deficient. For example, many children of normal intelligence may do poorly on orally administered spelling tests. A hearing-impaired child may not receive the stimulus input correctly, leading to an incorrect definition of the response needed. A dyslexic child may have chronic problems related to the use of letter symbols, affecting initial learning to spell and later success in many test situations. A blind child may have had less exposure to the test words. Similarly, a child from a foreign country may be less familiar with spelling words. Other children may be tired, uncooperative, disinterested, distracted, or unprepared on a given day, all of which may lower response accuracy. These children obviously fail for reasons specific to individual circumstances. Retarded children may also fail, but they are distinguished from these other children by the pervasiveness of their incompetence and by their limitations in developing effective problem-solving strategies. Since problems are not limited to formal academic settings, but arise in all aspects of everyday life (Brooks & Baumeister, 1977), retarded children should show deficits in social skills and social adaptation as well as in more clearly delineated intellectual situations.

The Social Ecology of Retarded Children's Lives

By definition, retarded children demand more care than most children in their society. Depending on a child's biological and behavioral profile, these demands may differ from those of nonretarded age-mates in quality, intensity, timing (age of onset and duration of need), complexity (number of simultaneous needs and rate of changing demands), or sequence. The impact of these child-specific demands and the family's responses to them will depend on the characteristics of individual family members, the functional relationships of family members to one another and to their extended social networks, and the number and type of resources available to the family and the child. There is no single scenario for a retarded child's life, but there are important and recurrent themes in social adaptation among retarded children.

Conclusion #1: The quality of environments cannot be judged without reference to the specific characteristics and needs of the children in those environments. For many years, investigators have noted that good environments, according to *a priori* standards, do not lead to positive outcomes for all children (Landesman-Dwyer, 1981; Windle, 1962). For example, in 1939 Kuenzel studied 82 "feebleminded" children in 42 foster-care settings. She concluded that homes with greater physical, social, and cognitive resources did not facilitate better adjustment for retarded children. In fact, she reported the opposite:

> A greater proportion of defectives, when placed in foster homes of average and inferior-average social standing, adjust [better] than when put into homes of superior caliber. The tempo of life in such homes and the goals set by these parents for attainment by defective children apparently correspond more closely to the rate of action and level of understanding of defectives than do homes and aspirations of superior foster parents. (p. 252)

Rautman (1949) also noted that the match between environmental expectations and children's abilities was an important factor in the personal adjustment of retarded children:

> Workers in the field have observed repeatedly that, in the case of mentally retarded children living at home, a retarded youngster who comes from a family where the standards are so low as to make his own retardation inconspicuous (sometimes because of retardation on the part of his parents) has a far more favorable educational and adjustmental profile than does a child who has equal or even less retardation but who comes from a family setting in which his intellectual handicap places him below the level of family aspirations. (p. 157)

Theoretically, a mismatch between children and environments increases the probability of negative social exchanges and decreases opportunities for a child to learn developmentally appropriate skills and strategies. If unreasonably high demands are placed on a child, the child will experience repeated failure, which may decrease his or her willingness to engage in challenging activities. An excessively demanding environment is less likely to provide information in ways that will maximize its assimilation and use (Haywood & Wachs, 1981). At the opposite extreme, an environment that fails to demand increasingly mature and independent responses from a child may also restrict development, by not providing problems to solve and by not offering feedback about the child's behavior.

Gazaway (1969) provided an example of a devastatingly undemanding environment in an isolated Appalachian setting. She observed:

> The youngsters are not required to fulfill obligations consistent with their abilites and age. Under these relaxed conditions they feel secure in a family setting where they are able to measure up to the low expectations of their clan. Parental and sibling approval is based on them as they are, not on their behavior or their contributions. . . . Parents are not distressed over a child's failure to develop in physical size, increasing skills, and complexity of function, if indeed they assess his progress at all. If undersized upon attaining school age, his introduction into formal education is delayed until he has

> "growed more." Behavioral patterns are likewise ignored, with no thought
> given to the many factors responsible for deviation. The youngsters, particu-
> larly the infants, are strikingly inadequate in motor coordination such as
> grasping and manipulating objects; language and speech development close-
> ly parallels that of parents and siblings; responsiveness and power to concen-
> trate are conspicuously absent. Deafness and poor vision go unnoticed. For
> the most part, the child must fend for himself as best as he can. (pp. 88–89)

By the standards of the outside world, most members of this Appalachian community are mentally deficient. Inbreeding, malnutrition,
primitive living conditions, widespread and untreated illnesses, and a
virtually barren physical environment contribute to low performance in
all aspects of adaptive behavior. Mutual social acceptance does not compensate for the inadequate cognitive and social challenges of the setting.
However, it must be acknowledged that a final determination of how
well matched an environment is to the needs of individuals depends on
the outcome measures selected and the values of those making the
judgment. Successful adaptation is not unidimensional, and a child's
progress in one domain, such as solving academic problems, is not
always predictive of other behavior, such as forming and maintaining
friendships.

Institutions clearly provide less child-specific stimulation than most
natural homes do. Yet contrary to popular belief, institutions do not
have uniformly negative effects, at least not soon after children enter
them. Zigler and Williams (1963) and Zigler, Balla, and Butterfield (1968)
found that children's preinstitutional experiences significantly influenced their initial responses to institutionalization. For children who
came from relatively stimulating homes, the institutions were socially
depriving. In contrast, changes in intelligence and motivation indicated
that the same environment was socially adequate or perhaps even enriched for children from socially deprived preinstitutional backgrounds.
The differences between the previously deprived and undeprived children decreased the longer they stayed in the same institutional environment. Even an environment that is well matched to the needs of children at one time must change as the children do, or the child-to-
environment match will be lost. Institutions seldom change.

Another example of the need to evaluate environments rela-
tivistically is provided in a prospective study of retarded individuals
who were randomly assigned to new living arrangements. Landesman-
Dwyer (in press) found that moving residents from traditional institutional wards that were large, crowded, understaffed, highly routinized,

and lacking in privacy to much smaller, more homelike duplexes did not promote general positive behavioral changes. Detailed observations of daily behavior and social interaction patterns revealed, however, that a few subgroups of individuals did benefit. Somewhat surprising was the finding that even comparatively minor changes in old institutional wards were sufficient to foster desired behavioral changes in some subjects. A post hoc functional analysis of the specific environmental demands, relative to the subjects' ability to perceive and to respond to these demands, led to better predictions of individual behavioral change over time and settings.

Conclusion #2: It is important to consider a child's mental or developmental age when evaluating his or her social environment. Early studies of mother–infant interactions overlooked the contribution of a child's mental age to observed patterns of behavior. For example, studies that focused on maternal speech patterns reported that mothers of retarded children were more controlling, less responsive, more stereotyped, or less challenging than mothers of normal children of the same chronological age (e.g., Buium, Rynders, & Turnure, 1974; Kogan, Wimberger, & Bobbitt, 1970; Marshall, Hegrenes, & Goldstein, 1973). Discussions of these differences often implied that mothers were behaving in restrictive or nonoptimal ways that partially accounted for their children's delay in language development. Later studies that controlled for the children's mental ages or levels of productive speech (e.g., Buckhalt, Rutherford, & Goldberg, 1978; Davis & Oliver, 1980; Rondal, 1978) found that mothers' behavior was in part a function of the children's characteristics. In other words, adults naturally tended to speak at levels matched to children's language abilities. Of course, not all group differences disappear when mental or developmental age is controlled, and the nature of the remaining differences varies with the technique used to describe interaction as well as with the ages, etiologies, and functional abilities of the retarded children.

This conclusion about social matters is consistent with our earlier one about learning and cognition. Retarded children's behavior is often similar to that of younger children, and the same general principles apply to both retarded and normal children's behavior. Once again, profoundly retarded individuals are the most likely to deviate from normative social patterns, sometimes by failing even to discriminate between social and nonsocial situations (Landesman-Dwyer, Berkson, & Romer, 1979).

Conclusion #3: Environmental variables account for much of the behavior of retarded children. Studies of retarded children in different types of home and school settings and in different social contexts indicate that environmental variables are highly correlated with behavioral patterns (e.g., Guralnick, 1981a,b; Hull & Thompson, 1980; Landesman-Dwyer *et al.*, 1979; McLain, Silverstein, Brownlee, & Hubbell, 1979; Seltzer & Seltzer, 1978; Stoneman & Brody, in press). That children show setting-specific behavior is not surprising. What is unexpected is the extent to which environmental variables can predict behavior. Environmental variables often provide more accurate bases for prediction than do a child's personal qualities. This observation does not imply direct causality, but it is consistent with the transactional hypothesis:

> Group × Place transactions encompass the processes by which groups are affected and, in turn, influence their physical milieu. A transactional approach to the study of settings highlights the active role taken by individuals and groups in creating and modifying their environments. Accordingly, the physical milieu of groups is construed not only as an antecedent of behavior but also as a sociocultural product, i.e., the material reflection of collective behavior and as a repository of shared social meanings. (Stokols, 1981, p. 395)

In an ethological study of 23 group homes, Landesman-Dwyer, Sackett, and Kleinman (1980) found that environmental variables such as the heterogeneity of the social group and the average intelligence of group members were associated with significant differences in daily activity patterns. Social behavior and friendship patterns also were influenced by environmental factors (Landesman-Dwyer *et al.*, 1979). Longitudinal studies of younger children in different social milieus provide further support for this conclusion (Guralnick, 1981a,b). Dunlop, Stoneman, and Cantrell (1980) observed preschoolers over a six-month period. Initially, children rated by their teachers as "likely to be referred to special education classes" differed significantly from other children in their social behavior. Over time, however, the two groups of children became less distinguishable in almost all aspects of social interaction. Other naturalistic experiments indicate that social behavior is modifiable by altering environmental conditions, even without specifically focusing on one-to-one training of social skills.

As we mentioned earlier, studies of early intervention programs confirm the importance of children's environments. Early intervention programs include a wide range of environmental modifications, from teaching parents new styles of interaction (cf. Baker & Clark, in press; Mitchell, 1980) to providing structured preschool programs designed to foster positive social interaction and to encourage the development of skills necessary for later adaptation to academic demands of school (cf. Bricker, Seibert, & Casuso, 1980; Guralnick, 1981a,b; Ramey & Haskins, 1981). Although many of the studies of early intervention lack adequate control conditions, random assignment of subjects, and precise control over or description of program content, they collectively provide convincing evidence that retarded children, like their nonretarded agemates, are highly responsive to their environments. What is less certain is the degree to which a child's developmental course remains altered after intervention ceases, whether the timing of interventions is critical, and what the key factors are in promoting behavioral change. Answering such questions requires a comparison of intervention programs run by different investigators, because individual programs have not been designed to answer them. Recently, Ramey and colleagues (Ramey & Bryant, 1982; Ramey, Sparling, Bryant, & Wasik, 1982) compared social and home variables of all controlled studies designed to prevent developmental retardation among children at high risk. They concluded that greater prevention results from more intense and extensive programming. We infer that there are greater social changes in children when they are required to assume active roles in organizing and responding to their environments. Theoretically, children should benefit the most from intervention when they are beginning to acquire independence and to develop relational concepts, and from intervention that is designed to change the behavior of key social agents in the children's worlds. Larger benefits should also come from interventions that teach general strategies for coping with and solving everyday problems, rather than training specific adaptive skills. This view is consistent with Greenspan's (1979, 1981) model of childhood social competence, which relies on a combination of cognitive and noncognitive factors.

Conclusion #4: To assess the impact of environmental variables, the child's total social repertoire and behavior in many situations should be considered. Changes in the environment often produce unanticipated behavioral consequences. These unexpected results occur for many reasons.

First, a change in the environment is seldom singular. Kogan, Tyler, and Turner (1974) taught mothers new skills for instructing their young handicapped children. Mothers clearly acquired and applied these skills, and they raised their expectations for their children's accomplishments. But from the children's perspective, their mothers also showed significantly less spontaneous warmth and physical affection. This side effect of training mothers was not anticipated, and in fact was noticed only during extensive review of videotaped mother–infant interactions before and after intervention. Had the investigators limited their outcome evaluation to the presence or absence of maternal teaching behavior and the number of new skills the children attained, they would have missed this other socially important consequence of their intervention.

Second, an individual's behavior is complex and multidimensional. The choice to act in one way may preclude or facilitate acting in some other ways. For example, a large increase in one activity will necessarily decrease the time spent in some other activity. Two children who are induced to increase their social interaction at school might do so at the expense of entirely different activities. One child could increase positive interactions by decreasing the amount of time spent staring out the window or disturbing classmates. The other child could increase social behavior at the expense of attending to academic tasks. The developmental consequences of teaching desirable social behavior would be entirely different for the two children. Similarly, when the objective of a program is to decrease or eliminate undesired behavior, new behavior must fill the emptied time. It has been shown frequently that the occurrence of repetitive, stereotyped, damaging, or bizarre behavior of institutionalized individuals can be reduced significantly by altering the consequences of such behavior. But when investigators evaluate the subjects' total behavioral repertoires, they frequently discover that subjects substitute equally undesired, albeit different, behavior for their eliminated stereotypies (Baumeister, 1978). If experimenters measure only the occurrence of the behavior to be eliminated, then they may judge the program's success on too narrow a basis. Unanticipated effects of environmental change may be quite unlike the behavior targeted for change. For example, Landesman-Dwyer and Sackett (1978) found that a simple intervention—consisting of upright positioning, physical peer contact, and exposure to highly salient objects—for multiply handicapped, nonambulatory, profoundly retarded children resulted in increased contact with objects and peers, decreased stereotyped "fixed action patterns," and elevated levels of visual alertness. Although these effects were most apparent during actual treatment, subjects continued to be more responsive in other settings for at least 6 months after treatment ended. The unanticipated findings were that treatment significantly altered the children's sleep–wake patterns and the quality of their performance on standardized infant assessment tools. Children who received treatment showed changes in their individual cycles of sleep and activity, while untreated controls did not. After the intervention, the pattern of sleep–wake cycles was more similar to that of older infants and more consistent across days. These changes were independent of the overall amount of sleep and wake activity. Similarly, treated subjects responded more reliably to test items on the Bayley Scales of Infant Development, even if their overall developmental level did not change. Such findings underscore the need to measure multiple aspects of behavioral adaptation, as well as to evaluate change of individuals. In this example, increased predictability of children's behavior probably had as much impact on their social environment as changing their overall level of performance or amount of wake time would have had.

Third, positive changes in a given behavior are not necessarily correlated with positive changes in closely related behaviors. An example is provided by the work of Walker, Greenwood, Hops, and Todd (1979). Socially withdrawn children were reinforced for either initiating or responding to social contact with others. In spite of successful acquisition of these skills, the children did not increase their overall social interaction. Accordingly, subjects were reinforced to maintain longer periods of social contact. Once again, children clearly acquired this ability, yet positive transfer effects did not occur; children continued to initiate interactions at an extremely low rate. Apparently, a program was needed to teach multiple aspects of positive social interaction, including the proper sequencing of the behavioral components. This social example is analogous to many from the experimental literature on cognition: Factors contributing to adaptive failure in the first place continue to affect children's behavior after they have been taught skills that seemingly should eliminate their problems. Similarly, programming and environmental provisions must be made to promote generalization of social learning (Stokes & Baer, 1977). One of the first full-scale training programs to prepare institutionalized adults for community integration was the Mimosa Cottage Experiment (Girardeau & Spradlin, 1964). After months of carefully implemented training, retarded people had learned many new skills and had lost many of their "institutional" behaviors (e.g., talking too loudly, poor posture, inappropriate expression of affect). Yet when the people took their first trip into the community, all of these institutional behaviors reappeared—immediately. The program had failed to teach for generalization of behaviors across settings, and the retarded people did not realize on their own the adaptive value of their new behavioral styles.

Since one intent of social ecological analysis is to learn how to improve conditions for optimal human development (Moos, 1973; Stokols, 1982), the outcome measures selected need to be broad and valid as indexes of a person's overall adaptation to his or her environment (Brooks & Baumeister, 1977). The challenge to develop sensitive and reliable measures of adaptive behavior and social competence has not yet resulted in tools that capture some of the most important features of everyday adaptation. Although researchers and clinicians have generated more than 300 adaptive behavior measures (Nihira, Foster, Shellhaas, & Leland, 1975; Research Group at Pacific State Hospital, 1977), the relationship between measured skills and success in different environments has been demonstrated in only a limited fashion. Given the tremendous practical and scientific constraints faced in conducting research in natural environments, we advocate choosing multiple measures and a broad perspective on children's lives. Then, similar findings from different studies, using different techniques to tap the same theoretical component of adaptation, should provide bases for sound recommendations about how to optimize conditions for human development.

In sum, retarded children show a variety of deficits and delays in their social development relative to normal children of the same chronological ages. Relative to normal children of the same mental ages, retarded children show fewer social differences. At least some of the differences are due to a mismatch between retarded children's social and learning environments and their cognitive capabilities. Theoretically, successful social adaptation is influenced by many personal qualities, such as self-confidence, physical attractiveness, communication style, consideration of others, willingness to change, and so forth. Many of these attributes can be modified by a combination of environmental manipulations, direct instruction, and provision of prosthetic aids.

Unless socially inadequate people are maintained permanently in special environments, their achievement of greater social success will depend on changes in their behavior. We propose that the most successful changes in social behavior will be accomplished by individuals who adopt a social problem-solving strategy that is parallel to the one suggested earlier for solving cognitive problems (see Table II). This strategy should permit individuals to control their social settings by establishing reasonable personal goals and by incorporating self-evaluation as a means of understanding their person–environment interactions. By reflecting on the probable outcomes of alternative social actions and comparing these with actual consequences, individuals could gain person-alized information to guide their behavior in subsequent social encount-ers. Individuals might choose to avoid environments that are too de-manding, and this choice could be highly adaptive. Clearly, this is the type of decision collectively made by the members of Duddie's Branch, the Appalachian hollow studied by Gazaway (1969). The members of this closed society recognize that "'We'd not make hit outta hyur'" (p. 150) and they consciously "don't accept responsibilities for anything of which they are not a part. . . . They don't have many problems in the complex world. Why? Because they don't let problems light, much less bore in" (p. 155). "The child from Duddie's Branch is not awed by the urgency of learning. That he cannot cope with academic pressures or any semblance of regular routine in no way lowers his estimation of himself" (p. 152). This separation of self-esteem from concrete accom-plishments may appear remarkable to those from highly competitive, achievement-oriented societies. From a social-ecological perspective, we would predict that implementing major changes in a setting such as Duddie's Branch would be difficult. If outside factors did succeed in changing either individuals or collectively agreed-on standards, the ef-fects of these factors should be manifold. Positive outcomes in some areas might well be countered by negative consequences in others.

WHY STUDY RETARDED CHILDREN?

The most straightforward reason for studying mentally retarded children is that many societies recognize them as a distinct group. Even though the concept *mental retardation* has not been created by all cul-tures, enough children in diverse societies are labeled mentally retarded to say that retarded children exist. In most action-oriented societies with sufficient resources and a commitment to children's welfare, the deci-sion has been made to assist mentally retarded children and their fami-lies. By itself, scientific inquiry will not change retarded children's prog-noses. However, science can help us judge the degree to which social programs and different environments do help prevent the negative out-comes associated with mental retardation. If evaluations of social pro-grams include careful analyses of the processes by which people change, then these evaluations will contribute to the discovery of general princi-ples of child–environment interaction. Such findings, in turn, will provide rational and empirically based reasons for improving social and educational policies.

As a result of federal legislation and appropriations, behavioral research in mental retardation increased dramatically in the United States during the 1960s. One hope of legislators, advocates, and researchers was that detailed description of retarded children's behavioral development would help elucidate normal development. The idea was that retarded children provide a natural instance of development "in slow motion." *If* retarded children were simply delayed in their rate of acquiring and demonstrating competencies, then they *would* represent a natural experiment in which behavioral change operated more slowly and accordingly might be more accessible to detailed description or selective manipulation. Unfortunately, this premise is flawed. Retarded children do not represent such an example because they deviate in too many biological and environmental ways from most nonretarded children. The hope of learning a great deal about normal development by studying retarded children has not been realized.

In some selected instances, the study of retarded children may lead to a more focused appreciation of the characteristics of normal development. As Haywood (1970) postulated, "Statistical deviation on a particular dimension may bring into sharp focus those individual differences that might otherwise be obscured in a 'normal' sample" (p. 5). In other cases, retarded children may represent a naturally occurring "control" or comparison condition for estimating the role of experience. Thus, if an investigator wished to assess the relative effects of experience and of cognitive level on a child's response to maternal separation (e.g., when first entering a preschool program), then observing both retarded and nonretarded children might be informative. The children could be selected to represent varying degrees of cognitive understanding of why they were being separated from their mothers. They could include children who differ widely in prior separation experiences and in the amount of exposure to adults and peers who are strangers. In such a study, including retarded children would expand the range of discrepancies between cognitive level and experience. For instance, a 4-year-old retarded child with the cognitive level of a 6-month-old normal infant may have had extensive experience with separation from his or her mother, as well as introduction into novel social situations. If experience were a significant component in mediating a child's response to starting preschool, then the 4-year-old retarded child should adapt more readily than the normal infant. If not, then the contribution of cognitive variables needs to be explored further. Alternatively, the actual nature of the retarded child's experiences in the past four years may have resulted in negative consequences that need to be evaluated in the current situation. So even in this example, retarded children would not necessarily provide a "clean" natural experiment.

Any comparative study of retarded children with other groups of children needs to be conducted with a sensitive appreciation of variables that may contribute to observed outcomes. To the degree that this is implemented, investigations of retarded children's development may yet clarify some of the influences on normal development. The value of scientific inquiry in mental retardation does not rely, however, on its implications for normal children or its applied social meaning. We concur with Baumeister (1967) that "if we aim to understand, predict, and control the behavior of retarded individuals, we need to know how they behave, not how they differ. Furthermore, a principle is no less important, meaningful, or reliable because it is established with reference to a group of [mentally retarded persons]" (p. 875).

Another hope of legislators and researchers during the 1960s was that mental retardation research would help elucidate basic brain–behavior relationships. This expectation seemed logical, because many forms of mental retardation are associated with neurological dysfunctioning and specific brain abnormalities (Warkany, 1971). However, careful review of the behavioral literature reveals little support for syndrome-specific patterns of behavior (Belmont, 1971; Berkson, 1973). Subsequent investigations have demonstrated that a few clinical syndromes are associated with particular deficiencies in cognitive performance (Rohr & Burr, 1978; Silbert, Wolff, & Lilienthal, 1977), although global cognitive functioning does not appear related to severity of brain damage when at least 60% of the brain mass is present (Lorber, 1980). Even when behavior does relate to neuroanatomical or neurological differences among children, the neurophysiological mechanisms responsible for the correlations have not been identified. Undoubtedly, retarded children will continue to provide clinical examples of brain abnormalities, some of which may be amenable to further descriptive study. Thus, Down's syndrome individuals are known to show accelerated biologic aging with concomitant brain changes in early adulthood that usually do not appear in the general population until old age. Young adults with Down's syndrome continue to learn and to show social and cognitive changes, in spite of decreasing neurological integrity. This attests to the lifelong plasticity of the human nervous system but does little to clarify the relationship between specific brain functions and behavioral outcomes. It remains to be determined by longitudinal study of retarded children whether anything precise can be learned about the complex interplay among brain function, observable behavior, and the role of the environment. We think that problems of measurement and ignorance about maturational changes in both behavior and brain processes make it unlikely that studying retarded people will substantially increase our understanding of brain–behavior interactions. Not all investigators concur with this judgment.

We are more optimistic about observing retarded children for the purpose of understanding how human beings adapt under diverse or atypical conditions. Mentally retarded children face much adversity and often seem to cope successfully with what appear to be overwhelming individual and environmental constraints. By conducting longitudinal studies of children who have identifiable sensory, motor, or social impairments, investigators may gain insights into uncommon solutions to everyday problems. When children are taught new ways of behaving, what effects will these newly acquired strategies have on other aspects of their behavior and on other individuals in their environment? What factors influence children's decisions to drop old ways of behaving (usually less mature and less successful) and adopt new strategies? Do children prefer to solve problems by the easiest possible route or by those routes that are most likely to maximize particular desired (and sometimes idiosyncratic) outcomes? As before, we conclude that comprehensive inquiry into how children cope with stress (Rutter, 1981), respond to everyday challenges, adapt to changing environmental conditions, and so forth, would proceed similarly for retarded and nonretarded children. Often, retarded children happen to be at higher-than-average risk for unusual combinations of person–environment variables. For instance, retarded children are at risk for being placed out of home, for being a minority or deviant member within a larger group, or for falling below the expectations of their parents or peers. Certainly, nonretarded children also face adverse environmental and personal conditions. Comparative studies may indicate the relative contribution of cognitive and noncognitive variables in determining styles of coping, attachment, social integration, successful problem solving, and so forth.

In sum, we view the life situations of most retarded infants, children, and adults as providing a valuable opportunity to study adaptive processes. Given society's commitment to providing effective intervention approaches, optimal residential environments, positive educational experiences, and structured learning opportunities for all children, social scientists have a diversity of natural developmental phenomena from which to select questions for further study.

How Adequate Are the Data?

Every scientific literature contains methodological and conceptual flaws. Although we believe that our general conclusions about retarded children's development are justified by the empirical literature, we would be remiss if we did not acknowledge some of the limits in the existing data base.

Understanding of retarded children is limited by several facts. First, the number of investigations of retarded children is small. There are hundreds of investigations of normal children for each investigation of retarded children. Second, some kinds of investigations have never been done with retarded children. Here are just two examples: There have been no comprehensive longitudinal investigations of either cognitive or social behavior of representative groups of retarded people, and there have been no investigations of the extent to which experimentally valid teaching procedures for producing generalized improvements in thinking operate with the variety of materials used in school curricula. Third, it requires exceptionally complex experiments to determine whether any behavior is controlled by experience, capacity, or some interaction of the two. If the question is about the role of past experience, no experiment can be definitive, because past experience cannot be observed or manipulated. If the question is about naturally occurring present experience, the problem is that no one knows how to characterize complex environments in objective terms that have known subjective meanings for children. Moreover, there are no psychometrically sound techniques for describing *in vivo* adaptations of children in terms that are comparable across different environments. If the question is about artificial or experimental environments, the problem is that there is no way to know in advance the range of behaviors that they might importantly change. While the question of relative contributions of experience and capacity is irrelevant to many studies of retarded people, it is central to others. No experiment has solved the foregoing complexities. Fourth, the technologies of data collection and analysis are so varied and relate so intricately to conflicting theoretical views that any experiment on human beings can be faulted on technical grounds. It would be a never-ending job to judge the implications of the faults for all of the interpretations of experiments about topics as broad as cognition and social adaptation.

Despite these limitations, we recommend our conclusions about retarded people to you. We are mindful of Sackett's homily, which he advanced while commenting on investigations reported at a conference on profound retardation:

> I would like to share a platitude with you that I use on my children and graduate students. This homily also seems applicable to [mental retardation] professionals and researchers, and almost certainly applies to the people who are the clients or subjects of the [mental retardation] discipline. It goes "You do the best you can with what you have." (1976, p. 156)

We have done the best we can, and we know that science is a self-correcting business. Investigations of mentally retarded children are now sounder than they used to be, and they will improve. The very process of studying complex problems such as mental retardation leads to improved methods for similar investigations with other populations, including normal people. Future investigations will provide stronger bases for valid conclusions about mental retardation in particular and about intellectual and social development in general.

ACKNOWLEDGMENTS

We greatly acknowledge the contribution of ideas from our many colleagues in the field of mental retardation. In particular, our collaborative research with Gene P. Sackett, John Belmont, and Gershon Berkson has influenced our notions about social and cognitive development.

REFERENCES

ABRAMOWICZ, H. K., & RICHARDSON, S. A. Epidemiology of severe mental retardation in children: Community studies. *American Journal of Mental Deficiency*, 1975, *80*, 18–39.
BAKER, B. L., & CLARK, D. B. Intervention with parents of developmentally disabled children. In S. Landesman-Dwyer & P. Vietze (Eds.), *The social ecology of residential environments*. Baltimore, Md.: University Park Press, in press.

BAUMEISTER, A. A. Problems in comparative studies of mental retardates and normals. *American Journal of Mental Deficiency*, 1967, 71, 869–875.

BAUMEISTER, A. A. Origins and control of stereotyped movements. In C. E. Meyers (Ed.), *Quality of life in severely and profoundly retarded people: Research foundations for improvement*. Washington, D.C.: American Association on Mental Deficiency, 1978.

BELMONT, J. M. Medical-behavioral research in retardation. In N. R. Ellis (Ed.), *International review of research in mental retardation* (Vol. 5). New York: Academic Press, 1971.

BELMONT, J. M., & BUTTERFIELD, E. C. The relations of short-term memory to development and intelligence. In L. Lipsitt & H. Reese (Eds.), *Advances in child development and behavior* (Vol. 4). New York: Academic Press, 1969.

BELMONT, J. M., & BUTTERFIELD, E. C. What the development of short-term memory is. *Human Development*, 1971, 14, 236–248.

BELMONT, J. M., BUTTERFIELD, E. C., & FERRETTI, R. P. To secure transfer of training, instruct self-management skills. In D. K. Detterman & R. J. Sternberg (Eds.), *How and how much can intelligence be improved*. Norwood, N.J.: Ablex Publishing, 1982.

BEREITER, C. E. An academically oriented preschool for culturally deprived children. In F. Hedinger (Ed.), *Preschool education today*. New York: Doubleday, 1966.

BERKSON, G. When exceptions obscure the rule. *Mental Retardation*, 1966, 4, 24–27.

BERKSON, G. Behavior. In J. Wortis (Ed.), *Mental retardation and developmental disabilities: An annual review* (Vol. 4). New York: Brunner/Mazel, 1973.

BERKSON, G., & LANDESMAN-DWYER, S. Behavioral research on severe and profound mental retardation (1955–1974). *American Journal of Mental Deficiency*, 1977, 81, 428–454.

BIJOU, S. W. Theory and research in mental (developmental) retardation. *Psychological Record*, 1963, 13, 95–110.

BIJOU, S. W. A functional analysis of retarded development. In N. R. Ellis (Ed.), *International review of research in mental retardation* (Vol. 1). New York: Academic Press, 1966.

BIRCH, H., RICHARDSON, S., BAIRD, D., HOROBIN, G., & ILLSLEY, R. *Mental subnormality in the community: A clinical and epidemiological study*. Baltimore, Md.: Williams & Wilkins, 1970.

BORKOWSKI, J. G., & CAVANAUGH, J. C. Maintenance and generalization of skills and strategies by the retarded. In N. R. Ellis (Ed.), *Handbook of mental deficiency, psychological theory and research* (2nd ed.). Hillsdale, N.J.: Lawrence Erlbaum Associates, 1979.

BRICKER, D., SEIBERT, J. M., & CASUSO, V. Early intervention. In J. Hogg & P. J. Mittler (Eds.), *Advances in mental handicap research* (Vol. 1). New York: Wiley, 1980.

BROOKS, P. H., & BAUMEISTER, A. A. A plea for consideration of ecological validity in the experimental psychology of mental retardation: A guest editorial. *American Journal of Mental Deficiency*, 1977, 81, 407–416.

BROWN, A. L. The development of memory: Knowing, knowing about knowing, and knowing how to know. In H. W. Reese (Ed.), *Advances in child development and behavior* (Vol. 10). New York: Academic Press, 1975.

BROWN, A. L. Knowing when, where, and how to remember: A problem in metacognition. In R. Glaser (Ed.), *Advances in instructional psychology*. Hillsdale, N.J.: Lawrence Erlbaum Associates, 1978.

BROWN, A. L., & BARCLAY, C. R. The effects of training specific mnemonics on the metamnemonic efficiency of retarded children. *Child Development*, 1976, 47, 70–80.

BROWN, A. L., CAMPIONE, J. C., & BARCLAY, C. R. Training self-checking routines for estimating test readiness: Generalization from list learning to prose recall. *Child Development*, 1979, 50, 501–512.

BUCKHALT, J. A., RUTHERFORD, R. B., & GOLDBERG, K. E. Verbal and nonverbal interaction of mothers with their Down's syndrome and nonretarded infants. *American Journal of Mental Deficiency*, 1978, *82*, 337–343.

BUIUM, N., RYNDERS, J., & TURNURE, J. Early maternal linguistic environment of normal and Down's syndrome language learning children. *American Journal of Mental Deficiency*, 1974, *79*, 52–58.

BUTTERFIELD, E. C. Instructional techniques that produce generalized improvements in cognition. In P. Mittler (Ed.), *Frontiers of knowledge in mental retardation*, (Vol. I): *Social, educational, and behavioral aspects*. Baltimore: University Park Press, 1981.

BUTTERFIELD, E. C. To cure cognitive deficits of mentally retarded persons. In F. J. Menolascino, R. Neman, & J. A. Stark (Eds.), *Curative aspects of mental retardation: Biomedical and behavioral advances*. Baltimore, Md.: Paul H. Brookes, 1983.

BUTTERFIELD, E. C., WAMBOLD, C., & BELMONT, J. M. On the theory and practice of improving short-term memory. *American Journal of Mental Deficiency*, 1973, *77*, 654–669.

CENTER FOR DISEASE CONTROL. *Congenital malformations surveillance (1982)*. Atlanta, Ga.: Author, U.S. Public Health Service, 1982.

CRISSEY, M. S. Mental retardation: Past, present, and future. *American Psychologist*, 1975, *30*, 800–809.

CRONBACH, L. J. Beyond the two disciplines of scientific psychology. *American Psychologist*, 1975, *30*, 116–127.

DAVIS, H., & OLIVER, B. A comparison of aspects of the maternal speech environment of retarded and nonretarded children. *Child: Care, Health, and Development*, 1980, *6*, 135–145.

DENNY, M. R. Research in learning performance. In H. Stevens & R. Heber (Eds.), *Mental retardation*. Chicago: Chicago University Press, 1964.

DEUTSCH, M. The disadvantaged child and the learning process. In A. H. Passow (Ed.), *Education in depressed areas*. New York: Columbia University Teacher's College Press, 1963.

DINGMAN, H. F., & TARJAN, G. Mental retardation and the normal distribution curve. *American Journal of Mental Deficiency*, 1960, *64*, 991–994.

DUNLOP, K. H., STONEMAN, Z., & CANTRELL, M. L. Social interaction of exceptional and other children in a mainstreamed preschool classroom. *Exceptional Children*, 1980, *47*, 132–141.

EDGERTON, R. B. *The cloak of competence: Stigma in the lives of the mentally retarded*. Berkeley: University of California Press, 1967.

EDGERTON, R. B., & BERCOVICI, S. M. The cloak of competence: Years later. *American Journal of Mental Deficiency*, 1976, *80*, 485–497.

ELLIS, N. R. The stimulus trace and behavioral inadequacy. In N. R. Ellis (Ed.), *Handbook of mental deficiency*. New York: McGraw-Hill, 1963.

ELLIS, N. R. A behavioral research strategy in mental retardation: Defense and critique. *American Journal of Mental Deficiency*, 1969, *73*, 557–566.

ELLIS, N. R., MCCARTNEY, J. F., FERRETTI, R. P., & CAVALIER, A. R. Recognition memory in mentally retarded persons. *Intelligence*, 1977, *3*, 310–317.

ESTES, W. K. *Learning theory and mental development*. New York: Academic Press, 1970.

FARBER, B. Effects of a severely retarded child on family integration. *Monographs of the Society for Research in Child Development*, 1959, 24(Whole No. 71).

FARBER, B. Family organization and crisis: Maintenance of integration in families with a severely retarded child. *Monographs of the Society for Research in Child Development*, 1960, 25(1, Serial No. 75).

Fourth Annual Report to Congress on the Implementation of Public Law 94-142: The Education for All Handicapped Children Act. U.S. Government Document, 1982.

GAZAWAY, R. *The longest mile.* New York: Doubleday, 1969.

GIRARDEAU, F. L., & SPRADLIN, J. E. Token rewards in a cottage program. *Mental Retardation,* 1964, *2,* 345–351.

GODDARD, H. H. Suggestions for prognostic classification of mental defectives. *Journal of Psycho-Asthenics,* 1909, *14,* 48–54.

GRAY, S. W., & KLAUS, R. A. An experimental preschool program for culturally deprived children. *Child Development,* 1965, *36,* 887–898.

GREENSPAN, S. Social intelligence in the retarded. In N. R. Ellis (Ed.), *Handbook of mental retardation: Psychological theory and research* (2nd ed.). Hillsdale, N.J.: Lawrence Erlbaum Associates, 1979.

GREENSPAN, S. Social competence and handicapped individuals: Practical implications of a proposed model. In B. K. Keogh (Ed.), *Advances in special education* (Vol. 3). Greenwich, Conn.: JAI Press, 1981.

GROSSMAN, H. J. (Ed.) *Manual on terminology and classification in mental retardation.* Washington, D.C.: American Association of Mental Deficiency, 1977.

GRUENBERG, E. M. Epidemiology. In H. A. Stevens & R. F. Heber (Eds.), *Mental retardation: A review of research.* Chicago: University of Chicago Press, 1964.

GRUNEWALD, K. Mentally retarded children and young people in Sweden. *Acta Paediatrica Scandinavica,* 1979, *Suppl. 275,* No. 75, 75–84.

GURALNICK, M. J. Programmatic factors affecting child-child social interactions in mainstreamed preschool programs. *Exceptional Education Quarterly,* 1981, *1,* 71–91. (a)

GURALNICK, M. J. The efficacy of integrating handicapped children in early education settings: Research implications. *Topics in Early Childhood Special Education,* 1981, *1,* 57–71. (b)

HAGBERG, B., HAGBERG, G., LEWERTH, A., & LINDBERG, U. Mild mental retardation in Swedish school children. I: Prevalence. *Acta Paediatrica Scandinavica,* 1981, *70,* 441–444.

HAYWOOD, H. C. Mental retardation as an extension of the developmental laboratory. *American Journal of Mental Deficiency,* 1970, *75,* 5–9.

HAYWOOD, H. C., & WACHS, T. D. Intelligence, cognition, and individual differences. In M. J. Begab, H. C. Haywood, & H. Garber (Eds.), *Psychosocial influences in retarded performance: Issues and theories in development.* Baltimore, Md.: University Park Press, 1981.

HENSHEL, A. *The forgotten ones.* Austin: University of Texas Press, 1972.

HULL, J. T., & THOMPSON, J. C. Predicting adaptive functioning of mentally retarded persons in community settings. *American Journal of Mental Deficiency,* 1980, *85,* 253–261.

HUNT, J. McV. *Intelligence and experience.* New York: Ronald Press, 1961.

HUNT, J. McV., & KIRK, G. E. Social aspects of intelligence: Evidence and issues. In R. Cancro (Ed.), *Intelligence: Genetic and environmental influences.* New York: Grune & Stratton, 1971.

KAUFFMAN, J. M. (Ed.). Are all children educable? *Analysis and Intervention in Developmental Disabilities,* 1981, *1* (entire issue).

KAZDIN, A. E. *Behavior modification in applied settings.* Homewood, Ill.: Dorsey, 1973.

KOGAN, K. L., WIMBERGER, H. C., & BOBBITT, R. A. The analysis of mother-child interaction in young mental retardates. *Child Development,* 1970, *14,* 205–220.

KOGAN, K. L., TYLER, N., & TURNER, P. The process of interpersonal adaptation between mothers and their cerebral palsied children. *Developmental Medicine and Child Neurology,* 1974, *16,* 518–527.

KUENZEL, M. W. Social status of foster families engaged in community care and training of mentally deficient children. *American Association on Mental Deficiency*, 1939, *44*, 244–253.

LANDESMAN-DWYER, S. Living in the community. *American Journal of Mental Deficiency*, 1981, *86*, 223–234.

LANDESMAN-DWYER, S. Residential environments and the social behavior of handicapped individuals. In M. Lewis (Ed.), *Beyond the dyad*. New York: Plenum Press, in press.

LANDESMAN-DWYER, S., & SACKETT, G. P. Behavioral changes in nonambulatory, profoundly mentally retarded individuals. In C. E. Myers (Ed.), *Quality of life in severely retarded people: Research foundations for improvement*. Washington, D.C.: American Association on Mental Deficiency, 1978.

LANDESMAN-DWYER, S., BERKSON, G., & ROMER, D. Affiliation and friendship of mentally retarded residents in group homes. *American Journal of Mental Deficiency*, 1979, *83*, 571–580.

LANDESMAN-DWYER, S., SACKETT, G. P., & KLEINMAN, J. S. Relationship of size to resident and staff behavior in small community residences. *American Journal of Mental Deficiency*, 1980, *85*, 6–17.

LARBRISSEAU, A., JEAN, P., MESSIER, B., & RICHER, C. Fragile X chromosome and X-linked mental retardation. *Canadian Medical Association Journal*, 1982, *127*, 123–127.

LORBER, J. Is your brain really necessary? *Science*, 1980, *210*, 1232–1234.

MARSHALL, N. R., HEGRENES, J. R., & GOLDSTEIN, S. Verbal interactions: Mothers and their non-retarded children. *American Journal of Mental Deficiency*, 1973, *77*, 415–419.

McLAIN, R. E., SILVERSTEIN, A. B., BROWNLEE, L., & HUBBELL, M. Attitudinal versus ecological approaches to the characterization of institutional treatment environments. *American Journal of Community Psychology*, 1979, *7*, 159–165.

MERCER, J. R. *Labeling the mentally retarded*. Berkeley: University of California Press, 1973.

MEYERS, C. E., NIHIRA, K., & ZETLIN, A. The measurement of adaptive behavior. In N. R. Ellis (Ed.), *Handbook of mental deficiency: Psychological theory and research* (2nd ed.). Hillsdale, N.J.: Lawrence Erlbaum Associates, 1979.

MILGRAM, N. A. The rationale and irrational in Zigler's motivational approach to mental retardation. *American Journal of Mental Deficiency*, 1969, *73*, 527–532.

MISCHEL, W. On the future of personality measurement. *American Psychologist*, 1977, *32*, 246–256.

MITCHELL, D. R. Down's syndrome children in structured dyadic communication situations with their parents. In J. Hogg & P. J. Mittler (Eds.), *Advances in mental handicap research* (Vol. 1). New York: Wiley, 1980.

MOOS, R. M. Conceptualizations of human environments. *American Psychologist*, 1973, *28*, 652–665.

NIHIRA, K., FOSTER, R., SHELLHAAS, M., & LELAND, H. *AAMD Adaptive Behavior Scale Manual*. Washington, D.C.: American Association on Mental Deficiency, 1975.

PENNINGTON, F. M., & LUSZCZ, M. A. Some functional properties of iconic storage in retarded and nonretarded subjects. *Memory and Cognition*, 1975, *3*, 295–301.

PERVIN, L. A., & LEWIS, M. (Eds.). *Perspectives in interactional psychology*. New York: Plenum Press, 1978.

PRESIDENT'S COMMITTEE ON MENTAL RETARDATION. *The six-hour retarded child*. Washington, D.C.: U.S. Government Printing Office, 1970.

RAMEY, C. T., & BRYANT, D. Evidence for primary prevention of developmental retardation. *Journal of the Division of Early Childhood*, 1982, *5*, 73–78.

RAMEY, C. T., & HASKINS, R. Early education, intellectual development, and school performance: A reply to Arthur Jensen and J. McV. Hunt. *Intelligence*, 1981, *5*, 41–48.

RAMEY, C. T., MacPHEE, D., & YEATES, K. O. Preventing developmental retardation: A general systems model. In L. Bond & Joffe (Eds.), *Facilitating infant and early childhood development*. Hanover, N.H.: University Press of New England, 1982.

RAMEY, C. T., SPARLING, J. J., BRYANT, D., & WASIK, B. Primary prevention of developmental retardation during infancy. In H. Moss, R. Hess, & C. Swift (Eds.), *Early intervention programs for infants*. New York: Hayworth Press, 1982.

RAUTMAN, A. Society's first responsibility to the mentally retarded. *American Journal of Mental Deficiency*, 1949, *54*, 155–162.

RESEARCH GROUP AT PACIFIC STATE HOSPITAL. *Performance measures of skill and adaptive competencies in the developmentally disabled: A listing with selected annotations of tests, instruments and scales reported as being used or considered for persons with developmental disabilities*. Pomona, Calif.: University of California, 1977.

ROBINSON, N. M. Mild mental retardation: Does it exist in the People's Republic of China? *Mental Retardation*, 1978, *16*, 295–299.

ROBINSON, N. M., & ROBINSON, H. B. *The mentally retarded child: A psychological approach* (2nd ed.). New York: McGraw-Hill, 1976.

ROHR, A., & BURR, D. B. Etiological differences in patterns of psycholinguistic development of children of IQ 30 to 60. *American Journal of Mental Deficiency*, 1978, *82*, 549–553.

RONDAL, J. A. Maternal speech to normal and Down's syndrome children matched for mean length of utterance. In C. E. Meyers (Ed.), *Quality of life in severely and profoundly retarded people: Research foundations for improvement*. Washington, D.C.: American Association on Mental Deficiency, 1978.

ROUTH, D. K. Learning sets—The Pittsburgh studies. In D. K. Routh (Ed.), *Learning, speech, and the complex effects of punishment*. New York: Plenum Press, 1982.

RUTTER, M. Stress, coping and development: Some issues and some questions. *Journal of Child Psychology and Psychiatry*, 1981, *22*, 323–356.

SACKETT, G. P. Conference summary. In C. C. Cleland, J. D. Swartz, & L. W. Talkington (Eds.), *The profoundly retarded: A conference proceedings*. Austin, Tx.: Western Research Conference and the Hogg Foundation, 1976.

SELTZER, M. M., & SELTZER, G. B. *Context for competence: A study of retarded adults living and working in the community*. Cambridge, Mass.: Educational Projects, 1978.

SILBERT, A., WOLFF, P. H., & LILIENTHAL, J. Spatial and temporal processing in patients with Turner's Syndrome. *Behavior Genetics*, 1977, *7*, 11–21.

SPRADLIN, J. E., & GIRARDEAU, F. L. The behavior of moderately and severely retarded persons. In N. R. Ellis (Ed.), *International review of research in mental retardation* (Vol. 1). New York: Academic Press, 1966.

STEIN, Z., & SUSSER, M. Public health and mental retardation: New power and new problems. In M. Begab & S. Richardson (Eds.), *The mentally retarded and society: A social science perspective*. Baltimore, Md.: University Park Press, 1975.

STOKES, T. F., & BAER, D. M. An implicit technology of generalization. *Journal of Applied Behavior Analysis*, 1977, *10*, 349–367.

STOKOLS, D. Group × place transactions: Some neglected issues in psychological research on settings. In D. Magnusson (Ed.), *Toward a psychology of situations: An interactional perspective*. Hillsdale, N.J.: Lawrence Erlbaum Associates, 1981.

STOKOLS, D. Environmental psychology: A coming of age. In A. Kraut (Ed.), *G. Stanley Hall Lecture Series* (Vol. 2). Washington, D.C.: American Psychological Association, 1982.

STONEMAN, Z., & BRODY, G. H. Observational research on retarded children, their parents, and their siblings. In S. Landesman-Dwyer & P. Vietze (Eds.), *The social ecology of residential environments*. Baltimore, Md.: University Park Press, in press.

STREISSGUTH, A. P., LANDESMAN-DWYER, S., MARTIN, J. C., & SMITH, D. W. Teratogenic effects of alcohol in humans and animals. *Science*, 1980, *209*, 353–361.

SUNDBERG, N. D., SNOWDEN, L. R., & REYNOLDS, W. M. Toward assessment of personal competence and incompetence in life situations. *Annual Review of Psychology*, 1978, *29*, 179–221.

SUSSER, M. W. *Community psychiatry.* New York: Random House, 1968.

TARJAN, G., DINGMAN, H. F., & MILLER, C. R. Statistical expectations of selected handicaps in the mentally retarded. *American Journal of Mental Deficiency*, 1960, *65*, 335–341.

TARJAN, G., WRIGHT, S. W., EYMAN, R. K., & KEERAN, C. V. Natural history of mental retardation: Some aspects of epidemiology. *American Journal of Mental Deficiency*, 1973, *77*, 369–379.

WALKER, H. M., GREENWOOD, C. R., HOPS, H., & TODD, N. M. Differential effects of reinforcing topographic components of social interaction: Analysis and direct replication. *Behavior Modification*, 1979, *3*, 291–321.

WARKANY, J. *Congenital malformations, notes and comments.* Chicago: Year Book Medical Publishers, 1971.

WECHSLER, D. Intelligence defined and undefined: A relativistic appraisal. *American Psychologist*, 1975, *30*, 135–139.

WINDLE, C. Prognosis of mental subnormals: A critical review of research. *American Journal of Mental Deficiency Monograph Supplement*, 1962, *66*.

YARROW, L. J. Maternal deprivation: Toward an empirical and conceptual reevaluation. *Psychological Bulletin*, 1961, *58*, 459–490.

ZEAMAN, D., & HOUSE, B. J. The role of attention in retardate discrimination learning. In N. R. Ellis (Ed.), *Handbook of mental deficiency: Psychological theory and research.* New York: McGraw-Hill, 1963.

ZIGLER, E. Familial and mental retardation: A continuing dilemma. *Science*, 1967, *155*, 292–298.

ZIGLER, E. Developmental versus difference theories of mental retardation and the problem of motivation. *American Journal of Mental Deficiency*, 1969, *73*, 536–556.

ZIGLER, E., & BUTTERFIELD, E. C. Motivational aspects of changes in IQ test performance of culturally deprived nursery school children. *Child Development*, 1968, *39*, 1–14.

ZIGLER, E., & WILLIAMS, J. Institutionalization and the effectiveness of social reinforcement: A three-year follow-up study. *Journal of Abnormal and Social Psychology*, 1963, *66*, 197–205.

ZIGLER, E., BALLA, D., & BUTTERFIELD, E. C. A longitudinal investigation of the relationship between preinstitutional social deprivation and social motivation in institutionalized retardates. *Journal of Personality and Social Psychology*, 1968, *10*, 437–445.

Author Index

Subject Index